AF520197

Molecular Physiology of Abiotic Stresses in Plant Productivity

Editor

A. Hemantaranjan

Department of Plant Physiology
Institute of Agricultural Sciences
Banaras Hindu University,
Varanasi – 221 005 INDIA

Published by:

SCIENTIFIC PUBLISHERS (INDIA)

Jodhpur	Delhi
–	–
5 A, New Pali Road	*4806/24, Ansari Road*
P.O. Box 91	*Daryaganj*
Jodhpur - 342 001	*New Delhi - 110 002*
INDIA	*INDIA*

E-mail: info@scientificpub.com
Website: www.scientificpub.com

Print: 2021

© 2017, Hemantaranjan, A.

All rights reserved. No part of this publication or the information contained herein may be reproduced, adapted, abridged, translated, stored in a retrieval system, computer system, photographic or other systems or transmitted in any form or by any means, electronic, mechanical, by photocopying, recording or otherwise, without written prior permission from the author and the publishers.

Disclaimer: Whereas every effort has been made to avoid errors and omissions, this publication is being sold on the understanding that neither the author nor the publishers nor the printers would be liable in any manner to any person either for an error or for an omission in this publication, or for any action to be taken on the basis of this work. Any inadvertent discrepancy noted may be brought to the attention of the publishers, for rectifying it in future editions, if published.

ISBN: 978-93-83692-46-0

Reprint from: Advances in Plant Physiology (Vol. 17)
Editor: A. Hemantaranjan
ISBN: 978-93-83692-47-7, 2017, 466 pages

Printed in India

PREFACE

The book on **Molecular Physiology of Abiotic Stresses in Plant Productivity** is the outcome of global dedication for researches at physiological and molecular levels that substantially deals with challenges of ongoing international concern over the abiotic stress research, which as the major environmental factors affects plant growth-development. On the other hand, this book also highlights focused researches of significance on image-based plant phenotyping; phenomics and its application in physiological breeding; trace elements; plant functions; physiological basis of yield variation; medicinal and aromatic plants and so on.

In this book over fourteen major chapters emphasize Abiotic stresses and plant productivity followed by seven more chapters in toto in different sections on plant trace elements in plant physiology; physiological basis of yield; medicinal & aromatic plant wealth and so on. The first chapter of this elucidates the molecular physiology of dehydration stress. This stress causes tremendous effect on plant antioxidant defense system that conveys for understanding mechanisms that trigger off physiological responses to dehydration stress and rehydration conditions. In second chapter, exercising of different promoters in ABA mediated pathways in crops species draws attention to a basic construct for a model system to satisfy the hypothesis for developing dehydration stress resistance in crop plants. Subsequently, with the aim to develop strategies for drought mitigation and management, a vibrant team of scientists portrays drought, agricultural practices and private sector lending for sustainability in India, as drought is a serious problem that inhibits nutrient acquisition through roots and limits both quantity and quality of crop productions in *Chapter 3*. Subsequently, in *Chapter 4*, the plant small heat shock proteins are particularly diverse and are a crucial component of the plant heat shock response. Authors consolidate considerable progress made in recent years revealing the structure, function, and evolution of the sHSPs. The *Chapter 5* authors explain physiological adaptations and dynamics for plant productivity under low light intensity highlighting shade tolerance as the maximization of light harvesting and efficient use of captured light in photosynthesis with decreased respiration costs for maintenance according to carbon gain hypothesis. Therefore, how low light intensity affects the morphology, anatomy, physiology and biochemistry of metabolism and the molecular changes in the plant has been thoroughly reviewed. Nevertheless, in *Chapter 6* authors have substantially focused on abiotic determinants of tree seedling growth in tropical dry forests, since information on the interactive effects of light, water and nitrogen on dry tropical tree seedlings is critical to formulating management practices to optimize the species and functional group specific resource availability. This is important for establishing the scientific basis of plantation and restoration programs in the dry tropical environment. Next to this in *Chapter 7,* the TNAU scientists have inclusively compiled major physiological adaptive traits of rice to anaerobic stress conditions caused by flooding during germination and early growth and its implications in crop improvement. This article further discusses the prospects of using knowledge gained in plant adaptation to anaerobic conditions for application in crop improvement programmes. **Of all non-essential heavy metals, cadmium (Cd) due to its potential toxicity to humans, and also its relative mobility in the soil–plant system** requires research for a better understanding of interactions between Cd and plant nutrients in soil-plant system. This has been nicely detailed by two experienced workers in *Chapter 8* as cadmium enrichment in soil: plant physiological manifestation. The *Chapter 9* reviews legume symbiosis under abiotic stresses, since legume–rhizobia association is critical for plant productivity in the view that soil nitrogen often limits plant growth. This symbiosis is affected by environmental stresses.

This book is novel beyond doubt by introducing phenomics and its application in physiological breeding in *Chapter 10* especially enriched through judicious compilation of vital importance in the present scenario by the brilliant team of researchers. This segment covers from simple to complex, focusing upon and discussing the genotyping and phenotyping; concept of a genotype–phenotype map; phenotyping system; plant phenotyping to phenomics; forward and reverse phenomics; phenomic tools; application of plant phenomics to trait-based physiological breeding; plant phenotyping platforms and phenomic projects; plant phenomics and data analysis; as well as plant phenomics in closing the 'gene to genotype' loop, which could be of outstanding interest to readers. The plant breeders have obtained valuable insights regarding heritability of traits, breeding value of the parents, influence of environment and selection approaches for genetically superior offspring using conventional tools like replication and sophisticated experimental designs. After this, in *Chapter 11,* another eminent team of scientists from NBPGR have devotedly endeavoured for contributing on image-based plant phenotyping, the latest consequential developments elucidating a candidate germplasm, which carries genes for targeted traits requires precise and accurate phenotyping of the germplasm. Authors draw attention that efforts are being made to develop new high throughput phenotyping techniques by using imaging, spectroscopy, robotics and high-performance computing. Further in *Chapter 12,* a brief insight into signal transduction pathways of some of the phytohormones based on plant physiological research done till now has been luminously compiled by vibrant workers of undersigned's department for providing enough insight into signal transduction pathways, which is of course an emerging challenge in plant physiological research.

Besides the above chapters, the last chapter of the first section presents an overview of various metal pollutants and their effects on key metabolic processes in plants, focusing mainly on uptake of metals, induction of oxidative stress, role of different components of enzymic and non enzymic antioxidants under metal toxicity conditions. Phytoremediation process as a measure to decontaminate metal polluted soils has also been commendably discussed by eminent scientists of Plant Biochemistry and out of two of them, illustrious Prof. R. S. Dubey is the Vice-Chancellor of two universities at present.

The ***Section II: Plant Trace Elements in Plant Physiology*** consists of two chapters. Notwithstanding, the importance of nickel depending enzymes in Archea, bacteria, plants and primitive eukaryotes is now very well documented so the scientist from Argentina reviewed the impact of nickel biochemistry and its role in plant physiology, as soluble and insoluble nickel compounds are also found in soils and in waters in *Chapter 15* correlating its impact on human health, hence significant an article in several respects. This chapter follows to *Chapter 16,* which gives an overview on iodine as a unique element in context with soil-plant-air system, since the distribution of iodine deficiency disorders (IDD) generally reflects the soil geochemistry of iodine, with areas remote from marine influence, such as central continental and rain shadow areas of high mountain ranges, being seriously affected. Therefore, producing iodine- enhanced vegetables is an attractive way to control IDD, as focused scientifically by devoted workers.

Besides the above, two fundamental chapters are contained in ***Section III: Plant Functions Research in Agricultural Progression.*** Author from Argentina, briefly presented in *Chapter 17* the most important aspects related to the structure and mechanism of action of alkaline phosphatases followed by discussion on purple acid phosphatases found in many organisms from bacteria to man. This is followed by an ideal article on plant lectins, agricultural advancements and mammalian toxicity in *Chapter 18* that scientifically describes plant lectins, the versatile proteins possessing cytotoxic, fungitoxic, anti-nematode and other properties and also promising candidates against many insect pests coupled with their capacity towards agricultural development. Subsequently, the *Chapter 19* of the ***Section IV: Physiological Basis of Yield*** provides precise information by a team of ICAR scientists about potential yield of a crop and the practical limitations to its realization in the field. This is essential for economically feasible meaningful research that ensures food security along with understanding of factors affecting yield are pivotal for agricultural research.

Eventually, the ***Section V*** on ***Nutraceuticals, Medicinal & Aromatic Plant Wealth,*** out of its two integral segments, in *Chapter 20* while illustrating plant wealth, focuses upon the general composition of identified and established microalgae as sources of nutraceuticals and pharmaceuticals of value based on pigments, protein and several bioactive compounds essential for human health and nutrition along with elucidating some side effects. This section finally concludes with a comprehensive review on medicinal and aromatic plants, the wealth of India at high and low altitudes where the single author portrays global biodiversity hotspots of Himalayas highlighting ecological, phyto-geographical and evolutionary factors that favour high species diversity amply understood significance of medicinal and aromatic plants. The author extensively and comprehensively depicts the subject by making emphasis upon quality assurance, industrial applications and biotechnological interventions with plant tissue culture approaches in splendid *Chapter 21* of this unique compilation.

Last but not the least, I am beholden to my family members for their blessings and good wishes for inception of this exclusive edited book on **Molecular Physiology of Abiotic Stresses in Plant Productivity.** Besides these, I am extremely thankful to the excellent and talented personnel of the Scientific Publishers, Jodhpur, India, for their authentic competence in the perfect printing of international standard and worldwide circulation.

A. Hemantaranjan
Professor & Editor in Chief
Advances in Plant Physiology Series

October 11, 2016

Department of Plant Physiology
Institute of Agricultural Sciences
Banaras Hindu University
Varanasi 221 005, INDIA
E-mail: hemantaranjan@gmail.com

CONTRIBUTORS

A. Hemantaranjan
Department of Plant Physiology, Institute of Agricultural Sciences, Banaras Hindu University, Varanasi 221 005, India

A. K. Trivedi
ICAR-Central Institute for Subtropical Horticulture, Rehmankhera, P.O. – Kakori, Lucknow – 226 101

A. Nishant Bhanu
Department of Genetics and Plant Breeding, Institute of Agricultural Sciences, Banaras Hindu University, Varanasi 221 005, India

A. Pandey
ICAR Research Complex – NEH Region, Umiam, Meghalaya – 793103

A. S. Raghubanshi
Institute of Environment & Sustainable Development (IESD), Banaras Hindu University, Varanasi-221005, India

Alok Shukla
Shri Guru Ram Rai Institute of Technology & Sciences, Dehradun 248001, India

Ankita Chandola
Department of Plant Physiology, Institute of Agricultural Sciences, Banaras Hindu University, Varanasi-221005 India

Ankita Kachhwaha
Algal Biotechnology Laboratory, Department of Post Graduate Studies and Research in Biological Science, Rani Durgavati University, Jabalpur (M.P.), India

Bhaben Tanti
Department of Botany, Guahati University, Guwahati 781014, Assam, India

C. Aochen
ICAR Research Complex – NEH Region, Umiam, Meghalaya – 793103

C. Vijayalakshmi
Department of Crop Physiology, Tamil Nadu Agricultural University, Coimbatore

D. Vijayalakshmi
Department of Crop Physiology, Tamil Nadu Agricultural University, Coimbatore

Dhiman Mukherjee
Department of Agronomy, Bidhan Chandra Krishi Viswavidyalaya, Kalayani, Nadia, W.B.-741235, India.

Enrique J. Baran
Centro de Química Inorgánica (CEQUINOR/CONICET, UNLP) Facultad de Ciencias Exactas, Universidad Nacional de La Plata, C. Correo 962, 1900-La Plata, Argentina

G. Pandey
ICAR-Central Institute for Subtropical Horticulture, Rehmankhera, P.O. – Kakori, Lucknow – 226 101

H. R. Bhandari, Department of Genetics and Plant Breeding, Institute of Agricultural Sciences, Banaras Hindu University, Varanasi 221 005, India

Hrishikesh Upadhyaya,
Department of Botany, Cotton College State University, Panbazar, Guwahati 781001, Assam, India.

J. Pohlong
ICAR Research Complex – NEH Region, Umiam, Meghalaya – 793103

Jyoti Kumari
Germplasm Evaluation Division, National Bureau of Plant Genetic Resources, New Delhi 110012, India

Kartikeya Srivastava
Department of Genetics and Plant Breeding, Institute of Agricultural Sciences, Banaras Hindu University, Varanasi 221 005, India

Kamla Dhyani
Shri Guru Ram Rai Institute of Technology & Sciences, Patel Nagar, Dehradun-248001 & G.B Pant University of Agriculture Science & Technology, U.S Nagar (Uttrakhand), India

Kavita Shah
Department of Biochemistry, Faculty of Science, & Institute of Environment and Sustainable Development, Banaras Hindu University, Varanasi-221005, India

M. N. Singh
Department of Genetics and Plant Breeding, Institute of Agricultural Sciences, Banaras Hindu University, Varanasi 221 005, India

M. Raveendran
Department of Plant Biotechnology, Tamil Nadu Agricultural University, Coimbatore

M. K. Adak
Physiology and Molecular Biology Research Laboratory, Department of Botany, University of Kalyani, Kalyani 74 1235, W.B., India

Manoj Kumar
Department of Biochemistry, Faculty of Science, Banaras Hindu University, Varanasi-221005, India

Md. Afjal Ahmad
Department of Plant Physiology, Institute of Agricultural Sciences, Banaras Hindu University,BHU, Varanasi - 221 005, India

Padmanabh Dwivedi
Department of Plant Physiology, Institute of Agricultural Sciences, Banaras Hindu University, BHU, Varanasi - 221 005, India

Prasann Kumar
Department of Plant Physiology, Institute of Agricultural Sciences, Banaras Hindu University,BHU, Varanasi - 221 005, India

Pratap Srivastava
Ecosystems Analysis Laboratory, Department of Botany, Institute of Science, Banaras Hindu University, Varanasi-221005, India

Pravin Prakash
Department of Plant Physiology, Institute of Agricultural Sciences, Banaras Hindu University,BHU, Varanasi - 221 005, India

Prerna Singh
Department of Biochemistry, Faculty of Science, Banaras Hindu University, Varanasi-221005, India

R. Naveen Kumar
Department of Sericulture, Tamil Nadu Agricultural University, Coimbatore

Rahul Bhadauria
Ecosystems Analysis Laboratory, Department of Botany, Institute of Science, Banaras Hindu University, Varanasi-221005, India

Ranju Gulati
Department of Botany, DAV College, Chandigarh, India

Rashmi Yadav
Germplasm Evaluation Division, National Bureau of Plant Genetic Resources, New Delhi 110012

Rishibha Dixit
Algal Biotechnology Laboratory, Department of Post Graduate Studies and Research in Biological Science, Rani Durgavati University, Jabalpur (M.P.), India

Rishikesh Singh
Institute of Environment & Sustainable Development (IESD), Banaras Hindu University, Varanasi-221005, India

Ruchi Bansal
Germplasm Evaluation Division, National Bureau of Plant Genetic Resources, New Delhi 110012

S. Aochen
Fazl Ali, College, Mokokchung, Nagaland - 798601

S. Pyngrope
ICAR Research Complex – NEH Region, Umiam, Meghalaya – 793103

S. S. Dutta
ICAR Research Complex – NEH Region, Umiam, Meghalaya – 793103

Sachchidanand Tripathi,
Department of Botany, Deen Dayal Upadhyay College, University of Delhi, New Delhi-110017, India

Sanjib Kumar Panda
Department of Life Science and Bioinformatics, Assam University , Silchar 788011, Assam, India

Savita Jangde
Department of Plant Physiology, Institute of Agricultural Sciences, Banaras Hindu University, BHU, Varanasi - 221 005, India

Shweta Sahni
Shri Guru Ram Rai Institute of Technology & Sciences, Dehradun 248001, India

Sundeep Kumar
Division of Genomic Resources, National Bureau of Plant Genetic Resources, New Delhi 110012

Surendra Singh
Algal Biotechnology Laboratory, Department of Post Graduate Studies and Research in Biological Science, Rani Durgavati University, Jabalpur (M.P.), India

Swati Agrawal
Faculty of Commerce, Banaras Hindu University, Varanasi-221005, India

Tapas Kumar Pandit
Regional Research Station (OAZ), U.B.K.V, Majhian, Patiram, Balurghat, Dakshin

V. K. Singh
ICAR-Central Institute for Subtropical Horticulture, Rehmankhera, P.O. – Kakori, Lucknow – 226 101

Vijai Pandurangam
Department of Plant Physiology, Institute of Agricultural Sciences, Banaras Hindu University, Varanasi - 221 005, India

CONTENTS

Preface *v*
Contributors *xi*

Section I:
Abiotic Stresses & Plant Productivity: Physiological & Molecular Perspectives

1. Molecular Physiology of Dehydration Stress Responses in Plant: A Review
— *Hrishikesh Upadhyaya, Bhaben Tanti and Sanjib Kumar Panda* 1

2. Exercise of Different Promoters in ABA Mediated Pathways Under Dehydration Stress In Crops Species
— *M.K. Adak* 25

3. Drought, Agricultural Practices and Private Sector Lending for Sustainability in India
— *Swati Agrawal, Manoj Kumar, Prerna Singh and Kavita Shah* 46

4. Small Heat Shock Proteins: The Thermal Stress Managers
— *Ankita Chandola, Vijai Pandurangam and Kamla Dhyani* 71

5. Physiological Approach of Heat Tolerance in Wheat: A Good Tool to Screen Wheat for Heat Tolerance
— *Kamla Dhyani, Ankita Chandola, Shweta Sahni and Alok Shukla* 81

6. Physiological Adaptations and Dynamics for Plant Productivity under Low Light Intensity
— *C. Aochen, J. Pohlong, S. S. Dutta, S. Pyngrope, S. Aochen and A. Pandey* 96

7. Abiotic Determinants of Tree Seedling Growth in Tropical Dry Forests
— *Sachchidanand Tripathi, Rahul Bhadauria, Pratap Srivastava, Rishikesh Singh and A. S. Raghubanshi* 119

8. Physiological Adaptation of Rice to Anaerobic Stress Conditions caused by Flooding During Germination and Early Growth and Its Implications in Crop Improvement

— D. Vijayalakshmi, R. Naveen Kumar, C. Vijayalakshmi and M. Raveendran 132

9. Cadmium Enrichment in Soil : Plant Physiological Manifestation

— Tapas Kumar Pandit and Dhiman Mukherjee 141

10. Legume Symbiosis under Abiotic Stresses- A Review

— Ranju Gulati 186

11. Introduction to Phenomics and Its Application in Physiological Breeding

— A. Nishant Bhanu, H.R. Bhandari, M.N. Singh and K. Srivastava 200

12. Image-based Plant Phenotyping

— Ruchi Bansal, Jyoti Kumari, Rashmi Yadav and Sundeep Kumar 227

13. An Insight in to Signal Transduction Pathways and the Roles of Phyto-hormones : An Emerging Challenge in Plant Physiological Research

— Md. Afjal Ahmad, Savita Jangde and Pravin Prakash 241

14. Metal Toxicity in Plants: Induction of Oxidative Stress, Antioxidative Defense System, Metabolic Alterations and Phytoremediation

— Arti Gautam and Rama Shanker Dubey 256

Section II:
Plant Trace Elements in Plant Physiology

15. Nickel and its Role in Plant Physiology

— Enrique J. Baran 291

16. Iodine: An Unique Element with Special Reference to Soil-Plant-Air System

— Prasann Kumar and A. Hemantaranjan 314

Section III:
Plant Functions Research in Agricultural Progression

17. Plant Purple Acid Phosphatases: Structure and Functions

— Enrique J. Baran 331

18. Plant Lectins, Agricultural Advancements and Mammalian Toxicity

— Prasann Kumar and Padmanabh Dwivedi 360

Section IV:
Physiological Basis of Yield

19. Physiological Basis of Yield Variation

— A. K. Trivedi, G. Pandey and V. K. Singh 379

Section V:
Nutraceuticals, Medicinal & Aromatic Plant Wealth

20. Microalalgae as a Potential Source of Nutraceuticals and Pharmaceuticals

— Surendra Singh, Ankita Kachhwaha and Rishibha Dixit 405

Comprehensive Review

21. Medicinal and Aromatic Plants: Wealth of India

— Dhiman Mukherjee 425

Subject Index 457

Section - I

Abiotic Stresses & Plant Productivity: Physiological & Molecular Perspectives

1

MOLECULAR PHYSIOLOGY OF DEHYDRATION STRESS RESPONSES IN PLANT: A REVIEW

***Hrishikesh Upadhyaya[1], Bhaben Tanti[2] and Sanjib Kumar Panda[3]**

[1]*Department of Botany, Cotton College State University, Panbazar, Guwahati- 781001, Assam, India.*
[2]*Department of Botany, Guahati University, Guwahati- 781014, Assam, India*
[3]*Department of Life Science and Bioinformatics, Assam University, Silchar 788011, Assam, India*
**Corresponding author:hkupbl_au@rediffmail.com*

INTRODUCTION

Dehydration stress is an important abiotic stress, which alters different physiological and biochemical processes in plant (Upadhyaya and Panda, 2004; Upadhyaya *et al.*, 2008; Upadhyaya *et al.*, 2016) leading to significant crop losses. Dehydration stress can be defined as water deficit conditions induced by either drought, salt stress, cold or high temperature stress that has adverse impact on plants growth. Dehydration stress is a significant limiting factor for agricultural productivity and generally inhibits plant growth through water absorption and nutrient uptake. Dehydration stress may involve accumulation of mineral elements in plant tissues by affecting root growth, nutrient mobility in soil and nutrient uptake and hence alter morphological, physiological, biochemical and molecular responses (Upadhyaya and Panda, 2004; Upadhyaya *et al.*, 2008; Bian and Jiang, 2009; Lu *et al.*, 2009; Basu *et al.*, 2010; Luo *et al.* 2010). Dehydration tolerance, the ability to withstand limited water contents is a trait found in some microorganisms like bacteria, algae, fungi, as well as animals and plants. Among plants, desiccation tolerance is common in ferns, mosses and most seeds and pollen of flowering plants (angiosperms). 'Resurrection plants', are able to withstand prolonged periods of dehydration and to recover within hours to a few days once water is available. The study of desiccation tolerance on resurrection plants reveals how they protect themselves against oxidative damage during dehydration, which is often accompanied by conditions of high irradiance (Bartels and Hussain, 2011; Farrant and Moore, 2011; Ding *et al.*, 2013; Zia *et al.*, 2016). Charuvi *et al.* (2015) reported that resurrection plant like *Craterostigma pumilum* can withstand dehydration stress by photoprotection of photosynthetic apparatus Plants respond to dehydration stress has been extensively studied. Such study by Ding *et al.*, (2013), revealed four distinct types of dehydration stress memory genes in *Arabidopsis thaliana*. It suggested implications for the cellular/organismal protective, adaptive, and survival functions encoded by the memory genes and added a new dimension to our understanding of plants' responses to dehydration stress and to current models for interactions between different signaling systems when adjusting to repeated spells of dehydration stress.

Plants under dehydration stress are highly regulated by components of osmoregulative and antioxidative systems and secondary metabolite contents as reported elsewhere (Jaleel *et al.*, 2007). Dehydration stress induces an increase in ROS production resulting in various degree of oxidative damage in different genotypes of crop plants including rice. However, recovery of stress induced damages after post-stress rehydration in terms of regaining antioxidative properties and photosynthetic functions has been reported earlier (Upadhyaya and Panda, 2004; Upadhyaya *et al.*, 2008). Crop plant like rice (*Oryza sativa* L.) is particularly susceptible to dehydration stress (Yang *et al.*, 2008). It is estimated that 50% of the world rice production is affected more or less by drought (Bouman *et al.*, 2005). Several studies on the dehydration stress effects and rehydration revealed physiological acclimation of plant by altered antioxidant responses (Upadhyaya and Panda, 2004; Upadhyaya *et al.*, 2008; Bian and Jiang 2009; Basu *et al.*, 2010; Luo *et al.*, 2010; Ding *et al.*, 2013; Bedi *et al.*, 2016; Zia *et al.*, 2016) and enhancement of recovery by hormone treatment (Vomaka and Pospisilava 2003). It is important to understand the mechanisms that trigger off physiological responses to dehydration stress and rehydration conditions. It has been well established that dehydration stress causes tremendous effect on plant antioxidant defense system. Dehydration stress tolerance and susceptibility has also been correlated with up and down regulation of antioxidant genes (Bian and Jiang, 2009). The current chapter gives an insight on our understanding of various aspects of molecular physiology of dehydration stress responses in plant.

MORPHOLOGICAL RESPONSES DURING DEHYDRATION STRESS

Plant responses to dehydration stress are usually monitored through various morphological parameters (Fig. 1) which have been proven to be good indicators of dehydration stress in different studies (Anjum *et al.,* 2011; Khosroshah *et al.,* 2014; Zaher-Ara *et al.,* 2016). Several morphological traits that can be observed in plants as a result of dehydration stress includes shrinkage in the size of leaves, decrease in the number of stomata, cutinization of leaf surface and thickening of leaf cell walls etc. In succulent plant, changes such as increase in the number of vessels, submersion of stomata, in xerotypes, induction of early senescence and formation of tube leaves in cereals has been observed in plants subjected to dehydration stress. Dehydration stress affects mitosis, cell elongation and expansion and resulted in reduced growth and yield traits (Hussain *et al.,* 2008). It reduces the size, longevity and number of leaves per plant as a consequence of low soil's water potential. Dehydration stress induced reduction in leaf area and consequently adversely effect crop plants due to the reduction in fresh and dry biomass production. Abdalla *et al.* (2015) found that shoot length declined in *Triticum aestivum* by increasing the severity and duration of dehydration stress as reported earlier (Farooq *et al.,* 2009). Such decrease in shoot length is due to decreased cell elongation resulting from water shortage which led to a decrease in cellular turgor. Dehydration stress induced significant reductions in total leaf area were recorded in the rice cultivar Nagina 22 and in wheat by Kadam *et al.* (2015). Plant biomass showed significant decrease in terms of fresh and dry weights of root and shoot during dehydration. Bolat *et al.* (2014) reported that dehydration-induced decreases of rootstocks in terms of fresh and dry weights in apple. Pre-anthesis dehydration treatment caused decrease in grain number and grain size in wheat as reported elsewhere(Dickens and Wright 2008). It was also reported that dehydration stress induced alteration in length, plant height, biomass and leaf area are associated with leaf senescence in various plants like *Abelmoschus esculentus* (Sankar *et al.,* 2008); in *Vigna unguiculata* (Manivannan *et al.,* 2007); in *Camellia sinensis* (Upadhyaya *et al.,* 2008, 2012, 2016) and in *Oryza sativa* and *Triticum estivum* (Kadam *et al.,* 2015).

DEHYDRATION STRESS INDUCED BIOCHEMICAL CHANGES

Dehydration caused by water deficit conditions alters various biochemical responses in plants (Fig. 1). Redox reaction, or the transfer of electrons, is a natural biochemical process for cellular metabolism during energy transduction in the inner mitochondrial and thylakoid membranes. The conditions that leads to loss of redox homeostasis as a consequent of the over accumulation of prooxidants may be referred to as oxidative stress. Redox modification of cellular compounds mediated by various abiotic stress factors including dehydration stress results the imbalance between prooxidant–antioxidant ratios which has been correlated with altered physiological conditions and many diseases in the plant kingdom. Redox regulation is the key in adjusting plant metabolism and development during abiotic stress in plants. Cellular organelles associated with energy metabolism like mitochondria and chloroplasts, have the ETC which is associated with carbon metabolism through NAD(P)H and ATP. Dehydration stress induced alterations in carbon and energy metabolism during stress have been reported in both these organelles. During oxidative stress a highly regulated system is needed to maintain coordination between these organelles, including reversible redox regulation of proteins by thiol–disulfide exchange, regulation of phosphoproteins, activation of signaling pathways by ROS-responsive regulatory genes, and ROS–antioxidant interactions (Foyer and Noctor 2009; Suzuki *et al.,* 2011). The ROS are produced in plant cells via a number of routes, and most of the cellular compartments have the potential to produce ROS (Table 1). Drought and salinity are the most important constraints that induce dehydration in crops and limits the average crop yield (Chen et al. 2013; Agarwal *et al.,* 2013). During dehydration stress, ROS can produce in different cell organelles that are very important for plant survival. In chloroplasts, photosystems I and II (PSI and PSII) are the major sites for the production of singlet oxygen ($^1O_2$) and superoxide ions ($O_2^{\cdot-}$). In mitochondria, complex I, ubiquinone, and complex III of the ETC are the major sites for the generation of $O_2^{\cdot-}$ (Gill and Tuteja 2010). The cell organelles follow different patterns for ROS metabolism (Miller et al. 2010). Dehydration stress lead to the generation of ROS in plants consequently damages carbohydrates, proteins, lipids, and DNA and results in oxidative stress (Gill and Tuteja 2010). Under normal conditions, ROS act as signal molecule which regulates expression of a number of genes useful in numerous physiological plant processes like root hair growth, stomatal closure, and senescence. In case ROS exceeds the normal level it can be removed by enzymatic (SOD, CAT, APX, GR, MDHAR, DHAR, GPX, GuPX, and GST) or non enzymatic (ASA, GSH, phenolic compounds, alkaloids, nonprotein amino acids, and α-tocopherols) antioxidant systems. Antioxidants and ROS initiate and control redox signal transduction and trigger differential expression of genes associated with the performance of the plant under salinity and dehydration stress is the most fascinating subject of future research in plant redox biology. Osmolyte mediated regulation of dehydration stress is due to accumulation of solutes such as proline, glycine betaine, mannitol etc. Decreased proline accumulation has been found to enhance the accumulation of ROS, which subsequently not only enhances plant sensitivity to salinity and dehydration stress but also increased oxidative damage (Miller et al. 2010). Exogenous GB application prevents dehydration-induced structural damage to ROS-producing organelles and suggests the role of proline and GB in the regulation of ROS metabolism during dehydration stress (Ashraf and Foolad 2007). During salinity induced dehydration stress, mannitol is accumulated in many plant species which mainly scavenges hydroxyl radicals from cells (Smirnoff, 2000).

Table 1: Localization of ROS and antioxidants in plants. Chl, Cyt, Mit, Per indicates chloroplast, cytoplasm, mitrochondria and peroxisome respectively

Antioxidants Mechanism	Reaction	Cellular locations	References
SOD (Superoxidase dismutase)	$O^-_2 \longrightarrow H_2O_2$	Chl, Cyt, Mit, Per	Nguyen & Donaldson 2005, del Rio et al., 2006
CAT (Catalase)	$H_2O_2 \longrightarrow H_2O$	Per	Nguyen & Donaldson 2005, del Rio et al., 2006 & Moller 2001
POX (Peroxidases)	$H_2O_2 \longrightarrow H_2O$	Many locations	Passardi et al ., 2005
Ascorbate/ Glutathione cycle (Asc, Glu , Apx &GR)	$H_2O_2 \longrightarrow H_2O$	Chl, Mit, Per	Nguyen & Donaldson 2005, del Rio et al., 2006 & Moller et al 2007
Carotenes and tocopherol	$^1O_2 \longrightarrow O_2$	Chl	Asada , 2006

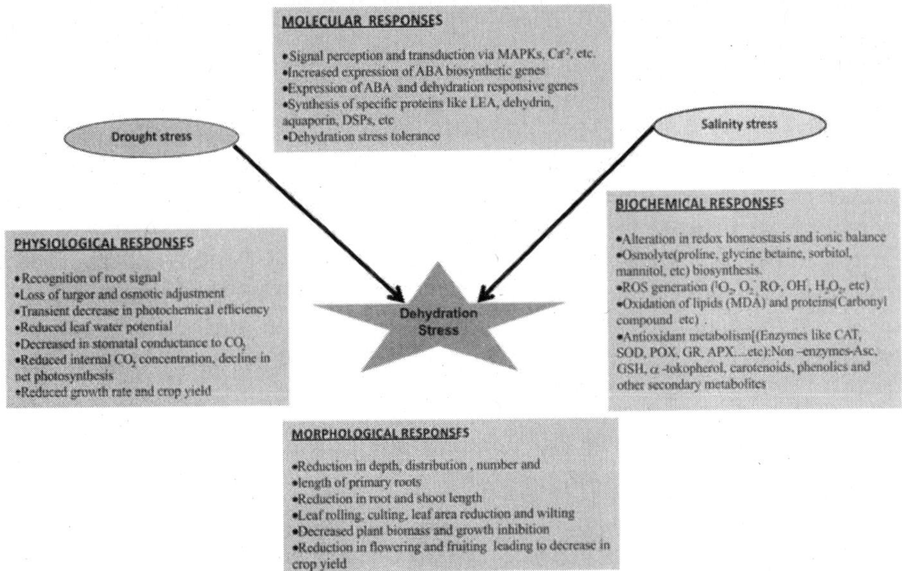

Fig. 1. Dehydration stress induces various morphological, physiological, biochemical and molecular responses in plants.

Budak *et al.* (2015) reported that proteomics analysis identified ABA 8-hydroxylase, MPK6, dehydrin, 30S ribosomal protein S1, and a 70 kDa HSP involved in ABA signaling in wheat. Recently, functional genomics has emerged as a power tool to identify the molecules that regulates dehydration signalling pathways. Transcriptomics analyses identified various genes including Hox22, bZIP TF, dehydrins, WRAB1, WCOR719,

HSPs, LEA, TaWRKY17, TaWRKY16, TaWRKY24, TaWRKY19-C, TaWRKY59, TaWRKY82, TaWRKY61, TaWLIP19, TaWRKY10, TaNAC69 and TaMYB33 for dehydration signaling pathways. As reported elsewhere (Budak *et al.,* 2015), proteomics and metabolomics studies identified several proteins like monomeric G-proteins, lipoxygenases, potassium channel β subunits, calnexin, LEAs, phosphatases and metabolites viz. proline, tryptophan, leucine, isoleucine, valine etc., are important in dehydration stress signaling in plants. Thus, highly efficient functional genomics tools have helped in identifying several important genes which can be exploited by breeders to develop dehydration resistant crop cultivars

PHYSIOLOGY OF PLANTS UNDER DEHYDRATION STRESS

Dehydration stress results when plant is subjected to limited water availability caused by several factors including drought and salinity and thus affects water relation in plants. Dehydration affects various physiological processes in plant and thus alters growth and productivity in crops (Fig. 1). Physiological traits like relative water content (RWC), leaf water potential, stomatal resistance, rate of transpiration, leaf temperature and canopy temperature greatly influence plant water relations. RWC of plant leaves is higher itially during leaf development and decreased as the dry matter accumulated. Dehydration stressed plants had lower RWC than non-stressed ones and decreased the leaf water potential and transpiration rate, with a concomitant increase in leaf temperature (Upadhyaya *et al.,* 2008, 2012, 2013; Farooq *et al.,* 2009). One of the important physiological responses to dehydration stress in plant is osmotic adjustment via osmolyte biosynthesis. Osmotic adjustment has been found in many species and has been implicated in the maintenance of stomatal conductance, photosynthesis, leaf water volume and growth in plants. Dehydration stress that induces a decrease in leaf water potential and in stomatal opening (Fig. 1), leading to the down-regulation of photosynthesis-related genes and reduced availability of CO_2, has been known as one of the major factors during stress.

Phytohormones particularly ABA, plays an important role in integrating dehydration stress signals and controlling downstream stress responses. ABA levels are continuously adjusted by plants in order to cope with abiotic stresses including drought, salinity and cold (Li *et al.,* 2016). During dehydration stress, cytokinin (CK) levels decrease, increasing shoot responses to ABA, leading to stomatal closure (Goicoechea et al. 1997). These stress-induced alterations in CKs and ABA level promote leaf senescence, which leads to leaf abscission and results in reduction of water loss (Pospíšilová *et al.,* 2000). Ethylene response factors (ERFs) were reported to enhance plant tolerance to dehydration stress. Transgenic tobacco expressing JERF3, an osmotic- and oxidative- stress responsive ERF, showed enhanced tolerance to drought and salinity stresses and decreased accumulation of ROS (Wu *et al.,* 2008). Auxin at physiological concentrations induces opening of stomata and may modulate the ABA effect on stomatal aperture (Tanaka *et al.,* 2006). It was proposed that in roots auxin and abscisic acid decrease resistance to waterflow. Therefore, an adaptation of plants to water stress may be an increase in root auxin concentration and a decrease in leaf auxin concentration, or auxin signalling, to minimize water loss. Indeed, Albacete *et al.* (2008) found an increase in root auxin content, whereas leaf auxin content of tomato stressed with 100 mm NaCl decreased. A decrease in auxin content has been shown in plants exposed to salt or drought stress. This has been reported for herbaceous plants, e.g., *Anastatica hierochuntica* and *Helianthus annuus*, but also in the developing xylem of poplar (Junghans *et al.,* 2006). Changes in auxin response can be achieved by regulation of auxin biosynthesis and auxin conjugation or by modulation of auxin signalling. In addition, local auxin concentration and auxin distribution may be regulated under stress by changes in auxin transport. It was reported that osmotic stress caused by increased salinity or drought has an impact on polar auxin transport, but response of the auxin transport machinery seems to be plant and tissue specific. Nevertheless, phytohormones play vital roles in dehydration tolerance of plants. Auxins induce new root formation by breaking root apical dominance induced by cytokinins. As a prolific root system is vital for dehydration tolerance, auxins have an indirect but key role in this regard. Drought stress limits the production of endogenous auxins, usually when contents of abscisic acid and ethylene increase. Dehydration stress and abscisic acid application enhance indole-3-butyric acid synthesis in maize. Müller (2007) reported that Indole-3-butyric acid synthetase from Arabidopsis is also dehydration-inducible. Song *et al.* (2009) observed that, out of the 31 predicted Aux/IAA genes in rice, 15 were induced by drought while three showed decreased expression.

Dehydration stress also alters nitrogen metabolism in plants. Naik *et al.* (2015) concluded that the nitrogen metabolism of *Portulaca oleracea* under water stress plays an important role in stress tolerance mechanism. Xu *et al.* (2015) observed that water stress regulate plant nitrogen metabolism by decreasing nitrate reductase activity. Glutamate dehydrogenase (GDH) may act as a stress-responsive enzyme in detoxification of high intracellular ammonia and production of glutamate for proline synthesis under stress conditions. Zhu *et al.* (2015) reported that over-expression of a glutamate dehydrogenase (GDH) gene confers dehydration tolerance in transgenic rice by preventing toxic accumulation of ammonia and increased production of glutamate, proline and soluble sugar.

Mineral nutrition in plants plays important physiological role in abiotic stress acclimation as it regulates cellular ionic homeostasis. Dehydration stress alters ionic homeostasis by causing reduced uptake of minerals. Different micro and macro nutrients are known to play dehydration stress ameliorative role. Nutrient uptake by crop plants greatly influenced by the overuse of the land in agricultural activities, climate change, precipitation regimes, root morphology, soil properties, quantity and quality of fertilizers, amount of irrigation. Dehydration stress caused reduction in uptake of minerals like, K, Ca, Cu, Zn, Mg, Mn, B etc. and may affect various physiological and biochemical processes in many plants (Upadhyaya *et al.,* 2012). Recently Ingram and Fujiwara (2016) emphasizes that mineral nutrition in plants have critical role in acclimation responses to various environmental stresses. Recent studies revealed a possible involvement of aquaporins in the regulation of hydraulic conductivity by nitrate nutrition (Li *et al.,* 2016; Ingram and Fujiwara 2016). Hydraulic conductivity affects overall nutrient transport in plants, and the possible link between the nitrate signalling and water transport provides a potential new mechanism allowing integration of nitrogen status with overall nutrient uptake and transport. Potassium (K^+) is the most important nutrient for plants and has roles in most aspects of plant physiology. K^+ is essential for a number of biochemical and physiological functions in plant cells including osmotic stress tolerance by regulating water potential. Upadhyaya *et al.* (2012) reported increased K^+ nutrition during dehydration stress improved stress tolerance and post stress recovery in *Camellia sinensis.* CIPK25 and OsAKT1 over expression resulted increased K^+ levels and lower sensitivity to osmotic/dehydration stress in transgenic tobacco and rice plant (Meena *et al.,* 2015; Ahmad *et al.,* 2016). Calcium is also known to mitigate dehydration stress effects and mediates post stress recovery efficiency in plants (Devi and Kar 2013; Upadhyaya *et al.,* 2013). Zinc is important for numerous physiological processes in plants. Zinc deficiency decreases growth, but beyond optimum level it is toxic and alters molecular physiology, biochemistry and anatomy of cell (Anderson 2007; Alloway 2013; Billard *et al.,* 2015; Mattiello *et al.,* 2015). Zinc also affects membrane integrity as it stabilises the molecular structure of cellular components and membranes. Zinc deficiency is one of the common nutrient deficiencies in plants that drastically reduce crop production. Zinc deficiency impairs plant cell by inhibition of growth, differentiation and development. Zinc fertilization mitigates dehydration stress effects and improve post stress recovery potential in plants (Upadhyaya et al. 2013)

Waraich *et al.,* (2011) reported that macro nutrients like N, K and Ca and micro- nutrients like Si, Zn and Mg decrease the toxicity of reactive oxygen species (ROS) by increasing the antioxidants levels such as superoxide dismutase (SOD) and thereby improve dehydration stress tolerance in plants. They also reported that micronutrients such as B and Cu reduce the adverse effects of dehydration stress by altering metabolic processes in plants. Zinc protects cells against ROS attack under drought condition by lowering the ROS generation. It is also reported by them that Boron help the plants to tolerate drought by improving flower retention, sugar transport, pollen formation and seed germination. Waraich *et al.*

(2011) observed that silicon nutrition reduces ROS generation and increases the antioxidants production thereby reducing the photooxidative damage and maintaining the integrity of chloroplast membrane and increases dehydration stress tolerance in plants. Cooke and Leishman (2016) reported that silicon addition during abiotic stress including dehydration alleviates abiotic stress responses. Currently, our lab is also engaged in understanding the importance of mineral nutrition particularly Ca, K and Zn in ameliorating dehydration stress in crop plant. Recent research on mineral nutrition emphasizing on application of nano nutrient or nano fertilizer which can be engineered to improve its function for agricultural applications (Adhikari *et al.*, 2016; Davarpanah *et al.*, 2016; Liu *et al.*, 2016; Khan *et al.*, 2016).

MOLECULAR RESPONSES UNDER DEHYDRATION STRESS

Dehydration induces a wide range of molecular responses in plants (Fig 1). Molecular studies on dehydration stress have identified multiple changes in gene expression and many genes that are dehydration induced, have been isolated. There is continues search for new genes and characterization of their regulatory elements to design innovative strategies for better plant adaptations. Molecular mechanism of dehydration stress signaling in plants have been well reviewed (Yamaguchi-Shinozaki & Shinozaki, 2006; Fujii and Zhu 2012; Patel and Hemantaranjan, 2013; Budak *et al.*, 2015). Studies on expressions patterns of dehydration induced gene suggested that most of the dehydration inducible genes responds to ABA treatment whereas some others do not (Shinozaki *et al.*, 2000; Osakabe *et al.*, 2014). There are ABA-independent as well as ABA-dependent regulatory system of gene expression under dehydration stress. Studies on ABA inducible genes in model plant showed that several genes require protein biosynthesis for their induction by ABA and established the fact that at least two independent pathways exist between the production of endogenous ABA and gene expression under stress condition. There are at least four signal transduction pathways two are ABA dependent (I & II) and two are ABA independent (III & IV). Protein synthesis is necessary for one of the ABA dependent signal pathways (I). Absicic acid responsive element (ABRE) is involved is one of the ABA-dependent pathway (II). In one of the ABA-independent pathways (IV), dehydration responsive elements (DRE) are involved in the regulations of genes not only by dehydration but also by salt and cold stress. Another ABA-independent pathway is controlled by drought and salt but not by cold.

Membrane localized receptor and sensor proteins play important roles in various stress signaling pathways. Plasma membrane signaling has been proposed to be involved in the initial process of dehydration stress perception outside the cell (Maathuis, 2013). In Arabidopsis, an AHK1, Arabidopsis histidine kinase (HK) function as an osmosensor and its overexpression enhanced dehydration tolerance in Arabidopsis (Tran *et al.*, 2007). Studies on ahk1 mutants revealed decreased sensitivity to ABA and the down regulation of ABA- and/or stress-responsive genes, suggesting that AHK1 as a positive regulator of osmotic-stress signaling (Tran *et al.*, 2007). In Arabidopsis, the cytokinin (CK) receptor HKs, AHK2, AHK3, and AHK4, have been known to negatively regulate ABA and dehydration signaling (Tran *et al.*, 2007, 2010). Studies on multiple mutants of ahk2, ahk3, and ahk4 display increased sensitivity to ABA and enhanced tolerance to dehydration (Tran *et al.*, 2007; Jeon *et al.*, 2010). Such findings indicate the existence of crosstalk among ABA, CK, and stress-signaling pathways (Nishiyama *et al.*, 2011; Ha *et al.*, 2012). In Arabidopsis, the receptor-like kinase (RLK) family includes more than 600 members, with the leucine rich-repeat (LRR)- and localized to the plasma membrane are known to be involved in dehydration induced osmotic-stress signaling in a variety of plant species (Osakabe *et al.*, 2013). These stress-related RLKs have different extracellular domains (e.g., LRR, an extensin-like domain, or a cysteine-rich domain) as suggesting their stimulation by stresses may activate RLK-mediated signaling pathways and convey the osmotic conditions outside of the cells as reported elsewhere (Bai *et al.*, 2009; Osakabe *et al.*, 2010; Tanaka *et al.*, 2012). RLKs that bind to cell-walls, such as cell wall-associated kinases (WAKs), the proline-rich extensin-like receptor kinase (PERKs) and the CrRLKs (*Catharanthus roseus* RLK1-like family) have been shown to be involved in the perception of turgor pressure (Christmann *et al.*, 2013; Osakabe *et al.*, 2014). Such findings would shed light on the signaling transduction system controlling dehydration stress tolerance and growth responses.

Transcriptomic and proteomic studies in various plant species have identified various processes associated with dehydration-responsive gene expression (Molina *et al.*, 2008; Jogaiah *et al.*, 2012; Le *et al.*, 2012; Utsumi *et al.*, 2012; Osakabe *et al.*, 2014; Budak *et al.*, 2015; Yadav *et al.*, 2016; Hichri *et al.*, 2016). Such studies have identified regulatory and functional dehydration-responsive genes that encodes various proteins which regulates biosynthesis of osmoprotectants and ABA, late embryogenesis abundant (LEA) and chaperone, ROS-related, ion homeostasis, and signaling proteins. In addition, several key transcription factors (TFs) such as MYB, MYC, DREB/CBF (drought-responsive cis-element binding protein/C-repeat-binding factor), ABF/AREB, NAC, and WRKY TFs which regulates transcription of drought/dehydration-responsive gene have also been identified (Yamaguchi-Shinozaki and Shinozaki, 2006; Osakabe *et al.*, 2014). ABA-responsive cis-element-mediated transcription via ABF/AREB is directly regulated by an ABA receptor complex involving SnRK2 that activate ABF/AREBs by phosphorylation (Umezawa *et al.*, 2010). The action of SnRK2 represents one of the important mechanisms regulating the rapid, adaptive response of plants to dehydration. DREB and AREB activate the transcription of various genes that are expressed in variety tissues. Additionally, TFs that regulates stomatal responses, have also been identified. DST (drought and salt tolerance), a C_2H_2-type TF, regulates the expression of genes involved in H_2O_2 homeostasis, and mediates ROS-induced stomatal closure and abiotic stress including dehydration stress tolerance in rice (Huang *et al.,* 2009). Li *et al.* (2008) reported dehydration-inducible nuclear TF, NFYA5, control stomatal aperture and play a role in drought induced dehydration tolerance in Arabidopsis. Overexpression of SNAC1 (STRESS-RESPONSIVE NAC1) in rice guard cells, enhanced ABA sensitivity and stomatal closure. The TFs like AtMYB60 and AtMYB61 are expressed mainly in guard cells, and play important role in regulating stomatal aperture and dehydration tolerance in plants (Cominelli *et al.,* 2005). Further studies to determine the molecular targets and signaling systems associated with these TFs in stomatal responses will increase our understanding of the regulatory networks controlling plant dehydration responses.

TRANSGENIC APPROACH IN DEHYDRATION STRESS TOLERANCE IN PLANTS

Dehydration stresses induced by drought, salinity etc., is a major challenge for sustainable food production as it may decrease the potential yields in crop plants by 70-80%. Scientist around the would have been continuously putting efforts to achieve successful crop breeding for developing stress adapted varieties by using novel biotechnological approaches, including molecular biology, genetics, plant and cell physiology and breeding. Functional genomics tools had been instrumental in identifying abiotic stress including dehydration stress responsive genes in plants in last few decades. In recent years, molecular biology tools helped to reveal the control mechanisms of dehydration-stress tolerance and enriched our knowledge on stress sensing, signalling including transcriptomic and proteomic regulatory networks in plants which is critical in developing transgenic crop for stress-tolerant crops is based on the expression of specific stress-related genes. Thus transgenic and molecular-marker approaches allow development of stress-tolerant germplasm. Transgenic plants for abiotic stress including dehydration tolerance are being developed, mainly by using gene transformation techniques. Transgenics plant carrying different genes relating to dehydration stress tolerance have been developed in crop plants like rice, wheat, maize, sugarcane, tobacco, Arabidopsis etc. Some of the genetic engineering approaches for developing dehydration stress tolerance in plants are depicted in Table 2. Su and Wu (2004) reported that when P5CS gene is cloned in *Oryza sativa* L., it showed significantly higher tolerance to dehydration stress induced by drought and salinity. In *A thaliana,* overexpression of P5CR results in increased root growth and decreased lipid peroxidation during salinity, drought and ABA mediated dehydration stress (Ma *et al.,* 2008). Transgenic Arabidopsis plants overexpressing deleted and active forms of AREB1 showed enhanced dehydration tolerance and ABA hypersensitivity (Fujita *et al.,* 2005). Overexpression of AREB1 also improved dehydration tolerance in rice and soybean (Barbosa *et al.,* 2013). Niu *et al.* (2014) have shown that overexpression of GSMT and DMT gene results dehydration and cold stress tolerance. Zhang *et al.* (2011) reported that BAD gene when cloned in *Solanum tuberosum* resulted drought and salinity induced dehydration stress tolerance. Ravikumar et al. (2014) developed a transgenic rice using the AtDREB1A transcription factor through Agrobacterium-mediated transformation. Expression of AtDREB1A was induced by drought induced dehydrationstress in transgenic rice lines, which were highly tolerant to dehydration stress in both the vegetative and reproductive stages without affecting their morphological or agronomic traits. Zhou *et al.* (2015) has shown that dehydration stress tolerance by preventing toxic accumulation of ammonia can be achieved by overexpression of BDH gene in *Oryza sativa* L. MYB15 gene confers salinity and dehydration stress tolerance by reducing oxidative damage and improved photosynthesis in transgenic tobacco (Shukla *et al.,* 2015). AaSMO1, gene isolated from *Artemisia annua* when overexpressed in tobacco resulted increased total sterols and improves tolerance to dehydration stress (Singh *et al.,* 2015). Recently, Li *et al.* (2016) reported that DREB A-5 overexpressed in *S. caninervis* confers dehydration and other multiple abiotic stress tolerance. NAC transcription factor from *Cicer arietinum* (CarNAC5) when overexpressed, *A thaliana,* develops tolerance for dehydration and other abiotic stress as evident from the work of Yu *et al.* (2016). Although the transfer of DREBs, MYBs, AREB and NAC have resulted in improved dehydration tolerance by modulating signaling pathways, further studies are required to improved whole downstream stress signaling cascade in transgenic plants.

Table 2: Some transgenic approaches for dehydration stress tolerance in plants

Gene	Gene origin	Target plant	Improved functions	References
P5CS	*V aconitifolia*	*Oryza sativa*	Plant showed significantly higher tolerance to dehydration stress induced by drought and salinity	Su and Wu (2004)
P5CR	*Triticum aestivum*	*A thaliana*	Increased root growth and decreased lipid peroxidation under salinity, drought and ABA mediated dehydration stress	Ma et al. (2008)
OAT	*A. thaliana*	*Oryza sativa*	Increased proline content and improved yield under dehydration stress caused by salt and drought	Wu et al. (2003)
betA	*Escherichia coli*	*Zea mays*	Increased glycine betain accumulation and confer tolerance to dehydration stress caused by salt and drought	Quan et al. (2004)

Gene	Gene origin	Target plant	Improved functions	References
GSMT and DMT	*Aphanothece halophytica*	*Oryza sativa*	Dehydration and cold stress tolerance	Niu et al. (2014)
BADH	*S. oleracea*	*Solanum tuberosum*	Drought and salinity induced dehydration stress tolerance	Zhang et al. (2011)
GDH	*Magnaporthe grisea*	*Oryza sativa*	Dehydration stress tolerance by preventing toxic accumulation of ammonia	Zhou et al. (2015)
MYB15	*Salicornia brachiata*	*Nicotiana tabacum*	Confers salinity and dehydration tolerance by reduced oxidative damage and improved photosynthesis	Shukla *et al.*, 2015
AaSMO1,	*Artemisia annua*	*Nicotiana tabacum*	Enhances total sterols and improves tolerance to dehydration stress	Singh et al. (2015)
DREB A-5	*Syntrichia caninervis*	*S. caninervis*	Confers dehydration and other multiple abiotic stress tolerance	Li et al. (2016)
Car NAC5	*Cicer arietinum*	*A thaliana*	Confers dehydration and other abiotic stress tolerance	Yu et al. (2016)

DEHYDRATION STRESS AND miRNAs IN PLANTS

MicroRNAs (miRNAs) are a class of small noncoding regulatory RNAs that regulate gene expression by guiding target mRNA cleavage or translational inhibition. Dearth of water is increasing worldwide and resulting into drought mediated dehydration stress. The dominant form of agriculture throughout the world is rain-fed agriculture and drought stress is a critical environmental factor which occurs frequently under such condition. By the application of high-throughput technologies such as genome wide gene expression and proteomics several genes are identified that are altered during dehydration condition (Bartels & Sunkar, 2005; Yamaguchi-Shinozaki & Shinozaki, 2006). It has been found that during dehydration stress condition expression of several micro RNAs (Table 3) altered gene expression (miRNAs) by up regulation and down regulation (Li *et al.,* 2008; Arenas-Huertero *et al.,* 2009). In plants like Arabidopsis, rice, and *Populus trichocarpa* miRNA-expression profiling has been performed under drought induced dehydration stress. The expression of *miR393*, *miR319* and *miR397* are up regulated in response to dehydration in Arabidopsis (Sunkar & Zhu, 2004) [Table 3]. In Arabidopsis, *miR396*, *miR168*, *miR167*, *miR165*, *miR319*, *miR159*, *miR394*, *miR156*, *miR393*, *miR171*, *miR158*, and *miR169* were shown to be drought responsive (Liu *et al.,* 2008). Several miRNAs are identified (miR157, miR167, miR168, miR171, miR408, miR393 and miR396) with the aid of a more comprehensive array based analysis technique, which were also up-regulated in drought-stressed Arabidopsis (Liu *et al.,* 2008). In rice, microarray analysis of miRNA expression profile under drought stress showed significant up-regulation of miR169g, the only member of miR169 family and the induction of miR-169g was more prominent in roots than in shoots. Sequence analysis revealed that miR-169g may be regulated directly by DRE binding transcriptional factors, while miR393 was transiently induced (Zhao *et al.,* 2007). In drought-challenged rice a genome-wide identification and analysis of miRNAs were carried out during different developmental stages, from tillering to inflorescence formation, using a microarray platform (Zhou *et al.,* 2010).The consequences of analysis showed that 16 miRNAs (*miR156*, *miR159*, *miR168*, *miR170*, *miR171*, *miR172*, *miR319*, *miR396*, *miR397*, *miR408*, *miR529*, *miR896*, *miR1030*, *miR1035*, *miR1050*, *miR1088*, and *miR1126*) were significantly down regulated in response to drought stress and also found that,14 miRNAs (*miR159*, *miR169*, *miR171*, *miR319*, *miR395*, *miR474*, *miR845*, *miR851*, *miR854*, *miR896*, *miR901*, *miR903*, *miR1026*, and *miR1125*) were significantly up-regulated under drought stress. It was found that the expression levels of *miR1446a-e, miR1444a, miR1447* and miR1450 were significantly reduced, while miR1711- a, miR482.2, miR530a, miR827, miR1445, and miR1448 were slightly downregulated in response to drought stress (Lu *et al.,* 2008). miRNAs expression patterns were investigated in drought-resistant wild emmer wheat (*Triticum turgidum)* using a plant miRNA microarray platform (Kantar *et al.,* 2010). In unstressed Triticum 205 miRNAs were used for the detection of miRNA expression levels while 438 miRNAs in drought-stressed were used for detection. Among all the miRNAs, 13miRNAs (*miR1867*, *miR896*, *miR398*, *miR528*, *miR474*, *miR1450*, *miR396*, *miR1881*, *miR894*, *miR156*, *miR1432*, *miR166*, and *miR171*) shown to be differentially regulated in drought induced dehydration condition (Kantar *et al.,* 2010; Frazier *et al.,* 2011; Li *et al.,* 2013; Gao *et al.,* 2015; Akdogan *et al.,* 2016). Thus physiological responses to dehydration stress are controlled by expression of a large number of genes, many of which are regulated by microRNAs.

Table 3: Some micro RNAs induced by dehydration stress and its function in plants. Others discussed in the text

MiRNA	Target gene	Plant species studied	Functions	References
MiR393	(F-box protein), AFB3	*A thaliana*	Regulates auxin dependent regulation process	Sunkar and Zhu (2004) Sunkar (2010) Si-Ammour et al. (2011)
MiR319	Genes encoding plant specific TFs like PCF5, PCF6, PCF8 and TCP14	*Agrostis stolonifera*	Increased leaf size, thickness, reduced water loss and confer dehydration tolerance	Zhou *et al.,* (2013)
MiR169	NFYA3	*Glycine max*	It directs NFYA3 mRNA cleavage and increased water loss and sensivity to dehydration	Ni et al. (2013)

MiRNA	Target gene	Plant species studied	Functions	References
MiR159	MYB transcription factor MYB33 and MYB101	*A thaliana* *Nicotiana tabacum* *Zea mays*	Confer dehydration tolerance by accumulation of plant hormone ABA	Reyes and Chua (2007) Frazier et al. (2011) Li et al. (2013) Akdogan et al. (2016)
MiR396	growth regulating factor 6 (OsGRF6) gene	*Oryza sativa*	Alters crop yield by regulating shape and architecture of inflorescence	Zhou et al. (2010) Gao et al. (2015)
MiR 172c	Glyma01g39520 (TF-APETALA-2)	*Glycine max* *Nicotiana tabacum*	Positive regulator of dehydration tolerance by reducing the water loss rate and increasing the expression of genes involved in the ABA-dependent and ABA-independent pathways; promote early flowering through modulation of the expression of flowering genes, such as FT and LFY during long days under dehydration stress	Wang et al. (2014) Li *et al.* (2015)

Hence, prior to coming to important conclusions, the overall theme of the write up, as described in preceding sections is being summarized for readers. "Dehydration stress is a serious threat to agriculture and crop production as it generates abiotic stress that interferes with plant growth and productivity. Dehydration stress reduces cell growth, cell elongation and cell expansion that results in retarded plant growth. Dehydration stress disrupts plants antioxidant function by promoting reactive oxygen species (ROS) accumulation. Its results in various alterations in morphological, biochemical, physiological and molecular responses as a consequence of up and down regulations of genes and proteins which functions have been revealed through various functional genomics , proteomics and metabolomic approaches. The current chapter gives an insight on various aspects of molecular physiology of dehydration stress responses in plant. Some of the keywords used throughout are **dehydration stress, growth , molecular physiology, ROS, plants."**

CONCLUSION AND PERSPECTIVE

Dehydration stress adversely affect plant growth and productivity by triggering a series of morphological, physiological, biochemical and molecular changes. Dehydration stress is caused by various factors such as drought, salinity, cold etc that limits water availability to plants. Plant functional genomics approaches including proteomic, transcriptomics and metabolomics along with other highthrouput tools and techniques had been promising and revealed dehydration regulated many genes, transcriptions factors and cellular signalling component which functions are critical in providing stress tolerance to plants. Recent studies on dehydration stress enriched our understanding on stress induced regulatory networks in plant and made it possible to develop transgenic for abiotic stress tolerance in major crops. Transgenic technologies have shown tremendous potential in producing dehydration stress tolerant plants. However, more attention is needed in understanding plants responses to dehydration stress and its interaction with other abiotic stress factors so that multiple stress tolerant crop can be developed in future which can withstand the changing climatic conditions and help to meet the food demand of growing human population.

Acknowledgements

The author (HU) thankfully acknowledge the financial support from SERB (Vide SO No: SB/YS/LS- 44/2013 dated 6th Sept, 2013), DST, Govt. of India HU and SKP also express their sincere thanks to the DBT, Govt of India, for funding DBT RGYI Project (SAN No.102/IFD/SAN/1716/2013-2014 dated July15 2013).

REFERENCES

Abdullah T, Pawar DD, Kale KD and Dingre S K 2015 Water and nutrient use efficiencies of wheat (*Triticum aestivum* L.) under drip fertigation. *Agriculture for Sustainable Development* **3(1)** 52-56.

Adhikari T, Kundu S and Rao AS 2016 Zinc delivery to plants through seed coating with nano-zinc oxide particles. *Journal of Plant Nutrition.* **39(1)** 136-46.

Agarwal PK, Shukla PS, Gupta K and Jha B 2013 Bioengineering for salinity tolerance in plants: state of the art. *Molecular biotechnology* **54(1)** 102-23.

Ahmad I, Mian A and Maathuis FJ 2016 Overexpression of the rice AKT1 potassium channel affects potassium nutrition and rice drought tolerance. *Journal of Experimental Botany.* **67 (9)** 2689-2698.

Akdogan G, Tufekci ED, Uranbey S and Unver T 2016 miRNA-based drought regulation in wheat. *Functional & integrative genomics.* **16(3)** 221-33.

Albacete A, Ghanem ME, Martínez-Andújar C, Acosta M, Sánchez-Bravo J, Martínez V, Lutts S, Dodd IC and Pérez-Alfocea F 2008 Hormonal changes in relation to biomass partitioning and shoot growth impairment in salinized tomato (*Solanum lycopersicum* L.) plants. *Journal of Experimental Botany.* **59(15)** 4119-31.

Alloway BJ 2013 Bioavailability of elements in soil. In: Selinus, O. (Ed.), Essent. Med. Geol. Springer, Netherlands, pp. 351e373.

Andersen P 2007 A review of micronutrient problems in the cultivated soil of Nepal. *Mt. Res. Dev.* **27** 331-335.

Anjum SA, Xie XY, Wang LC, Saleem MF, Man C and Lei W 2011 Morphological, physiological and biochemical responses of plants to drought stress. *African Journal of Agricultural Research*. **6(9)** 2026-32.

Arenas-Huertero C, Pérez B, Rabanal F, Blanco-Melo D, De la Rosa C, Estrada-Navarrete G, Sanchez F, Covarrubias AA and Reyes JL 2009 Conserved and novel miRNAs in the legume Phaseolus vulgaris in response to stress. *Plant molecular biology*. **70(4)** 385-401.

Asada K 2006 Production and scavenging of reactive oxygen species in chloroplasts and their functions. *Plant Physiol*. **141** 391-396.

Ashraf M and Foolad M 2007 Roles of glycine betaine and proline in improving plant abiotic stress resistance. *Environmental and Experimental Botany*. **59(2)** 206-16.

Bai, L, Zhang G, Zhou Y, Zhang Z, Wang W, Du Y, *et al.*, 2009 Plasma membrane-associated proline-rich extension-like receptor kinase 1, a novel regulator of Ca signalling, is required for abscisic acid responses in *Arabidopsis thaliana*. *Plant J*. **60** 314–327.

Barbosa EGG, Leite JP, Marin SRR, Marinho JP, Fátima Corrêa Carvalho, J, Fuganti-Pagliarini, R, *et al.*, 2013. Overexpression of the ABA-dependent *AREB1* transcription factor from *Arabidopsis thaliana* improves soybean tolerance to water deficit. *Plant Mol. Biol. Rep*. **31** 719–730.

Bartels D and Hussain SS 2011 Resurrection plants: physiology and molecular biology. In Plant Desiccation Tolerance, pp. 339-364, Springer Berlin Heidelberg.

Bartels D and Sunkar R 2005 Drought and salt tolerance in plants. Critical reviews in Plant Sciences. **24(1)** 23-58.

Basu S, Roychoudhury A, Saha PP, Sengupta DN 2010 Differential antioxidative responses of indica rice cultivars to drought stress. *Plant Growth Regulation*. **60(1)** 51-9.

Bedi S, Sengupta S, Ray A and Chaudhuri RN 2016 ABI3 mediates dehydration stress recovery response in *Arabidopsis thaliana* by regulating expression of downstream genes. *Plant Science*. **250** 125-140.

Bian S and Jiang Y 2009 Reactive oxygen species, antioxidant enzyme activities and gene expression patterns in leaves and roots of Kentucky bluegrass in response to drought stress and recovery. *Scientia Horticulturae*. **120(2)** 264-70.

Billard V, Maillard A, Garnica M, Cruz F, Garcia Mina JM, Yvin JC, Ourry A, Etinne P 2015 Zn deficiency in *Brassica napus* induces Mo and Mn accumulation associated with chloroplast proteins variation without Zn remobilization. *Plant Physiol Biochem*. **86** 66-71.

Bolat I, Dikilitas M, Ercisli S, Ikinci A, and Tonkaz T 2014 . The Effect of Water Stress on Some Morphological, Physiological and Biochemical Characteristics and Bud Success on Apple and Quince Rootstocks. *The Scientific World Journal*, 769732. http://doi.org/10.1155/2014/769732

Bouman BA, Peng S, Castaneda AR and Visperas RM 2005 Yield and water use of irrigated tropical aerobic rice systems. *Agricultural Water Management*. **74(2)** 87-105.

Budak H, Hussain B, Khan Z, Ozturk NZ, and Ullah N 2015 From Genetics to Functional Genomics: Improvement in Drought Signaling and Tolerance in

Wheat. *Frontiers in Plant Science* **6** 1012. http://doi.org/10.3389/fpls.2015.01012.

Charuvi D, Nevo R, Shimoni E, Naveh L, Zia A, Adam Z, Farrant JM, Kirchhoff H and Reich Z 2015 Photoprotection conferred by changes in photosynthetic protein levels and organization during dehydration of a homoiochlorophyllous resurrection plant. *Plant physiology* **167(4)** 1554-65.

Chen Q, Tao S, Bi X, Xu X, Wang L and Li X 2013 Research progress in physiological and molecular biology mechanism of drought resistance in rice. *Am J Mol Biol.* **3(2)** 102–7

Christmann A, Grill E, and Huang J 2013 Hydraulic signals in long-distance signaling. *Curr. Opin. Plant Biol.* **16** 293–300.

Cominelli E, Galbiati M, Vavasseur A, Conti L, Sala T, Vuylsteke, M, et al. 2005 A guard-cell-specific MYB transcription factor regulates stomatal movements and plant drought tolerance. *Curr. Biol.* **15** 1196–1200

Cooke J and Leishman MR 2016 Consistent alleviation of abiotic stress with silicon addition: a meta-analysis. *Functional Ecology.* **30(8)** 1340-57.

Davarpanah S, Tehranifar A, Davarynejad G, Abadía J and Khorasani R 2016 Effects of foliar applications of zinc and boron nano-fertilizers on pomegranate (*Punica granatum* cv. Ardestani) fruit yield and quality. *Scientia Horticulturae.* **210** 57-64.

Devi SH and Kar M 2013 Amelioration of moisture stress effect by $CaCl_2$ pre-treatment in upland rice. *Ind J Plant Physiol* **18(4)** 384–387.

Dickin E and Wright D 2008 The effects of winter waterlogging and summer drought on the growth and yield of winter wheat (*Triticum aestivum* L.). *European Journal of Agronomy.* **28(3)** 234-44.

Ding Y, Tao Y and Zhu C 2013 Emerging roles of microRNAs in the mediation of drought stress response in plants. Journal of Experimental Botany. **64(11)** 3077-86.

Elnitsky MA, Hayward SA, Rinehart JP, Denlinger DL and Lee RE 2008 Cryoprotective dehydration and the resistance to inoculative freezing in the *Antarctic midge, Belgica antarctica. Journal of Experimental Biology.* **211(4)** 524-30.

Farooq M, Wahid A and Basra SM. Improving water relations and gas exchange with brassinosteroids in rice under drought stress. *Journal of Agronomy and Crop Science.* **195(4)** 262-9.

Farrant JM and Moore JP 2011 Programming desiccation-tolerance: from plants to seeds to resurrection plants. *Current opinion in plant biology.* **14(3)** 340-5.

Foyer CH and Noctor G 2009 Redox regulation in photosynthetic organisms: signaling, acclimation and practical implications. *Antioxid Redox Signal.* **11** 861–905.

Frazier T, Sun G, Burklew C and Zhang B 2011 Salt and drought stresses induce the aberrant expression of microRNA genes in tobacco. *Mol Biotechnol* **49** 159–165.

Fujii H and Zhu JK 2012 Osmotic stress signaling via protein kinases. *Cellular and Molecular Life Sciences.* **69(19)** 3165-73.

Gao F, Wang K, Liu Y, Chen Y, Chen P, Shi Z, Luo J, Jiang D, Fan F, Zhu Y and Li S 2015 Blocking miR396 increases rice yield by shaping inflorescence architecture. *Nature plants.* **2** :15196.

Gill SS and Tuteja N 2010 Reactive oxygen species and antioxidant machinery in abiotic stress tolerance in crop plants. *Plant Physiol Biochem.* **48(12)** 909–30.

Goicoechea N, Antolin MC, Sánchez-Díaz M 1997 Gas exchange is related to the hormone balance in mycorrhizal or nitrogen-fixing alfalfa subjected to drought. *Physiol Plant.* **100** 989–97.

Ha S, Vankova R, Yamaguchi-Shinozaki K, Shinozaki K and Tran LS 2012. Cytokinins: metabolism and function in plant adaptation to environmental stresses. *Trends Plant Sci.* **17** 172–179.

Hichri I, Muhovski Y, Clippe A, Žižková E, Dobrev PI, Motyka V and Lutts S 2016 SlDREB2, a tomato dehydration-responsive element-binding 2 transcription factor, mediates salt stress tolerance in tomato and Arabidopsis. *Plant, Cell & Environment.* **39(1)** 62-79.

Huang XY, Chao DY, Gao JP, Zhu MZ, Shi M and Lin HX 2009 A previously unknown zinc finger protein, DST, regulates drought and salt tolerance in rice via stomatal aperture control. *Genes Dev.* **23** 1805–1817.

Hussain M, Malik MA, Farooq M, Ashraf MY and Cheema MA2008 Improving drought tolerance by exogenous application of glycinebetaine and salicylic acid in sunflower. *Journal of Agronomy and Crop Science.* **194(3)** 193-9.

Ingram GC and Fujiwara T 2016 Special focus issue on plant responses to the environment. *Plant and Cell Physiology*. pcw058.

Jaleel CA, Manivannan P, Sankar B, Kishorekumar A, Gopi R, Somasundaram R and Panneerselvam R 2007 Induction of drought stress tolerance by ketoconazole in *Catharanthus roseus* is mediated by enhanced antioxidant potentials and secondary metabolite accumulation. *Colloids and surfaces B: Biointerfaces.* **60(2)** 201-6.

Jeon J, Kim NY, Kim S, Kang NY, Novák O, Ku SJ, *et al.,* 2010. A subset of cytokinin two-component signaling system plays a role in cold temperature stress response in Arabidopsis. *J. Biol. Chem.* **285** 23371–23386.

Jogaiah S, Govind S R and Tran LS 2012 Systems biology-based approaches toward understanding drought tolerance in food crops. *Crit. Rev. Biotechnol.* **33** 23–39.

Junghans U, Polle A, Düchting P, Weiler E, Kuhlman B, Gruber F and Teichmann T 2006 Adaptation to high salinity in poplar involves changes in xylem anatomy and auxin physiology. *Plant, Cell & Environment.* **29(8)**1519-31.

Kadam N, Yin X, Bindraban P, Struik P and Jagadish K 2015 Does morphological and anatomical plasticity during the vegetative stage make wheat more tolerant of water-deficit stress than rice? Plant physiology **22** pp-114.

Kantar M, Unver T and Budak H 2010 Regulation of barley miRNAs upon dehydration stress correlated with target gene expression. *Functional & Integrative Genomics.* **10(4)** 493-507.

Khan MN, Mobin M, Abbas ZK, AlMutairi KA and Siddiqui ZH 2016 Role of nanomaterials in plants under challenging environments. *Plant Physiology and Biochemistry* 10.1016/j.plaphy.2016.05.038.

Le DT, Nishiyama R, Watanabe Y, Tanaka M, Seki M, Ham LH, et al. (2012). Differential gene expression in soybean leaf tissues at late developmental stages under drought stress revealed by genome-wide transcriptome analysis. PLoS ONE **7(11)** 1-10

Li G, Tillard P, Gojon A and Maurel C 2016 Dual regulation of root hydraulic conductivity and plasma membrane aquaporins by plant nitrate accumulation and high-affinity nitrate transporter NRT2.1. *Plant Cell Physiol* **57** 733–742.

Li H, Zhang D, Li X, Guan K and Yang H 2016 Novel DREB A-5 subgroup transcription factors from desert moss (*Syntrichia caninervis*) confers multiple abiotic stress tolerance to yeast. *Journal of Plant Physiology* **194** 45-53.

Li J-S, Fu F-L, An M, Zhou SF, She YH and Li WC 2013 Differential expression of micro RNAs in response to drought stress in maize. *J Integrated Agriculture* **12** 1414–1422.

Li WX, Oono, Y, Zhu J, He XJ, Wu, JM, Iida K *et al.,* 2008. The Arabidopsis NFYA5 transcription factor is regulated transcriptionally and post transcriptionally to promote drought resistance. *Plant Cell* **20** 2238–2251.

Liu R, Zhang H, Lal R 2016. Effects of Stabilized Nanoparticles of Copper, Zinc, Manganese, and Iron Oxides in Low Concentrations on Lettuce (*Lactuca sativa*) Seed Germination: Nanotoxicants or Nanonutrients? *Water, Air & Soil Pollution.* **227(1)** 1-4.

Lu G, Gao C, Zheng X, Han B 2009 Identification of OsbZIP72 as a positive regulator of ABA response and drought tolerance in rice. *Planta.* **229(3)** 605-15.

Ludwig-Müller J 2007 Indole-3-butyric acid synthesis in ecotypes and mutants of Arabidopsis thaliana under different growth conditions. *Journal of Plant Physiology.* **164(1)** 47-59.

Luo LJ 2010 Breeding for water-saving and drought-resistance rice (WDR) in China. *Journal of Experimental Botany* **61 (13)** 3509-3517.

Maathuis FJ 2013 Sodium in plants: perception, signalling, and regulation of sodium fluxes. *J. Exp. Bot.* **65** 846–858.

Manivannan P, Abdul Jaleel C, Kishorekumar A, Sankar B, Somasundaram R, Sridharan R and Panneerselvam R 2007 Changes in antioxidant metabolism of *Vigna unguiculata* (L.) Walp. By propiconazole under water deficit stress. *Colloids and Surfaces Biointerfaces,* **57** 69–74.

Mattiello, EM, Ruiz HA, Neves JCL, Ventrella MC and Araújo WL 2015 Zinc deficiency affects physiological and anatomical characteristics in maize leaves. *J Plant Physiol.* **183** 138-143

Meena MK, Ghawana S, Dwivedi V, Roy A, Chattopadhyay D 2015 Expression of chickpea CIPK25 enhances root growth and tolerance to dehydration and salt stress in transgenic tobacco. *Frontiers in plant science.* **6** 683

Miller GA, Suzuki N, CIFTCI-YILMAZ SU, Mittler RO 2010 Reactive oxygen species homeostasis and signalling during drought and salinity stresses. *Plant, Cell & Environment.* **33(4)** 453-67.

Molina C, Rotter B, Horres R, Udupa SM, Besser B, Bellarmino L, *et al.,* 2008. Super SAGE: the drought stress-responsive transcriptome of chickpea roots. *BMC Genomics* **9** 553.

Moller IM 2001 Plant mitochondria and oxidative stress: electron transport, NADPH turnover, and metabolism of reactive oxygen species. *Annual Review of Plant Biology.* **52(1)** 561-91.

Moller IM, Jenesen PE and Hanssen A 2007 oxidative modification to cellular components in plants. *Annu. Rev. Plant Biol.* **58** 459-481.

Naik KP, Krishnamurthy N and Ramachandra C 2015 Effect of nutrient sources on grain yield, methane emission and water productivity of rice (*Oryza sativa*) under different methods of cultivation. *Indian Journal of Agronomy* **60(2)** 249-54.

Nguyen AT and Donaldson RP 2005 Metal-catalyzed oxidation induces carbonylation of peroxisomal proteins and loss of enzymatic activities. *Archives of Biochemistry and Biophysics.* **439(1)** 25-31.

Ni Z, Hu Z, Jiang Q, Zhang H. GmNFYA3, a target gene of miR169, is a positive regulator of plant tolerance to drought stress. *Plant Molecular Biology* **82(1-2)** 113-29.

Nishiyama R, Watanabe Y, Fujita Y, Le DT, Kojima M, Werner T, Vankova R, Yamaguchi-Shinozaki K, Shinozaki K, Kakimoto T, Sakakibara H 2011 Analysis of cytokinin mutants and regulation of cytokinin metabolic genes reveals important regulatory roles of cytokinins in drought, salt and abscisic acid responses, and abscisic acid biosynthesis. *The Plant Cell* **23(6)** 2169-83.

Niu X, Xiong F, Liu J, Sui Y, Zeng Z, Lu BR and Liu Y 2014 Co-expression of ApGSMT and ApDMT promotes biosynthesis of glycine betaine in rice (*Oryza sativa* L.) and enhances salt and cold tolerance. *Environmental and Experimental Botany.* **104** 16-25.

Osakabe Y, Yamaguchi-Shinozaki K, Shinozaki K and Phan Tran LS 2013. Sensing the environment: key roles of membrane-localized kinases in plant perception and response to abiotic stress. *J. Exp. Bot.* **64** 445–458.

Osakabe Y, Mizuno S, Tanaka H, Maruyama K, Osakabe K, Todaka D, *et al.*, 2010. Overproduction of the membrane-bound receptor-like protein kinase 1, RPK1, enhances abiotic stress tolerance in Arabidopsis. *J. Biol. Chem.* **285** 9190–9201.

Osakabe Y, Osakabe K, Shinozaki K and Tran LS 2014 Response of plants to water stress. *Frontiers in plant science.* **5** 86.

Passardi F, Cosio C, Penel C and Dunand C 2005 Peroxidases have more functions than a Swiss army knife. *Plant Cell Reports* **24** 255-265

Patel PK and Hemantaranjan A 2013 Differential sensitivity of chickpea genotypes to salicylic acid and drought stress during preanthesis: effects on total chlorophyll, phenolics, seed protein and protein profiling. *The Bioscan.* **8(2)** 569-74.

Pospíšilová J, Synkova H and Rulcova J 2000 Cytokinins and water stress. *Biologia Plantarum.* **43(3)** 321-8.

Ravikumar G, Manimaran P, Voleti SR, Subrahmanyam D, Sundaram RM, Bansal KC, Viraktamath BC, Balachandran SM 2014 Stress-inducible expression of AtDREB1A transcription factor greatly improves drought stress tolerance in transgenic indica rice. *Transgenic Research.* **23(3)** 421-39.

Reyes JL and Chua NH 2007 ABA induction of miR159 controls transcript levels of two MYB factors during Arabidopsis seed germination. *Plant J:Cell Mol Biol* **49** 592–606.

Sankar B, Jaleel CA, Manivannan P, Kishorekumar A, Somasundaram R, Panneerselvam R 2008 Relative efficacy of water use in five varieties of *Abelmoschus esculentus* (L.) Moench. under water-limited conditions. *Colloids and Surfaces B: Biointerfaces.* **62(1)** 125-9.

Shinozaki K and Yamaguchi-Shinozaki K 2000 Molecular responses to dehydration and low temperature: differences and cross-talk between two stress signaling pathways. *Current Opinion in Plant Biology.* **3(3)** 217-23.

Shukla PS, Gupta K, Agarwal P, Jha B and Agarwal PK (2015) Overexpression of a novel SbMYB15 from *Salicornia brachiata* confers salinity and dehydration tolerance by reduced oxidative damage and improved photosynthesis in transgenic tobacco. *Planta.* **242(6)** 1291-308.

Si-Ammour A, Windels D, Arn-Bouldoires E, Kutter C, Ailhas J, Meins F and Vazquez F 2011 miR393 and secondary siRNAs regulate expression of the TIR1/AFB2 auxin receptor clade and auxin-related development of Arabidopsis leaves. *Plant Physiology.* **157(2)** 683-91.

Singh A, Jindal S, Longchar B, Khan F and Gupta V 2015 Overexpression of *Artemisia annua* sterol C-4 methyl oxidase gene, AaSMO1, enhances total sterols and improves tolerance to dehydration stress in tobacco. *Plant Cell, Tissue and Organ Culture* **21(1)** 167-81.

Smirnoff N 2000 Ascorbic acid: metabolism and functions of a multi-faceted molecule. *Curr Opin Plant Biol* **3** 229–235

Song Y, You J and Xiong L(2009) Characterization of OsIAA1 gene, a member of rice Aux/IAA family involved in auxin and brassinosteroid hormone responses and plant morphogenesis. *Plant Molecular Biology* **70(3)** 297-309.

Su J and Wu R 2004 Stress-inducible synthesis of proline in transgenic rice confers faster growth under stress conditions than that with constitutive synthesis. *Plant Science.* **166(4)** 941-8.

Sunkar R 2010 Micro RNAs with macro-effects on plant stress responses. In Seminars in cell & developmental biology (Vol. 21, No. 8, pp. 805-811). Academic Press.

Sunkar, R. and Zhu, J-K. (2004) Novel and stress-regulated microRNAs and other small RNAs from Arabidopsis. Plant Cell **16** 2001–2019

Suzuki N, Miller G, Morales J, Shulaev V, Torres MA and Mittler R 2011 Respiratory burst oxidases: the engines of ROS signaling. *Current Opinion in Plant Biology* **14(6)** 691-699.

Tanaka H, Osakabe Y, Katsura S, Mizuno S, Maruyama K, Kusakabe K, *et al.,* 2012 Abiotic stress-inducible receptor-like kinases negatively control ABA signaling in Arabidopsis. *Plant J.* **70** 599–613.

Tanaka Y, Sano T, Tamaoki M, Nakajima N, Kondo N and Hasezawa S 2006 Cytokinin and auxin inhibit abscisic acid-induced stomatal closure by enhancing ethylene production in Arabidopsis. *Journal of Experimental Botany* **57(10)** 2259-66.

Tran LS, Urao T, Qin F, Maruyama K, Kakimoto T, Shinozaki K., *et al.,* 2007. Functional analysis of AHK1/ATHK1 and cytokinin receptor histidine kinases in response to abscisic acid, drought, and salt stress in Arabidopsis. *Proc. Natl. Acad. Sci. U.S.A.* **104** 20623–20628.

Tran, L. S., Shinozaki, K., and Yamaguchi-Shinozaki, K 2010 Role of cytokinin responsive two-component system in ABA and osmotic stress signalings. *Plant Signal. Behav.* **5** 148–150

Umezawa T, Nakashima K, Miyakawa T, Kuromori T, Tanokura M, Shinozaki K., et al. 2010. Molecular basis of the core regulatory network in ABA responses: sensing, signaling and transport. *Plant Cell Physiol.* **51** 1821–1839.

Upadhayaya H, Panda SK, Dutta BK 2008 Variation of physiological and antioxidative responses in tea cultivars subjected to elevated water stress followed by rehydration recovery. *Acta Physiologia Plantarum.* **30** 457–468,

Upadhyaya H, Dutta BK and Panda SK 2013 Zinc modulates drought induced biochemical damages in tea [*Camellia sinensis* (L) O Kuntze]. J. Agric. Food Chem. **61** 6660–6670

Upadhyaya H and Panda SK 2004 Responses of *Camellia sinensis* to drought and rehydration . *Biol. Plant.*,**48 (4)** 597-600.

Upadhyaya H, Dutta B.K and Panda SK 2016 Drought Induced Physiological and Biochemical Changes in Leaves of Developing Seedlings of Tea [*Camellia sinensis* (L) O Kuntze] Cultivars. *Journal of Tea Science Research*, **6(12)** 1-11

Upadhyaya H, Dutta BK, Sahoo L and Panda SK 2012 Comparative Effect of Ca, K, Mn and B on Post-Drought Stress Recovery in Tea [*Camellia sinensis* (L.) O Kuntze]. *Am J Plant Sci* **3** 443-460.

Utsumi Y, Tanaka M, Morosawa T, Kurotani A, Yoshida T, Mochida K, *et al.*, 2012. Transcriptome analysis using a high-density oligomicroarray under drought stress in various genotypes of cassava: an important tropical crop. *DNA Res.* **19** 335–345.

Vomacka L and Pospisilava J 2003 Rehydration of Sugar Beet Plants after Water Stress: Effect of Cytokinins *Biologia Plantarum* **46(1)** 57-62.

Wang Y, Wang L, Zou Y, Chen L, Cai Z, Zhang S, Zhao F, Tian Y, Jiang Q, Ferguson BJ and Gresshoff PM 2014 Soybean miR172c targets the repressive AP2 transcription factor NNC1 to activate ENOD40 expression and regulate nodule initiation. *The Plant Cell.* **26(12)** 4782-801.

Waraich EA, Ahmad R and Ashraf MY 2011 Role of mineral nutrition in alleviation of drought stress in plants. *Australian Journal of Crop Science.* **5(6)** 764.

Waraich EA, Ahmad R, Ashraf MY 2011 Role of mineral nutrition in alleviation of drought stress in plants. *Australian Journal of Crop Science* **5(6)** 764.

Wu QS, Xia RX, Zou YN 2008 Improved soil structure and citrus growth after inoculation with three arbuscular mycorrhizal fungi under drought stress. *European journal of soil biology.* **44(1)** 122-8.

Yadav A, Khan Y and Prasad M 2016 Dehydration-responsive miRNAs in foxtail millet: genome-wide identification, characterization and expression profiling. *Planta.* **243(3)** 749-66.

Yamaguchi-Shinozaki K and Shinozaki K 2006 Transcriptional regulatory networks in cellular responses and tolerance to dehydration and cold stresses. *Annu. Rev. Plant Biol.* **57** 781-803.

Yang T, Chaudhuri S, Yang L, Du L, and Poovaiah BW 2010. A calcium/calmodulin-regulated member of the receptor-like kinase family confers cold tolerance in plants. *J. Biol. Chem.* **285** 7119–7126.

Yu X, Liu Y, Wang S, Tao Y, Wang Z, Mijiti A, Wang Z, Zhang H and Ma H 2016 A chickpea stress-responsive NAC transcription factor, CarNAC5, confers enhanced tolerance to drought stress in transgenic Arabidopsis. *Plant Growth Regulation.* **79(2)** 187-97.

Zaher-Ara T, Boroomand N and Sadat-Hosseini M 2016 Physiological and morphological response to drought stress in seedlings of ten citrus. Trees. **30(3)** 985-93.

Zdzislaw Attila Tátrai, Rabab Sanoubar, Zsuzsanna Pluhár, Silvia Mancarella, Francesco Orsini, and Giorgio Gianquinto, "Morphological and Physiological Plant Responses to Drought Stress in *Thymus citriodorus*," *International Journal of Agronomy*, doi:10.1155/2016/4165750

Zhang X, Zou Z, Gong P, Zhang J, Ziaf K, Li H, Xiao F and Ye Z 2011 Over-expression of microRNA169 confers enhanced drought tolerance to tomato. Biotechnology letters. **33(2)** 403-9.

Zhao J, Ren W, Zhi D, Wang L, Xia G 2007 Arabidopsis DREB1A/CBF3 bestowed transgenic tall fescue increased tolerance to drought stress. Plant Cell Reports. **26(9)** 1521-8.

Zhou L, Liu Y, Liu Z, Kong D, Duan M and Luo L 2010 Genome-wide identification and analysis of drought-responsive microRNAs in *Oryza sativa*. *Journal of Experimental Botany*. **61(15)** 4157-68.

Zhou M, Li D, Li Z, Hu Q, Yang C, Zhu L and Luo H 2013 Constitutive expression of a miR319 gene alters plant development and enhances salt and drought tolerance in transgenic creeping bentgrass. *Plant Physiology*. **161(3)** 1375-91.

Zhou Y, Zhang C, Lin J, Yang Y, Peng Y, Tang D, Zhao X, Zhu Y and Liu X 2015 Over-expression of a glutamate dehydrogenase gene, MgGDH, from Magnaporthe grisea confers tolerance to dehydration stress in transgenic rice. *Planta*. **241(3)** 727-40.

Zia A, Walker BJ, Oung HM, Charuvi D, Jahns P, Cousins AB, Farrant JM and Reich Z 2016 Kirchhoff H. Protection of the photosynthetic apparatus against dehydration stress in the resurrection plant *Craterostigma pumilum*. *The Plant Journal*. doi: 10.1111/tpj.13227.

Zokaee-Khosroshahi M, Esna-Ashari M, Ershadi A and Imani A 2014 Morphological Changes in Response to Drought Stress in Cultivated and Wild Almond Species. *International Journal of Horticultural Science and Technology*. **1(1)** 79-92.

■■■■

2

EXERCISE OF DIFFERENT PROMOTERS IN ABA MEDIATED PATHWAYS UNDER DEHYDRATION STRESS IN CROPS SPECIES

M.K. Adak

Physiology and Molecular Biology Research Laboratory, Department of Botany, University of Kalyani, Kalyani 74 1235, West Bengal, INDIA
E. mail: mkadak09@gmail.com, mkadakjoymaa@gmail.com

INTRODUCTION

It is an irony of fate that water deficit in plant's life is an inevitable providence irrespective the growth phases, organs or tissues. It is now needless to say that water deficit in any forms may ensure the drought stress to plants and regardless of its intensity, duration it happens to be everlasting problem in modern agriculture. In more recent statistics, it reveals that around 12 per cent fall in corn production as compared to earlier year was observed out of agricultural drought in USA in 2013 (USDA, 2014). Therefore, it became obvious to be a threat for the sustainability of existing crop and its production coupled with good drought tolerant cultivars. It might be an overestimate, still, alarming that by the year of 2050 and onwards the arable fallows under worldwide cultivation would be affected with salinity coupled with water deficit. Ironically, world population may be challenged with two billion people to feed. Therefore, new insights along with strategies with sustainable crop productivity ought to be framed to meet this confront for extreme harsh environments (Todaka *et al.,* 2015). Now, plants response is diverse and still remains unsolved due to complexity as well as unpredictability of stress exposure. In instance, stress might singly be occurring in multiple phases of plant's development but its interface might also be unequal (Hanin *et al.,* 2011). On the other hand the interaction of multiple stress factors also occur same time or different thus making more a complex net work. The improvement of water use efficiency (WUE) is set as most prime focus under water deficit stress where stomatal regulation is the pivotal path adjusting the tissue dehydration (Jiang *et al.,* 2012). This is more illustrated with the citation of rice (*Oryza sativa* L), an important cereal along with wheat, maize, sorghum etc. In recent report on 2013 the rice has been cultivated in 124 countries in the world with the production of 745 million ton (FAOSTAT). Rice in comparison to other cereals is earlier sensitive to water deficit with its inherent ecological nature since its domestication. Soils saturated with moisture well before permanent wilting percentage along with very less (almost nil) metric potential characterize the rice genotypes to be a semi aquatic species (Karaba *et al.,* 2007). With this it could be predicted that in comparison to other cereals, rice obliges around 3000-5000 cm^3 of water to produce 1 kg grain. For maize the requirement of water is almost half for equivalent amount of seed. Therefore, improvement of WUE signifies the sustainable agricultures in advocating the water conservation, particularly, under drought prone zone. Soil salinity is also drawn with other impacts water stress where a low negative water potential (μ_w) may ensure the constant demand for water for thriving plant genotypes. Under hardy, dry as well as saline soil the imposition of water deficit induces the mechanical constraints on roots growth (Deinlein *et al.,* 2014). Osmotic stress causes cellular dehydration due spillage of water into apoplastic spaces and finally an oxidative burst is ensured with elevated level of reactive oxygen species (ROS) (Hasegawa *et al.,* 2000, Mittler *et al.,* 2004). An estimate documents about 1/3 of arable land worldwide is in constant and chronologically victim irrespective of agriculture regions in virtual and thereby the fall in yield is mostly mate with drought. In another approach the transgenic cereals with their proper acclimatization have opened up a new avenue for launching of developed varieties in drought and saline prone areas. Under *in vitro* condition, the performances of those transgenic plants are to perform well and demonstrate their mode of tolerance in terms of cellular, physiological as well as phenological levels and thereby become the keys to understand plants responses (Xiong *et al.,* 2002). Undoubtedly, the immediate behavior of plants might not be necessarily correlated with actual attainment and adjustment of plants to the water deficit at successive stages of growth and development. Still, to decipher the genes functioning in support to over express and/or down regulate particular paths are most welcome with transgenic plants. This is more prevalent with the signaling cascade for particular elicitors in response to water deficit in soil (Duan, *et al.,* 2012) This is followed in down-stream movement through the cytoplasmic receptor proteins and finally to reach into nucleus (Golldack *et al.,* 2014). Moreover, the co-ordination between a wide array of regulatory proteins, their diversification in different taxa and finally decipher the precise roles in a particular up/down regulation are also more privileged with transgenic plants. Plant growth regulators or hormones are also exciting in elucidation of signaling and has been conceiving a lot of cross talks in this regard (Wu Le *et al.*, 2008). A concomitant changes in hormonal profiles along with variation in specific genes are mostly offered by plants to communicate signaling approving plant hormones as secondary messengers. Absicic acid (ABA), ethylene (C_2H_2) are most readily cited by the plants in this signaling cascade as scored in many cereals (Wu Le *et al.*, 2008, Huang, *et al.*, 2012). In the present review, some highlights with reference to molecular signaling in water stress perception, its coordination with different transcription factors and their precise functions have been elucidated with attention to cereal crop species. The expression of specific genes directly affecting the biosynthesis and functionalities of those hormones in relation to the signaling net work under the ongoing dehydration of cytosol has also been incorporated. With high expectation this present appraisal or review would be a basic construct for a model system to satisfy the hypothesis and thus may also be suggestive in developing stress resistance in crop plants (Wang *et al.*, 2013). The information are still lacking to be confirmed as well as conclusive

It is still in dilemma whether specific hormone/hormones is /are responsible for gene expression or the genes are induced to over-produce hormonal paths under drought stress in modulation of physiological responses. Any particular physiological process may or must be not solely mediate the recovery of plants in linear way; rather it is the conscious activity of many cellular events in concert (Pinheiro and Chaves 2011). It should also be well thought-out that gene expressions under moisture deficit in cytosol surely overlie the hormone regulated gene expression. In most of the cases the information have been enriched from the data under investigations for direct drought exposure through attaining the a higher negative water potential ($\ddot{Y}_w$) along with fall in turgor (Hasegawa *et al.*, 2000) and indirectly osmotic impairment (elevation the level of compatible solutes). In another approaches the interference of elicitors based changes in signaling and thereby induction of genes are also most exciting sites for further research. Therefore, it is really difficult to be precise and conclusive to adopt a particular strand and in so doing to concentrate the research for better plant's tolerance. The chemical signaling and its interference had been most possible with the role for stomatal regulation in plants. Classically, ABA has the unique opportunity to place itself as prime concern with direct interaction in signaling for desiccation from roots to shoots (Huang, *et al.*, 2012). During dehydration, in a complex net work ABA is exercised in a short-term response for plants under dehydration (Zhang *et al.*, 2006). In addition, ABA under depleted water deficit in soil comes in the category of under long-term responses through expression of genes for rapid and deeper root system. Still, ABA not being solely can attribute for stomatal signaling with jasonic acid (JA) and nitric oxide (NO) for the short term responses. Again, for long term gene expression shares the induction of counter genes induced by ethylene (C_2H_2), cytokinin (CK) and auxin also. Gene expression with ABA under drought stress came under understanding from stomatal closure with changes of allied leaf character after exogenous application of the ABA (Finkelstein and Rock, 2002). Physiologically, a fall in ¥$_w$ in roots starts initially starts the accumulation of ABA in the leaf mesophyll cells, more accurately in chloroplast. A complex, however, yet to be deciphered networks with transcriptional regulation has been the focus for long days in ABA dependant pathways. By different molecular studies the key but candidate genes have been elucidating for analysis of responses from tolerant plants, mostly transformed with stress tolerance (Shinizaki *et al.,* 2006). Two already established and well characterized paths for transcriptional directive in water/salt stress responses in *Arabidopsis* deciphered in hormones-gene interaction. The first one hold a direct involvement of ABA or ABA sensitive/ dependant pathway and the second involves a non dependant of ABA mediation rather regulatory net work mediated by dehydration response elements (DRE) (Golldack *et al.,* 2014). So far the most targeted gene is concerned in ABA mediated dehydaration tolerance; it is the 9 cis-epoxy carotenoid dioxydrogenase (NCED). A number of variants exist of the NCED and out of those NCED-3 has most thoroughly been studied in desiccation tolerance for *Arabidopsis* and it recorded that there was an over-fold expression irrespective of any abiotic stress (Behnam *et al.,* 2013). In more analytical study it also revealed that a specific DNA sequence as cis–element appears for up regulation of NCED gene. This sequence is located on up-stream of G box (-2.248bp) from start site and induced not only by water stress but also with other abiotic components (Kuromori, 2014). Not only synthesis or recognition of ABA sequence, but also the transport of ABA across the cellular membrane is also essential for initiation of ABA dependant gene responses under water/salinity stress. In the mesophyll tissues of leaf or in chloroplast of guard cells ABA is automatically synthesized and with polar transport through H^+ gradient under photosynthetic irradiance reaches the cytosol. In case of hydro-active stomatal movement ABA could also be accumulated in the guard cells by H^+/ATP ase functioning and may coupled with influx of K^+ transitory complex. Thus, the depletion of K^+ in ionic form may substantially reduced the stomatal aperture with more -¥$_w$ in the guard cells (Behnam *et al.,* 2013).

MOLECULAR CHARACTERIZATION OF ABA TRANSPORT SYSTEM

Some of the important findings have been emerged with molecular regulation of ABA transport, more specifically to the protein receptor cascade, still, how far it is applicable regardless of species may open the question. A series of membrane spanning proteins are activated with their *in vitro* phosphorylation activity with induction of ABA and in downstream phenomena of stomatal closure. In more recent study by Bauer et al. (2013). Amongst those the most studied are Pyrabactin resistance (PYR), Pyrabactin resistance like (PYR1-L), Regulatory component of ABA response (RCAR), protein phosphatase- 2C(PP2C), Sucrose non fermenting protein kinase 2(SnRK2). In their studies supported by other the signaling of ABA is dependent on a series of conformational changes through phosphorylation pattern of secondary messengers through cytosol and finally transducer into nucleus for gene induction. As for e.g. the dephosphorylation of SnRK2 triggers the repression of PP2C when plants are under moderate to insignificant stress with water deficit. The isolation of those membrane bound signaling proteins are also responded negatively in phophorylation without ABA *in vitro* system (Miyakawa *et al.,* 2013). But in presence of ABA, the PYL/ PYR/RCARs with bound ABA could distinguish and bind the PP2C. This would release the SnRK2 from PP2C and thus the repression is over. The released SnRK2 then can phosphorylate the different proteins associated in the up-stream of ABA inducing genes (Wang *et al.,* 2011). Those proteins essentially include ABA-responsive cis element binding proteins or ABA binding proteins (ABA-BP). A typical sequence for ABA responsive elements (ABRE) is characterized by a core sequence : PyACGTGG/TC, which is flanked by other recognition sequences like MYCRS; CANNTG, MYB recognition sequence: MYBRS; C /TAACNA/G). Interestingly, the core ABRE is also compatible to other water stress response elements like that of drought responsive elements (DRE). Another DRE element of a drought responsive gene with 5,284 bp have been elucidated which shows compatibility to the other sequence of drought tolerant *indica* sub species of rice. This sequence also confers correlation in expression of α-linoleic acid metabolism as well expressed in drought resistant genotype. Other genes adhered to the sensitivity of ABA tolerance including the senescence related degradative genes (SDG) and photosynthetic genes were down regulated in water stress resistant genotypes. With transcript profiling along with comprehensive data there recorded a few marker transcripts for assigning drought tolerant types from some rice cultivars (Degenkolbe *et al.,* 2013). Almost the markers recorded a significant correlation with induced water or even salt induced water stress tolerance. In fact, the transcription regulation along with ABA involvement is much complex when it encounters with other regulatory elements under dehydration or related abiotic factors inducing water stress. The ABA transcription factors can bind in its phosphorylated state with cis element which span the core promoter region of drought inducible genes. In fact ABA binding protein (AREB), a *bzip* class transcription factor, has four conserved domains for phosphorylation by SnRK2. Despite a number of transcription factors amenable in plants, still, ABFs and AREB are the pivotal factors to accommodate the transcriptional operation into a common cluster moderated with ABA dependant path.

Plants can Accomplish ABA Independent Path for Water/ Salinity Stress

This is in addition to ABA induced gene regulation where other sequences in sensitive to ABA paths are followed in plants. A number of plants showing their ubiquitous exercise of ABA insensitivity through some cis elements that are allowed to bind factors under dehydration. Dehydration responsive proteins (DREB 2) is such a type of protein that might originate as the evolutionary descendant of AP2/ERF family, a specific transcription factor (Sharma *et al.,* 2013). A number of variants are demarcated according to their conserve domains in DNA binding sites and mostly cultivated are DREB2A and DREB2B in *Arabidopsis.* Another domain which can negatively regulate the DREB2A amino acid sequence to be also interactive under ABA independent pathway (Mizoi *et al.,* 2012). This got more information from the molecular ubiqutinization by C3HC4 ring domain in a protein called DREB2A interacting proteins 1 & DREB2A interacting proteins 2 (DRIP1 & DRIP2). However, this is essentially happened when the plants are not under any water/ salinity or any other stressful condition. In more analysis of this pathway it reveals that a growth inhibiting factor GRF7 can repress the growth under non stressed conditions of plants. In fact it is the protein that finds binding site to the DREB1/2 amino acid sequence through ABA independent pathway (Kim *et al.,* 2012). On the other hand DREB2A has also been documented to be involved in heat tolerance under water stress condition and thus suggest possible involvement of ABA impendent pathway for cross tolerance in other stress factors (Yoshida *et al.,* 2011). In more details it reveals that a heat shock factor A1, a protein is over expressed in few mutants of Arabidopsis when the plants were stressed under dehydration following elevated ranges of temperature. Now, DREB has also been found to interconnect with cold inducing factors under cold stress. Under such condition the cold induced factors like DREB1-CBF, a class of b-zip has attended for involvement in genes under low temperature stress. In transgenic *Arabidopsis* the over-expressed *DREB1-CBF* has improved the resistance for freezing tolerance in terms of both cellular stability as well as anti-oxidation (Yamaguchi-Shinozaki and Shinozaki, 2006). Besides this a number of other transcription factors like HOS1, ICE1, SIZ1, MYb15 are also attributed cold tolerance of plant species as well as cross tolerance to salinity and water stress also. Several transcription factors, exclusively for some regulatory genes have been dictatorial for different developmental processes in plants. *CBF,* is belonging to the class of those genes conferring tolerance for both woody and herbaceous plants (Wisniewski *et al.,* 2014). Interestingly, the precise roles of *CBF* do also vary in woody and herbaceous plants in relation to binding affinity of transcription factors to those cis element in gene concerned. More so, *CBF* can also accommodate the growth processes of plants like timing and development in flowering in few plant species. Regardless the cases these all are also involved with ABA independent path in relation to DREB1/CBF interaction under dehydration/salinity stress. Collectively, all these facts and figures have been the supplementary for the complexity of CBF mediated dehydration tolerance (Mizoi *et al.,* 2012). It also allows the cross-talk as well as some degree of overlie may subsist in CBF and/or DREB mediated response to dehydration stress.

Recognition of Signal, Cellular Transduction and Responses at Molecular Level under Water/Salinity Stress

Out of a number of plant/crop species the gene pool so far studied, rice and *Arabidopsis* have been more exercising. In case of the first crop, a total number of 5000 and 6000 genes are up and down regulated respectively under drought as well as salinity induced drought stress (Maruyama *et al.,* 2014). With this expression of genes in any way the essential proteins and metabolites are featured to compensate the cellular anomalies under drought stress. However, these are highly species specific. To substantiate it is recorded a high expression for TCA cycle regulatory enzymes like isocitrate lyase and malate synthatase in rice to replenish the keto acids in energy yielding paths. But, in *Arabidopsis* these expressions were found albeit under similar condition of stress. Similarly, in case of rice also the down regulated activities of P450 735A was found to be accompanied in declining trend with cytokinin content in leaf tissues, however it remained inconsistent in root (Maruyama *et al.,* 2014). But this was lacking with *Arabidopsis* sp. Still, another transcription factor OREB1-bZIP, orthologous to *Arabidopsis* have found to be similarly interacted when rice seedlings were treated with ABA for assessment of tolerance for drought stress (Hong *et al.,* 2011). It was already established that rice registers a significant interaction taking DREB as a transfactor in ABA-independent pathway. With further analysis scanning the whole genome sequence rice contains at least five DREB1 genes, of which *OsDREB2A* and *OsDREB2B* are quite consistent with drought stress. While the role of others the sequence homology though matches well, still, functional modules are not consistent with those of *OSDREB2A and OSDREB2B*. In reference to rice the abundance and expression specificity also varies according to stressors. With an example transgenic rice expressing those two genes bestowed drought tolerance (Cui *et al.,* 2011). *OsDREB2A* has two mRNA to transcribe: *OsDREB2A1* and *OsDREB2A2*. The former interestingly encodes a non functional protein of a mini open reading frame. But the second one encodes a long ORF which matches AP2/ ERF DNA domain from *Arabidopsis*. Moreover, this transcript is abundant in regardless of abiotic stressors but *OsDREB2A1* are not expressed under cold stress. This draws a conclusion that *OsDREB2A2* may be regulated by an alternative splicing mechanism. The expression of other transcripts like *OsDREB1A* and *OsDREB2A* are regulated by cold treatments where as *OsDREB1F* and *OsDREB1G* are also induced by dehydration stress (Wang *et al.,* 2008). Though DREB1 does not participate in any stress tolerance, OSBDERB1F can bind as factor to the regulatory step for ABA independent pathway in rice. In transgenic rice the over expression of *OsDREB1F* can accommodate for cold resistance by co expression of Os1EC1, Os1EC2, Os1Z1 and Os1Z2 etc (Nakamura *et al.,* 2011).

Availability of Transcription Factors under Water / Salinity Stress in Plants

In plant system the availability and abundance of transcription factors has been a major exercising domain to find the possible paths for gene regulation. Basic leucine zipper (b Zip), the family of transcription factor that find homology of many other proteins in regulation of ABA dependant pathway for water/salinity tolerance (Antoni *et al.,* 2011). In its molecular structure one part of this protein contributes in DNA binding domain to a specific *cis* element and other part has the affinity between two polypeptide chains. The basic amino acids like arginine, leucine can share the DNA binding domain function and constitute the transcription factors. b-zip factors are supposedly the most conserved DNA binding protein that from the evolutionary point of view stands to be most primitive type. In more details there are four genes to produce the protein for b-zip factors and those are very common ancestral of all recent plants (Ishiguro *et al.,* 1994). In recent study it revealed that in rice an Osb-Zip23 shares some homology between ABF /AREB of *Arabidopsis* in regulation of ABA dependant pathway. Moreover, the expression was more acute in germinating seedlings than post germination period thereby desiccation tolerance is imparted by these homologous class of factors. In fact, Osb-Zip23 acts as a activator to transcription rather than a direct transcription factor. Two distinct domains both at N and C terminal (1-59) and (210-240) respectively are more important to lead it as transcriptional activator (Zou *et al.,* 2010) In more recent analysis it is clear that Osb-Zip23, a common activator for related genes under ABA dependant path and on downstream more than hundred genes are activated in that category. Mostly proteins adhering to desiccation tolerance (LEA, dehydrin), transcription factor (ERBP, AP1) and protein kinases are most important. Another transcriptional factor OsbZip46 has drawn more attention in relation to the family of OsbZip23 mediated interaction of *cis* elements in rice. In this member of OsbZip23, few amino acid residues (122-219) have a negative interaction on transactivation activity. Therefore, by simple deletion mutation technique of this amino acid domain a constitutive mutant was developed entitled OsbZip46(*CA1*) (Tang *et al.,* 2012). This when over-expressed in rice recorded several fold higher activity in other genes related to drought insensitivity. On detection through micro array analysis these down regulated genes found to be different from OsbZip23 down regulatory genes. Thereby it could be predicted that OsbZip (CA1) controls for those genes different by OsbZip23. It is interesting to note that plant height remains hardly sensitive on over-expression of OsbZip46(CA) than control. However, on transgenic plant with application of ABA made a significant shortening of shoot. This was equally compatible with other growth attributes on over-expressed plants with OsbZip23. This may be conclusive that ABA dependant path is equally shared by both OsbZip 23 and OsbZip 46 in relation to growth suppression under drought stress. Physiologically the subdued growth status of plants under water deficit is accomplished to conserve more water through mediation of ABA dependant pathway also (Huang *et al.,* 2012). Therefore, transgenic rice over-expressing the OsbZip23 is trailed to realize adequate growth and vigor despite of rendering the stress tolerance under soil moisture stress. Another homologue, ABF3 of b-Zip class of protein found *Arabidopsis* shares the homology with OsbZip23 and OsbZip46. In this plant also transgenesis with ABF3 showed an increased tolerance to salinity coupled with osmotic stress but appreciably with no growth alteration (Kang *et al.,* 2002). Another class of bZip family in rice is exampled by OsbZip71 and OsbZip16 (belonging to the family IV) which are both different from different amino acid domain from ABF3. Those are more different from AREB, another family in *Arabidopsis.* A very contradictory and distinct expression pattern was found in over-expressed rice with OsbZip 16. Interestingly, plants showed induced tolerance to stress but no significant changes in growth attributes with ABA application (Chen *et al.,* 2012). On the other hand a multi stress resistance variety was developed with transgenesis by OsbZip 71 insert cloned under 35S CaMV or dehydration inducible RD29A promoter. Still, under control condition the recombinant plants had hardly scored any changes in growth attributes as compared to stressed condition. This all together probe the possibility of transgenic varieties with triple accomplishment with ABA content of plants, the cis element in ABA sensitivity and finally the bZip class of transcription factors finding homology amongst themselves. Still, more attention and investigation are required prior to draw any analogy for the possible stress mitigation with over expression of those factors. Because compromising the growth and development in expense of transgenic expression, however, not commutable for stress tolerance. An increase in ABA content under drought and allied stress in almost the cases is an integral feature and thereby transgenesis might remain overburden for plant breeders.

A number of promoters both constitutive and inducible ones have been reported with high to moderate efficiency in rice transformation. Likewise taking *Arabidopsis* as model, DREB1A has set the most common drought inducible factors in this species. Few numbers of groups have reported with the use of ubiquitin promoter in rice also the over-expression of DREB1A gene and recorded variable degrees of stress tolerance (Oh *et al.,* 2005). Interestingly, these transformed plants have shown to be insensitive in growth retardation under non stressed condition. Thereby it suggests that DREB1A might have modulated the stress tolerance in such a way that plants are to compromise its growth under salinity. On the contrary, it makes confusion for some transformed plants with their higher yield potential under stressed condition while using the RD29 promoter for DREB1A over-expression (Datta *et al.,* 2012). The same transformation has enriched the potential of plants for muti stress tolerance like cold, heat and water stress also (2006). In most of the cases plants were characterized physiologically with higher accumulation in proline, sugars and other osmolytes content responsible for osmoprotectant. With DNA micro array analysis, the transgenic plants were remarkably shown to up regulate some downstream genes like ß-amylse and dehydrin, a LEA class of protein (Oh *et al.,* 2005). On the contrary, in some cases it recorded some retardation of transformed plants with this gene showed some discrepancies with others through some observations of stunted growth of rice shoots (Ito *et al.,* 2006). In this study a series of transgenic plants were developed with different inserts taking rice homogue of DREB1A, *Arabidopsis* paralogue DREB1B and DREB1C, rice orthologues *Os* DREB1A as well as *Os* DREB1B are the most frequently used constitutive promoters like Chl a/b binding proteins and ubiquitin. Efforts have also highlighted the possibility of success for drought tolerance in hybrid rice also. The interspecific hybrid species (*Oryza glaberrima* crossed with *Oryza sativa*), NERICA1 when transformed with *Os* DREB1C recorded a significant drought tolerance but partial fall in yield. It is well coordinated with the interplay of different transcription factors, even specific for a particular stressors, still, showed some cross protection against drought stress (Ishizaki, *et al.,* 2012). Likewise, from barley a factor, *HVCB4* sensitive to cold reaction had promised to confer drought tolerance when transformed in rice. In fact, HVCB4 is a family of *Arabidopsis* DREB1A family. In this case a multi stress tolerance has characterized the plants establishing the regulation in a more complex web with different trans factors induced under dehydration (Oh *et al.,* 2007). Among the transfactors the *Arabidopsis* DREB2A has been widely studied in relation to diverse stress tolerance in many crop species. This has been a master regulator for both drought and related heat stress induced factors in few crop species (Mizoi *et al.,* 2012). More so, the regulatory paths for DREB2 under ABA independent regulation has shared almost similar cascade in perception of stress signal. Even in rice there recorded at least five DREB2 family variants: *OsDERB2A, Os DREB2B, OsDREB2C, OsDREB2F* and *OsABI4.* Arabidopsis transgenic with *OsDREBA* has shown more tolerance to depleted moisture level in soil (Matsukura *et al.,* 2010). Over expression with *DREB2B* in *indica* rice under ABRE promoter has also been successful in retention of tissue moisture by overproducing soluble sugars and other compatible solutes. ABRE is the promoter sequence of those gene under the induction of ABA under osmotic shock. In a study of Cui *et al.* (2011) the plants were more tolerant at booting stages following moisture deficit and producing more effective tillers bearing panicles. In addition to DREB factors other ones, most commonly animal in origin are also coordinating the stress response in ABA independent path. These especially include AP2/ERF and alike factors contribute the stress tolerance in rice plants.

Ethylene responding transcription factors are next in abundance to b-Zip classes

In addition to ABA another plant growth regulator subjected for salinity/drought tolerance is ethylene (C_2H_2). A number of cellular events are modified in a similar way for plant's responding to salinity stress by ethylene in a similar as ethylene. Along with DREB1A and DREB2, a few transcription factors like AP2/ERF type and AP2/ERF like factors confer tolerance to salinity in many transgenic plants like rice also (Matsukura *et al.,* 2010). In *Arabidopsis* the detection of one gene coding the AP2/ERF like transcription factor is *HARDY* gene. This gene has immense activity in reduction of photorespiration as well as enhanced photosynthetic activities when transformed with HARDY gene. An significant increase in root and shoot biomass also included as another advantages for such transformation. In more details way a transcription factor OsDERF1 binds the CGG box of the ~300bp of upstream of *Os*ERF3 and *Os*AP2-39 gene of rice. A mutation with *Os*ERF3 leads to increase an ethylene production along with drought tolerance in rice (Zhang *et al.,* 2013). On the contrary, an over expression of OsERF3/OsERF4a had minimized the repressor protein and simultaneously increased the defense resistance, shoot growth as well as water saturation at tissue level. Likewise an over expression of ethylene response factor from tomato increased the osmotic stress tolerance without any changes of growth and development. An antisense through *OsERF3* in knockdown process causes the increase of C_2H_2 biosynthesis as well as stress resistance. This definitely may suggest that water stress tolerance also involves C_2H_2 in involvement, however, in other paths. Rice transformed with *OSERF3* also recorded a fall in drought tolerance (Zhang *et al.,* 2013). On the contrary, transgenic with double inserts of OsERF3/OsERF4a can indirectly induce the stress tolerance by inactivating a repressor protein that may down regulate the dehydration stress like decrease in foliages and vegetative growth. The transgenic with over-expression of ethylene related factors (e.g., ERF gene from other plants) can interact other transcription factors like MYC/MYB from maize. Other analogues of ERF like *JERF*$_3$ can interact the genes related to ABA biosynthesis, proline metabolism and photosynthesis (Quan *et al.,* 2010). Transcription factors like AP2/ ERF in nature, as for e.g., *AP37* when over-expressed in rice shown enhanced tolerance to salinity, drought and cold. The genes can regulate the most important down -stream proteins like zinc fingers and membrane transporter for metals like $Fe^{+3/+2}$. There are also some cold binding factors may interact with ethylene response factors for drought tolerance as reported from rice. Likewise, for AP2/ERF factors few genes (*DREB1/CBF*) have been reported for development of transgenic plants in various species like

Pisum (Bhatnagar-Mathur *et al.,* 2013), *Glycine* (Polizel *et al.,* 2011), *Nicotiana* (Kasuga *et al.,* 2004) etc.

Molecular Probing for Traits of Salinity Tolerance with Regulatory Elements

The regulatory elements for gene transcription are paid a special attention for the probing the physiological traits adhered to salinity. In most of the cases the physiological alterations to invade the salinity impact is mediated by some interaction of cis element and trans factors. In rice and other cereals a number of such cis elements have been identified and analyzed with their homology and phylogenic lineages with other higher plants. With regards to such elements, some of the plants are also possessing ABRE and showed ABA sensitivity also (Weiner *et al.,* 2010). Even unrelated transcription factors like Arabidopsis EDT1/HDG11 encoding a homeodomain leucine zipper on over-expression in rice recorded activation of ABA inducing genes by interaction with ABRE. An enhanced root length couple with thick cuticle necessarily imparts tolerance to water deficit under salinity also (Yu *et al.,* 2013). In this transformed plants related antioxidation capacity, compatible solutes like glycine betaine, proline are also associated traits for water stress resistance with several fold increase in ABA content. Interestingly, the interaction of other stress trans-factors like SNAC1, SNAC2, NCED3, OsbZIP23 were also more important to be induced with adequate grain yield also (Zhao *et al.,* 2014). Maize is another potential source for cloning of some important transcription factor to transfer traits onto other cereals like rice also. Likewise, *Zmhgz10,* a homeodomain transcription factor when made in transgenesis in rice it recorded a significant inhibition of growth inhibition with application of exogenous ABA. This shows that there must have some negative affinity for similar DREBP factor with other genes that down regulate the insensitivity to ABA. In rice other MYB factors (*OsMYB*2) had experienced similar result when cloned under ABRE promoter (Yang *et al.,* 2012). Another most important factor is WRKY. WRKY factors contain 2-3 WRKY domain. The WRKY domain contains 6—70 amino acids chain for DNA binding (Zhang *et al.,* 2005.). This domain is characterized with a highly conserved core sequence WRKYGQK and a zinc finger protein. The two essential domains are histidine and cystein containing $CX_{4\text{-}5}CX_{22\text{-}23}HXH$ or $CX_7CX_{23}HXC$ (X= any amino acids). The interaction between WRKY domains happens in the major grove of DNA with its typical sequence (T/A)TGAC(T/A) cis-element (W-box). These transcription factors strictly recognize a typical cis element : (T/A)TGAC(T/A), also to be called as W-box. Apart from recognition of plants in abiotic as well as biotic stress, WRKY factor also can sense other few modules of plants development and some secondary metabolism also. These functionales are interacted through the DNA binding affinity with other DNA binding or non-DNA binding protein domains in plants (Shen *et al.,* 2012). A number of molecular events are recognized in respect to

WRKY mediated stress tolerance: variation in cis element recognition, protein –protein interaction, phosporylation, proteosomeal degradation etc (Wang *et al.,* 2007). The most fascinating factor for WRKY interaction in plant stress response is based on diversity in cis element recognition. From the earlier literatures the W-box contains the motif of (T/A) TGAC(T/A). This refers to the Early work indicated that the WRKY family could bind W-box). The non specificity of WRKY factors was established from the observation that SUSIBA2, a factor from Hordeum could bind the cis element (TAAAGATTACTAATAGGAA) for sugar response. Still, the core sequence for WRKY is assigned to have GAC with flanking sequence for DNA–Protein interaction. Still, the differences among the WRKYs are on protein side as well as downstream arginine and lysine residues (Cui *et al.,* 2011). These two amino acids residues may be liable for interaction of phosphate backbones in consecutive seven DNA base pairs of a strand. Any changes for amino acids mentioned (e.g. tryptophan, tyrosine or even lysine of WRKYGQK motif could turn over the binding of DNA at least partially. The flanking sequence helps to dictate the binding of DNA strength or DNA binding motif. On the other hand, the substitution with alanine for arginine, glutamine or glycine had not mate with any sort of significant changes in DNA binding abilities. In brief, this WRKY protein-DNA interaction can result in different gene activation/ induction under various abiotic stresses.

The evolution of WRKY factors bears some relevance in analysis phylogeny of different crop species. There recorded some diffused and patchy relationship amongst the plant lineage and even outside of it. The later includes few diplomonads, amoebae, fungi and even some amoebozoa also (Degenkolbe *et al.,* 2009). This type of distribution may be suggestive of the fact that there might have some lateral gene transfer by which the update formation of WRKY groups appeared in past. With this concept, in brief, a well defined 7 groups in their discrete lineages had appeared in the form of Groups I + IIc, Groups IIa + IIb, Groups IId + IIe, and Group III. In another school of thoughts in angiosperms posses some proteins with domains typified for both resistances (R) proteins having some lineages with WRKYGQK motif (Dubouzet *et al.,* 2003). Flowering plants also contain proteins with domains typical for both resistance (R) proteins and WRKY transcription factors. In studies it also revealed that R protein-WRKY genes had evolved a number of times in angiosperms. It is interesting to note more that each of those were to some precise angiospemic plants lineages. These heteromeric assemblage of proteins contain some unique combination. These chimeric proteins contain not only novel combinations of protein domains but also novel combinations and numbers of WRKY domains.

Other Transcription Factors Related to Abiotic Stress

Most of the cases these include the protein kinases related to phosphorylation and dephosphorylation in imparting the stress tolerant of plants.

As for e.g.*OsCPK4*, a protein kinase with ATP dependency showed a combination of gene induction for drought and salinity tolerance (Campo *et al.,* 2014). This was found in transgenic rice plants with over-expression with other genes like those of lipid degradation. These cover lipid binding proteins in lipid tranferases reactivity from cell membrane to sub-cellular fractions and some lipases (Jiang *et al.,* 2012). Protein kinases with references to transfer of inorganic phosphate (Pi) to those of oxidative stress rsponses like anti-oxidative enzymes of peoxidase groups, catalases, GST were te most important (Hoetal *et al.,* 2013). These are all in support for membrane protection of ROS under induced salinity as well as water deficit in crops. Transgenic rice plants are most important and significant in over-expressing of some those like *OsCDPK1* to consider those tolerance attributes for salinity. Other important proteins like those share with PR protein in plants like 14-3-3 protein, GF14C etc for their over-expression in rice plants. It is assumed that plants expressing GF14C is mediated by phosphorylation with OsCPK4 enhancing the activation of some important proteins in antioxidant cascade (Mizoi *et al.,* 2012). In another study it also revealed that a maize transcription factor *ZmGF14-6,* a homologue of 14-3-3 protein showed an over-production of those enzymes. In their study it also revealed that other genes for salinity tolerance like *Rab 21, Dip 1* when expressed in transgenic rice showed a lower tolerance to fungal invasion also suggesting the co -linearity to both abiotic and biotic stress. Therefore, the *ZmGF14-6* has a dual roles in conferring the tolerance for abiotic and biotic stress, however, both share the commonness through oxidative stress. This has been contradicted by others that the same gene, ZmGF14-6 has improved the tolerance to other abiotic stressors like chilling, drought and salinity (Xiang *et al.,* 2007). The homologue of *OsCDPK1* like *OsCIPK12* was found to increase drought tolerance in a similar way for rice as an affirmative modulator for stress tolerance. Physiologically plants were evaluated with the enhanced activities of proline and soluble sugar accumulation as a common marker for stress tolerance.

EXPRESSION OF TRANSGENIC WITH LEA PROMOTER

Higher or angiospermic plants are well adapted to terrestrial ecosystem overcoming every sort of their hazardous in soil and environment. This is more prelude with the scarcity or excess of water inputs essential to make hydration of the tissues. Proteins are there even in the developing embryos in seed also contributory to desiccation tolerance in the early phases of seed development. Late embryogenic abundant proteins (LEA) belongs to class of proteins that constitute the basic leucine zipper class with high fidelities of basic amino acids like arginine and lysine (Tuteja *et al.,* 2013). The evolution of LEA proteins encompasses the possible changes which has been accommodating the lineage between cellular dehydration resistance and lowering the osmotic potential under soil moisture deficit. Therefore, it is advantageous to decipher the functionalism of their promoter efficiencies under ongoing dehydration in gene expression for LEA and other related dehydrin class of proteins. Predominantly, to ensure food security it is the two prime elements, the crop sustainability and grain production mostly concerned with drought tolerance. So, it is the collection of proteins that interfere with the selection of species focusing the molecular mechanism in induction of the LEA related/allied genes. In the crop species the LEA ranges 10 KD to 300 KD depending on tissues specific localization. Irrespective of plant organs LEA are exclusively localized in the cytoplasm and nuclear organizing zones. Based on the mRNA homologies and alignment of amino acids in different taxa, LEA are categorized within 1-5 groups (Sharma *et al.*, 2013).

From the cis element analysis almost the LEA promoters are responsive to ABA induction. The homology between RD 29 (Responsive to dehydration 29) gene is also having with the equivalent core region of promoter but varying flanking sequence also shows ABA dependency in induction under water stress. Many of LEA protein genes can be induced by ABA. *Arabidopsis* thaliana L. drought-induced gene RD22 is regulated by ABA [42]. Under severe stress, two cis elements (MYC and MYB-like element) were identified, which were cooperative to ABA-induced-gene expression. From the homology model it is assumed that the DNA binding protein encoded by gene rd22 BP1 has the characteristics feature resembling MYC-like sequence. More so, this is also induced by drought and ABA like other *Rab* class of genes (Kang *et al.,* 2002). One of the feature of the promoter of these genes that those are in concomitant in expression as a function of duration in water deficit stress. These expressions attained their peak with a decrease in the progressive stress out of water deficit. On recovery phase the expression was transitory with the above level. Therefore, it is confirmed that short but rapid exposure is hardly able to make any significant changes in gene expression for those of LEA proteins. In most of the LEA promoter, it is the ABRE that governs the sensitivity of ABA in plants under desiccation condition with a core sequence CACGTG and allied a G Box. The promoter ligated with wheat Emla and rice Rab 16A gene could confer the tolerance in other crops in transformation experiment tested with 35S promoter as well as other binary vector like pcambia1201. At a glance regardless of crops there recorded three predominant genes: PRD8, PRD14 and PRD22, expressed under water deficit. However, hardly those remain sensitive with cold or heat stress. Still, PRD and PRD14 are expressed under exogenous induction with ABA which is not with PRD22. This may possibly be the indicative of the fact that there should be different transduction pathways for PRD either with ABA mediation or non ABA mediated channel. In addition the promoters for mutant genes like *Rab*, *ABAI* responded well both with drought or salinity, however, suggesting the possibility for other substitute for ABA non-dependency. Related other genes and their cis elements acted for drought are communal with to some level with LEA related promoter. This is more established at least, on their flanking sequences (Kang *et al.,* 2002). Likewise, rd29A, lti78 and cor78 have showed their matching in homologies of cis elements as well as the binding factors to it. With this concept it is prevalent that promoter responses might be conferred with two possible elements and theirs interplay or interchange. The ABREs and non ABREs are great caring in their homologies regularly in core sequences as well as flanking sequences. Further dissection in fine structure of such promoter with reference to *rd29* suggests it is with an essential sequence: TACCGACAT, more closely to DRE motif (Kasuga *et al.,* 2004). A fusion between DRE of *rd29A* with GUS and in use for *Arabidopsis* transformation led to development in their mutual behavior under heat, drought and even cold stress alsom (Kasuga *et al.,* 2004, Polizel *et al.,* 2011).

CONCLUSION

Therefore, it is very much prudent that the key element for a gene to express in any physiological features is the transcription factors. From the ongoing discussion of the different facets of transgenes to be over-expressed in crop species it comes in question how a foreign DNA sequence has been the potential to accommodate within the genome space. Moreover, it remains still in some sort of enigma that *cis* elements are though variable in their tandem repeat, however, has the relatedness on their flanking/core sequences in almost the cultivated crop species. This gives the impetus that hardly crops are there in cultural practices where domestication of any gene/genes under the same regulatory elements is not the effects of mutualism of those sequences. Different crops withstanding any sort of abiotic stress are the variation of gene expression, in quantitative, however, not in qualitative manner. The expression potential depends on the copy number of transgenes, its stable integration as well as synergistic expression with other allied genes (Deinlein *et al.,* 2014). Understanding well the mechanism of abiotic stress resistance it becomes prudent that multiple transgenics ought to be more realistic than to single gene transfer, if the crops are accommodated at al. It's still remains subtle that a transcription factor/ factors are really the driver for a particular gene to over-express subjected to the choice of stressors. Whether the regulatory elements already present on a gene needs to be over expressed in association with a strong promoter in agro-vector or a direct cloning of Trans factors on harbor on *cis* element. It should be taken in account the metabolic traffic load of a metabolite/ metabolites in any physiological path to circumvent in accomplishment of stress tolerance (Parent *et al.,* 2010) Another insight must be probing in transgenic assessment with regards to agronomic screening is the stableness and heritability of a transcription factors in over-expression of a particular gene product. The behavior of a particular gene for long/ short term responses and regardless of edhapic condition allows more critical or stringent probing for declaration of a stress tolerance variety under either natural or simulated condition. Therefore, assuming that the highest efficiency is extracted by certain promoters it is still yet to understand which transcription factors has unanimous binding ability over a wide range of *cis* element for a number genes concerned to abiotic stress tolerance. Still, it is extremely required to fill the knowledge gap among a number of variables with genomic constitution and functinolaes, diversity in total gene pools, coverage in gene expressivity and corroboration of those into physiological activities. Those all together may assist in selection of complex traits more specifically to develop suitable quantitative trait loci (QTL) to select from existing crop population or even to select a transgenic with stronger phenotypic traits.

REFERENCES

Antoni, R., L., Rodriguez, M. Gonzalez-Guzman, GA. Pizzio and PL. Rodriguez. 2011. News on ABA transport, protein degradation, and ABFs/WRKYs in ABA signaling. *Curr Opin Plant Biol* 14: 547–553

Bauer, H., P. Ache, S. Lautner, J. Fromm, W. Hartung and K. A. S. Al-Rasheid. 2013. The stomatal response to reduced relative humidity requires guard cellautonomous ABA synthesis. *Curr. Biol.* 23, 53–57. doi: 10.1016/j.cub.2012. 11.022

Behnam, B., S. Iuchi, M. Fujita, Y. Fujita, H. Takasaki, and Y. Osakabe. 2013. Characterization of the promoter region of an *Arabidopsis* gene for 9-cisepoxycarotenoid dioxygenase involved in dehydration-inducible transcription. *DNA Res.* 20, 315–324. doi: 10.1093/dnares/dst012.

Bhatnagar-Mathur, P., J. S. Rao, V. Vadez, S. R. Dumbala, A. Rathore and K. Yamaguchi Shinozaki. 2013. Transgenic peanut overexpressing the DREB1A transcription factor has higher yields under drought stress. *Mol. Breeding* 33, 327–340. doi: 10.1007/s11032-013-9952-7

Campo, S., P. Baldrich, J. Messeguer, E. Lalanne, M. Coca, and B. SanSegundo. 2014. Over expression of a calcium-dependent protein kinase confers salt and drought tolerance in rice by preventing membrane lipid per oxidation. *Plant Physiol.* 165, 688–704.doi:10.1104/pp.113.230268

Chen, H., W. Chen, J. Zhou, H. He, L. Chen and H. Chen. 2012. Basic leucine zipper transcription factor OsbZIP16 positively regulates drought resistance in rice. *Plant Sci.* 193–194, 8–17. doi: 10.1016/j.plantsci.2012.05.003

Cui, M., W. Zhang, Q. Zhang, Z. Xu, Z. Zhu and F. Duan. 2011. Induced overexpression of the transcription factor OsDREB2A improves drought tolerance inn rice. *Plant Physiol. Biochem.* 49, 1384–1391. doi: 10.1016/j.plaphy. 2011.09.012

Datta, K., N. Baisakh, M. Ganguly, S. Krishnan, K. Yamaguchi Shinozaki and S. K. Datta. 2012. Overexpression of Arabidopsis and rice stress genes' inducible transcription factor confers drought and salinity tolerance to rice. *Plant Biotechnol. J.* 10, 579–586. doi: 10.1111/j.1467-7652.2012. 00688.x

Degenkolbe, T., P. T. Do, J. Kopka, E. Zuther, D. K. Hincha and K. I. Köhl. 2013. Identification of drought tolerance markers in a diverse population of rice cultivars by expression and metabolite profiling. *PLoS ONE* 8:e63637. doi: 10.1371/journal.pone.0063637

Deinlein, U., A.B. Stephan, T. Horie, W. Luo, G. Xu and J.I. Schroeder. 2014. Plant salt-tolerance mechanisms. *Trends Plant Sci.* 19, 371–379. doi: 10.1016/j.tplants.2014.02.001

Duan, J. and W. Cai. 2012. OsLEA3-2, an abiotic stress induced gene of rice plays a key role in salt and drought tolerance. *PLoS ONE* 7:e45117. doi: 10.1371/journal.pone.0045117

Dubouzet, J. G., Y. Sakuma, Y. Ito, M. Kasuga, E. G. Dubouzet, and S. Miura. 2003. OsDREB genes in rice, *Oryza sativa* L., encode transcription activators that function in drought-high-salt-, and cold-responsive gene expression. *Plant J.* 33, 751–763.doi:10.1046/j.1365-313X.2003.01661.x

Golldack, D., C. Li, H. Mohan and N. Probst. 2014. Tolerance to drought and salt stress in plants: unraveling the signaling networks. *Front. Plant Sci.* 5:151. doi: 10.3389/fpls.2014.00151

Hanin, M., F. Brini, C. Ebel, Y. Toda, S. Takeda, S., and K. Masmoudi. 2011. Plant dehydrins and stress tolerance: versatile proteins for complex mechanisms. *Plant Signal. Behav.* 6, 1503–1509. doi: 10.4161/psb.6.10. 17088.

Hasegawa, P.M., R.A. Bressan, J.K. Zhu and H.J. Bohnert. 2000. Plant cellular and molecular responses to high salinity. Annu Rev Plant Physiol Plant Molec Biol 51: 463–499.

Hong, J. Y., M. J. Chae, I. S. Lee, Y. N. Lee, M. H. Nam, and D. Y. Kim. 2011. Phosphorylation-mediated regulation of a rice ABA responsive element binding factor. *Phytochemistry* 72, 27–36. doi: 10.1016/j.phytochem. 2010.10.005

Huang, J., S. Sun, D. Xu, H. Lan, H. Sun and Z. Wang. 2012. A TFIIIA-type zinc finger protein confers multiple abiotic stress tolerances in transgenic rice (*Oryza sativa* L.). *Plant Mol. Biol.* 80, 337–350. doi: 10.1007/s11103-012-9955-5

Ishiguro S. and K. Nakamura K. 1994. Characterization of a cDNA encoding a novel DNA-binding protein, SPF1, that recognizes SP8 sequences in the 5' upstream regions of genes coding for sporamin and beta-amylase from sweet potato. *Mol Gen Genet* 244: 563–571.

Ishizaki, T., K. Maruyama, M. Obara, A. Fukutani, K. Yamaguchi-Shinozaki and Y. Ito. 2012. Expression of *Arabidopsis* DREB1C improves survival, growth, and yield of upland New Rice for Africa (NERICA) under drought. *Mol. Breed.* 31, 255–264. doi: 10.1007/s11032-012-9785-9

Ito, Y., K. Katsura, K. Maruyama, T. Taji, M. Kobayashi and M. Seki. 2006. Functional analysis of rice DREB1/CBF-type transcription factors involved in cold-responsive gene expression in transgenic rice. *Plant Cell Physiol.* 47, 141–153. doi: 10.1093/pcp/pci230

Jiang, S.Y., R. Bhalla, R. Ramamoorthy, H.-F. Luan, P.N. Venkatesh and M. Cai. 2012. Over-expression of OSRIP18 increases drought and salt tolerance in transgenic rice plants. *Transgenic Res.* 21, 785–795. doi: 10.1007/s11248-011-9568-9.

Kang, J., H. Choi, M. Im and S.Y. Kim. 2002. *Arabidopsis* basic leucine zipper proteins that mediate stress-responsive abscisic acid signaling. *Plant Cell* 14, 343– 357. doi: 10.1105/tpc.010362.tase

Karaba, A., S. Dixit, R. Greco, A. Aharoni, K.R. Trijatmiko and N. Marsch-Martinez. 2007. Improvement of water use efficiency in rice by expression of HARDY, an *Arabidopsis* drought and salt tolerance gene. *Proc. Natl. Acad. Sci. U.S.A.* 104, 15270–15275. doi: 10.1073/pnas.0707294104

Kasuga, S. Miura, K. Shinozaki and K. Yamaguchi-Shinozaki. 2004. A combination of the *Arabidopsis* DREB1A gene and stress inducible rd29A promoter improved drought- and low-temperature stress tolerance in tobacco by gene transfer. *Plant Cell Physiol.* 45, 346–350. doi: 10.1093/pcp/pch037

Kim, H., H. Hwang, J.-W. Hong, Y.-N. Lee, I.P. Ahn and I.S. Yoon. 2012b. A rice orthologue of the ABA receptor, OsPYL/RCAR5, is a positive regulator of the ABA signal transduction pathway in seed germination and early seedling growth. *J. Exp. Bot.* 63, 1013–1024. doi: 10.1093/jxb/err338

Kim, J.S., J. Mizoi, S. Kidokoro, K. Maruyama, J. Nakajima and K. Nakashima. 2012a. Arabidopsis growth-regulating factor7 functions as a transcriptional repressor of abscisic acid- and osmotic stress-responsive genes, including DREB2A. *Plant Cell* 24, 3393–3405. doi: 10.1105/tpc.112.100933

Kuromori, T., E. Sugimoto, and K. Shinozaki. 2014. Intertissue signal transfer of abscisic acid from vascular cells to guard cells. *Plant Physiol.* 164, 1587–1592. doi: 10.1104/pp.114.235556

Maruyama, K., K. Urano, K. Yoshiwara, Y. Morishita, N. Sakurai and H. Suzuki. 2014. Integrated analysis of the effects of cold and dehydration on rice metabolites, phytohormones, and genetranscripts. *Plant Physiol.* 164,1759–1771. doi: 10.1104/pp.113.231720

Matsukura, S., J. Mizoi, T. Yoshida, D. Todaka, Y. Ito and K. Maruyama. 2010. Comprehensive analysis of rice DREB2-type genes that encode transcription factors involved in the expression of abiotic stress-responsive genes. *Mol. Genet. Genomics* 283, 185–196. doi: 10.1007/s00438-009-0506-y

Mittler, R., S. Vanderauwera, M. Gollery and F. Van Breusegem. 2004. Reactive oxygen gene network of plants. *Trends in Plant Sci* 9: 490–498.

Miyakawa, T., Y. Fujita, K. Yamaguchi-Shinozaki and M. Tanokura. 2013. Structure and function of abscisic acid receptors. *Trends Plant Sci.* 18, 259–266. doi: 10.1016/j.tplants.2012.11.002

Mizoi, J., K. Shinozaki and K. Yamaguchi-Shinozaki. 2012. AP2/ERF family transcription factors in plant abiotic stress responses. *Biochim. Biophys. Acta* 1819, 86–96. doi: 10.1016/j.bbagrm.2011.08.004

Oh, S., S.I. Song, Y.S. Kim, H. Jang, S.Y. Kim, and M. Kim. 2005. *Arabidopsis* CBF3 / DREB1A and ABF3 in transgenic rice increased tolerance to abiotic stress without stunting growth. *Plant Physiol.* 138, 341–351. doi:10.1104/pp.104.059147.1

Oh, S.J., C-W. Kwon, D-W. Choi, S.I. Song and J-K. Kim. 2007. Expression of barley HvCBF4 enhances tolerance to abiotic stress in transgenic rice. *Plant Biotechnol. J.* 5, 646–656. doi: 10.1111/j.1467-7652.2007.00272.x

Parent, B., B. Suard, R. Serraj and F. Tardieu 2010. Rice leaf growth and water potential are resilient to evaporative demand and soil water deficit once the effects of root system are neutralized. *Plant Cell Environ.* 33, 1256–1267.doi: 10.1111/j.1365-3040.2010.02145.x

Polizel, A.M., M.E. Medri, K. Nakashima, N. Yamanaka, J. R. B. Farias and M. C. N. de Oliveira. 2011. Molecular, anatomical, and physiological properties of a genetically modified soybean line transformed with rd29A: AtDREB1A for the improvement of drought tolerance. *Genet. Mol. Res.* 10, 3641–3656. doi: 10.4238/2011.October.21.4

Quan, R., S. Hu, Z. Zhang, H. Zhang, Z. Zhang and R. Huang. 2010. Over-expression of an ERF transcription factor TSRF1 improves rice drought tolerance. *Plant Biotechnol. J.* 8, 476–488. doi: 10.1111/j.1467-7652.2009.00492.x

Shen, H., C. Liu, Y. Zhang, X. Meng, X. Zhou and C. Chu, C. 2012. OsWRKY30 is activated by MAP kinases to confer drought tolerance in rice. *Plant Mol. Biol.* 80, 241–253. doi: 10.1007/s11103-012-9941-y

Tang, N., H. Zhang, X. Li, J. Xiao and L. Xiong. 2012. Constitutive activation of transcription factor OsbZIP46 improves drought tolerance in rice. *Plant Physiol.* 158, 1755–1768. doi: 10.1104/pp.111.190389

Todaka, D., K.Shinozaki and K.Yamaguchi-Shinozaki. 2015. Recent advances in the dissection of drought-stress regulatory networks and strategies for development of drought-tolerant transgenic rice plants. *Frontiers in Plant Science*, Doi: 10.3389/flps,vol. 6, Article 84.

Tuteja, N., R. K. Sahoo, B. Garg and R. Tuteja 2013. OsSUV3dualhelicase functions in salinity stress tolerance by maintaining photosynthesis and antioxidant machinery in rice (*Oryza sativa* L. cv. IR64). *Plant J.* 76, 115–127.doi: 10.1111/tpj.12277

Wang, C., P. Deng, L. Chen, X. Wang and H. Ma. 2013. A Wheat WRKY Transcription Factor TaWRKY10 Confers Tolerance to Multiple Abiotic Stresses in Transgenic Tobacco. *PLoS ONE.* 8(6): e65120. doi:10.1371/journal.pone.0065120

Wang, D., Y. Pan, X. Zhao, L. Zhu, B. Fu and Z. Li. 2011. Genome-wide temporal-spatial gene expression profiling of drought responsiveness in rice. *BMC Genomics* 12:149. doi: 10.1186/1471-2164-12-149

Wang, H.H., J.J. Hao, X.J. Chen, Z.N. Hao and X. Wang. 2007. Overexpression of rice WRKY89 enhances ultraviolet B tolerance and disease resistance in rice plants. *Plant Mol Biol* 65: 799–815.

Wang, Q., Y. Guan, Y. Wu, H. Chen, F. Chen and C. Chu. 2008. Overexpression of a rice OsDREB1F gene increases salt, drought, and low temperature tolerance in both *Arabidopsis* and rice. *Plant Mol. Biol.* 67, 589–602. doi: 10.1007/s11103- 008-9340-6

Weiner, J.J., F.C. Peterson, B.F. Volkman and S.R. Cutler. 2010. Structural and functional insights into core ABA signaling. *Curr. Opin. Plant Biol.* 13, 495–502. doi: 10.1016/j.pbi.2010.09.007

Wu, L., H. Zhang, X-C. Zhang, X-C. Wang and R. Huang. 2008. Transcriptional modulation of ethylene response factor protein JERF3 in the oxidative stress response enhances tolerance of tobacco seedlings to salt, drought, and freezing. Plant Physiol 148: 1953–1963.

Xiang Y., Y. Huang and L. Xiong. 2007. Characterization of stress-responsive CIPK genes in rice for stress tolerance improvement. *Plant Physiol.* 144, 1416– 1428. doi:10.1104/pp.107.101295

Xiong, L. and J.K. Zhu. 2002. Molecular and genetic aspects of plant responses to osmotic stress. *Plant Cell Environ* 25: 131–139.

Yamaguchi-Shinozaki, K. and K. Shinozaki 2006. Transcriptional regulatory networks in cellular responses and tolerance to dehydration and cold stresses. *Annu. Rev. Plant Biol.* 57, 781–803. doi: 10.1146/annurev.arplant.57.032905.105444

Yang, A., X. Dai, and W.H. Zhang. 2012. A R2R3-type MYB gene, OsMYB2, is involved in salt, cold and dehydration tolerance in rice. *J.Exp. Bot.* 63,2541–2556. doi: 10.1093/jxb/err431

Yoshida, T., N. Ohama, J. Nakajima, S. Kidokoro, J. Mizoi and K. Nakashima. 2011. *Arabidopsis* HsfA1 transcription factors function as the main positive regulators in heat shock-responsive gene expression. *Mol. Genet. Genomics* 286, 321–332. doi: 10.1007/s00438-011-0647-7

Yu, L., X. Chen, Z. Wang, S. Wang, Y. Wang and Q. Zhu. 2013. *Arabidopsis* enhanced drought tolerance1/HOMEODOMAIN GLABROUS11 confers drought tolerance in transgenic rice without yield penalty. *Plant Physiol.* 162, 1378–1391. doi: 10.1104/pp.113.217596

Zhang Y.J. and L.J. Wang. 2005. The WRKY transcription factor super family: its origin in eukaryotes and expansion in plants. *BMC Evol Biol* 5: 1–12.

Zhang, H., J. Zhang, R. Quan, X. Pan, L. Wan and R. Huang. 2013. EAR motif mutation of rice OsERF3 alters the regulation of ethylene biosynthesis and drought tolerance. *Planta* 237, 1443–1451. doi: 10.1007/s00425-013-1852-x

Zhao, Y., Q. Ma, X. Jin, X. Peng, J. Liu, L. Deng. 2014. A novel maize homeodomain-leucine zipper (HD-Zip) I gene, Zmhdz10, positively regulates drought, and salt tolerance in both rice and *Arabidopsis*. *Plant Cell Physiol.* 55, 1142–1156. doi: 10.1093/pcp/pcu054.

■■■■■

3

DROUGHT, AGRICULTURAL PRACTICES AND PRIVATE SECTOR LENDING FOR SUSTAINABILITY IN INDIA

Swati Agrawal[a], Manoj Kumar[b], Prerna Singh[b] and Kavita Shah[b,c]

[a]*Faculty of Commerce, Banaras Hindu University, Varanasi-221005, India*
[b]*Department of Biochemistry, Faculty of Science,*
Banaras Hindu University, Varanasi-221005, India
[c]*Institute of Environment and Sustainable Development,*
Banaras Hindu University, Varanasi-221005, India
Corresponding author: Prof. Kavita Shah; E-mail: kavitashah@bhu.ac.in

INTRODUCTION

Recently the plant breeding techniques have assisted to some extent, with the selection and development of crop cultivars adapted to drought. Prolonged drought causes starvation, famine, epidemics and death of lives and force mass migration of human and animals. Under drought stress the physiological and biochemical processes in plants are decreased. (Flexas *et al.,* 2004; Valentovic *et al.,* 2006; Westgate and Boyer, 1986). Almost half of the terrestrial land surface in the world comprises of dry lands and are distributed among all the different regions of the earth (Fig. 1).

Indian agriculture is heavily dependent on climate: a favourable southwest summer monsoon is critical in securing water for irrigating crops. Droughts, combined with policy failures, have periodically led to major Indian famines, including the Bengal famine of 1770, the Chalisa famine, the Doji bara famine, the Great Famine of 1876–78, and the Bengal famine of 1943. Some commentators have identified British government inaction as contributing factors to the severity of famines during the time India was under British rule. Famine largely ended by the start of 20th century with the Bengal famine of 1943 being an exception related to complications during World War II. The 1883 Indian Famine Codes, transportation improvements, and changes following independence have been identified as furthering famine relief. In India, traditionally, agricultural labourers and rural artisans have been the primary victims of famines. In the worst famines, cultivators have also been susceptible. Fig 2 shows major drought affected areas in India in the year 2016.

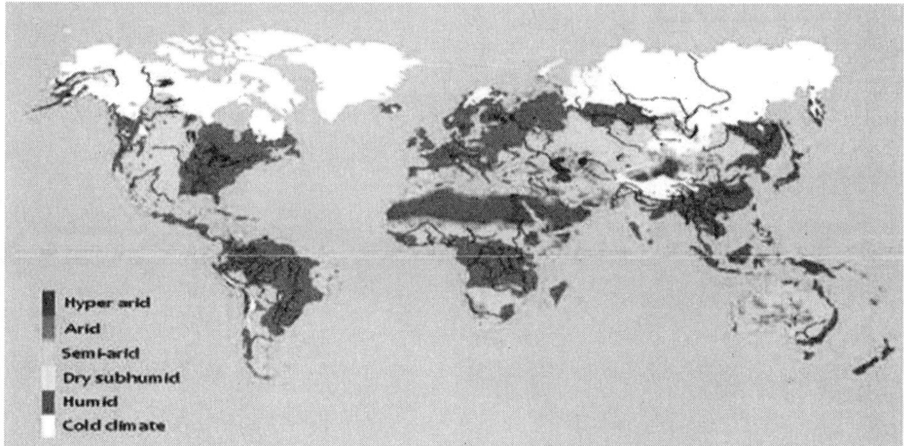

Fig. 1. Geographical distribution of drought affected areas in the world (Karim *et al.,* 2015)

The Indian economy is divided into have and have nots. Most of the farmers live below poverty line. Many drought stricken farmers are bound to commit suicide. They do not have property to mortgage as collateral security for taking loans. Thus, they are deprived of banking facilities. This further created a big gap between have and have nots.

Drought is a serious problem as it inhibits nutrient acquisition through roots and limits both quantity and quality of crop productions. Table 1 shows 10 drought-hit states in India. Drought stress is often accompanied by a number of other environmental factors including temperature, high solar radiation and wind. Water and nutrition are two of the major components of environmental variations and together provide limitations to successful crop production.

Mineral nutrients are essential for plant growth and development through their fundamental roles in plant metabolism. Many studies revealed that balanced fertilization can improve photosynthetic activity by stabilizing biochemicals *i.e.* superoxide dismutase (SOD) activity, improve proline, abscisic acid (ABA) and make the crop drought tolerant with efficient root system and finally improve crop yield. Several reports have demonstrated that proline and abscisic acid (ABA) accumulation in plants can enhance tolerance to abiotic stresses (Bray, 1997; De Ronde *et al.*, 2004; Hmida-Sayari *et al.*, 2005).

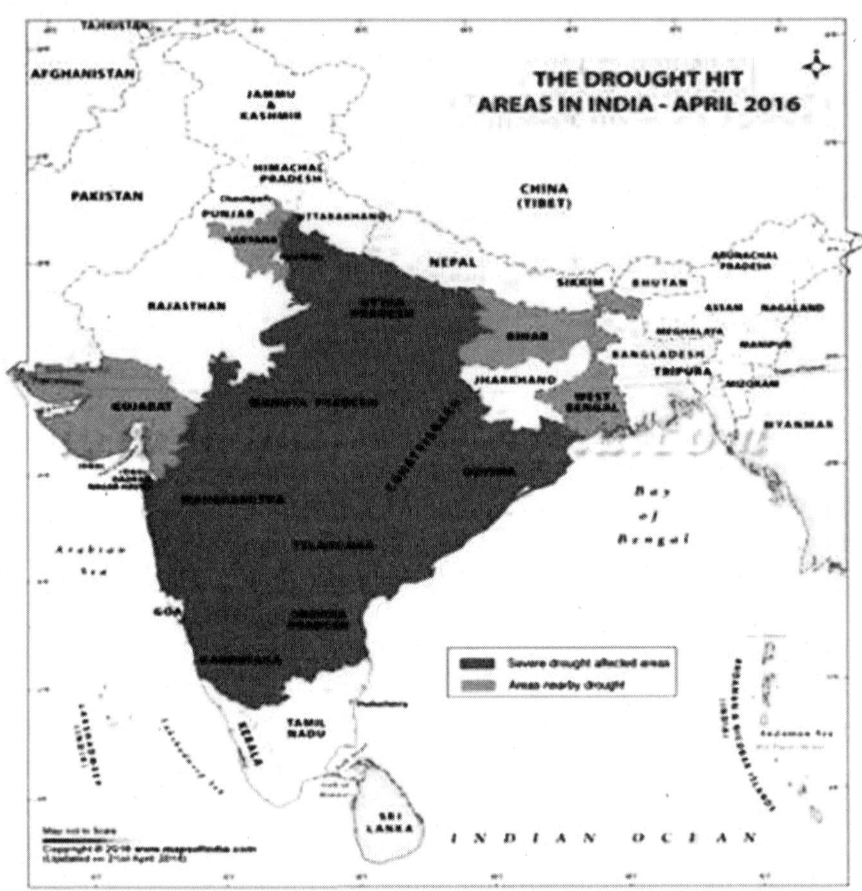

Fig. 2. Map of drought affected area in 2016 in India

Various roles of micro- and macro-nutrients as an essential plant nutrient, and some of the better-understood mechanisms responsible for genotypic variations in their efficiency are described. The rate of photosynthesis, pollen viability, number of fertile spikes and number of grains per spike are increased by late foliar application of zinc (Zn), boron (B) and manganese (Mn) micronutrients (Karim *et al.*, 2015). These nutrients also play a key role in the maintenance of photosynthetic activity (Brown *et al.*, 1993; Karim *et al.*, 2012), pollen viability (Karim *et al.*, 2012; Sharma *et al.*, 1990), the preservation of membrane integrity (Bettger and O'Dell, 1981) and enzyme activities (Cakmak and Marschner, 1988), as well as being an important factor in a plant's defense against reactive oxygen species, which proliferate under various stress conditions, including drought stress (Cakmak, 2000).

Table 1: 10 Major drought affected states of India

States	Total no. of districts	No. of drought affected districts	Government Funds promised for drought relief
Maharashtra	36	21	-
Odisha	30	16	1,000-crore in 16 districts
Telangana	10	07	-
Rajasthan	33	19	Rs. 50 lakh to each district
M.P.	51	46	Rs. 38,500 crore
Jharkhand	24	22	Rs. 143.25 crore
U.P.	75	50	Rs. 1,304 crore
Karnataka	30	27	-
Andhra Pradesh	09	07	Rs. 10,000 crore
Chhattisgarh	27	25	Rs. 150 crore for the seeds Rs. 223 crore for drought relief

Many physiological processes in plants are impaired by drought stress, including photosynthesis, enzyme activity, membrane stability, pollen viability and ultimately growth (Flexas *et al.,* 2004; Valentovic *et al.,* 2006). Rice (*Oryza sativa* L.), maize (*Zea mays* L.) and wheat (*Triticum aestivum* L.), are major three cereals that feed the world, with an estimated global production of 470,854 and 655 million tons and consumption of 469,866 and 673 million tons of rice, maize and wheat, respectively in 2012–2013 (International Grain Council, 2013).

Drought has a direct impact on the photosynthetic apparatus, essentially by disrupting all major components of photosynthesis including the thylakoid electron transport, the carbon reduction cycle and the stomatal control of the CO_2 supply, together with an increased accumulation of carbohydrates, peroxidative destruction of lipids and disturbance of water balance. Drought decreased photosynthetic rate and high temperature hastened the decline in photosynthetic rate in wheat. Many studies have shown that the decreased photosynthetic activity in wheat under drought stress due to stomatal or non-stomatal mechanisms (Ahmadi, 1998; Del Blanco *et al.,* 2000). Stomata are the entrance of water loss and CO_2 absorbability and stomatal closure is one of the first responses to drought stress which result in declined rate of photosynthesis. Stomatal closure deprives the leaves of CO_2 and photosynthetic carbon assimilation is decreased in favor of photorespiration.

Down regulation of Rubisco large subunit has been observed in drought stressed susceptible wheat lines (Bota *et al.* 2004, Demirevska *et al.* 2009) showing its involvement in drought tolerance mechanism. The activity of photosynthetic electron chain is finely tuned to the availability of CO_2 in the plant and photosystem II (PS II) often declines in parallel under drought conditions. Both the chlorophyll a and b are prone to soil dehydration (Farooq *et al.* 2009).

India's Drought - Climate Relationship

Cracked soils, farmer suicides and desperate migration are at odds with the country's image as an emerging economic and technological power. India is in the grip of a severe drought as a result of two successive weak monsoons and a searing heat wave. Its reservoirs dipped to less than a fifth of their total capacity in May, and a quarter of the country's 1.1 billion people are estimated to be affected in some way.

Reports of parched, cracked soils, farmers' suicides and desperate migration from Marathwada in the west of the country – one of the worst-hit regions – are at odds with the country's image as an emerging economic and technological power, aspiring towards a trillion-dollar economy "with no poverty" by 2032.

India's economy still depends largely on monsoon rains, with two-thirds of its agricultural land fed by rain. Other parts of the country are irrigated but this is costing the country dearly, leading to rapidly depleting groundwater and declining water tables.

An analysis carried out for the World Bank in 2013 found that India is already feeling the effects of a warming climate. If global average temperatures rise by 2°C, it predicts that unprecedented spells of hot weather will occur far more frequently and cover much larger areas.

GROWTH RESPONSES ON CEREALS DURING DROUGHT STRESS

Depending on the cereal species and on the geographical location of plant cultivation, drought may occur during the phase of vegetative/generative transition in the shoot apical meristem (Bidinger *et al.,* 1987). Three main mechanisms reduce crop yield by soil water deficit: (i) reduced canopy absorption of photosynthetically active radiation, (ii) decreased radiation-use efficiency and (iii) reduced harvest index (Earl and Davis, 2003). The rate of photosynthesis is reduced mainly by stomatal closure, membrane damage, and disturbed activity of various enzymes, especially those involved in ATP synthesis (Xiong *et al.,* 2006). Drought stress during flower induction and inflorescence development leads to a delay in flowering (anthesis), or even to complete inhibition (Wopereis *et al.,* 1996).

Root is a constitutive trait for plant growth that makes better use of water and minerals which is an important component of drought tolerance at different growth stages (Blum, 1996; Weerathaworn *et al.,* 1992). The potential quantity of accumulated water depends on the extent of root proliferation in the soil volume. When drought stress occurs during early growth stages considerable changes occurs in the root/shoot ratio (Nielsen and Hinkle, 1996). Recent research at the International Maize and Wheat

Improvement Center (CIMMYT) has investigated the possibility of measuring the root capacity to assess the absorptive area of the roots and to use this as a selection parameter for enhanced drought tolerance (Edmeades *et al.,* 2000).

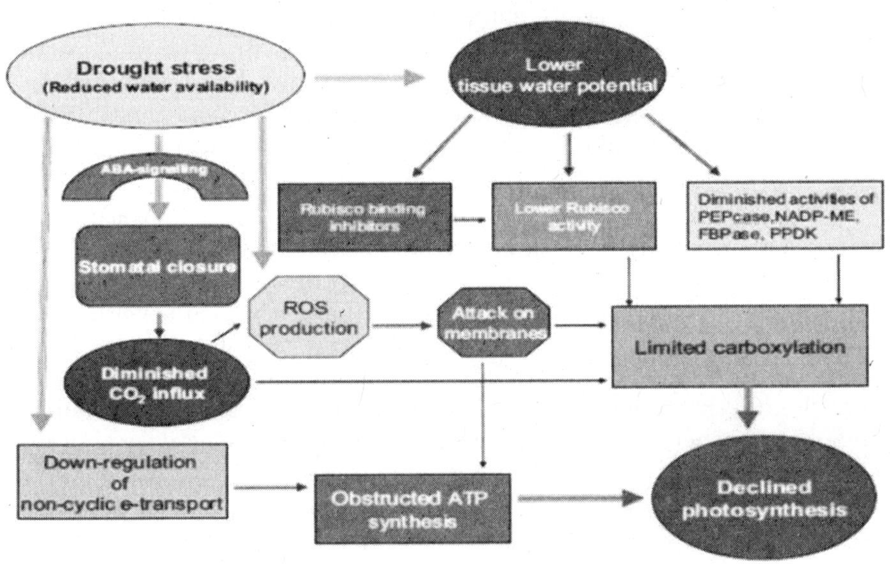

Fig. 3. Photosynthesis under drought stress (Farooq *et al.,* 2009)

During water stress, activities of the phosphoenolpyruvate carboxylase, nicotinamide adenine dinucleotide phosphatemalic enzyme, Rubisco, fructose-1, 6-bisphosphatase and pyruvate orthophosphate dikinase decreased linearly with lowered leaf water potential. Figure 3, shows that the photosynthesis decreased under drought stresses. Drought stress disturbs the balance between the production of reactive oxygen species and the antioxidant defense, causing accumulation of reactive oxygen species, which induces oxidative stress. Upon reduction in the amount of available water, plants close their stomata (plausibly *via* ABA signaling), which decreases the CO_2 influx. Reduction in CO_2 not only reduces the carboxylation directly but also directs more electrons to form reactive oxygen species. Severe drought conditions limit photosynthesis due to a decrease in the activities of ribulose-1, 5-bisphosphate carboxylase/oxygenase (Rubisco), phosphoenolpyruvate carboxylase (PEPCase), NADP-malic enzyme (NADP-ME), fructose-1, 6-bisphosphatase (FBPase) and pyruvate orthophosphate dikinase (PPDK). Reduced tissue water contents also increase the activity of Rubisco binding inhibitors. Moreover, non-cyclic electron transport is down-regulated to match the reduced requirements of NADPH production and thus reduces the ATP synthesis.

Plant Nutrition and Drought Tolerance

The nutrient deficiencies in soils are common problem under arid and semi-arid conditions, like in different parts of Afghanistan, Nepal, Bhutan, Myanmar, Laos Cambodia, India, Pakistan, China, Philippine and also in Bangladesh resulting in severe decreases in grain yield (Alloway, 2008; Karim *et al.,* 2006; Misra *et al.,* 1992). Therefore, development of genotypes with high tolerance to both drought and nutrient deficiency stresses is a high priority research area, and combination of these traits with high grain micro- and macro-nutrient concentrations would be a most desirable breeding goal especially for the arid and semi-arid regions (Bagci *et al.,* 2007).

Drought stress reduces the availability, uptake, translocation and metabolism of nutrients. A reduced transpiration rate due to water deficit reduces the nutrient absorption and efficiency of their utilization. Studies of genotypic variation in drought tolerance have revealed several criteria including the maintenance of photosynthesis activity under drought stress, improved membrane stability and maintenance of enzyme activity. Coincidentally or otherwise, Zn, B and Mn are also involved in maintaining photosynthetic activity (including maintaining the activity of drought sensitive enzymes), and preserving membrane integrity. Although nutrient deficiency is known to affect a wide range of crops in many parts of the world, genotypic differences between species render some crops more susceptible to deficiency than others (Martens and Westermann, 1991).

Nutrients deficiency effects to the physiological aspect of plants resulting from Zn, B and Mn associated with the disruption of normal enzyme activity in certain tissues. Reduced enzyme activity as a result of Zn, B and Mn deficiency can lead to impaired carbohydrate metabolism (a reduction in photosynthesis), decreased chlorophyll content and the abnormal structure of chloroplasts, induced impairments in cellular function and integrity, and a depression of male fertility (Karim *et al.,* 2012; Khan *et al.,* 2004; Wei *et al.,* 2005).

Oxidative Damage under Drought Stress

Various organelles including chloroplasts, mitochondria and peroxisomes are the seats as well as first target of reactive oxygen species produced under drought stress. The environmental stresses leads to the generation of reactive oxygen species, including superoxide anion radicals ($O^{-\cdot}$), hydroxyl radicals ($OH^{\cdot}$), hydrogen peroxide (H_2O_2), alkoxy radicals ($RO^{\cdot}$) and singlet oxygen ($O_2{}^1$) (Munné-Bosch and Penuelas, 2003). Reactive oxygen species may react with proteins, lipids and deoxyribonucleic acid, causing oxidative damage and impairing the normal functions of cells (Foyer and Fletcher, 2001). Many cell compartments produce reactive oxygen species; of these, chloroplasts are a potentially important source because excited pigments in thylakoid membranes may interact with O_2 to form strong oxidants such as $O^{-\cdot}$ or $O_2{}^1$ (Reddy *et al.,* 2004). Further downstream reactions produce other reactive oxygen species such as H_2O_2 and OH^-. The interaction of O_2 with reduced components of the electron transport chain in mitochondria can lead to reactive oxygen species formation (Möller, 2001), and peroxisomes produce H_2O_2 when glycolate is oxidized into glyoxylic acid during photorespiration (Fazeli *et al.,* 2007).

At very low O_2 concentrations, plant terminal oxidases and the formation of reactive oxygen species *via* the mitochondrial electron transport chain still remain functional. Among enzymatic sources of reactive oxygen species is xanthine oxidase (Apel and Hirt, 2004).

The enzymatic step is the dismutation of the superoxide anion by superoxide dismutase to yield H_2O_2 (Lamb and Dixon, 1997). Peroxidases and catalases also play an important role in the fine regulation of reactive oxygen species in the cell through activation and deactivation of H_2O_2 (Sairam *et al.,* 2005). Several apoplastic enzymes may also generate reactive oxygen species under normal and stressful conditions (Apel and Hirt, 2004).

Reactive oxygen species are formed as by-products in the electron transport chains of chloroplasts (Apel and Hirt, 2004), mitochondria and the plasma membrane (Sairam *et al.,* 2005). The plant mitochondrial electron transport chain, with its redox-active electron carriers, is considered as the most probable candidate for intracellular reactive oxygen species formation. Mitochondria can produce reactive oxygen species due to the electron leakage at the ubiquinone site – the ubiquinone: cytochrome *b* region (Gille and Nohl, 2001) and at the matrix side of complex I (NADH dehydrogenase) (Möller, 2001).

Superoxide radical and its reduction product H_2O_2 are potentially toxic compounds, and can also combine by the Haber-Weiss reaction to form the highly toxic OH^- (Sairam *et al.,* 1998). Many reports show the deleterious effects of reactive oxygen species, whose production is stimulated under water stress (Blokhina *et al.,* 2003). Reactive oxygen species cause lipid peroxidation and consequently membrane injuries, protein degradation and enzyme inactivation (Sairam *et al.,* 2005). Under biochemical and molecular investigation it was also found that over-expression of the BcZAT12 gene transcripts in tomato as well as by largely enhancing the ROS-scavenging capacity and up regulation of Hsp transcripts. Therefore, transformed tomato lines have improved electrolyte leakage, relative water content and chlorophyll levels with an enhanced activities of antioxidant enzymes viz. superoxide dismutase, catalase, ascorbate peroxidase and glutathione reductase when exposed to drought stress (Shah *et al.,* 2013).

Alleviation and Risk Management of Drought for Increasing Cereal Productions

The best way to management of drought for cereal productions by helping the farmers to grow the crops under this unfavorable climatic condition such as drought. To succeed over the long term, following key policies are recommended for drought risk management to increase cereal productions (1) Research and development; (2) Agricultural policies; (3) Access and market system; (4) Integration of science, policy makers and stakeholders.

GOVERNMENT STEPS TOWARDS MANAGEMENT AND MITIGATION OF DROUGHT

India's attempts to get its agriculture into a state where drought has no negative impact on the economy – are, at best, patchy. In 2013, the country signed up to policy recommendations thrashed out at a United Nations meeting of the World Meteorological Organization. The aim was to develop strategies for drought mitigation and management.

India has national crop insurance schemes for drought-hit farmers but there has been little take-up as many are too poor to pay the premium, and even if they do, the process required to verify crop losses can be too cumbersome to complete. Similarly, rainwater harvesting schemes are neglected despite policies to make them mandatory.

And while experts recommend that the government should encourage farmers to grow crops that need little water such as hardy millets, water-guzzling crops like sugarcane continue to be subsidised to keep the politically powerful sugar barons on side.

There have been no serious attempts to utilise local crop diversity, conserve water or recharge depleted aquifers. Farmers need money to buy seed or livestock and the soil damage hampers the improvement of water table due to disability to absorb.

Like its people, the Indian government, which has been criticised by the country's Supreme Court for its "ostrich-like attitude", is praying that this year's monsoon will set things right. The rains are predicted to be bolstered by the weakening El Niño. But this won't give impoverished farmers the money they need to buy seeds or livestock, and it may not do much for the water table – if the soils are so damaged that their ability to absorb the heavy rains is hampered.

Either way, if a rising economic power like India can't manage its crops and water, it doesn't bode well for the region in a future, warmer world. This was realized by the Government and the RBI in India way back. Accordingly, the two also visualized a very important role for banks to play. Having developed a firm belief in this philosophy, and thus in order to bridge the gap between have and have nots, they perceived and emphasized that the Indian banking system should play an active role in the financing of those deprived sectors of the economy. For this purpose, the Government with the help of RBI initiated priority sector lending.

Priority Sector Lending

Priority sector lending is a scheme which is proposed to give loans to the important sectors of the economy i.e. agriculture, small scale industries, education loan, housing loan, weaker sections. It is an active instrument of our financial policy which aims to restore sectoral balance with in credit disbursement and to develop the weaker sections within these sectors (Rani & Garg 2015).

Early years of Priority Sector Lending and RBI

Social control on banks was instituted through Banking Laws (Amendment) Bill 1967, which was introduced in the Lok Sabha on December 23, 1967. Through social control, banks were directed to align their operations in line with the national objectives. It was aimed at imposition of certain restrictions on banks to give advances to small industries, agriculture etc. It also facilitates the government to have control over the affairs of the bank management. The National Credit Council (NCC) has been set up in February 1968 with the responsibility of estimating the demand for bank credit from different sectors of the economy. The report stressed the need for co-ordination between co-operative banks and commercial banks in order to achieve optimum utilization of resources and to substitute the usurious money lenders and indigenous bankers (Report of RBI 2006-08).

The social control on banks has led the commercial banks to move towards mass banking. Despite commercial banks, lending to agriculture under (a) direct financing and (b) Indirect financing, the lending towards agriculture did not exceed two per cent of the total credit. This shows that social control did not show much impact on rural lending and still, much of the credit went to big business houses. Bank credit to hitherto neglected areas was still a dream. In order to correct the situation, 14 major commercial banks were nationalized on July 19, 1969 through an ordinance and this was a milestone in banking history of India. The broad objectives of nationalization were:

a. To provide adequate and timely credit for agriculture, small industries and exports
b. To give professional bent to bank management
c. To encourage new class of entrepreneurs (Shekhar & Shekhar 2008).

After nationalization of banks in 1969, the broad categories under priority sector were Agriculture, small scale industries and exports (Kanagasabai, 1999). This is the genesis of priority sector lending, which is actually directed lending imposed on the Indian Banks in the name of "social control". There was increase in the availability of credit but the situation was still in favour of big industries.

There have been persistent complaints that several priority sectors such as agriculture, small scale industries and exports have not been receiving their due share of bank credit. This appears to be the first occasion that the term priority sector was used (Vimla, 2002).

Thus the origin of priority sector prescriptions for banks in India can be traced to the Credit Policy of the year 1967-68. In that policy it was emphasized that commercial banks should increase their involvement in the financing of priority sectors, viz., agriculture, exports and small-scale industries, as a matter of urgency. However, the concept of the priority sector was properly formalized in the year 1972 on the basis of the report submitted by the Informal Study Group on Statistics Relating to Advances to the Priority Sectors constituted by the Reserve Bank in May 1971. On the basis of this report, the Reserve Bank prescribed a modified return for reporting priority sector advances and certain guidelines were issued in this regard in February 1972, indicating the scope of the items to be included under various categories of priority sector (Report of RBI, 1967-68).

The guidelines issued indicated only the general description of the advances to be included and no ceilings were fixed, except in the case of small-scale industry and road and water transport operators where ceilings on the value of original investments were indicated. The scope and extent of priority sector has undergone several changes since then with several new areas and sectors being brought within the purview of this sector.

Formalisaton of Priority Sector lending

In 1969, under the influence of 1967-68 Credit Policy, the priority sector procured a total Rs. 505 crore as loans. Out of this Rs. 188 crore was for Agricultural sector and Rs. 286 crore was for small scale industries. The percentage share of priority sector lending in total credit thus worked out was merely 14%. It revealed that at that time of the total bank credit only a small portion was received by priority sectors. There was a need to give clear and specific definition of different components of priority sector because some of the banks were not clear about the exact scope of agricultural lending (Joshi 1972). Keeping this in mind, in 1972, the definition of priority sector was given a formal shape and in 1974, banks were directed to provide one-third of total credit to priority sectors by 1979. Later on this limit was extended to 40%. Out of 40 percent, 10% was to be lent to weaker sections for their financial upliftment.

Reforms in Priority Sector Lending in 1991

Over the years the stipulation for lending to the priority sector has been retained, more or less the same, even though its scope and definition was fine tuned by including new items. There came a major change when the Indian economy and the banking system witnessed a major reform process as enunciated in the year 1991. Under the influence of reform process the priority sector policy also witnessed a major change. Based on the recommendations of the working group, set up by Reserve Bank, the priority sector norms were revised in April 2007 (RBI Reports, various issues). As per the revised norms, priority sectors which have been retained are:

Those sectors that impact large sections of the population, the weaker sections and the sectors that are employment-intensive such as agriculture, and tiny and small enterprises. The broad sectors under the revised norms include agriculture (both direct and indirect), small enterprises (direct and indirect), retail trade in essential commodities and consumer cooperatives stores, micro credit, education loans and housing loans.

The weaker sections of the society include following:-

Small and marginal farmers with land holding of 5 acres and less and landless labourers, tenant farmers and share croppers.

- Artisans, village and cottage industries where individual credit limits do not exceed Rs. 50,000/-
- Beneficiaries of Swarnjayanti Gram Swarojgar Yojana (SGSY)
- Scheduled Castes and Scheduled Tribes
- Beneficiaries of Differential Rate of Interest (DRI) scheme
- Beneficiaries under Swarna Jayanti Shahari Rojgar Yojana (SJSRY)
- Beneficiaries under the Scheme for Liberation and Rehabilitation of Scavangers (SLRS).
- Self Help Groups (SHGs)
- The sectors which are employment intensive such as agriculture (both direct and indirect)
- Micro and small enterprises (direct and indirect)
- Retail trade in essential commodities and consumer cooperatives stores.
- Education loans and housing loans up to Rs. 20 lakh.

As the constituents of the priority sector were redefined, so was done with the targets fixed for banks of different categories. The targets and sub-targets set under priority sector lending for domestic and foreign banks operating in India as revised under reform process are furnished in Table 2.

Under the revised norms, the overall priority sector lending targets at 40 percent and 32 percent for the domestic and foreign banks, respectively, as also other sub-targets, have been retained unchanged. Individual farmers, under priority sector advances, can avail of up to Rs. 10 lakh against agricultural produce for a period not exceeding 12 months. Indirect credit to agriculture should not exceed 4.5 percent of adjusted net bank credit or 25 percent of overall credit to agriculture.

Under the priority sector norms direct institutional credit to the agricultural sector includes loans sanctioned for small and marginal farmers for purchase of land for agricultural purposes. Distressed farmers indebted to non-institutional lenders can obtain loans against appropriate collateral and group security.

Table 2: Private sector lending through banks in India

	Domestic banks (both public sector and private sector banks)	Foreign banks operating in India
Total Priority Sector Advances	40 percent of NBC	32 percent of NBC
Total Agricultural Advances	18 percent of NBC	No target
SSI Advances	No target	10 percent of NBC
Export Credit	Export credit does not form part of priority sector	12 percent of NBC
Advances to Weaker Sections	10 percent of NBC	No target

Note: NBC denotes net bank credit

In the revised guidelines, with the aim of overcoming the crowing out effect against small loans, particularly to agriculture, only one-third of amount of loans/advances are allowed to be classified as direct agriculture. In case, the domestic scheduled commercial banks are unable to meet the priority sector targets and/or agriculture target, they would be allocated amounts for contribution to the Rural Infrastructure Development Fund (RIDF) established with NABARD or funds with other financial institutions, as specified by the Reserve Bank. The foreign banks with a shortfall in priority sector lending as per stipulated targets are required to contribute to Small Enterprises Development Fund (SEDF) established with Small Industries Development Bank of India (SIDBI).

Priority Sector Lending - Pre 1990 Scenario

Data on priority sector lending by banks in pre 1990 period is presented in Table 3.1. A significant trend is visible from the Table 3.1 which is about priority sector lending in pre 1990 period. The Table shows that priority sector lending in the year 1969 was Rs. 505 crore which increased by 325.76% to Rs. 2150.1 crore in the year 1975. The percentage share which was 14% in the year 1969 increased to 25% in the year 1975. The increase in the share was due to nationalization of banks in 1969. In the year 1969 the share of agriculture in total bank lending was 37.23% which fell to 36.51% in 1975. During the same period, share of small scale industries dipped from 56.63% to 48.5%. This fall in the share was due to increase in the share of other priority sectors (RBI Reports, various issues).

Table 3.1: Credit Provided by SCB to Priority Sector (1969-1990) (In Crore Rs.)

	1969	1970	1971	1972	1973	1974	1975	1976	1977
1. Agricultural Sector	188 (37.23)	-	382 (37.23)	439 (36.34)	567 (36.35)	598.8 (33.23)	785 (36.51)	1092 (38.79)	1381 (39.62)
2. Small Scale Industries	286 (56.63)	-	500 (48.73)	597 (49.42)	750 (48.08)	926.6 (51.42)	1042.9 (48.5)	1222 (43.41)	1460 (41.88)
3. Other Priority Sector	31 (61.14)	-	144 (14.04)	172 (14.24)	243 (15.58)	276.6 (15.35)	322.2 (14.99)	501 (17.8)	645 (18.5)
4. Total Priority Sector (1+2+3)	505	-	1026	1208	1560	1802	2150	2815	3486
5. % of Priority Sector Out of Total Credit	14.0	-	-	21.0	23.1	24.2	25	24.5	25.5

	1978	1979	1980	1981	1982	1983	1984
1. Agricultural Sector	1726 (39.82)	2288 (40.30)	2767 (41.11)	3584 (42.14)	4615 (43.23)	5375 (42.53)	6551 (41.22)
2. Small Scale Industries	1756 (40.52)	2252 (39.67)	2635 (39.15)	3229 (37.97)	3901 (36.54)	4584 (36.27)	5750 (36.18)
3. Other Priority Sector	852 (19.66)	1137 (20.03)	1328 (19.73)	1691 (19.88)	2160 (20.23)	2678 (21.19)	3593 (22.61)
4. Total Priority Sector (1+2+3)	4334	5677	6730	8504	8504	12637	15894
5. % of Priority Sector Out of Total Credit	28.6	30.9	33	35.6	35.6	36.1	38.1

	1985	1986	1987	1988	1989	1990
1.Agricultural Sector	7978 (41.53)	9391 (42.11)	10852 (41.77)	12285 (41.16)	14133 (40.10)	16939 (41.28)
2. Small Scale Industries	6956 (36.21)	8101 (36.32)	9525 (36.66)	11164 (37.41)	3677 (38.81)	15672 (38.20)
3. Other Priority Sector	4274 (22.25)	4810 (2187)	5603 (21.57)	6396 (21.43)	7432 (21.09)	8419 (20.52)
4. Total Priority Sector (1+2+3)	19208	22302	25980	29845	35242	41030
5. % of Priority Sector Out of Total Credit	39.9	41.0	42.9	43.8	42.6	40.7

Note: Data given in brackets are % of total credit to priority sector.

Source: 1. Report Related to Nature and Progress of Banking in India, RBI. 2. Banking Statistics, RBI

In the year 1969, other priority sector had a share of 6.14%, it increased in the year 1975 to 14.99%. This change was in accordance with the policies of government. Although the share of priority sector increased, yet it was not as per demand at that time.

Keeping in mind the requirements of priority sector, in the year 1976, Regional Rural Bank Act was passed. The main objectives of setting up Regional Rural Bank were as follows:

- Development of Agriculture
- To provide loans to small and marginal farmers
- To provide loans to agricultural labourers, craftsmen and other small workers.
- To make agricultural economy stronger.

Because of these policies, there was tremendous growth in priority sector lending. The flow of credit to priority sector was Rs. 2150.1 crore in the year 1975. This improved remarkably by about 213% to Rs. 6730 crore in the year 1980. In percentage share it increased from 25% to 33%. Thus, banks fulfilled the goal of one-third of total credit. During this period share of agricultural sector increased from 36.51% in the year 1975 to 41.11% in the year 1980. This was also due to nationalization of banks in 1980.

The Table reveals that during first half of 1980s, the share of priority sector in total bank credit kept on increasing. It rose sharply from Rs. 6730 crore in the year 1980 to Rs. 19,208 crore (185.4% that is 2.85 times growth) in the year 1985. It further had a hike from Rs. 19,208 crore to Rs. 41,030 crore (113.6% that is 2.14 times growth) in the year 1990. The percentage share of credit to priority sector was 33% in the year 1980, which also rose sharply to 39.9% in the year 1985 and 40.7% in the year 1990 respectively.

Thus, we can see that according to the social necessity loans were provided by Scheduled Commercial Banks and consequently the share of priority sector in total credit increased from 14% in the year 1969 to a high rate of 40.7% in the year 1990.

Priority Sector Lending – Post 1990 Scenario

While studying about new reforms in the banking sector as also our study in previous section we have seen that banks in the beginning had extended major part of their credit to industries in developed cities. Banks were not interested in providing credit to agricultural and small scale industries. Consequently, Government of India and RBI adopted strict policy to provide credit to priority sector. Accordingly the banks started extending more and more credit to the priority sector. Due to the policy and banking reforms there was a remarkable change in the scenario.

Now let us study the trends in priority sector lending in post 1990 period. The data of pre 1990 period has also been included to understand

and relate the situation of priority sector lending in post 1990 period. It is presented in Table 3.2.

Table 3.2: Priority Sector Lending by Scheduled Commercial Banks (Amount in Rs. Crore)

Year	Agriculture	Small Industries	Other Priority Sector	Total Lending	Share of Priority Sector in Total Credit (in %)
1969	188	286	31	505	14.0
	(37.23)	(56.63)	(6.14)	(100)	
1975	785	1042.9	322.2	2150.1	25.0
	(36.51)	(48.50)	(14.99)	(100)	
1980	2767	2635	1328	6730	33.0
	(41.11)	(39.15)	(19.73)	(100)	
1985	7978	6956	4274	19208	39.9
	(41.53)	(36.21)	(22.25)	(100)	
1990	16939	15672	8419	41030	40.7
	(41.28)	(38.20)	(20.52)	(100)	
1995-96	27044	31884	14401	73329	33.7
	(36.88)	(43.48)	(19.64)	(100)	
2000-01	51922	56002	46490	154414	35.4
	(33.62)	(36.27)	(30.11)	(100)	
2005-06	173972	91212	245554	510738	36.7
	(34.06)	(17.86)	(48.08)	(100)	
2007-08	273658	155804	309224	738686	34.3
	(37.05)	(21.09)	(41.86)	(100)	

Note: Figures in Parenthesis denote percentage share in total lending by banks.

Source: Statistical Tables, Banking Statistics and Report on Trend and Progress of Banking in India, RBI, Various Issues

A perusal of the table reveals that in the year 1969, priority sector had merely 14% share in total credit extended by banks. It meant that banks provided credit up to 86% to industries. Looking at this situation, in the year 1974, banks were directed to increase the share of priority sector to one-third of total credit. Based on the instructions of the Government, the banks started more credit to priority sector. In the year 1975, the share of priority sector increased from 14% to 25%. It further increased to 33% in the year 1980. This was one-third of total credit. Thus, banks achieved the target (Report of RBI, 2003).

In this respect the Government of India and RBI further increased the target for banks and directed banks to provide credit up to 40%. Due to the influence of the directives issued by government, the share of priority sector lending increased to 39.9% by 1985. It continuously increased for 3 years and in the year 1988 it had gone up to its maximum level of 43.8%. After that it started declining and in the year 1990 it fell down to 40.7%. Then after the share of priority sector in total bank credit registered a continuous decline.

While the share of priority sector was 40.7% in 1990, it fell down to 34.3% in 2007-08. It declined further to 31.2% in the year 2009-10.This falling trend continued and it reached 29.1% in 2013-14.

The decline in the priority sector lending was because of new orientation of banks towards commercial principles of banking more poised and focused on prudence and profit under the reform process (Ghosh, 1992).

We have also done the intra-sector wise analysis of priority sector lending by banks. A detailed intra-sector wise breakup of priority sector lending by commercial banks is presented in Table 3.3. A perusal of the table reveals that in the post 1990 period the share of agricultural sector has declined from 39.97% in 1991-92 to 34.58% in 1998-99 and further to 32.83% in 2004-05. While the share of agriculture was declining, the share of others was increasing. The share of other priority sectors increased from about 20% in 1990 to 47.6% in 2004-05.

In post 2005 period, the priority sector lending was again in favour of agricultural sector so the percentage share increased from 32.83% in 2004-05 to 37.05% in 2007-08.It showed an increasing trend till 2009-10 with 42.42%. It slightly had a downward trend with 40.99% in 2013-14. While the share of agriculture was increasing, the share of others was declining. The share of other priority sector declined from 47.6% in 2004-05 to 41.8% in 2007-08. It further declined to 13.96% in 2013-14. This clearly shows that agriculture and SSI sector have been given importance in recent years.

A perusal of the Table 3.3 further reveals that in post 1990 share of priority sector lending declined from 40.7% to 35% and remained at this level till 2007. But in recent years it came down to 31.2% and further to 29.1% in 2013-14 (Refer Chart 3.5, 3.6). The positive effect of it could be seen on the profitability and financial health of banks. Now a large amount of credit is being made available to those sectors from where banks could earn at higher rates. This fact is reflected from the increase in share of personal loans.

Table 3.3: Sectorwise Breakup of Priority Sector Lending by Commercial Banks

Year	Agriculture	Small Scale Industries	Other Priority Sector	Total Priority Sector receiving credit	% Share of Priority Sector in total Credit Provided
1991-92	18,157 (39.97)	18,150 (39.96)	9,118 (20.07)	45,425 (100)	37.7
1992-93	19,963 (40.06)	20,026 (40.19)	9,843 (19.75)	49,832 (100)	37.1
1993-94	21,208 (39.36)	22,617 (41.98)	10,055 (18.66)	53,880 (100)	34.4

Year	Agriculture	Small Scale Industries	Other Priority Sector	Total Priority Sector receiving credit	% Share of Priority Sector in total Credit Provided
1994-95	23,983	27,638	12,540	64,161	36.5
	(37.38)	(43.48)	(19.64)	(100)	
1995-96	27,044	31,884	14,401	73,329	33.7
	(36.88)	(43.48)	(19.64)	(100)	
1996-97	31,442	35,944	17,494	84,880	38.3
	(37.04)	(42.35)	(20.61)	(100)	
1999-2000	44,381	52,814	34,632	1,31,827	35.3
	(33.67)	(40.06)	(26.27)	(100)	
2000-01	51,922	56,002	46,490	1,54,414	35.4
	(33.62)	(36.27)	(30.11)	(100)	
2001-02	60,761	57,199	57,299	1,75,259	31.0
	(34.67)	(32.64)	(32.69)	(100)	
2002-03	73,518	60,394	77,697	2,11,609	34.8
	(34.74)	(28.54)	(36.72)	(100)	
2003-04	90,541	65,855	1,07,438	2,63,834	35.1
	(34.32)	(24.96)	(40.72)	(100)	
2004-05	1,25,250	74,588	1,81,638	3,81,476	34.5
	(32.83)	(19.55)	(47.61)	(100)	
2005-06	1,73,972	91,212	2,45,554	510738	36.7
	(34.06)	(17.86)	(48.08)	(100)	
2006-07	2,30,398	1,17,880	2,85,864	6,34,142	35.3
	(36.33)	(18.59)	(45.08)	(100)	
2007-08	2,73,658	1,55,804	3,09,224	7,38,686	34.3
	(37.05)	(21.09)	(41.86)	(100)	
2008-09	375595	256128	300737	932460	30.3
	(40.28)	(27.47)	(32.25)	(100)	
2009-10	463321	362291	266668	1092280	31.2
	(42.42)	(33.17)	(24.41)	(100)	
2010-11	507176	478530	253684	1239390	32.65
	(40.92)	(38.61)	(20.47)	(100)	
2011-12	583343	527685	310017	1421045	34.86
	(41.05)	(37.13)	(21.82)	(100)	
2012-13	645433	687210	207153	1539796	32.06
	(41.92)	(44.63)	(13.45)	(100)	
2013-14	769800	846130	262170	1878100	29.10
	(40.99)	(45.05)	(13.96)	(100)	

Note: Figures in parenthesis are percentage to total.

Source: (1) Handbook of Statistical on the Indian Economy, RBI, Various Issues.

(2) Banking Statistics, RBI, Various Issues.

Analysis of priority sector lending by commercial banks reveals another important change. This change came as the increase in the share of personal loans in total lending. It increased from 6.42% in the year 1990 to 22% in the year 2008. The increased share of personal and professional lending by banks helped them earn returns at higher rates (RBI Bulletin, various issues).

Table 3.4(A): Bank Group-wise Break-up of Priority Sector Lending (Amount in Rs. Crore)

Year	Public Sector Banks	Private Bank	Foreign Bank	All Banks
1991	-	-	-	-
1992	43,034	1,812	741	45,587
1993	44,742	2,383	947	48,072
1994	45,932	2,876	2,234	51,042
1995	52,405	4,226	3,136	59,767
1996	61,809	6,049	4,470	72,328
1997	68,931	8,859	5,038	82,828
1998	80,574	9,966	5,417	95,957
1999	93,712	12,054	6,516	1,12,282
2000	1,08,209	14,761	7,614	1,30,584
2001	1,49,116	21,567	11,572	1,82,255
2002	1,71,484	24,184	9,936	2,05,604
2003	1,99,786	36,648	14,555	2,50,989
2004	2,44,456	48,920	17,960	3,11,336
2005	3,07,046	69,886	23,843	4,00,775
2006	4,09,748	1,06,586	30,439	5,46,773
2007	5,21,376	1,44,549	37,831	7,03,756
2008	6,08,963	1,63,223	50,301	8,22,487
2009	7,20,083	1,90,207	55,483	9,65,773
2010	8,64,564	2,15,552	60,290	11,40,406
2011	1021500	249100	66700	1337300
2012	1129993	286419	80559	1496971
2013	1282200	327400	84854	1694454
2014	1619000	464500	90700	2174200
2015	1751200	530300	97000	2378500

'-' Unavailability

Source: (i) Handbook of Indian Banking Statistics, RBI, Various Issues.
(ii) Annual Report, RBI, Various issues.

Priority Sector Lending – Bank Groupwise Breakup

We have also analysed the priority sector lending under the break-up of data on three broad categories of banks. Data in this regard are presented in Table 3.4(A), Table 3.4(B). On the basis of our analysis in the said tables on various categories of banks providing credit, it has emerged out very clearly that Public Sector Banks still provided a large share of credit to priority sectors, inspite of the fact that in post 1990 period, there is major reduction in the share of priority sectors. This has been depicted in the Tables 3.4(A) and 3.4(B).

A perusal of the Table reveals that in the year 1992 PSBs accounted for 94.40% of total credit extended to priority sectors by all banks. With the passage of time, the share of PSBs has gone down and the share of private banks has gone up. The share of PSBs has come down to 75% and 73.63% in the year 2010 and 2015 respectively and the corresponding increase has been claimed by the private banks. The share of private banks has gone up to 19% in the 2010 and 22.3%% in 2015 (See Table 3.4B).

Table 3.4(B) Percentage Share of Three Groups of Banks in Total Priority

Year	Public Sector Banks	Private Banks	Foreign Banks	Total Banks
1992	94.40	3.97	1.63	100
1995	87.68	7.07	5.25	100
2000	82.87	11.30	5.83	100
2005	76.61	17.44	5.95	100
2008	74.04	19.85	6.11	100
2009	74.56	19.69	5.75	100
2010	75.81	18.90	5.29	100
2011	76.39	18.63	4.98	100
2012	75.49	19.13	5.38	100
2013	75.67	19.32	5.01	100
2014	74.46	21.36	4.18	100
2015	73.63	22.30	4.07	100

Sector Credit by all Banks (in Percentage)
Source: Based on Table 3.4 (A)

In the year 1992, private banks and foreign banks accounted for only 3.97% and 1.63% of total credit extended by all banks to priority sector, while in the year 2010, the share of private banks and foreign banks went up to 18.9% and 5.2% (Refer Chart 3.7). This reveals that the centralized pressure on Public Sector Banks to provide credit was decentralized to private and foreign banks only to very small extent. Still the burden social philosophy is being shouldered by the PSBs in incomparable proportions.

Thus we can see that before 1969, commercial banks neglected priority sector lending as they were owned by and controlled by big industrialists. But after nationalization, commercial banks started financing to priority sectors like agriculture, small industry and business, retail trade, professional and self-employed persons, education, housing loans for weaker sections and consumption loans. The scenario in this regard has changed and the change is a pleasing and welcome one. Although different categories of banks have achieved the overall target for priority sector lending by March 2010, but there are a few individual bank defaulters. 3 out of 27 PSBs, 2 out of 22 private sector banks and 4 out of 28 foreign banks have not achieved the priority sector lending targets of 40% and 32% respectively. They too should be brought to book and made to conform to norms like others banks.

Priority Sector Lending –Revised Guidelines 2015

Drawing upon the recommendations of an Internal Working Group, revised guidelines on priority sector lending were issued in April 2015. With the intention of giving a thrust to the MSME sector, medium enterprises were included in the priority sector. Additionally, a specific target was introduced for lending to micro enterprises. In addition to medium enterprises, two new categories, viz., social infrastructure and renewable energy were included under the priority sector (RBI Report, 2014-15). The salient features of the revised priority sector lending guidelines are:

- Separate targets of 8 per cent for small and marginal farmers (within the agriculture target of 18 per cent) and 7.5 per cent for micro industries have been prescribed to be achieved in a phased manner by 2017. These targets will be made applicable to foreign banks with 20 branches and above post 2018 after a review in 2017.
- Priority sector widened to include medium enterprises, social infrastructure and renewable energy.
- Monitoring of priority sector lending compliance on a 'quarterly' average basis at the end of the respective year from 2016-17.
- Priority Sector Lending Certificates (PSLCs) will be an eligible tradable instrument for achieving priority sector targets. The buyer (a deficient bank) will pay a 'price/ fee' to the seller bank (a bank which has over achieved its PSL requirements) for purchasing a specified amount of PSL obligation applicable for a particular date.
- Educational loans (including loans for vocational courses) up to 1 million irrespective of the sanctioned amount to be considered as eligible for the priority sector.
- Export credit up to 32 per cent of adjusted net bank credit (ANBC) or credit equivalent amount of off-balance sheet exposure (CEOBE), whichever is higher, to be eligible as part of the priority sector for foreign banks with less than 20 branches. For domestic and foreign banks with year will be reckoned up to 2 per

cent of ANBC or CEOBE (whichever is higher), as part of the priority sector.

- Foreign banks with less than 20 branches are required to achieve the total priority sector target of 40 per cent of ANBC or CEOBE, whichever is higher, in a phased manner by 2020.

CONCLUSIONS

This study has also identified a number of gaps in our knowledge on plant nutrition and drought tolerance. Drought stress and micronutrient deficiency are serious abiotic stress factors inhibiting micronutrient absorption by roots and limiting crop productions.

It shown that Zn and B nutrition can provide cereal plants with a level of tolerance of drought and it's relation with their efficiency. Therefore, it can be concluded that soil application is a promising approach to alleviate drought stress. Another attractive environmental friendly approach is to select efficient and drought tolerant genotypes with a more efficient root system.

Acknowledgement

Authors gratefully acknowledge BHU where the work was carried out. Authors are grateful to CSIR and UGC for funding the research work of PS and MK. The present work is a part of doctoral thesis of SA. Our apologies to those workers whose work could not be included herein due to space limitations.

REFERENCES

Ahmadi, AA. 1998. Effect of post-anthesis water stress on yield regulating processes in wheat (*Triticum aestivum* L.). Ph.D. Thesis. University of London, Ashford, U.K.

Apel, K.; Hirt, H. 2004. Reactive oxygen species: metabolism, oxidative stress, and signal transduction, *Annu. Rev. Plant Biol,*. 55: 373–99.

Bettger, WJ; O'Dell, BL. 1981. A critical physiological role of zinc in the structure and function of biomembranes. *Life Sci.,* 28: 1425.

Bidinger, FR; Mahalakshmi, V; Rao, GDP. 1987. Assessment of drought resistance in pearl millet [*Pennisetum americanum* (L.) Leeke]. I. Factors affecting yields under stress. *Aust. J. Agric. Res.,* 3: 37–48.

Blokhina, O; Virolainen, E; Fagerstedt, KV. 2003. Antioxidants, oxidative damage and oxygen deprivation stress: a review. *Ann. Bot.,* 91: 179–194.

Bota, J; Flexas, J; Medrano, H. 2004. Is photosynthesis limited by decreased Rubisco activity and RuBP content under progressive water stress? *New Phytol.,* 162: 671-681.

Bray, EA. 1997. Plant responses to water deficit. *Trend Plant Sci.,* 2: 48–54.

Brown, PH; Cakmak, I; Zhang, Q; Robson, AD. 1993. Form and function of zinc plants. *Dev. Plant Soil Sci.,* 55: 93 (93).

Cakmak I, Marschner H (1988). Zinc-dependent changes in ESR signals, NADPH oxidase and plasma membrane permeability in cotton roots. *Physiol. Plant.,* 73: 182–186.

Cakmak, I. 2000. Tansley review no. 111. Possible roles of zinc in protecting plant cells from damage by reactive oxygen species. New Phytol. 146, 185–205.

De Ronde, JA; Cress, WA; Krüger, GHJ; Strasser, RJ; Van Staden, J. 2004. Photosynthetic response of transgenic soybean plants, containing an Arabidopsis P5CR gene, during heat and drought stress. *J. Plant Physiol.,* 161: 1211–1224.

Del Blanco, IA; Rajaram, S; Kronstad, WE; Reynolds, MP. 2000. Physiological performance of synthetic hexaploid wheat derived populations. *Crop Sci.,* 40:1257-1263.

Demirevska, K; Zasheva, D; Dimitrov, R; Simova-Stoilova, L; Stamenova, M; Feller, U. 2009. Drought stress effects on Rubisco in wheat: changes in the Rubisco large subunit. *Acta Physiol. Plant.,* 31: 1129-1138.

Du, YC; Kawamitsu, Y; Nose, A; Hiyane, S; Murayama, S; Wasano, K; Uchida, Y. 1996. Effects of water stress on carbon exchange rate and activities of photosynthetic enzymes in leaves of sugarcane (*Saccharum* Sp.), *Aust. J. Plant Physiol.,* 23: 719–726.

Earl, H; Davis, RF. 2003. Effect of drought stress on leaf and whole canopy radiation use efficiency and yield of maize. *Agron. J.,* 95: 688–696.

Farooq, M; Wahid, A; Kobayashi, N; Fujita, D; Basra, SMA. 2009. Plant drought stress: effects, mechanisms and management. *Agron. Sustain. Dev.,* 29:185-212.

Fazeli, F; Ghorbanli, M; Niknam, V. 2007. Effect of drought on biomass, protein content, lipid peroxidation and antioxidant enzymes in two sesame cultivars. *Biol. Plant.,* 51: 98–103.

Flexas, J; Bota, J; Loreto, F; Cornic, G; Sharkey, TD. 2004. Diffusive and metabolic limitations to photosynthesis under drought and salinity in C3 plants. *Plant Biol.,* 6: 269–279.

Foyer, CH; Fletcher, JM. 2001. Plant antioxidants: colour me healthy, *Biologist,* 48: 115–120.

Ghosh, DN. 1992. Bank Profitability and Priority Sector Lending, Economics and Political Weekly, Vol. 27 (8).

Gille, L; Nohl, H. 2001. The ubiquinol/bc1 redox couple regulates mitochondrial oxygen radical formation, *Arch. Biochem. Biophys.,* 388: 34–38.

Hmida-Sayari, A; Gargouri-Bouzid, R; Bidani, A; Jaoua, L; Savouré, A; Jaoua, S. 2005. Overexpression of Δ1-pyrroline-5-carboxylate synthetase increases proline production and confers salt tolerance in transgenic potato plants. *Plant Sci.,* 169: 746–752.

International Grain Council, 2013. Grain market report (GMR no. 414).

Joshi, PN.1972. Financing of Priority Sectors by Commercial Banks, The Journal of Indian Institute of Bankers, XLIII(1):27-34.

Kanagasabai, S. 1999. An Economic Analysis of Priority Sector Lending under Lead Bank Scheme in Union Territory of Pondicherry, Ph.D Thesis, Pondicherry University, Pondicherry, pp.122 94

Karim, MR; Zhang, YQ; Zhao, RR; Chen, XP; Zhang, FS; Zou, CQ. 2012. Alleviation of drought stress in winter wheat by late foliar application of zinc, boron, and manganese. *J. Plant Nutr. Soil Sci.,* 175: 142–151.

Khan, HR; McDonald, GK; Rengel, Z. 2004. Zinc fertilization and water stress affects plant water relations, stomatal conductance and osmotic adjustment in chickpea (*Cicer arientinum* L.). *Plant Soil,* 267: 271–284.

Lamb, C; Dixon, RA. 1997. The oxidative burst in plant disease resistance. *Annu. Rev. Plant Phys.,* 48: 251–275.

Martens, DC; Westermann, DT. 1991. Fertilizer application for correcting micronutrient deficiencies. In: Mortvedt, JJ, Cox; FR, Schuman; LM, Welch; RM. (Eds.), Micronutrients in Agriculture, 2nd ed. Soil Science Society of America, Madison, Wisc, pp. 549–592 (Book Series No. 4).

Möller, IM. 2001. Plant mitochondria and oxidative stress: electron transport, NADPH turnover, and metabolism of reactive oxygen species. *Annu. Rev. Plant Phys.,* 52: 561–591.

Munné-Bosch, S; Penuelas, J. 2003. Photo and antioxidative protection, and a role for salicylic acid during drought and recovery in fieldgrown *Phillyrea angustifolia* plants. *Planta,* 217:758 766.

Rani, S; Garg, D. 2015. Priority Sector Lending: Trends, Issues and Strategies, International Journal of Management and Social sciences Research, Vol.4

Rao, C. Bhujanga 2014. Appraisal of Priority Sector Lending by Commercial Banks in India, Monograph 27, Madras School of Economics, Chennai

Reddy, AR; Chaitanya, KV; Vivekanandan, M. 2004. Drought-induced responses of photosynthesis and antioxidant metabolism in higher plants. *J. Plant Physiol.,* 161: 1189–1202.

Reserve Bank of India (1967-68), Report on Currency and Finance, Evolution of Banking in India, RBI publications, Vol.4 , Chapter III, pp.96

Reserve Bank of India (2003), *Report on Trends and Progress of Banking in India,* Mumbai.

Reserve Bank of India (2006-07), The Handbook of Statistics on the Indian Economy, RBI publications

Reserve Bank of India (2006-08), Report on Currency and Finance, Lending and Investment operations of Banks, RBI publications, Vol II

Reserve Bank of India (2014-15), Report on Trend and Progress in India, Operations and Performance of Scheduled Commercial Banks, RBI Publications

Reserve Bank of India (2014-15), The handbook of Statistics on the Indian Economy, RBI publications

Reserve Bank of India, Statistical Tables Relating to Banks in India, Various Issues

Reserve Bank of India: Banking Statistics - Banking Statistical Returns, Various Issues.

Sairam, RK; Deshmukh, PS; Saxena, DC. 1998. Role of antioxidant systems in wheat genotypes tolerance to water stress. *Biol. Plant.,* 41: 387–394.

Sairam, RK; Srivastava, GC; Agarwal, S; Meena, RC. 2005. Differences in antioxidant activity in response to salinity stress in tolerant and susceptible wheat genotypes. *Biol. Plant.,* 49: 85–91.

Shah, K; Singh, M; Rai, AC. 2013. Effect of heat-shock induced oxidative stress is suppressed in *BcZAT12* expressing drought tolerant tomato. *Phytochemistry,* 95: 109–117.

Sharma, PN; Chatterjee, C; Agarwala, SC; Sharma, CP. 1990. Zinc deficiency and pollen fertility in maize (*Zea mays*). *Plant Soil,* 124: 221–225.

Shekhar, KC; Lekshmy, Shekhar. 2008. Banking Theory and Practice, Vikas Publishing House Pvt. Ltd, New Delhi, 9th edition, pp.151-153.

Singla, A; Arora, RS. 2005. Financial Performance of Public Sector Banks: A Comparative Study of Canara Bank and Indian Bank, Punjab Journal of Business Studies, Vol.1 (1), April-September, pp1-13

Valentovic, P; Luxova, M; Kolarovic, L; Gasparikova, O. 2006. Effect of osmotic stress on compatible solutes content, membrane stability and water relations in two maize cultivars. *Plant Soil Environ.,* 52: 184.

Vimala, P A. 2002. Study on the Priority Sector Lending by commercial banks in Kerala, Ph.D Thesis, Mahatma Gandhi University, Kerala

Westgate, ME; Boyer, JS. 1986. Reproduction at low and pollen water potentials in maize. *Crop Sci.,* 26951–26956.

Wopereis, MCS; Kropff, MJ; Maligaya, AR; Tuong, TP. 1996. Drought-stress responses of two lowland rice cultivars to soil water status. *Field Crop Res.,* 46: 21–39.

Xiong, L;Wang, R; Mao, G; Koczan, JM. 2006. Identification of drought tolerance determinants by genetic analysis of root response to drought stress and abscisic acid. *Plant Physiol.,* 142; 1065–1074.

■■■■

4

SMALL HEAT SHOCK PROTEINS: THE THERMAL STRESS MANAGERS

***Ankita Chandola[1], Vijai Pandurangam[1] and Kamla Dhyani[2]**

[1] Department of Plant Physiology, Institute of Agricultural Sciences, Banaras Hindu University, Varanasi-221005, India
[2]Shri Guru Ram Rai Institute of Technology & Sciences, Patel Nagar, Dehradun-248001
**Corresponding author: ankycd@gmail.com*

INTRODUCTION

All organisms respond to high temperatures or heat stress by altering their gene expression patterns and turning on the heat shock genes. These events are referred to as the heat shock response (Linquist and Craig, 1988; Nover, 1991; Vierling, 1991; Kotak *et al.,* 2007; Richter *et al.,* 2010). The products of the heat shock genes are the heat shock proteins or HSPs. The HSPs include a number of conserved protein families: the HSP100s, HSP90s, HSP70s, HSP60s, and the HSP20s or the small HSPs. Despite their close working relationship in the cell, the individual HSP families are evolutionarily unrelated to each other. The HSPs were first identified based on their up-regulation during heat shock. Small heat shock proteins (sHSPs) are the most ubiquitous molecular chaperones subgroup with molecular weights ranging from 12 to 42 kDa that can prevent the irreversible aggregation of denaturing proteins. The number of sHSP genes increases along the evolutionary scale (Haslbeck *et al.,* 2005). Single-celled organisms such as bacteria have only one or two sHSP, whereas multicellular organisms have many sHSP in the genome. The expression of genes for these sHSPs is limited in the absence of environmental stress and occurs in some stages of growth and development of plants such as embryogenesis, germination, development of pollen grains, and fruit ripening (Sun *et al.,* 2002). However, sHSPs are usually undetectable in plant cells under physiological conditions, are induced upon stress and plant tolerance to stress, including drought, salinity, oxidized species, and low temperatures (Lo¨w *et al.,* 2000; Hamilton and Heckathorn, 2001; Scharf *et al.,* 2001; Zhang *et al.,* 2008). The ability of sHSPs to assemble into heat shock granules (HSGs) and their disintegration is a prerequisite for survival of plant cells under continuous stress conditions at sublethal temperatures (Miroshnichenko *et al.,* 2005). HSP70 and HSP90 mRNAs can increase ten-fold, while low-molecular weight (LMW) HSPs can increase as much as 200-fold. Small heat shock proteins are involved in stress tolerance mainly in heat stress. Kim *et al.* (2012) previously isolated and characterized a rice cDNA clone, *Oshsp26*, encoding a chloroplast-localized small heat shock protein that is expressed following oxidative or heat stress. The plant small heat shock proteins are particularly diverse and are a crucial component of the plant heat shock response. In recent years, considerable progress has been made in elucidating the structure, function, and evolution of the sHSPs (Waters, 2013).

STRUCTURE

Most sHSPs are in the range of 15-22 kDa thus, these proteins are also sometimes referred to as the HSP20 family. The primary sHSP amino acid sequence includes a variable N-terminal region, a more conserved C-terminal region, and a C-terminal extension (Fig. 1). The highly conserved C-terminal region is often referred to as the α-crystallin domain, or the HSP20 domain. The sHSPs that function in specific cellular organelles or compartments have N-terminal transit, leader, or signal sequences needed to get the sHSP to the proper cellular compartment. The C-terminal extension can be quite variable in both length and sequence, and may also contain organelle specific retention amino acid motifs. Guan *et al.,* 2004 defined the model for sHSP chaperone activity based on studies of a cytosolic sHSP family named as Class I sHSPs (sHSP-CI), which represent the most abundant sHSP in plants. The first two high-resolution crystal structures determined for sHSPs were Hsp16.5 from an archaea, *Methanocaldococcus (Methanococcus) jannaschii* (Kim *et al.,* 1998), and Hsp16.9 from wheat (*Triticum aestivum*) (Montfort *et al.,* 2002). In solution, wHSP16.9 can dissociate into smaller oligomeric states in a temperature dependent manner (Montfort *et al.,* 2001). The basic building block of wHSP16.9 is a dimer, which further assembles as a 12-mer consisting of two trimers of dimers.

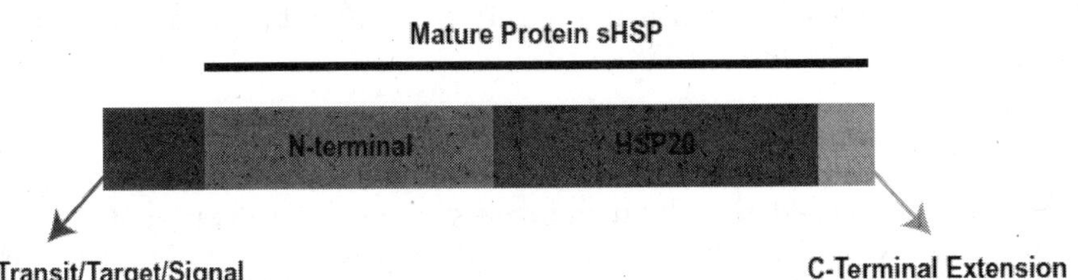

Fig. 1. Small heat shock protein domains. The sHSP monomer contains a variable N-terminal domain (light blue), the conserved HSP20 or α-crystallin domain (red), and a variable C-terminal extension (grey). The sHSPs that localize to the organelles, chloroplast, mitochondrion, endoplasmic reticulum, and peroxisome, have the necessary transit, targeting, or signal sequences (dark blue) needed for successful entry into those individual organelles. These sequences are not part of the mature sHSP. The C-terminal extension is variable in length and contains the endoplasmic reticulum (KDEL) or peroxisome (SKL) retention signals in those sHSPs that function in those cellular compartments (Source: Waters, 2013).

CATEGORIZATION AND LOCALIZATION

Small heat shock proteins are required for life and are found in even the smallest genomes (Waters *et al.,* 2003). The categorization is based on the sequence similarity and the location of these proteins in the cell. Higher plants are characterized by the presence of at least 20 types of sHSPs, but one species could contain 40 types of these sHSPs (Vierling, 1991). Particularly, ten separate families of small heat shock proteins have been recognized to be conserved in both monocot and dicot plants, indicating the potential for diversity in sHSP mechanisms (Siddique *et al.,* 2008; Scharf *et al.,* 2001). Small heat shock proteins encoded by four of these families localize to the cytoplasm and there are two classes of proteins (Class I and Class II) in the cytoplasm encoded by two groups of genes. Those encoded by the other six families localize to cellular organelles including nucleus, chloroplasts, mitochondria, endoplasmic reticulum, and peroxisomes (Basha *et al.,* 2010). An example of this diversification of sHSPs in plant species with their location is listed in Fig. 3. Furthermore, Scharf *et al.* (2001) and Siddique *et al.* (2003) divided the sHSPs of *A. thaliana* and *Lycopersicon esculentum* into three subclasses. These included:

- Subclass CI represented by six proteins in *A. thaliana* and five proteins in *L. esculentum*
- Subclass CII represented by two genes in both plants
- Subclass CIII represented by one gene in both plants

Siddique *et al.* (2008) reported the presence of other groups in the cytoplasm of *A. thaliana* cells and could be categorized into subclasses:

CIV, CV, CVI, and CVII. It is believed that all this diversification and abundance of the sHSPs in a plant reflect an adaptation of the plant to heat stress (Waters *et al.,* 1996).

MECHANISM

It has been proposed that heat-induced oligomer dissociation is a major mechanism by which plant sHSPs can expose normally in accessible, hydrophobic client binding surfaces (Haslbeck *et al.,* 2005). It is likely that heat-induced dissociation of sHSP oligomers may expose the hydrophobic patches buried in the oligomeric interface, resulting in binding and stabilization of denatured proteins (Montfort *et al.,* 2001; Haslbeck *et al.,* 2005). The model suggests that sHSP assembles into a large homo-oligomer, which binds denatured proteins in an ATP-independent manner, keeping them in a folding competent state. Then, it cooperates with ATP-dependent molecular chaperones, such as HSP70 and HSP90, to refold those proteins. Notably, sHSP has a much larger binding stoichiometry than other molecular chaperones, which has led to the speculation that sHSP functions as a reservoir to stabilize the flood of denatured proteins in response to stress (Lee and Vierling, 2000; Ehrnsperger *et al.,* 1997). Nevertheless, the details about the interactions between sHSP and non-native proteins and how these non-native proteins are subsequently refolded are still lacking. This is partially due to limited knowledge on the molecular structure of sHSPs (Haslbeck *et al.,* 2005).

ROLE

Investigations of sHSPs from mammals, yeast, plants and bacteria have unveiled their role in thermotolerance such that the over-expression of eukaryotic recombinant sHSPs could increase the thermotolerance of E. coli cells (Landry *et al.,* 1989; Soto *et al.,* 1999; Torok *et al.,* 1999). The sHSPs have been observed to exist as large homo-oligomers and the reversible dissociation of their oligomers has been reported to be important for their enhanced chaperone activity at high temperatures (Montfort *et al.,* 2001; Haslbeck *et al.,* 1999; Gu *et al.,* 2002; Haslbeck *et al.,* 2008; Jaya *et al.,* 2009; Stengel *et al.,* 2010; Jiao *et al.,* 2005). The model for the role of sHSPs in the chaperone network is given in Fig. 2 (Waters, 2013). Small heat shock proteins complex with a variety of non-native proteins in an ATP-independent manner and, in the context of the stress response, form a first line of defence against protein aggregation in order to maintain protein homeostasis (Haslbeck and Vierling, 2015). One of the characteristic functions of this class is the degradation of the proteins that have unsuitable folding. Another characteristic that distinguishes these sHSPs from other chaperone classes is that their activity is independent from ATP (Miernyk, 1999). These two functional features of the sHSPs stand out among other HSP chaperones (Haslbeck *et al.,* 2005; Nakamoto and Vigh, 2007; McHaourab *et al.,* 2009; Eyles and

Gierasch, 2010; Tyedmers *et al.*, 2010). Small HSPs may play structural roles in maintaining cell membrane integrity also. The special importance of small HSPs in plants is suggested by their unusual abundance and diversity.

Several reports have been listed below showing the different roles of small heat shock proteins in different plants.

- It was noticed that there was a positive qualitative relation between the accumulation of sHSPs in the plastids and thermotolerance of heat shock (from 28 to 40°C) in six divergent Anthophyta species, including C3, C4, CAM, monocot, and dicot species Downs and Heckathorn (1998).
- Another study by Downs and Heckathorn (1998), indicated that the mitochondrial sHSP protected NADH: ubiquinone oxidoreductase (complex I) during heat stress in apple fruit of *Pyrus pumila* (P. Mill.) K. Koch var. McIntosh. This information might indicate some role of these proteins in adaptation of plants to heat stress.
- Localization of small HSPs in chloroplast membranes further suggested that these proteins protect the PSII from adverse effects of heat stress and play a role in photosynthetic electron transport (Barua *et al.*, 2003). Kim *et al.* (2012) also reported that OsHSP26 protein plays an important role in the protection of PSII during heat and oxidative stress in vivo.
- Mogk *et al.* (2003) showed that the sHSP 18.1 isolated from *Pisum sativum*, as well as the sHSP 16.6 from *Synechocystis* sp. PCC6803 under in vitro conditions, binds to unfolded proteins and allows further refolding by HSP70/ HSP100 complexes.
- In their review, Nakamoto and Vigh (2007) concluded that there were some indications that small heat shock proteins play an important role in membrane quality control and thereby potentially contribute to the maintenance of membrane integrity especially under stress conditions.
- Acunzo *et al.* (2012) reported that HSP27, αB-crystallin and HSP22 are ubiquitous small heat shock proteins whose expression is induced in response to a wide variety of unfavourable physiological and environmental conditions. These sHSPs protect cells from otherwise lethal conditions mainly by their involvement in cell death pathways such as necrosis, apoptosis or autophagy. Hence, HSP27, αB-crystallin and HSP22 are appropriate targets for modulating cell death pathways. At a molecular level, the mechanisms accounting for sHSP functions in cell death are
    - prevention of denatured proteins aggregation,
    - regulation of caspase activity,

- regulation of the intracellular redox state,
- function in actins polymerization and cytoskeleton integrity and
- proteasome-mediated degradation of selected proteins.

- Glatz *et al.* (2016) reported that sHSP15.8 and sHSP16 were induced after prolonged high-temperature treatment but with different kinetic profiles. Both sHSPs are able to bind to different lipids isolated from S. pombe, whose interaction might provide a powerful protection against heat-induced damages of the membranes.

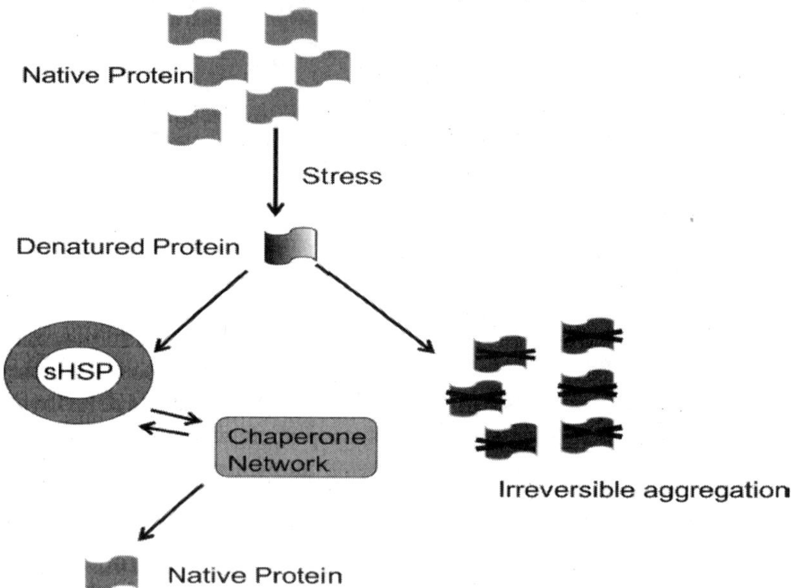

Fig. 2. The sHSPs are an important part of the chaperone network. The sHSPs bind denaturing proteins, protecting them from irreversible aggregation. The substrate proteins can then be refolded by other parts of the chaperone network (Waters, 2013).

CONCLUSIONS

Summing up we conclude that the sHSPs are a fascinating group of proteins. They are ancient and ubiquitous, yet highly variable. In addition, as chaperones, they have a crucial and fundamental role in plant cell biology. Many experimental results indicated that the small heat shock proteins play an important role in protection of cells from different stresses. Induction of small heat shock proteins gene expression and protein accumulation upon environmental stresses point to the hypothesis that these proteins play an important role in stress tolerance. The sHSPs null mutants were also reported to display a weak, but significant, heat sensitivity indicating their importance in the thermal stress management. Thus, studies of the sHSPs have a broader importance because they connect cellular function to organismal-level traits such as tolerance to heat stress.

Species	Protein	DNA accession number
Chloroplast-localized proteins		
Arabidopsis thaliana	HSP21	X54102
Glycine max	HSP22	X07188
Petunia hybrida	HSP21	X54103
Pisum sativium	HSP21	X07187
Triticum aestivum	HSP26A	X58280
Triticum aestivum	HSP26B	X67328
Zea mays	HSP26	L28712
Mitochondrial-localized protein		
Chenopodium rubrum	HSP23	X15333
Endoplasmic reticulum-localized proteins		
Arabidopsis thaliana	HSP22	U11501
Glycine max	HSP22	X63198
Pisum sativum	HSP22	M33898
Class I Cytocolically localized proteins		
Arabidopsis thaliana	HSP17.6	X16076
Arabidopsis thaliana	HSP17.4	X17293
Arabidopsis thaliana	HSP18.2	X17295
Chenopodium rubrum	HSP18.3	X53870
Daucus carota	HSP18.0	X53852
Daucus carota	HSP17.8	X53851
Glycine max	HSP17.5	M11318
Glycine max	HSP17.6	MI1317
Glycine max	HSP18.5	X07160
Helianthus annuus	HSP17.6	X59701
Lycopersicon esculentum	HSP17.8	X56138
Medicago sativa	HSP18.1	X58710
Medicago sativa	HSP18.2	X5871 1
Oryza sativa	HSP16.9	X60820
Oryza sativa	HSP17.4	D 12635
Pisum sativum	HSP18.1	M33899
Triticum aestivum	HSP16.9A	X13431
Triticum aestivum	HSP16.9B	X64618
Triticum aestivum	HSP16.9C	L 14444
Zea mays	HSP17.2	X65725
Class II Cytocolically localized proteins		
Arabidopsis thaliana	HSP17.6	X63443
Glycine max	HSP17.9	X07159
Ipomea nil (Pharbatis nil)	HSP18.8	M99430
Ipomea nil (Pharbatis nil)	HSP17.2	M99429
Lilium longiflorum	HSP17.6	D21816
Lilium longiflorum	HSP16.5	D21818
Pisum sativum	HSP17.7	M33901
Triticum aestivum	HSP17.3	X58279
Zea mays	HSP17.5	X54076
Zea mays	HSP17.8	X54075

Fig. 3. Location of sHSPs types in plant cell and the DNA accession number (after Waters *et al.*, 1996)

FUTURE DIRECTIONS

Studies of sHSPs have clearly made great strides in recent years, but many important questions can still be addressed. First, while we have a general understanding of the function and structure of the sHSPs, we do not yet know how the diversity in primary sequence reflects differences in structure or function both within the plant subfamilies and between subfamilies. In addition, little is known about higher level structural variation among plant sHSPs. Additional high-resolution structures of plant sHSPs will provide crucial information and will elucidate the differences in structure and function between and within the diverse plant subfamilies. Analysis of the interaction of sHSPs with biological membranes and membrane proteins is also necessary. Joint utilization of biochemical and cell biological methods will provide new information important for a detailed description of the mechanism of action of sHSPs in the cell.

REFERENCES

Acunzo J, Katsogiannou M and Rocchi P 2012 Small heat shock proteins HSP27 (HspB1), αB-crystallin (HspB5) and HSP22 (HspB8) as regulators of cell death. *The International Journal of Biochemistry and Cell Biology* **44** 1622-1631.

Barua D, Downs CA and Heckathorn SA 2003 Variation in chloroplast small heat-shock protein function is a major determinant of variation in thermotolerance of photosynthetic electron transport among ecotypes of *Chenopodium album. Funct. Plant Biol.* **30** 1071-1079.

Basha E, Jones C, ysocki VW and Vierling E 2010 Mechanistic differences between two conserved classes of small heat shock proteins found in the plant cytosol. *Journal of Biological Chemistry* **15** 11489-11497

Downs CA and Heckathorn SA 1998 The mitochondrial small heat shock protein protects NADH : ubiquinone oxidoreductase of the electron transport chain during heat stress in plants. *Febs Lett.* **430** 246-250

Ehrnsperger M, Gr¨aber S, Gaestel M and Buchner J 1997 Binding of non-native protein to Hsp25 during heat shock creates a reservoir of folding intermediates for reactivation. *EMBO Journal* **16** 221-229

Eyles SJ and Gierasch LM 2010 Nature's molecular sponges: small heat shock proteins grow into their chaperone roles. *Proceedings of the National Academy of Sciences, USA* **107** 2727-2728

Gu L, Abulimiti A, Li W and Chang Z 2002 Monodisperse Hsp16.3 nonamer exhibits dynamic dissociation and reassociation, with the nonamer dissociation prerequisite for chaperone-like activity. *Journal of Molecular Biology* **319** 517-526

Guan JC, Jinn TL, Yeh CH, Feng SP, Chen YM and Lin CY 2004 Characterization of the genomic structures and selective expression profiles of nine class I small heat shock protein genes clustered on two chromosomes in rice (*Oryza sativa* L.). *Plant Molecular Biology* **5** 795- 809

Hamilton EW and Heckathorn SA 2001 Mitochondrial adaptations to NaCl. Complex I is protected by anti-oxidants and small heat shock proteins, whereas complex II is protected by proline and betaine. *Plant Physiol.* ***126*** 1266-1274

Haslbeck M, Franzmann T, Weinfurtner D and Buchner J 2005 Some like it hot: the structure and function of small heat-shock proteins. *Nature Structural and Molecular Biology* **12** 842-846

Haslbeck M, Kastenmuller A, Buchner J, Weinkauf S and Braun N 2008 Structural dynamics of archaeal small heat shock proteins. *Journal of Molecular Biology* **378** 362-374

Haslbeck M, Walke S, Stromer T, Ehrnsperger M, White HE, Chen S, Saibil H and Buchner J 1999 Hsp26: a temperature-regulated chaperone. *EMBO J.* **18** 6744-6751

Haslbeck M and Vierling E 2015 A First Line of Stress Defence: Small Heat Shock Proteins and Their Function in Protein Homeostasis. *Journal of Molecular Biology* **427** 1537-1548

Jaya N, Garcia V and Vierling E 2009 Substrate binding site flexibility of the small heat shock protein molecular chaperones. *Proc. Natl. Acad. Sci. U S A* **106** 15604-15609

Jiao W, Qian M, Li P, Zhao L and Chang Z 2005 The essential role of the flexible termini in the temperature-responsiveness of the oligomeric state and chaperone like activity for the polydisperse small heat shock protein IbpB from *Escherichia coli*. *Journal of Molecular Biology* **347** 871-884

Kim KH, Alam I, Kim YG, Sharmin SA, Lee KW, Lee SH and Lee BH 2012 Overexpression of a chloroplast-localized small heat shock protein OsHSP26 confers enhanced tolerance against oxidative and heat stresses in tall fescue. *Biotechnol Lett* **34** 371-377

Kotak S, Larkindale J, Lee U, von Koskull-Doring P, Vierling E and Scharf KD 2007 Complexity of the heat stress response in plants. *Current Opinion in Plant Biology* **10** 310-316

Landry J, Chretien P, Lambert H, Hickey E and Weber LA 1989 Heat shock resistance conferred by expression of the human HSP27 gene in rodent cells. *J Cell Biol.* **109** 7-15

Lee GJ and Vierling E 2000 A small heat shock protein cooperates with heat shock protein 70 systems to reactivate a heat-denatured protein. *Plant Physiology* **122** 189-197

Lindquist S and Craig EA 1988 The heat-shock proteins. *Annual Review of Genetics* **22** 631-677

Lo¨w D, Bra¨ndle K, Nover L and Forreiter C 2000 Cytosolic heatstress proteins Hsp17.7 class I and Hsp17.3 class II of tomato act as molecular chaperones in vivo. *Planta* **211** 575-582

McHaourab HS, Godar JA and Stewart PL 2009 Structure and mechanism of protein stability sensors: chaperone activity of small heat shock proteins. *Biochemistry* **48** 3828-3837

Miernyk JA 1999 Protein folding in the plant cell. *Plant Physiol.* **121** 695-703

Miroshnichenko S, Tripp J, Nieden U, Neumann D, Conrad U and Manteuffel R 2005 Immunomodulation of function of small heat shock proteins prevents their assembly into heat stress granules and results in cell death at sublethal temperatures. *Plant J.* **41** 269-281

Mogk A, Schlieker C, Friedrich KL, Scho¨nfeld HJ, Vierling E and Bukau B 2003 Refolding of substrates bound to small Hsps relies on a disaggregation reaction mediated most efficiently by ClpB/DnaK. *J. Biol. Chem.* **278** 31033-31042

Montfort RL, Basha E, Friedrich KL, Slingsby C and Vierling E 2001 Crystal structure and assembly of a eukaryotic small heat shock protein. *Nat. Struct. Biol.* **8** 1025-1030

Nakamoto H and Vigh L 2007 The small heat shock proteins and their clients. *Cellular and Molecular Life Sciences* **64** 294-306

Nover L 1991 Heat shock response. *Boca Raton, FL: CRC Press*

Richter K, Haslbeck M and Buchner J 2010 The heat shock response: life on the verge of death. *Molecular Cell* **40** 253-266

Scharf KD, Siddique M and Vierling E 2001 The expanding family of Arabidopsis thaliana small heat stress proteins and a new family of proteins containing α-crystallin domains (ACD proteins). *Cell Stress and Chaperones* **3** 225-237

Siddique M, Gernhard S, Von Koskull-D¨oring P, Vierling E and Scharf KD 2008 The plant sHSP superfamily: five new members in Arabidopsis thaliana with unexpected properties. *Cell Stress and Chaperones* **13** 183-197

Siddique M, Port M, Tripp J, Weber C, Zielinski D, Calligaris R, Winkelhaus S and Scharf KD 2003 Tomato heat stress protein Hsp16.1-CIII represents a member of a new class of nucleocytoplasmic small heat stress proteins in plants. *Cell Stress Chaperon* **8** 381-394

Soto A, Allona I, Collada C, Guevara MA, Casado R, Cerezo ER, Aragoncillo C and Gomez L 1999 Heterologous expression of a plant small heat-shock protein enhances Escherichia coli viability under heat and cold stress. *Plant Physiol.* **120** 521-528

Stengel F, Baldwin AJ, Painter AJ, Jaya N, Basha E, Kay LE, Vierling E, Robinson CV and Benesch JLP 2010 Quaternary dynamics and plasticity underlie small heat shock protein chaperone function. *Proceedings of the National Academy of Sciences, USA* **107** 2007-2012

Sun W, Motangu MV and Verbruggen N 2002. Small heat shock proteins and stress tolerance in plants. *Biochim. Biophys. Acta* **15** 1-9

To¨ro¨k Z, Goloubinoff P, Horva´th I, Tsvetkova NM, Glatz A, Balogh G and Varvasovszki V 2001 Synechocystis HSP17 is an amphitropic protein that stabilizes heat-stressed membranes and binds denatured proteins for subsequent chaperone-mediated refolding. *Proc Natl Acad Sci USA* **98** 3098-3103

Tyedmers J, Mogk A and Bukau B 2010 Cellular strategies for controlling protein aggregation. *Nature Reviews Molecular Cell Biology* **11** 777-788

Vierling E 1991 The heat shock response in plants. Annual Review of Plant Physiology and Plant Molecular Biology **42** 579-620

Waters ER 2013 The evolution, function, structure, and expression of the plant sHSPs. Journal of Experimental Botany **64** 391-403

Waters ER, Lee GJ and Vierling E 1996 Evolution, structure and function of the small heat shock proteins in plants. *J. Exp. Bot.* **47** 325-338

Waters E, Hohn MJ, Ahel I, Graham DE, Adams MD, Barnstead M, Beeson KY, Bibbs L, Bolanos R, Keller M, *et al.,* 2003 The genome of Nanoarchaeum equitans: Insights into early archaeal evolution and derived parasitism. *Proc Natl Acad Sci USA* **100** 12984-12988

Zhang JH, Wang LJ, Pa, QH, Wang YZ, Zhan JC and Huang WD 2008 Accumulation and subcellular localization of heat shock proteins in young grape leaves during cross-adaptation to temperature stresses. *Scientia Horticulturae* **117** 231-240.

■■■■

5

PHYSIOLOGICAL APPROACHES FOR HEAT TOLERANCE IN WHEAT: A GOOD TOOL TO SCREEN WHEAT FOR HEAT TOLERANCE

Kamla Dhyani[1]*, Ankita Chandola[2], Shweta Sahni[3] and Alok Shukla[3]

*1*G.B Pant University of Agriculture Science & Technology, U.S Nagar (Uttrakhand), India*
[2]Department of Plant Physiology, Institute of Agricultural Sciences, Banaras Hindu University, Varanasi 221 005, India
[3]Shri Guru Ram Rai Institute of Technology & Sciences, Dehradun, India
Corresponding Author- dhyani.k@rediffmail.com*

INTRODUCTION

Wheat (*Triticum aestivum*), is second most important cereal after rice, it represents about 30% of the world's cereal area and is cultivated world-wide. Climate change and global warming is a matter of concern as wheat growth is greatly affected by heat stress (HS) various developmental stage. Stress due to high-temperature has emerged as a major constraint for the successful wheat production worldwide (Hays *et al*. 2007; Kumar *et al*. 2012).

Annual yield loss in wheat due to global warming is expected to be 7.7 billion dollars, and by 2025, this would be around 18 billion dollars. The average global temperature is reported to be increasing at a rate of 0.18°C every decade (Hansen *et al.,* 2012). To deliver food security for the 9 billion population in 2050, a 70% increase in world food supply will be required, climate change models suggests further increase in average temperature, large warming in 2005-2015 exists in major region of the world (Hansen *et al.,* 2016). Yield losses are associated with both high temperatures from initial growth stages (18°C–25°C and up to 32°C during grain filling, as well as heat shocks (temperatures greater than 32°C) at reproductive stage.

As current situation of weather is characterized by changing local weather events and increasing the frequency of weather hazards in near future, is a major challenge to plant breeders, in increasing yield potential (Semenov *et al.,* 2014), multidisciplinary approach is helpful to understand how plants respond to abiotic stresses and required application of this knowledge in the context of climate change (Martre *et al.,* 2015). Scientists recently expressed alarm at the record has baked the planet in the first few months of 2016, which follows a year that was the warmest on record. It has been shown by various workers that wheat is particularly sensitive to extreme cold and hot temperatures during the reproductive stage (Alghabari *et al.,* 2014; Vara Prasad and Djanaguiraman, 2014). Middle latitude countries have near-optimum temperature and limited impact from projected temperature change, but, in contrast, warmer countries, such as Indonesia, India and Nigeria are on a steep slope with rapidly declining productivity as temperature rises (Burke *et al.,* 2015) and a tendency for increasing heat waves and strong droughts to coincide (Mazdiyasni and Agha-Kouchak, 2015) are expected consequences of increased global warming.

The present chapter considers the physiological aspects and traits for which there is evidence that genetic improvement could improve wheat adaptation to Heat Stress.

DEVELOPMENTAL STAGES

Developmental stages of wheat greatly affected by heat and plays very important role to obtain high yield in high temperature. Days to heading, days taking to anthesis and days taken to physiological maturity are the important parameters to cope up with high temperature stress, and these traits decides the yield and grain size of the crop, for example, if a genotype good in all traits like early ground cover, chlorophyll and others it is not necessary it will be a good performer in case of yield. With other physiological characters plants developmental studies also required to study .Yield ultimately is a combined effect of various morphological, developmental, physiological and genetical factors. Days to heading and days to physiological maturity which generally decides the grain filling duration which is an important heat tolerant trait which is commonly heritable in nature, it is also explore by various workers. Some early maturing varieties are also heat tolerant in case of yield. Thorough study of developmental stages of wheat and its relation with yield could be a good tool to asses heat tolerance.

RAPID GROUND COVER AND CANOPY STRUCTURE

Initial active development (Rapid Ground Cover) or early vigour decides the capacity of genotypes to develop leaf area (canopy) or above ground biomass. Rapid ground cover is associated with difference in seed germination behaviour of crop (Emergance rate, specific leaf area, grain, embryo size (Matthew & Reynolds, 2014, Richards and Lukacs, 2002), such characteristics are heritable, are good to make them breeding targets (Rebetzke *et al.,* 2008). Rapid ground cover is a genetical diverse character and has been exploited by various workers (Mullan and Reynolds, 2010). Canopy at vegetative stage and reproductive stage plays very important role in photosynthesis, less leaf area index is responsible for less photosynthesis wheat shows great diversity in canopy architecture, and canopy plays very important role to optimized light distribution, and it could improve radiation use efficiency (Murchie *et al.,* 2009), and in some research it was found that tolerant genotype have more LAI as compare to susceptible one. Optimum LAI is a characterstic feature of a tolerant genotype (Dhyani *et al.,* 2012) genetically, canopy architecture may not be easy to address, since the canopy is a highly complex structure. But now some instruements like Plant Canopy Analyzer and Leaf Area Meter are available with the help of those, one can estimate the Plant Canopy parameters like (LAI, MTA, DIFN). Identification of gene responsible of leaf area is required further study. Comparative study of leaf area index and other physiological and agronomical traits require more research.

LATE SENESCENCE (STAY-GREEN ABILITY OF A GENOTYPE)

Stay-green ability of a plant desides, to delayed senescence during post-anthesis stages of plant development (Thomas and Howarth, 2000). Stay-green may refers to either the delayed onset of senescence or a slower rate of senescence (Thomas and Smart, 1993). Contribution of Photosynthesis under optimum temperature is 80–90% of assimilates for grain filling (Evans *et al.,* 1975). Therefore, senescence before, maturity and the rate of senescence may be important factors that influence yield potential under stress (Thomas and Howarth, 2000). In many crop species, stay-green plants have better foliage quality, high chlorophyll content, and greater resistance to insects, pests and diseases (Xu *et al.,* 2000). Most studies across plant species indicate that stay-green is a quantitative trait.

Leaf senescence is important developmental phenomena in crop development which is characterized by structural changes initially and followed by a controlled vacuolar collapse, and a final loss of integrity of leaf, stay green is also effected by various other biotic and abiotic factors including heat stress, this ability is due to delay expression of senescence-related genes (Lim *et al.,* 2007). Stay Green ability which is quite related to long time for physiological maturity is recognized as an important physiological traits for stress conditions, but only this stay green trait is not helpful in heat tolerance. During senescence generally chlorosis occurs and it reduce the photosynthetic area and mobilization of nitrogen to the maturing grain (Vijayalakshmi *et al.,* 2010). Although three components chlorophyll at anthesis duration of senescence, and rate of senescence determine the stay-green ability during heat stress, the rate of senescence not the start of senescence, is important component of stay-green (Harris *et al.,* 2007).

Recent studies by Jespersen (2016) on some perennial grasses shows that the enzymatic activity or gene expression of several major chlorophyll-synthesizing and chlorophyll-degrading enzymes (chlorophyllase, and peroxidase) indicated heat-induced decline in leaf chlorophyll content was mainly due to accelerated chlorophyll degradation, as manifested by increased gene expression levels of chlorophyllase and pheophytinase, and the activity of pheophytinase (PPH), while chlorophyll-synthesizing genes and enzymatic activities were not differentially altered by heat stress in the two genotypes.

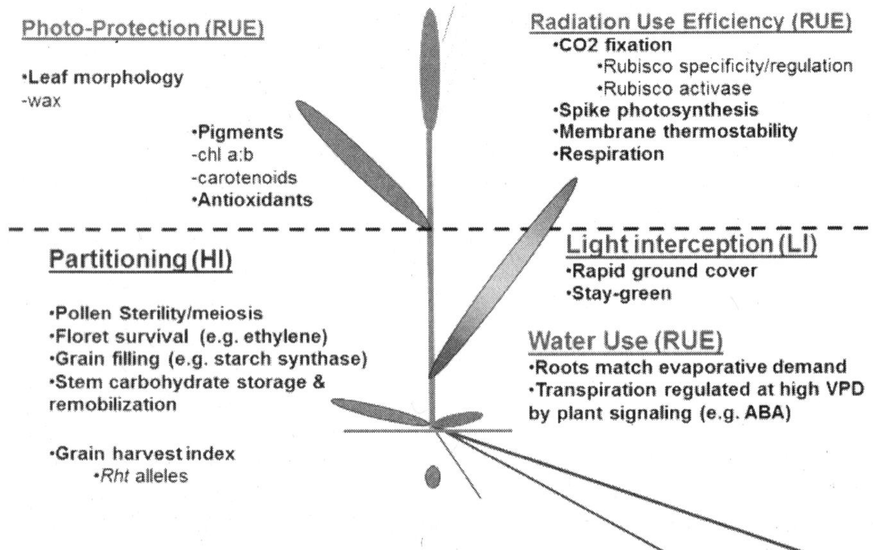

Fig. 1- Model of Heat-adaptive traits grouped by three main drivers of yield (Li, Rue and Hi) in the absence of water limitations (Adapted from Reynolds *et al.,* 2007).

RADIATION USE EFFICIENCY (RUE)

Crop biomass is depends on net radiation use efficiency, because plant use sun light for photosynthesis which is result of total carban assimilation minus losses associated with growth (structural process), maintenance (damages due to diseases), and repair (stress), metabolic inefficiencies (mainly photorespiration in C_3 species), and a range of photoprotective strategies. Plant generally used CO_2 from the atmosphere and convert it in to carbohydrate, but all the carbohydrate is not stored in the grain, storage is depends on different physiological processes of plants mainly partitioning of the reserve, some genotypes are seems to be good in other morphological characters but are low yielder.

Improvements in photosynthetic capacity and efficiency are key targets for increasing the yield potential of crops. The main objectives, namely increased affinity of Rubisco for CO_2 versus oxygen and a better catalytis as it is a very important enzyme of photosynthesis (Parry *et al.*, 2011), are likely to be particularly beneficial at warmer temperatures. Rubisco's affinity for CO_2 decreases with temperatures (Jordan and Ogren, 1984).

PHOTOSYNTHESIS AND CHLOROPHYLL CONTENT

Photosynthesis is the most sensitive physiological process effected by elevated temperature and light (Wahid *et al.* 2007) and photosynthesis affects growth and grain yield of wheat. Heat stress disrupts the function of chloroplast and reductions in chlorophyll content. Oxidative stress may induce lipid peroxidation leading to protein degradation, membrane rupture and enzyme inactivation (Sairam *et al.* 2000). Rubisco is more sensitive to increased temperatures than the rest of the enzymes involved in carboxylation. Prasad *et al.* (2008) reported that the most important reasons for PSII sensitivity to high temperature are heat-induced increase in thylakoid membrane fluidity and electron-transport dependent integrity of PSII. The inhibition of PSII electron transport under heat stress is often indicated by a sharp increase in the basal level of chlorophyll fluorescence that corresponds to photosynthetic inhibition (Ristic *et al.* 2007). The Fv/Fm and SPAD measurements are used to measure the photosynthetic efficiency in plants indirectly. Light absorbed by the leaf at a optimum physiological temperature is used in the photochemical act and the balance is dissipated as heat or reemitted as chlorophyll fluorescence. Chlorophyll fluorescence is estimated as Fv/Fm and reflects thylakoid membrane integrity. Fv (variable fluorescence) is the difference between Fo and Fm where, Fo is the fluorescence at ground state which rises to its maximum value (Fm) upon an application of a saturated flash of light. In healthy leaves, the Fv/Fm ratio is close to 0.8. A lower value indicates that a proportion of PSII reaction centers are damaged. This phenomenon is called as photo-inhibition, and is observed in the plants under stress conditions (Fracheboud *et al.* 2002). Heat stress induced damage and disruption of the integrity of thylakoid membranes also causes the photophosphorylation to cease (Dias and Lidon, 2009).

PHOTOPROTECTIVE METABOLITES

The photosynthetic machinery is very sensitive to damage under abiotic stress, in excess light condition there is formation of Reactive Oxygen Species. Photoprotective molecules include carotenoids, flavonoids, glutathione, ascorbate, and tocopherols are scavengers of ROS, these molecules scavange ROS prior to their diffusion from the site of generation and protects complex molecules. Enzymatic antioxidants were substantially increased in heat-tolerant genotypes of wheat and have been correlated with heat the capacity to acquire thermo-tolerance (Sairam *et al.* 2000; Mittler, 2002; Almeselmani *et al.* 2009). The main scavenging molecules are found throughout the cell and its organelles. Depending on location, the main functions of photoprotective molecules are:

(1) They protects photooxidative damage to membranes and proteins caused by excessive light
(2) Scavenge singlet oxygen and dissipate extra energy via xanthophyll-mediated non photochemical quenching
(3) Interact with the membrane lipids acting like energy receptors and in down regulation processes,
(4) Provide direct protection by quenching or chemically scavenging of $1O_2$, O_2 2, and OH.
(5) Before damaging the cells they neutralize free radicals and ROS, under high irradiance levels
(6) Redox signaling (Murchie and Niyogi, 2011).

MEMBRANE THERMOSTABILITY

Membrane thermostability is good trait to asses heat tolerance, Heat shock increases cell membrane permeability, which inhibit cellular functions, unsaturation of fatty acids and denaturation of membrane proteins takes place during heat shock conditions that disrupt water, ion, and organic solute movement across membranes. Thylakoid membranes which is a site of light reaction in plants typically show swelling, increased leakiness and separation of the chlorophyll molecules from LHCII (light harvesting complex II) from the PSII core complex, and it disrupts PSII-mediated electron transfer (Ristic *et al.,* 2008). Reactive oxygen species (ROS) accumulation during HS also damages membranes (Abdul-Razack and Tarpley, 2009). Monogalactosyldiacylglycerol (50%) and galactosyldiacylglycerol (20%) are the main composition of membrane with the rest including phosphatidylglycerol and sulfoquinovosyldiacylglycerol, increased ratio of digalactosyldiacylglycerol to monogalactosyldiacyl glycerol favours thermotolerance with high level of saturation. However, more research is needed to explain genetic variation in MT and its relationship with saturation/unsaturation of fatty acids (Essemine *et al.,* 2011).

OSMOLYTE ACCUMULATION

Osmotic adjustment (osmoregulation) is a remarkable part of physiology of plants by which they respond to water deficits. This process is generally related with water deficit condition, In this process, cellular osmotic potential is decreased by the accumulation of solutes. These compounds include proline, glutamate, glycine-betaine, mannitol, sorbitol, polyols, trehalose, sucrose, oligosaccharides and inorganic ions like K+. These compounds help the cells to maintain their hydrated state and therefore function to provide resistance against drought and cellular dehydration (Ramanjulu and Bartels, 2002; Chaves *et al.* 2003). Osmolytes in low accumulation function in protecting macromolecules either by stabilizing the tertiary structure of protein or by scavenging ROS produced in response to drought (Zhu, 2001).

Proline is one of the amino acids, which appear most commonly in response to stress. Wheat is marked by low level of these compatible solutes and the accumulation and mobilization of proline was observed to enhance tolerance to heat stress (Dhyani *et al* , 2012, Ali Sajid 2016). Proline can act as a signalling molecule to modulate mitochondrial functions, influence cell proliferation or cell death and trigger specific gene expression, which can be essential for plant recovery from stress (Szabados and Savoure, 2010). Hong-Bo *et al.* 2006 investigated the role of proline as a wheat anti-drought defence protein under drought.

QUANTITATIVE TRAITS LOCI (QTL) ANALYSIS

Recently, several QTLs have been identified in wheat for heat tolerance during the reproductive phase. Byrne *et al.* (2002) detected QTLs for heat tolerance under hot and dry conditions on chromosomes 2B and 5B in a spring wheat population.

Dhanda and Munjal (2006) reported the presence of both additive and dominant types of gene action in wheat as well as QTL and associated simple sequence repeat markers (Ciuca and Petcu, 2009). The QTL for thousand grain weight (TGW) was dissected in to single Mendelian gene and mapped on short arm of chromosome 7D by Röder *et al.* (2008). Nine QTLs across the wheat genome for effective tiller per plant were mapped by Li *et al.* (2010) using a set of 16 doubled haploid (DH) lines. Paliwal *et al.* (2012) used the parameter heat susceptibility index (HSI) for thousand grain weight.

QTLs identified for heat tolerance in wheat during the reproductive phase

Trait	**Chromosome**	**Reference**
Rate of senescence	2A , 6A, 6B	Vijayalakshmi *et al.* (2010)
Greenness at maximum		

Senescence	4B, 5D, 3A, 6B	
SPAD chlorophyll content	7B	
Fv/Fm chlorophyll fluorescence	7A	
Yield	4A	
Canopy temperature	4A	Pinto *et al.* (2010)
Number of grains	1A, 2A, 3B, 4A, 5B	
Grain weight	1B, 2B, 3B, 5A, 6D	
Days to flowering	2D, 7D	Mason *et al.* (2010)
Stay-green	1A, 3B, 7D	Kumar *et al.* (2010)

Some traits associated with heat tolerance:

YIELD TRAITS

Plot yield	Fisher et al. 1998
Thousand grains weight (TGW)	Shpiler and Blum 1991
Grain filling duration (GFD)	Randall and Moss 1990
Number of effective tiller per plant	Richards 1996

MORPHOLOGICAL TRAITS

Early ground cover	Richards 1996
Stay green	Zhao *et al.* 2007
Epicuticular wax/leaf glaucousness	Richards 1996
Leaf rolling	Araus 1996
Biomass	Reynolds *et al.* 2001

PHYSIOLOGICAL TRAITS

Canopy temperature	Reynolds *et al.* 1994
Photosynthetic rate	Rijven 1986
Chlorophyll content	Al-Khatib and Paulsen 1984
Stomata conductance	Reynolds *et al.* 1994
Stem reserve	Mohammadi *et al.* 2009
Membrane thermostability	Shanahan *et al.* 1990
Transpirational cooling	Reynolds and Rebetzke, 2011; Pinto *et al.* 2010
Stomatal conductance	Reynolds *et al.* 1994
Early maturation	Tewolde *et al.* 2006
Antioxidant defense	Almeselmani *et al.,* 2009; Dhyani et al 2013

ASSIMILATE TRANSLOCATION (PARTITIONING)

Water-soluble carbohydrates stored in the stems and leaf sheaths of winter cereals provide an important source of assimilate for remobilization during grain-filling. Consequently, WSC are a major contributor to wheat grain yield and grain size in all environments but especially where photosynthesis is compromised as occurs where water is limiting. Breeding programs targeting greater water soluble carbohydrates should provide improved varieties with greater and more stable yields in stress environments.

The main objective of heat tolerance breeding is to select such characters/traits which are could be helpful to obtain such genotype which are tolerant to heat, or can produce more in high temperature environment. Assimilate partitioning is a important phenomena which decides the yield of wheat crop, During grain filling, assimilates generally transfer either from current assimilation or from pre-anthesis stored stem reserves. Under ideal conditions, 90 to 95% of carbon required for grain filling is transferred from current carbon assimilation (Kobata *et al.,* 1992). Under water and heat stress, the relative contribution of pre-anthesis stored reserves and current assimilation changes substantially Palta *et al.,* 1994). When the photosynthetic source of assimilates is reduced by heat stress, the alternative source for grain filling is remobilized stem reserves. The demand for stem reserves under heat stress dramatically increases, ranging from 6 to 100% depending on the heat-induced reduction in photosynthesis (Blum, 1998). However, under heat stress genotypic variation exists for the contribution of stem reserves for grain filling (Yang *et al.,* 2002). Tolerance to heat stress is associated with genotypes with stable photosynthesis, but also those with high capacity to store stem reserves (Blum *et al.,* 1994). At high temperatures, assimilate translocation which occurs through both symplastic and apoplastic pathways, is substantially reduced. Assimilate transport from flag leaf to grain is substantially reduced by temperatures above 30°C. This indicates that, in wheat, the effect of heat stress on assimilate translocation is indirect even though heat stress reduces the rate of assimilate transport from vegetative organs to grain (Plaut *et al.,* 2004).

PROTEOMICS

The expression of plant genes during growth, development and exposure to environmental variations is regulated by changes at the level of various enzymes and or proteins of various metabolic pathways (Hakeem et al. 2012). Thus to understand the molecular basis of heat tolerance in some wheat cultivars, the analysis of wheat proteome is very important. Proteomics has emerged as a powerful tool in discovering genes and pathways involved in abiotic stress response in crop plants (Chen and Harmon 2006; Kosová et al. 2011). Heat tolerant varieties of wheat could be designed by targeting key proteins or enzymes and metabolic pathways identified from tolerant wheat lines. The effect of heat stress on hexaploid wheat grain proteome using two-dimensional electrophoresis (2-DE) and matrix-assisted laser desorption/ionization-time of flight-mass spectrometry (MALDI-TOF-MS) was analysed by Majoul et al. (2004). A total of 43 proteins differentially expressed were identified under heat stress. It consisted of enzymes of carbohydrate metabolism, the ATP synthase ß-chain that was related to four heat-decreased proteins, heat shock proteins and some other defense proteins. The diverse group of proteins expressed included HSPs, and energy metabolism associated proteins, enzymes of carbohydrate metabolism, as well as stress related proteins (Laino et al. 2010). HSPs perform some well established functions like preventing denaturation of existing protein or misfolding of newly synthesized proteins (Vierling 1991; Timperio et al. 2008). The exact role of many HSPs in wheat under heat stress is still not clearly understood but certain HSPs like HSP70, HSP90 (high molecular weight HSPs) and HSP26 (small molecular weight HSPs) are shown to be induced by heat stress (Laino et al. 2010). Amandeep et al (2016), worked on differential expression of smaller HSP (hsp 20 and hsp 26.5) and found that expression of small hsp was high in barley as compare to wheat so that barley is more tolerant to heat as compare to wheat HSP70 has been found to be linked with acquired thermo-tolerance and involved in folding and oligomerization of protein to prevent irreversible aggregation (Hendrick and Hartl 1993).

Enzymes of carbohydrate metabolism are also differentially expressed under heat stress. They could be involved in either degradation (susceptible genotypes) or synthesis (tolerant genotypes) of starch at grain filling. The relevance of two key enzymes, glucose-1-phosphate adenyl-transferase and granule bound starch synthase can be understood by the fact that reduction in the level of these enzymes at grain filling stages leads to decrease in yield under heat stress (Majoul et al. 2003, 2004). Thus, proteomics play a key role in getting detailed insight of metabolic pathways of wheat affected from heat stress and identifying candidate enzymes as potential target for designing heat tolerant wheat cannot be over ruled.

METABOLOMICS

Plants have the ability to adapt to adverse environmental changes by altering composition and concentration of metabolites so that they can survive in abiotic stress conditions. Through global metabolite profiling simultaneous monitoring of precursors, intermediates, and products of all the metabolic pathways of any organism, including plants, can be done (Kaplan *et al.,* 2011). Thus, metabolic profiling has accelerated discovery of stress signal transduction molecules and compounds that are integral part of plant response to abiotic stresses (Shulaev *et al.,* 2008). But, metabolic profiling in wheat is not reported much and none of the reports are related to heat stress (Matthews *et al.,* 2012). Majority of past studies have mainly focused only on proteins and carbohydrates rather than focusing on whole metabolite change in response to heat stress. It is necessary to analyze global metabolite profile under heat stresses involving heterogeneous group of signal molecules involved in complex metabolic networks.

MANAGEMENT STRATEGIES

With physiological and molecular strategies, agronomic strategies are very important for mediating future increases in ambient temperature. Agronomical traits includes practices that conserve water, fertilization during critical growth stages and timing of sowing. For example, optimum water supply to heat stressed wheat helped to maintain grain-filling rate, duration and size (Dupont *et al.,* 2006). Of course, this is not possible in rainfed wheat growing regions if it doesn't rain. Application of nitrogen, phosphorus and potassium improve plant growth under moderate heat stress (Dupont *et al.,* 2006). When nitrogen, phosphorous and potassium are applied post anthesis, more protein accumulates in the grain at day/night. Time of sowing, timely nutrient application is needed to protect wheat from heat stress.

CONCLUSIONS

As we all know that heat stress is a major constrains which limits wheat productivity all over the world, Demands for the agricultural products is expected to increase at 1.1 per cent per year from 2005 to 2050 (Alexandratos and Bruinsma, 2012), so there is a great need to research in the right direction in this field. There is a great need to develope varieties not high productive but high productive in changing environment. Integrating physiology and biotechnological tools with conventional breeding techniques will help to develop wheat varieties with better grain yield under stress during reproductive and grain-filling phases. Thus the physiological phenotyping alongwith gene discovery can be valuable to pin down desired alleles and understand their genetic mechanism. Recent advances in molecular biology, functional genomics, and comparative tools opened new opportunities for the molecular improvement of modern wheat. The genetic basis of heat adaptation is poorly understood. Till date no "heat tolerance" genes have been cloned. In the present situation of heat tolerance research, physiological traits (PTs) associated with heat adaptation constitute the best available "tool" for genetic improvement of crops, as they represent important factors favourable combinations of alleles. But recently developed techniques are helpful in faster identification and characterization of heat related gene and gene region. There are many old and modern variants harbour a large reservoir of potential stress related genes and hold a tremendous potential for wheat improvement. More work is needed to screen such type of ancient and modern cultivars in wheat, after identification introduction of these traits of wheat can be performed either with breeding through marker-assisted selection or transgenic methods. Transgenic strategies with enhanced transformation and selection methods are currently being developed. There is lot of instruments are now available for physiological studies like Canopy structure, growth analysis and photosynthetic performance which can be measured non-destructively in model plant and cereal crops with the help of those one can study accurately the physiological traits.

Further research on enzymes which are responsible for the loss of chlorophyll during heat-induced senescence is a important part of development of genotypes with stay-green traits either through MAS or transgenic methods.

Agronomical traits also cannot be neglected to make wheat tolerant to high temperature. Genetic engineering can be a good tool but a plenty of research work is needed in this aspect Until now, transcription factors have been the most appealing targets for transgenic wheat improvement, due to their role in multiple stress-related pathways.

REFERENCES

Alghabari F, Lukac M, Jones HE, Gooding MJ (2014). Effect of Rht alleles on the tolerance of wheat grain set to high temperature and drought stress during booting and anthesis. *Journal of Agronomy and Crop Science* 200, 36–45.

Ali S, Khan SU, Gurmani AR and Khan A (2016) Agronomic and physiological response of wheat (*Triticum aestivum* L.) genotypes under terminal heat stress condition, International *Journal of Biosciences* 8, No. 5, p. 1-7, 2016

Burke M, Hsiang SM, Miguel E (2015) Global non linear effect of temperature on economic production. *Nature.* doi:10.1038/nature15725

Dhanda S.S, Munjal R (2012) Heat tolerance in relation to acquired thermo tolerance for membrane lipids in bread wheat. *Field Crops Res* 135: 30–37

Dhyan K, Ansari WM, Yalaga RR, Verma RS, Shukla A and Tuteja N 2013 Comparative physiological response of wheat genotypes in late sown condition, Plant Signalling and Behavior. *Plant Signaling & Behavior* 8:6, e24564; June 2013

Eberhart SA, Russell WA (1966) Stability parameters for comparing genotypes. *Crop Science* 6: 36-40.

Hakeem KR, Chandna R, Ahmad P, Iqbal M, Ozturk M (2012) Relevance of proteomic investigations in plant abiotic stress physiology. OMICS doi:10.1089/omi.2012.0041

Hall AE, Antony E (1992) Breeding for heat tolerance. In: Jules Janick (ed) Off prints From *Plant Breeding Review*. Rockville, Maryland, USA.

Hays D, Mason E, Hwa Do J, Menz M, Reynolds M (2007). Expression quantitative trait loci mapping heat tolerance during reproductive development in wheat (*T. aestivum*). In: Buck HT, Nisi JE, Salomo'n N (eds.), Wheat production in stressed environments. Springer, Amsterdam, pp 373-382.

Hays DB, Do JH, Mason RE, Morgan G, Finlayson SA (2007) Heat stress induced ethylene production in developing wheat grains induces kernel abortion and increased maturation in a susceptible cultivar. *Plant Sci* 172: 1113–1123

Jespersen D, Zhang J, Huang, B (2016). Chlorophyll loss associated with heat induced senescence in bentgrass. *J Plant Science* 249, 1-12, doi: 10.1016/j.plantsci.2016.04.016. Epub 2016 May 3

Kaur A, Gupta OP, and Sharma P (2016) Comparative expression analysis of *HSP* genes in wheat and barley under heat stress". *Journal of Wheat Research* 8(1):63-64 Homepage: http://epubs.icar.org.in/ejournal/index.php/JWR

Kosova K, Vítámvás P, Prášil IT, Renaut J (2011) Plant proteome changes under abiotic stress-contribution of proteomics studies to understanding plant stress response. J Proteomics 74(8):1301– 22

Kumar A, Li C, Portis AR Jr (2009) Arabidopsis thaliana expressing a thermostable chimeric Rubisco activase exhibits enhanced growth and higher rates of photosynthesis at moderately high temperatures. *Photosynth Res* 100: 143–153

Kumar U, Joshi A, Kumari M, Paliwal R, Kumar S, Röder M (2010) Identification of QTLs for stay green trait in wheat (*Triticum aestivum* L.) in the 'Chirya 3' 3 'Sonalika' population. *Euphytica* 174: 437–445

Laino P, Shelton D, Finnie C, De Leonardis AM, Mastrangelo AM, Svensson B, Lafiandra D, Masci S (2010) Comparative proteome analysis of metabolic proteins from seeds of durum wheat subjected to heat stress. *J Proteom* 10:2359–2368

Lim PO, Kim HJ, Nam HG (2007) Leaf senescence. *Annu Rev Plant Biol* 58: 115–136

Majoul T, Bancel E, Triboi E, Hamida JB, Branlard G (2004) Proteomic analysis of the effect of heat stress on hexaploid wheat grain: characterization of heat-responsive proteins from non-prolamins fraction. *Proteomics* 4:505–513

Martre P, Jamieson PD, Semenov MA, Zyskowski RF, Porter JR, Triboï E. 2006. Modelling protein content and composition in relation to crop nitrogen dynamics for wheat. *European Journal of Agronomy* 25: 138–154

Mason RE, Mondal S, Beecher FW, Pacheco A and Hay DB (2010) QTL associated with heat susceptibility index in wheat (*Triticum aestivum* L.) under short-term reproductive stage heat stress, *BMC Genetics*201415:97 DOI: 10.1186/s12863-014-0097-4

Mohammadi V, Qannadha MR, Zali AA, Yazdi-Samadi B (2004) Effect of Post Anthesis Hear Stress on Head Traits of Wheat. T*International Journal of Agriculture & Biology*. 1: 42-44.v

Mullan DJ, Reynolds MP (2010) Quantifying genetic effects of ground cover on soil water evaporation using digital imaging. *Funct Plant Biol* 37: 703–712

Murchie EH, Niyogi KK (2011) Manipulation of photoprotection to improve plant photosynthesis. *Plant Physiol* 155: 86–92

Murchie EH, Pinto M, Horton P (2009) Agriculture and the new challenges for photosynthesis research. *New Phytol* 181: 532–552

Ollinger SV (2011) Sources of variability in canopy reflectance and the convergent properties of plants. *New Phytol* 189: 375–394

Parry MAJ, Reynolds M, Salvucci ME, Raines C, Andralojc PJ, Zhu X-G, Price GD, Condon AG, Furbank RT (2011) Raising yield potential of wheat. II. Increasing photosynthetic capacity and efficiency. *J Exp Bot* 62: 453–467

Pinto RS, Reynolds MP, Mathews KL, McIntyre CL, Olivares Villegas J J, Chapman SC (2010). Heat and drought adaptive QTL in a wheat population designed to minimize confounding agronomic effects. *Theo. Appl Genet* 121:1001-1021.

Porch TG (2006) Application of stress indices for heat tolerance screening of common bean. *Journal of Agronomy & Crop Science* 192: 390-394.

Radmehr M, Kajbaf A, Lotfaliaeineh Gh A (1996) Effect of heat stress on agronomic traits, grain yield and its components on twenty-five genotypes of bread wheat. *Seed and Plant Journal* 12(1): 13-23.

Rebetzke G, Rattey A, Farquhar G, Richards R, Condon T (2012). Genomic regions for canopy temperature and their genetic association with stomatal conductance and grain yield in wheat. *Funct Plant Biol* 40:14-33.

Reddy KR (1995) Effect of temperature. A review 149-150 in: George C (ed), KUO, Adaptation of Food Crops to Temperature and Water Stress. *CABI Publishing*, New York, USA.

Reynolds MP, Manes Y, Izanloo A, Langridge P (2009) Phenotyping approaches for physiological breeding and gene discovery in wheat. *Ann Appl Biol* 155: 309–320

Reynolds MP, Pierre CS, Saad ASI, Vargas M, Condon AG (2007) Evaluating potential genetic gains in wheat associated with stress-adaptive trait expression in elite genetic resources under drought and heat stress. *Crop Sci* 47: S-172–S-189

Reynolds MP, Rebetzke G (2011) Application of plant physiology in wheat breeding. In AP Bonjean, WJ Angus, M Van Ginkel, eds, The World Wheat Book: A History of Wheat Breeding, Vol 2. Editions TEC, Paris, pp 877–906

Reynolds MP, Rebetzke G (2011). Application of plant Kosová K, Vítámvás P, Prášil IT, Renaut J (2011). Plant proteome changes under abiotic stress-contribution of proteomics studies to understanding plant stress response. *J. Proteomics* 74(8):1301-22. Physiology in wheat breeding. In: Bonjean AP, Angus WJ, Van Ginkel M (eds.), The World Wheat Book: A History of Wheat Breeding, Vol 2. Editions TEC, Paris, pp 877-906.

Reynolds. M, Balota. M, Delgado M, Amani I, Fischer R (1994) Physiological and morphological traits associated with spring wheat yield under hot, irrigated conditions. *Funct Plant Biol* 21: 717–730

Richards RA, Lukacs Z (2002) Seedling vigour in wheat: sources of variation for genetic and agronomic improvement. Aust J Agric Res 53: 41–50 Hansen J, Sato M, Ruedy R (2012) Perception of climate change. *Proc Natl Acad Sci* USA (in press)

Ristic Z, Bukovnik U, Vara Prasad PV, West M (2008) A model for prediction of heat stability of photosynthetic membranes. *Crop Sci* 48: 1513–1522

Semenov MA, Halford NG (2009). Identifying target traits and molecular mechanisms for wheat breeding under a changing climate. *J Exp Bot* 60:2791-2804.

Semenov MA, Stratonovitch P, Alghabari F, Gooding MJ. (2014). Adapting wheat in Europe for climate change. *Journal of Cereal Science* 59, 245–256

Szabados L, Savoure A (2010). Proline: a multifunctional amino acid. *Trends Plant Sci* 15(2):89-97.

Thomas H, Howarth CJ (2000) Five ways to stay green. *Journal of Exp Bot* 51:329–337

Timperio AM, Egidi MG, Zolla L (2008) Proteomics applied on plant abiotic stresses: role of heat shock proteins (HSP). *J Proteomics* 71:391–411

Tuinstra MR, Grote EM, Goldsbrough PB, Ejeta G (1997) Genetic analysis of post-flowering drought tolerance and components of grain development in *Sorghum bicolor* (L.) Moench. *Mol Breed* 3:439–448

Ugarte C, Calderini DF, Slafer GA (2007) Grain weight and grain number responsiveness to pre-anthesis temperature in wheat, barley and triticale. *Field Crops Res* 100: 240-248.

Vara Prasad PV, Djanaguiraman M (2014). Response of floret fertility and individual grain weight of wheat to high temperature stress: sensitive stages and thresholds for temperature and duration. *Functional Plant Biology* 41, 1261–1269

Vijayalakshmi K, Fritz A, Paulsen G, Bai G, Pandravada S, Gill B (2010) Modeling and mapping QTL for senescence-related traits in winter wheat under high temperature. *Mol Breed* 26: 163–175

Wardlaw IF, Wrigley CW (1994) Heat tolerance in temperate cereals: an overview. *Aust J Plant Physiol* 21: 695–703

Wollenweber B, Porter JR, Schellberg J (2003). Lack of interaction between extreme high-temperature events at vegetative and reproductive growth stages in wheat. *J. Agron. Crop Sci* 189:142-150.

■■■■

6

PHYSIOLOGICAL ADAPTATIONS AND DYNAMICS FOR PLANT PRODUCTIVITY UNDER LOW LIGHT INTENSITY

C. Aochen, J. Pohlong, S. S. Dutta, S. Pyngrope, S. Aochen* and A. Pandey

ICAR Research Complex – NEH Region, Umiam, Meghalaya – 793103
**Fazl Ali, College, Mokokchung, Nagaland - 798601*

INTRODUCTION

Light is a critical natural resource which essentially controls morphogenesis and production in crop plants. It has a major role in photosynthesis, photoperiodism, and photonasty. It is an indispensable feature for plants, being the energy source for their growth and development. Besides being essential to photosynthesis, it also acts as an environmental signal which, when perceived, triggers changes in plant metabolism and development (Jiao *et al.,* 2007).

Low light intensity - diffused light or cloudiness is a general phenomenon. Light intensity fluctuates highly within a day, within a month, and among the seasons. Its intensity increases from sunrise to the middle of the day and then gradually decreases toward sunset. The failure of the monsoon is invariably associated with higher radiation, and an overcast sky with lower radiation. High-rainfall areas are also usually associated with lower incident radiation caused by cloudiness. At higher latitudes rains occur even without cloudiness. Wet season (May to Nov/Dec in the northern hemisphere) is characterized by gradual decreasing temperatures, decreasing photoperiod, shorter sunshine hours, and lower solar radiation because of several rainy, cloudy, partly cloudy, and overcast days, and higher relative humidity. The amount of light reaching the earth's surface and the length of the day are functions of latitude and are greatly affected by the presence of clouds, smoke, dust, gas molecules, organic materials, insect debris, and water vapor in the atmosphere. Moreover, with the rapid economic development and urbanization happening, high-rise buildings and structures are emerging at a quicker pace, so shade environments are produced more widely.

Solar radiation penetrating the earth's atmosphere is reflected, absorbed, and scattered. Of the total solar energy entering the earth's atmosphere only 75% reaches a height of 1800 m while only 24-50% reaches sea level. Photosynthesis in green leaves uses solar energy in wavelengths from 0.4 to 0.5 μm, referred to as photosynthetically active radiation (PAR); 100% irradiance corresponds to PAR of 1500 ± 30 μmol $m^{-2}s^{-1}$. PAR is estimated to be about 43% to 50% of total radiation. Runkle (2006) commended the use of light intensity unit μmol $m^{-2}s^{-1}$ or mol $m^{-2}d^{-1}$ in plant response studies since these provide information on the photon content of light. Because plant growth proceeds by the addition of new modules (e.g. leaves, stems, branches), each plant usually experiences different light conditions during its growth and development due to its sedentary lifestyle (Bazzaz, 1996). Solar radiation intercepted by the leaf is either reflected, absorbed, or transmitted. Approximately in a single leaf, 30% is reflected, 20% transmitted, and the remaining 50% absorbed (Birkebak and Birkebak, 1964).

The intensity and spectral quality of light exhibits large temporal and spatial variation. For example, the daily photon flux available for plants grown in the deep shade of the tropical forest is more than 100 times less than that available for plants exposed to direct sunlight. In rice-growing tropical and sub-tropical climates, lower productivity was attributed mainly to lower incident radiation rather than to temperature. Reduced light intensity coupled with low sunshine hours is considered to be a major environmental constraint for productivity of rice during wet season (Murty and Sahu, 1987). Processes in plants that are affected by shade are photosynthesis, transpiration, respiration, nitrate reduction, protein synthesis, hormone production and photosynthate translocation, root growth and mineral absorption. Conversely, shade also can reduce the loss of soil moisture, retain nutrients, suppress weeds, reduce soil temperature and plant during the day, raising the night air temperature, reduce water runoff, as well as the removal of humidity and CO_2.

Low light tolerance of shade tolerance refers to the capacity of a given photosynthetic organism to tolerate low light levels and it is typically characterized by a set of morphological and physiological traits such as decrease in growth rate, light compensation point, dark respiration rate, net photosynthetic rate and Chl a:b, increase in quantum yield, chlorophyll content and carbohydrate storage together with many other traits (Valladares and Niinemets, 2008). These traits represent the most fundamental criteria for shade tolerance. While high light responses occur to maximize light-saturated rates of photosynthesis, low-light responses occur to enhance the efficiency of photon capture. Carbon exchange rate (CER) is strongly dependent on irradiance, absorption, and utilization of photon energy, and low irradiance influences CER directly by reducing photon energy utilization, thus decreasing productivity (Gregoriu *et al.*, 2007). Givnish (1988), in his 'carbon gain hypothesis', explains shade tolerance as the maximization of light harvesting and efficient use of captured light in photosynthesis with decreased respiration costs for maintenance. According to this hypothesis any trait that enhances the light use efficiency and hence the carbon gain, would increase the shade tolerance of a species.

A review of how low light intensity affects the morphology, anatomy, the physiology and biochemistry of metabolism and the molecular changes in the plant is, therefore, presented.

LOW LIGHT INTENSITY AND GROWTH ANALYSIS

Low light at vegetative stage in rice is characterized by high tiller mortality, and reductions in dry matter production, panicle number, relative growth rate (RGR), net assimilation rate (NAR) and specific leaf weight (SLW) (Venkateswarlu *et al.,* 1977), whereas height, leaf area index (LAI), leaf area ratio (LAR) increase (Janardhan and Murty, 1980). Wider and thinner leaf development is also considered an adaptive phenomenon to capture insufficient irradiance more efficiently with less investment in terms of N and C (Sims and Pearcy, 1994).

Rosette plants can orient their leaves to a more vertical position, sometimes referred to as hyponasty, and redirect accumulation of biomass to stems and petioles rather than leaf blades (Maliakal *et al.,* 1999). In cotton-growing regions of Central India, partial to full cloudy conditions pertaining to active monsoon season (July to September) cause low light stress affecting growth and development of cotton; longer low light duration during reproductive phase lead to increase in physiological shedding of fruiting parts thus affecting final yield due to lower biomass generation and reduction in SLW. It responds to low light stress mainly by persistent increase in source area and chlorophyll content for effective light interception (CICR Technical Bulletin No 2). In bottle gourd cv. High-green (hybrid) grown under four levels of light (100, 75, 50 and 25% PAR), main stem length, internode length and individual leaf area increased, whereas main stem diameter and numbers of leaves per plant decreased. In popular landscape interior plant *Schefflera arboricola,* reduced light intensity significantly reduced fresh and dry weight, elongated internodes in young shoots and very small leaves (Kubatsch and Grüneberg, 2007).

Long-term exposure of olive leaves to various PAR levels caused a number of serious anatomical and morphological changes in leaves that might be permanent. Long-term exposure to shade increased shoot length, internodal length, and leaf area, but decreased the number of flower buds and fruit set in three olive cultivars (Tombesi and Cartechini 1986). Shading of legumes generally causes elongated growth, reduced SLW, and increased LAR (Bowes *et al.,* 1972). Chu and Robertson (1974) found that shading reduced both nodule number and size in white clover (*Trifolium repens* L.). Turf grass exhibited a decline in root numbers and tilling, a shorter rhizome, higher and thinner stem with longer internodes, lighter colored and thinner leaf with flatter leaf angles and slower growth (Beard, 1997; Bell and Danneberger, 1999). Shading did not affect plant heights of cowpea and bushbean except for some petiole elongation in bushbean (Eriksen and Whitney, 1984).

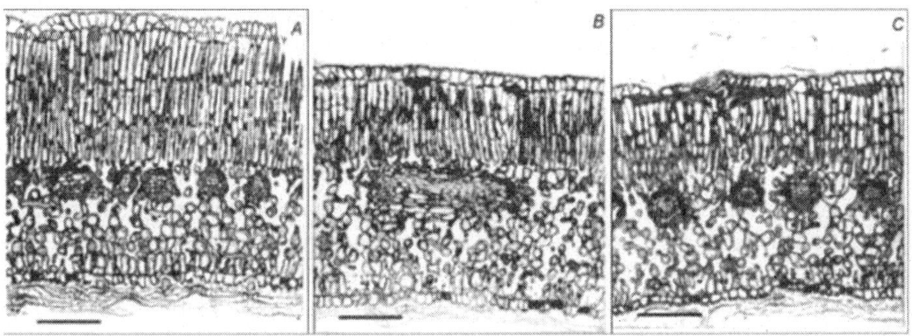

Fig. 1: Light micrographs showing the leaf structure and thickness of olive leaves in full daylight (a) 30% shade (b) 60% shade (c). Bars = 50μm and 60% shade (c). Bars = 7μm (Gregoriou *et al.,* 2007)

LOW LIGHT AND LEAF ANATOMY

The leaf anatomical structure changes enhance the cell ability to capture light, which is an advantage of the light penetrating leaf epidermis reaching mesophyll, or the photochemical reaction process in the leaf epidermis that directly improves photosynthetic capacity. Shading produces thinner mesophyll of rice leaves, smaller individual cell volume, and smaller cell number per unit area (Chonan 1967). In *Tradescantia pallid* cv. Purpurea plants subjected to light intensities of 100, 50 and 25% of the normal, the hypodermal tissue was constituted of 2-3 cell layers, whereas in plants under 0 and 5% irradiance, the hypodermal layer was reduced to 1-2 cell layers along with the size of the cells. A consideration is that the hypodermis together with the epidermis functions in focusing and concentrating light. The cells of epidermis on the abaxial surface had reduced volume, when compared to the adaxial surface. These variations in epicuticular layers thickening were clear responses to environmental variations, especially light intensities (Paiva *et al.,* 2003). The leaf anatomy and vein density of *Flavia bidentis* plants grown at 150 μmol quanta $m^{-2}s^{-1}$ and medium intensity 500 μmol quanta $m^{-2}s^{-1}$ were different in several respects. Leaf mesophyll thickness was greater, and bundle sheath cell area and vascular bundle area were higher in the former resulting in larger vascular bodies (Pengelly *et al.,* 2010). In olives, leaf thickness was significantly reduced in proportion to shade, citing it mainly to reduction of both palisade and spongy parenchyma (Fig. 1). Epidermis morphological structure turn out to be cell convex thoroughly, with reduced layers, increased size, thinner cell walls, thinner epidermis cutin membrane or no cutin membrane in *Alchornea triplinervia* (Roacaas and Scarano, 2001).

LOW LIGHT INTENSITY AND PHOTOSYNTHETIC PIGMENTS

Leaf thickness and architecture affect the content of photosynthetic component per unit leaf area. Chlorophyll a (Chl a) and Chlorophyll b (Chl b) are important pigments involved in the absorption and transmission of solar energy with part of Chl a involved in converting solar energy into electrochemical energy. An increase in accessory chlorophyll results in maximized light absorption rate (Evans, 1989). Chlorophyll content is considered a limiting factor in photosynthetic rate under weak light and hence cultivars with high chlorophyll concentration had higher photosynthesis under low light intensity. Recent study by Restrepo and Garcés (2013) showed that the leaf chlorophyll content was higher in rice leaves under low irradiance. Shade leaves contain a higher proportion of Chl b relative to Chl a, and therefore, a lower Chl a:b ratio. Chl a:b ratio indicates changes in the composition of photosynthetic complexes. The absorption of chlorophyll is very low in green light between 480 nm and 550 nm. The so-called green gap of Chl a is closed partially by absorption of the accessory pigment Chl b which exhibits absorption maxima between 350-450 nm. Under low light the relative proportion of blue light is higher, and a higher Chl b translates into its higher efficiency in utilizing blue light. Therefore, light absorption by the antenna system decreases Chl a:b ratio at low light levels. Rice cultivars with higher Chl b, thereby, have advantage to adapt to low light (Nayak *et al.,* 1978). Meanwhile reduction of light intensity beyond 50% of the normal light inhibited chlorophyll synthesis resulting in decreased chlorophyll content (Nayak and Murthy, 1980).

Beneragama and Goto (2010) reported for the first time that the shade tolerance of photosynthetic flagellate *Euglena gracilis* is characterized by the increased Chl a:b ratio, contradicting the generally accepted hypothesis of decreased Chl a:b in shade tolerance response. As the Chl b content was remarkably low compared to Chl a, Chl a:b ratio appears to be determined solely by the amount of Chl a in *Euglena.* The study demonstrated that *Euglena* maximizes light capturing by increasing chlorophylls under low light. Moreover, the reduced light compensation point, which is a simple measure of shade tolerance and low dark respiration rates under low light have assisted *Euglena* to be efficient by reducing the losses. High chlorophyll contents under low light situations found in *Euglena* can, therefore, be considered as a first line of adaptive response to reduced light, and changes in Chl a:b ratios depending on the light environment might be a characteristic of species themselves.

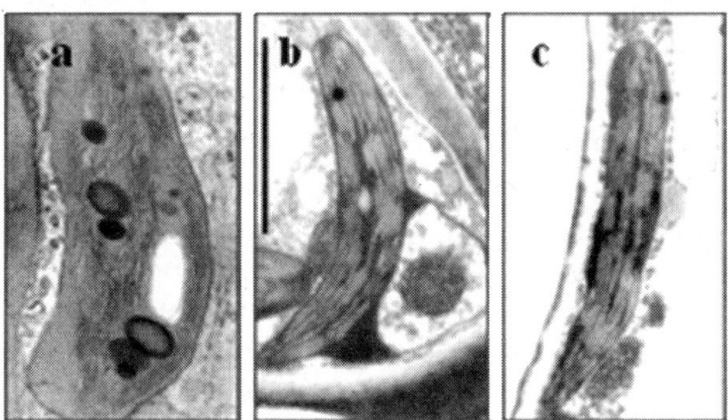

Fig. 2: Transmission electron micrographs of olive chloroplasts in full daylight (a) 30% shade (b) and 60% shade (c). Bars = 7μm (Gregoriou *et al.,* 2007)

LOW LIGHT INTENSITY AND PHOTOSYNTHETIC PROCESS

Photosynthesis is a complex physiological procedure in plants encompassing light absorption, energy conversion, electron transfer, ATP synthesis, key regulating enzyme activities etc. Leaves grown in low irradiance often have lower rates of photosynthesis due to a lower content per unit leaf area of most photosynthetic components, including Rubisco and components of electron transport and ATP synthesis. Shi *et al.* (2006) noted that Rubisco activity in chloroplasts decline dramatically under low light conditions. According to Chapman and Carter (1976), the minimum limit for the process of photosynthesis in most plants is between 100 and 200 footcandle (fc*). Below a minimum light intensity, the plant falls below compensation point, which is the metabolic point at which the rate of photosynthesis equals respiration. At this point the leaves do not gain or lose dry matter. Photosynthesis significantly slows down or ceases while respiration continues. Decreased photosynthetic energy flow at low light intensities could not be buffered by a down-regulation of energy-consuming processes (Pörs, 1999). Differences in plants adapted or acclimated to sun and shade adapted from is represented in Table 1.

At the molecular level, the light harvesting antenna complex of PSII (LHCII), which collects the most significant part of the light energy, was found to play a central regulatory role by finely controlling the amount of energy delivered to the reaction centers. Under low light starvation, the photosynthetic yield will depend on both - the efficiency of light energy capture by antenna pigments and its delivery to the reaction centers, and the captured light energy redistribution between PSI and PSII, which operate in series to form the electron transfer chain from water to NADP. Shade plants grown under light filtered on the forest canopy (more far-red light) will possess a higher PSII/PSI ratio to compensate for the reduction in the amount of red light, which is required to excite PSII (Anderson and Osmond, 2001). Usually PSII antenna has more pigments than that of PSI. Whilst the PSI reaction center absorbs at 700 nm, the PSII reaction center absorbs at 680 nm. Therefore, to match the energy of reaction center, PSI antenna absorbs longer wavelength light than PSII. This is due to enrichment in chlorophyll b in PSII on one hand, and the presence of the long wavelength chlorophyll a in PSI on the other.

Net photosynthetic rate (P_N) is greatly dependent on irradiance, absorption and utilization of photon energy (Jackson 1980). Low irradiance affects net photosynthesis (P_N) directly by reducing the utilization of photon energy, but this effect differs amongst plants and is dependent on their saturation irradiance. Reduced P_N was partly due to the lower PAR and partly to the morphological changes in leaves *viz.,* stomatal density, leaf area and width, palisade cells, areal leaf mass (ALM) and chloroplasts (Gregoriou *et al.,* 2007). Ma *et al.,* 2015 observed that decreases in P_N associated with the reduced transpiration rate, stomatal conductance (g_s), and intercellular CO_2 concentration (C_i) under 25%, 50%, and 100% irradiance indicated that stomatal limitation occurred, which was consistent with lower values in stomatal length, width, apertures, and number of stoma per unit LA. The lower stomatal number and aperture per unit LA help reduce the burden of the photosynthetic reaction center and protect plants from light deficiency (Lawson *et al.,* 2011).

At low irradiance photosynthesis rises linearly, giving a highly conserved quantum yield (Ogren and Evans, 1993). Changes in light intensity lead to rearrangement of the light harvesting antenna complexes and changes in composition of the two photosystems. Growth under low light promotes large PSI and PSII antenna size while high light generates a small photosynthetic unit. Moreover, plants grown under low light conditions have more peripheral LHCs per PSII reaction centre, a higher amount of Rubisco per unit chlorophyll, and cytochrome b/f complex per unit chlorophyll (Anderson *et al.,* 1995; Murchie and Horton, 1998).

Whilst the low light molecular adaptations balance the energy capture rate and distribution into two photosystems, the high light adaptations act to avoid absorption and to safely deal with the captured energy. The most documented adaptation to the light quality resulting in the imbalance of the energy absorbed by photosystems is known as the State Transitions. This process occurs within minutes and is effective at low/moderate light intensity, for example when light intensity is not limiting photosynthesis. The molecular mechanism involved is designed to restore this energy imbalance, utilizing the redox state of the plasto-quinone pool or/and cytochrome b as a sensor/transducer, which are involved in activation of a protein kinase. When the energy imbalance is in favor of PSII, this kinase phosphorylates some polypeptides of LHCII complex, targeting *Thr* or *Ser* residues of the stroma-exposed N-terminus.

A so-called "mobile" LHCII (originating exclusively from the outer-LHCII pool) becomes dissociated from the PSII under the influence of the negative charge, introduced by phosphate, thereby reducing the PSII antenna. Phosphorylation leads to a partial unstacking of the grana and the detached phosphor-LHCII can migrate into those regions of the thylakoid membrane, which are enriched in PSI (grana margins and stroma lamellae), and interact with the PSI complex.

Table 1: Characteristic differences between plants adapted or acclimated to sunny and shady extremes in irradiance level (Givnish, 1988)

Trait	Sun	Shade
Leaf-level		
Photosynthetic light response		
Light-saturated rate	High	Low
Compensation irradiance	High	Low
Saturation irradiance	High	Low
Biochemistry		
N, Rubisco, and soluble protein content / mass	High	Slightly lower
Chlorophyll *a* / chlorophyll *b* ratio	High	Low
Chlorophyll / soluble protein ratio	Low	High
Anatomy and ultrastructure		
Chloroplast size	Small	Large
Thylakoid / grana ratio	Low	High
Morphology		
Leaf mass / area	High	Low
Leaf thickness	High	Low
Stomatal size	Small	Large
Stomatal density	High	Low
Palisade / spongy mesophyll ratio	High	Low
Mesophyll cell surface / leaf area ratio	High	Low
Leaf orientation	Erect	Horizontal
Iridescence, lens-shaped epidermal cells	None	Rare
Reddish leaf undersides	Very rare	Infrequent
Canopy-level		
Leaf area index	High to low	Low
Phyllotaxis	Spiral	Distichous
Twig orientation	Erect	± Horizontal
Asymmetric leaf bases	Very rare	Infrequent
Plant-level		
Fractional allocation to leaves	Low	High
Fractional allocation to roots	High	Low
Reproductive effort	High	Low

Shade tolerance can importantly affect plant response to global change. Elevated CO_2 may facilitate the spread of tolerant species, whereas warming and ecosystem fragmentation and degradation negatively affect shade-tolerant species. Low light negatively affects g_s and enhance the concentrations of intercellular CO_2 in rice leaves (Yang *et al.*, 2011; Liu *et al.*, 2014). Intercellular space was increased by shade; this change might alter the CO_2 conductance from the sub-stomatal cavities to the sites of carboxylation in chloroplasts, thus restricting photosynthetic rate (Syvertsen *et al.,* 1995). Powles and Critchley (1980) reported bean plants grown under low light had lower rates of photosynthetic electron transport and CO_2 assimilation, leading to reduced growth in *Tabebuia chrysotricha*. The respiration rate decreases more than P_N, thus resulting in a higher ratio of respiration to P_N (Sato and Kim, 1980).

At the chloroplast level, the ratio of PSII to PSI has been shown to vary according to irradiance level (Murchie and Horton, 1998; Yamazaki *et al.,* 1999). Land plants usually add more antenna chlorophyll to PSII (increased antenna size) or increase the PSII reaction centers (RCs) relative to PS I RCs in response to shade to enhance light capture and energy transfer (Ehleringer, 2006); a reverse trend has also been reported by Chow *et al.* (1988). Falkowski and Owens (1980) identified two contrasting strategies in low-light adaptations in marine phytoplanktons; increase in the size of PS I units in *Skeletonoma* sp., a diatom, and increase in the number of PS I units in *Dunaliella*, a chlorophyte. Green algae *Scenedesmus* and *Chlorella* responded to low light in such a way that number of RCs of both PS II and PSI increased while the antenna size remain unaltered (Humbeck *et al.,* 1988). It can, therefore, be suggested that *Euglena* increases the PSII reaction centersand decreases the antenna size of PSII under low light conditions. This suggests, *Euglena* either increase the size or increase the number of PS I units under low light conditions givingrise to an increased total chlorophylls (Beneragama and Goto, 2010). Increase in PS I units (size or number) explains the increased Chl a:b ratio under low light, as PS I often does not contain Chl b (Hirashima *et al.,* 2006). On the other hand, increased number of PS II RCs also supports the higher Chl a:b ratio under low light, as ~85% of Chl b in *Euglena* is confined to antenna (Cunningham and Schiff, 1986). As a whole, the shade tolerance response of *Euglena* in relation to photosynthetic units is characterized by increase in number of RCs in PS II, decrease in antenna size of PS II, and increase in either number of PS I RCs or the antenna size of PS I (Fig. 3).

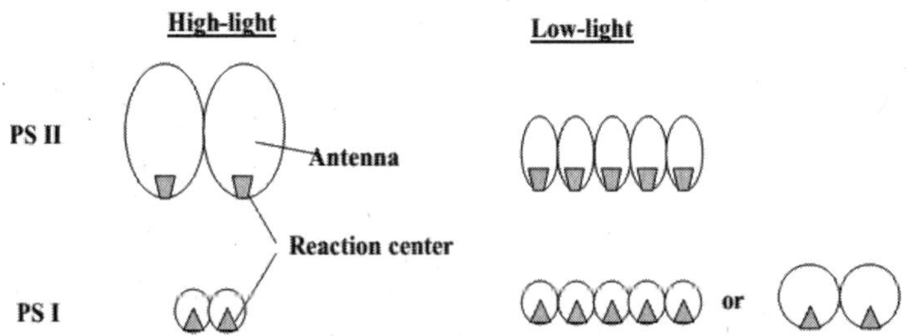

Fig. 3: Schematic representation of the organization of the photosynthetic apparatus in *Euglena* adapted to high and low light conditions (Beneragama and Goto, 2010).

Shading also increased thylakoid number in grana and stroma, but reduced trichome density, plastoglobuli, suggesting another means of photosynthesis limitation in shaded leaves (Gregoriou *et al.*, 2007). Grana number of each chloroplast in turf grass *Festuca arundinacea*, which is not resistant to shade, increased sharply when transferred from sunshine to shade condition (Wherley *et al.*, 2005).

LOW LIGHT AND PHOTOPERIODISM

The intensity of light, an indication of the strength of a light source at a particular wavelength, is also important in determining the flowering process in many plants. Flowering plant mainly uses phytochromes to regulate the time of flowering based on the length of day and night and to set circadian rhythms. While Baloch *et al.* (2012) reported that many long day ornamental plants responded better to more intense light sources, the intensity of irradiance vary from plant to plant such as flowering plants require 6,000-10,000 lux (74-124 μmol $m^{-2}s^{-1}$), flowering bulbs need 500-1000 lux (6-12 μmol $m^{-2}s^{-1}$) and most foliage plants need 1,000-6,000 lux (12-74 μmol$m^{-2}s^{-1}$) light irradiance (Thomas and Vince-Prue, 1997). For cyclamen's growth, leaf development and rate of flowering, 320 μmol $m^{-2}s^{-1}$ light intensity is more important than the length of day or photoperiod (Karlsson, 2001). As rice is a short day plant which subject to face long photoperiods of 14h after floral induction by short day cycle led to pollen sterility, mainly in the later formed florets under low light condition. Although *Saintpaulia* flowered at 5,000-13,000 lux (62-161 μmol$m^{-2}s^{-1}$) but the rate of flowering is higher at increased light intensity (Hildrum and Kristoffersen, 1969).

LOW LIGHT INTENSITY AND DRY MATTER ACCUMULATION, TRANSLOCATION AND DISTRIBUTION

Photosynthetic apparatus of organisms adapt to low light environments allowing coordinated allocation of resources not only to achieve and maintain optimal rates of photosynthesis, but also to function efficiently under limited light (Anderson *et al.*, 1995). Most of the dry matter produced sustains the growth of leaves, culm and sheaths rather than being allocated to panicles, which results in decrease of dry matter weight in the panicles (Ren *et al.*, 2003a). The post-anthesis photosynthetic production in leaves and pre-anthesis dry matter stored in the culm and sheaths are the main sources of nutrients for developing rice panicles which is about 60% and 40% of the total dry matter in the panicles, respectively (Cao *et al.*, 2001). The rate of dry matter weight in shoots to total dry matter (shoots + roots) in rice plants increases under low light conditions, suggesting that dry matter weight in roots decreases more than that of shoots (Yamamoto *et al.*, 1995).Moreover, exportation and translocation rates of culm-sheath dry matter decrease, which leads to an increase in the rate of culm-sheath dry matter weight compared with the total dry matter weight of shoots as well as a reduction in the dry matter weight of panicles (Sun *et al.,* 2012). A reduced leaf P_N caused by shading might decrease the saccharide contents in olive leaves and consequently reduce inflorescent bud initiation (Proietti, 2000).

Low light conditions results in lesser amount of N allocated to rice panicles; also the amount of N used for the development of leaves and culm-sheath increases under low light conditions, thus, leading to enhanced grain protein content under low light (Ren *et al.,* 2003c). This situation implies that applying of more N fertilizer during low light conditions is not productive and reduces rice culture to a non-profitable enterprise (Venkateswarlu and Visperas, 1987).

Under the condition of low light, on one hand, turf grass gets fewer illuminations and reduces the production of photosynthesis, hence, less energy and carbohydrates are accumulated. On the other hand, mowing of turf grass frequently make the plants need more energy for leaves regeneration. Energy stored in the root or in the stolons is transferred to the aboveground thereby reducing the carbohydrates stored in the grass. Therefore non-structural carbohydrates and photosynthetic efficiency of the turf grass are greatly reduced in low light conditions (Qian and Engelke, 1999; Ervin, 2003).

LOW LIGHT INTENSITY AND SUCROSE-STARCH METABOLIC PATHWAY

Shading not only affects the filling rate, carbohydrate accumulation of grain, and dry matter transportation in stem-sheath, but it also affects starch synthase and related enzyme activities. Decline of the sucrose content was one of the important reasons for the decrease of starch synthesis under low light intensity. In wheat, low light intensity during grain filling period depressed grain growth and starch production by about 20%. In rice, starch synthase I activity was depressed by shading during the later stage of grain development; reduced enzyme activity rather than insufficient supply of assimilates accounted for poor starch production (Mengel and Judel, 2006). Decreased starch synthase activity in grains, therefore, enhanced the occurrence of chalky rice (Li *et al.,* 2006a). Activities of leaf sucrose-phosphate synthase (SPS), and kernel sucrose synthase (SS), soluble starch synthase (SSS), granule-bound starch synthase (GBSS), ADP-glucose pyrophosphorylase (AGPase), and UDP-glucose pyrophosphorylase (UGPase) of maize decreased significantly after shading, and their activities decreased with different shading phases in the order, flowering to maturity > joining to flowering > seedling to jointing (Zhang *et al.,* 2008). The decrease is attributed to the decrease in translocation and distribution of photosynthates from leaf to grain.

Under weak light (50% shading) in rice, starch, amylose and sucrose contents decreased. Among the enzymes of sucrose-starch metabolic pathway, there was no significant change in AGPase activity, and while the activities of SSS and GBSS decreased, the activities of both soluble starch branching enzyme (SSBE) and granule bound starch branching enzyme (GBSBE) increased. Both AGPase and SSBE activity were positively and significantly correlated with starch accumulation rate. The increase in activity of starch branching enzyme promoted the conversion of amylose to amylopectin, thereby, decreasing the ratio of amylose to total starch under weak light (Li *et al.,* 2006b). Their previous study had also demonstrated that increased activity of SSBE impairs the accumulation of amylose in grains under low light (Li *et al.,* 2005).

LOW LIGHT INTENSITY AND YIELD

It is estimated that a cumulative solar radiation of 14,000 g cal/cm^2 or 200 h bright sunshine during the 30 d before harvest would be optimum for grain yield in rice (Murty and Sahu, 1987). Rice grown under low light produces lower grain yield because they have smaller photosynthetic capacity and lesser accumulation of assimilates. When the influence of low light intensity was examined at different growth stages of rice, progressive yield reduction was recorded with successive growth stages, from 9.5 t ha^{-1} in the vegetative phase to 4.4 t ha^{-1}during ripening phase. The critical stages were the reproductive and ripening phases; low light intensity affected spikelet number m^{-2} at reproductive stage, while at ripening phase it affected grain percentage (Venkateswarlu, 1977).

The incident radiation at crucial growth stages of rice is more important than the mere total radiation. During wet season, low light is commonly experienced by short and medium duration rice varieties. The major constraints to yield are low grain numbers per panicle in short-duration varieties, high spikelet sterility in medium-duration varieties and lesser productive panicles in long duration varieties (Murty and Rao, 1982). Its effect during the reproductive phase is irreversible while reversible during the vegetative phase. Before heading, low light gives rise to a pronounced decrease in fertile panicles; after heading shading cause impairment of P_N as well as lower dry matter accumulation and sink capacity in rice plants, thereby significant reductions in the number of filled grains and 1000-grain weight (Deng *et al.,* 2009). From initial heading to maturity stages, decreases in the seed-setting rate and 1000-grain weight decreased grain yield (Liu *et al.,* 2014). While evaluating two rice cultivars, a traditionally tall local variety and a modern semi-tall variety, under 40-60% light intensity, Singh (2005) recorded significant increase in plant height and spikelet sterility, and therefore, decreased number of grains per panicle. Reduced spikelet number per panicle has been attributed to shading effects on cell division (Matsushima, 1957).

Depressed starch production under low light conditions (without a decrease in starch phosphorylase activity) has been reported in wheat (Mengel and Judel, 2006). Grain yield of soybean reduced under low light because of reduction in dry matter production, and total pod number per plant (Panhwar *et al.,* 2004). Wahua and Miller (1978) also found that number of pods per plant was the main component of yield in soybean altered by shading, thereby affecting yield.

LOW LIGHT INTENSITY AND YIELD/PRODUCT QUALITY

When rice is grown under low light for 32 d, starting from the initial heading stage, yield of brown rice, milled rice and head rice, as well as their Shading decreased grain amylose content and gel consistency while the percentage of chalky kernels and grain protein increase, resulting in poor appearance and milling qualities of rice grains (Ren *et al.,* 2003b).

2-acetyl-1-proline (2-AP) is the major compound responsible for the aromatic character of fragrant rice. Proline is a precursor of 2-AP in rice (Yoshihashi *et al.,* 2002), and γ-aminobutyraldehyde (GABald) is likely to be the direct precursor to 2-AP production via BADH2; Bradbury *et al.* (2008) proposed that functional BADH2 catalyses the conversion of GABald to γ-aminobutyric acid (GABA), while non-functional BADH2 in fragrant rice leads to cyclisation of GABald to Δ1-pyrroline, with the addition of an acetyl group forming 2-AP. The inactivation of BADH2, encoding an aminoaldehyde dehydrogenase, is responsible for the accumulation of 2-AP in fragrant rice varieties (Bradbury *et al.,* 2005). Shading (67% reduction of natural light) increased 2-AP production in two fragrant rice varieties, significantly increased accumulation of the plant stress response molecules prolineand gamma-aminobutyric acid (GABA), which are implicated in pathways leading to 2-AP production, under all shading treatments (Mo *et al.,* 2015).While an increased content of 2-AP as a result of the shading effect of air pollution in may be desirable, yield loss and loss of grain quality would clearly be detrimental.

Shading at earlier growth stages had less effect on starch than did shading at later growth stages (Wang *et al.,* 2013). In this study, later growth stages were marked by decreased peak viscosity and breakdown, but increased pasting temperature. IIyou 498 and Gangyou 906 had higher endurance to shading before heading, while Dexiang 4103 could maintain high quality when exposed to shade after heading. These differences in shade endurance among rice varieties can offer a theoretical foundation for selecting and breeding shade-tolerant rice. Therefore, the relationship between key enzyme activity of starch and starch pasting characteristics, and the technique of rice breeding and cultivation to improve shade endurance require further investigation.

LOW LIGHT INTENSITY AND ANTIOXIDATIVE DEFENSE SYSTEM

Under low light conditions, the promotion of antioxidative enzyme activity and osmotic regulation for low light tolerant varieties can help maintain the scavenging of reactive oxygen species and the water potential in cells, and it can also minimize the adverse effects of low light on plant physiological metabolism. Light intensity at 600 µmol $m^{-2}s^{-1}$ increased rapidly the activities of SOD, CAT and POD indicating that a mild light stress had occurred in lettuce (Fu *et al.,* 2012). SOD, pyridoxine, and 2-caffeicacido-methyltransferase (involved in the biosynthesis of phenols) were up-regulated in low-light leaves in *P. oceanica* (Mazzuca *et al.,* 2009). In seedlings of *Camptotheca acuminata,* a potential medicinal plant, the less pronounced plant growth under 100% and 25% irradiance is associated with a decline in photosynthetic capacity and photochemical efficiency, with increases in the activity of SOD, CAT and POD, and malondialdehyde (MDA) content and relative electrolyte conductivity

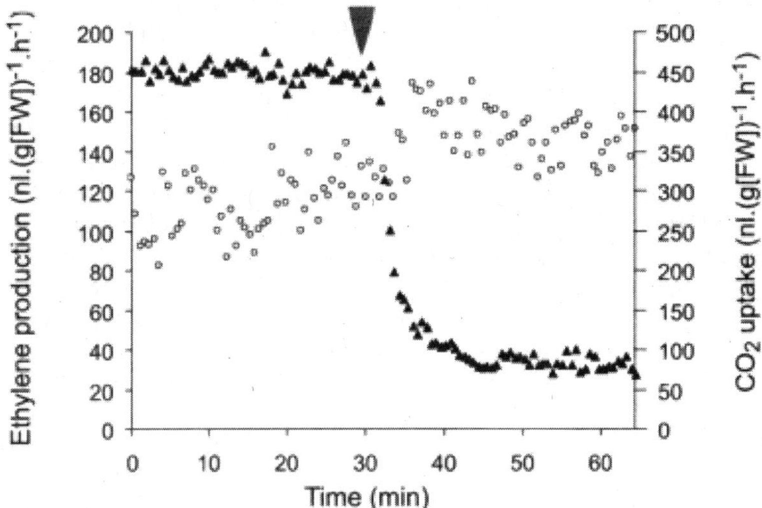

Fig. 4: Changes in ethylene production in 2-week-old ACC-treated rosette *Arabidopsis* upon a switch in light intensity from 125 to 30 µmol $m^{-2}s^{-1}$. Black triangles - CO_2 uptake; white circles -emanated ethylene; Arrow head - Time point of light switch; FW - Fresh weight (Vandenbussche *et al.,* 2003)

LOW LIGHT INTENSITY AND HORMONAL EFFECT

Morphological responses of plants to shading have long been studied as a function of light quality, in particular the ratio of red to far red light that affects phytochrome activity. Decreased light intensities coincided with rapid increased ethylene production but decreased CO_2 uptake in Arabidopsis rosettes (Fig. 4). In addition, ethylene- and auxin-insensitive mutants are impaired in their reaction to shading, which is reflected by a defect in leaf elevation and an aberrant leaf biomass allocation. At the molecular level, several auxin-inducible genes are up-regulated in wild-type Arabidopsis in response to a reduction in light intensity, including the primary auxin response gene IAA3 and a protein with similarity to

AUX22 and the 1-aminocyclopropane-1-carboxylic acid synthase genes ACS6, ACS8, and ACS9 that are involved in ethylene biosynthesis. Ethylene and auxin signaling are, therefore, required for the response to low light intensities. The greater leaf size in shaded leaves is also observed to be caused by an increase in auxins and gibberellins within leaves under low irradiance (Salisbury and Ross, 1978).

LOW LIGHT INTENSITY AND NITROGEN FIXATION AND ASSIMILATION

Sprent (1973) found that shading decreased N_2-fixation activity in lupin in proportion to the logarithm of relative irradiance. This was largely due to reduced nodule mass under shade, while the specific nodule activity was only slightly reduced. Sprent and Bradford (1977) found that shading of fababean delayed nodule senescence and prolonged total nodule activity as a result of it. Seasonal and diurnal variations in N_2 fixation indicate that the process is quite sensitive to the supply of photosynthetic assimilates (Sloger *et al.,* 1975). Nitrogen uptake is inhibited by low light during post flowering period in semi-dwarf rice varieties (Janardhan *et al.,* 1980). Higher nitrate reductase activity for a longer duration is essential for efficient N metabolism under low light (Murty *et al.,* 1975). Low light also led to reduction of nitrate reductases activity reduction in the blade, whole activity of nitrate reductases in the root decreased in greater manner. This makes plants to absorb more nitrate in order to satisfy the demand of nitrogen (Takahashi and Sherman, 1993; Gouia *et al.,* 2000).

LOW LIGHT INTENSITY AND PROTEOMIC AND TRANSCRIPTOME STUDIES

Acclimation to different light conditions is also manifested by changes in organization and/or level of protein complexes in thylakoid membranes and by different contents of stromal components involved in carbon fixation. Protein expression in the very valuable coastal endemic seagrass *Posidonia oceanica,* collected from turbid waters (low-light), was evaluated. More than 2600 proteins were detected in leaves, among which, 26 proteins were differentially expressed in low-light conditions, 12 of which were identified through MASCOT analyses. A 30% down-regulation of Rubisco large subunit in low-light acclimated leaves was also observed. Enzymes involved in carbohydrate cleavage (1-fructose-bisphosphate aldolase, nucleoside diphosphate kinase, and beta-amylase) were also upregulated. Furthermore, under lowlight, four components of the ubiquitin/mediated proteolysis pathway (26S proteasome regulatory, proteasome beta type 1, proteasome 7 D beta type, and proteasome alpha 7), and the perchloric acid soluble translation inhibitor protein, were upregulated. This suggests that, in *P. oceanica* leaves, enhanced protein turnover mediates acclimation to low-light conditions (Mazzuca *et al.,* 2009).

Polypeptide composition of all Chl a/b-binding LHCs and reaction centre (RC) polypeptides (PsbD and PsaD) in mesophyll (M) and bundle sheath (BS) chloroplasts of maize plants grown at high light (HL), moderate light (ML) and low light (LL) was analyzed by immunodetection with specific antibodies. A decrease in Chl a/b ratio was observed in LL leaves indicating that more light-harvesting complexes were present in chloroplasts of these plants because Chl b is mainly bound to LHCI and LHCII antennas. The amount of some polypeptides, like Lhcb1, Lhcb4, Lhcb5 increased in LL plants more than other polypeptides, indicating changes in composition of PSII, with more prominence in BS chloroplasts. At low irradiance the level of LHCII increased also in BS chloroplasts although they are agranal with reduced amount and electron transferring activity of PSII. This has led to suggest that agranal maize BS chloroplasts might be also the place where acclimation processes to light irradiance is realized (Drozak and Romanowska, 2006).

Suzuki *et al.* (2001) showed that the amount of leaf Rubisco RNA (large subunit and small subunit) declined throughout the lifetime of the leaf in parallel with the amount of Rubisco protein. A decline in RNA content was also observed in low-light leaves only, between day 0 and day 12. An increase in the expression of genes encoding ADP glucose pyrophosphorylase, starch debranching enzyme, and starch branching enzyme under high light intensity might represent an increase in the use of starch as a storage carbohydrate. This was possibly caused by the added photosynthate availability (Murchie *et al.*, 2005).

Umate (2010) conducted genome-wide analysis of the light-harvesting chlorophyll a/b-binding proteins and light-inducible proteins in Arabidopsis and rice. They identified 17 candidate loci which encodes for chlorophyll a/b-binding proteins in rice. These are light-harvesting complex genes of PSI and PSII in rice. Three-helix early light-inducible proteins (ELIPs), two-helix stress-enhanced proteins (SEPs) and one-helix high light-inducible proteins [one-helix proteins (OH Ps)] are identified in rice. Interestingly, two independent genomic loci (LOC_Os01g64960, LOC_Os04g59440) encoding PsbS protein are also identified with implications on additional mode of non-photochemical quenching (NPQ) mechanism in rice. A few additional LHC related genes are also identified in rice (LOC_Os09g12540, LOC_Os02g03330). This is the first report of identification of light-harvesting complex genes and light-inducible genes in rice.

LOW LIGHT INTENSITY AND FUTURE CHALLENGES

Research on low light intensity through the years highlights areas which needs emphasis for yield improvement. Traits conferring shade tolerance require novel experiments that explicitly quantify irradiance (not merely as a fraction of sunlight) and light quality (red-far red ratio, fraction of diffuse light, lateral versus vertical light) affecting individual plants.

Moreover such experiments should factor in control for ontogenetic effects, effects of seed size, limited rooting space (e.g., pot effects), and interacting environmental drivers (Valladares and Niinemets, 2008). Improving photosynthetic efficiency has been a targeted role in improving crop yield till date. However, increasing the photosynthetic efficiency has to be supported by increased assimilation of carbon dioxide (i.e. supply of RuBP must be in steady state), as well as proper partitioning into sink to realize more production. Light-harvesting chlorophyll a/b-binding protein (LHCP) collect and transfer light energy to photosynthetic reaction centers. However, the roles of different LHCPs in light-harvesting antenna systems remain obscure. Exploration of nucleotide variation in the genes encoding LHCP can facilitate a better understanding of the functions of LHCP.

Genes related to tolerance for transient low light stress in rice are mainly located in the cytoplasm. However in some situations nuclear genes from both maternal and paternal lines interact with cytoplasmic genes to make tolerance stronger or weaker. The positive correlation between photosynthesis and accumulation of dry matter will make this genetic behavior of shade tolerance in rice useful for breeding for high yield. In fruits like olives, the effect of shade on fruit yield, seeming not to be an effect of reduced P_N alone, indicates that other factors such as flower initiation, fruit set, or other morphogenetic characteristics warrant further investigation.

Global change generates novel combinations of factors that can crucially alter shade tolerance and affect both the role of species early life-history traits and the resilience of established individuals. Research, therefore, also needs focus on modifications of shade tolerance by and the role of shade tolerance in mediating community and ecosystem responses to global change drivers.

REFERENCES

Anderson, J.M., Chow, W.S. and Park, Y.I. 1995. The grand design of photosynthesis: Acclimation of the photosynthetic apparatus to environmental cues. *Photosyn. Res.,* **46**: 129-39

Anderson, J.M. and Osmond, B. 2001 Sun-shade responses: Compromises between acclimation and photoinhibition. In: Photoinhibition. Kyle, D.J., Osmond, B., Arntzen, C.J. (Eds.). Amsterdam: Elsevier, pp. 1-38

Baloch, J.U.D., Munir, M., Iqbal, M. and Abid, M. 2012. Effects of varied irradiance on flowering time of facultative long day ornamental annuals. *Pak. J. Bot.,* **44(1)**: 111- 117

Bazzaz, F.A. 1997. Allocation of resources in plants: state of the science and critical questions. In: Plant resource allocation, Bazzaz, F.A. and Grace, J. (eds). New York: Academic Press, pp. 1–37.

Beard J.B. 1997. Shade stress and adaption mechanisms of turf grasses. *Turfgrass,* **8**: 1186–1195

Bell, G.E. and Danneberger, T.K. 1998. Temporal shade on creeping bentgrass turf. *Crop Sci. Soc. Ame.r,* **39(4)**: 1142-1146

Beneragama, C.K. and Goto, K. 2010. Chlorophyll a:b ratio increases under low-light in `shade-tolerant' *Euglena gracilis. Tropical Agril. Res.*, **22(1)**: 12-25

Birkebak, R. and Birkebak, R. 1964. Solar radiation characteristics of tree leaves. *Ecology,* **45(3)**: 646-649

Bowes, G., Ogren, W.J. and Hageman, R.H. 1972. Light saturation photosynthesis rate, RuDP carboxylase activity and specific leaf weight in soybean grown under different light intensities. *Crop Sci.,* **12**: 77 - 79

Bradbury, L.M., Gillies, S.A., Brushett, D.J., Waters, D.L. and Henry, R.J. 2008. Inactivation of an aminoaldehyde dehydrogenase is responsible for fragrance in rice. *Plant Mol. Biol.,* **68(4–5)**: 439–449

Bradbury, L.M.T., Fitzgerald, T.L., Henry, R.J., Jin, Q. and Waters, D.L. 2005. The gene for fragrance in rice. *Plant Biotech. J.,* **3(3)**: 363–370

Cao, S.Q., Zhai, H.Q., Yang, T.N., Zhang, R.X. and Kuang, T.Y. 2001.Studies on photosynthetic rate and function duration of rice germplasm resources. *Chin. J. Rice Sci.,* **15(1)**: 29– 34

Chonan, N. 1967. Studies on the photosynthetic tissues in the leaves of cereal crops. III. The mesophyll structure of rice leaves inserted at different levels of the shoot. *Jap. J. Crop Sci.,* **36**: 291–296

Chow, W.S., Anderson, J.A. and Hope, A.B. 1988. Variable stoichiometries of photosystem II to photosystem I reaction centers. *Photosyn. Res.,* **17B**: 277-281

Chu, A.C.P. and Robertson, A.G. 1974. The effects of shading and defoliation and nodulation and nitrogen fixation by white clover. *Plant Soil,* **41**: 509-519

CICR Technical Bulletin No 2. Abiotic stresses in cotton: A physiological approach. www.cicr.org.in /pdf/abiotic_stress.pdf

Cunningham, F.X.Jr. and Schiff, J.A. 1986. Chlorophyll-Protein complexes from *Euglena gracilis* and mutants deficient in Chlorophyll *b*1. *Plant Physiol.,* **80(1)**: 231-238

Deng, F., Wang, L., Yao, X., Wang, J.J., Ren, W.J. and Yang, W.Y. 2009.Effects of different-growing-stage shading on rice grain-filling and yield. *J. Sichuan Agric. Univ.*, **27(3)**: 265 269

Drozak, A. and Romanowska, E. 2006. Acclimation of mesophyll and bundle sheath chloroplasts of maize to different irradiances during growth. Biochimicaet Biophysica Acta. *Bioenergetics,* **1757(11)**: 1539-1546

Ehleringer, J. 2006. Photosynthesis: Physiological and ecological considerations. In: Plant Physiology, 4th Edition, Taiz, L. and Zeiger, E. (Eds.), Sinauer Associates, Inc., Publishers, Sunderland, MA. pp. 197-220

Eriksen, F.I. and Whitney, A.S. 1984. Effects of solar radiation regimes on growth and N2 fixation of soybean, cowpea, and bushbean. *Agron J,* **76**: 529– 535

Evans, J.R. 1989. Partitioning of nitrogen between and within leaves grown under different irradiances. *Aus. J. Plant Physiol.,* **16(6)**: 533 - 548

Falkowski, P.G. and Owens, T.G. 1980. Light-Shade adaptation 1. Two strategies in marine phytoplankton. *Plant Physiol.,* **66(4)**: 592–595

Givnish, T.G. 1988. Adaptation to sun and shade: a whole-plant perspective. *Australian J Plant Physiol.*, **15(2)**: 63-92

Gouia, Houda, Ghorbal, M.H. and Meyer, C. 2000. Effects of cadmium on activity of nitrate reductase and on other enzymes of the nitrate assimilation pathway in bean. *Plant Physiol. Biochem.,* **38(7)**: 629-638

Gregoriou, K., Pontikis, K. and Vemmos, S. 2007. Effects of reduced irradiance on leaf morphology, photosynthetic capacity, and fruit yield in olive (*Olea europaea* L.). *Photosynthetica,* **45(2)**: 172- 181

Hildrum, H. and Kristoffersen, T. 1969. The effect of temperature and light intensity on flowering in *Saintpaulia ionantha* Wendl. *Acta Hort.,* **14**: 249-259

Hirashima, M., Satoh, S., Tanaka, R., and Tanaka, A. 2006. Pigment shuffling in antenna systems achieved by expressing prokaryotic Chlorophyllide *a* Oxigenase in *Arabidopsis*. *J. Biol. Chem.,* **281(22)**: 15385-15393

Humbeck, K., Hoffmann, B. and Senger, H. 1998. Influence of energy flux and quality of light on the molecular organization of the photosynthetic apparatus in Scenedesmus. *Planta,* **173(2)**: 205-212

Jackson, J.E. 1980. Light interception and utilization by orchard systems. *Hort. Rev.* **2**: 208-267

Janardhan K.V., Murty K.S. and Dash, N.B. 1980. Effect of low light during ripening period on grain yield and translocation of assimilates in rice varieties. *Ind. J. Plant Physiol.,* **23**: 163- 168

Janardhan K.V, and Murty, K.S. 1980. Effect of low light during vegetative stage on photosynthesis and growth attributes in rice. *Ind. J. Plant Physiol.,* **23**:156-162

Jiao,Y., Lau, O.S. and Deng, X.W. 2007. Light-regulated transcriptional networks in higher plants. *Nat. Rev. Genet.,* 8(3): 217 - 230

Kakiuchi, J. and Kobata, T. 2004. Shading and thinning effects on seed and shoot dry matter increase in determinate soybean during the seed-filling period. *Agron. J.,* **96**: 398-405

Karlsson, M. 2001. Recent findings may make you rethink cyclamen. In: Bedding Plants, Greenhouse Product News, **11(3)**: 22-24

Khush, G.S. 2003. Productivity improvements in rice. *Nutrition Rev.,* **6**: 6.

Kubatsch, A. and Grüneberg, H. 2007. The effect of low light intensity and temperature on growth of *Schefflera arboricola* in interior landscapes. *Hort. Sci.,* **42(1)**: 65–67

Lawson, T., von Caemmerer, S. and Baroli, I. 2011. Photosynthesis and stomatal behavior. In: Progress in Botany, Vol. 72, Lüttge, U.E. Beyschlag, W., Büdel, B. and Francis, D. (Eds), Berlin, Heidelberg: Springer, pp 265–304

Li, T., Ohsugi, R., Yamagishi, T. and Sasaki, H. 2005. Effects of weak light on rice starch accumulation and starch synthesis enzyme activities at grain filling stage. *Chin. J. Rice Sci.,* **19(6)**: 545– 550

Li, T., Ohsugi, R., Yamagishi, T. and Sasaki, H. 2006a. Effects of weak light on rice sucrose content and sucrose degradation enzyme activities at grain filling stage. *Acta Agron. Sin.,* **32(6)**: 943-945

Li, T., Ohsugi, R., Yamagishi, T. and Sasaki, H. 2006b Effects of weak light on starch accumulation and starch synthesis enzyme activities in rice at the grain filling stage. *Rice Sci.,* **13(1)**: 51–58

Liu, Q.H., Xiu, W.U. Chen, B.C. and Jie, G.A.O. 2014. Effects of low light on agronomic and physiological characteristics of rice including grain yield and quality. *Rice Sci.,* **21(5)**: 243-251

Ma, X., Song, L., Yu, W., Hu, Y., Wu, J., and Ying, Y. 2015. Growth, physiological, and biochemical responses of *Camptotheca acuminata* seedlings to different light environments. *Frontiers in Plant Sci.,* 6: Article 321

Maliakal, S.K., McDonnell K., Dudley, S.A. and Schmitt, J. 1999. Effects of red to far- red ratio and plant density on biomass allocation and gas exchange in *Impatiens capensis. Int. J. Plant Sci.,* **160**: 723

Matsushima, S. 1957. Analysis of developmental factors determining yield and yield prediction in lowland rice. *Bull. Natl. Inst. Agric. Sci.,* **A5**: 1-271

Mazzuca, S., Spadafora, A., Filadoro, D., Vannini, C., Marsoni, M., Cozza, R., Bracale, M., Pangaro, T. and Innocenti, A.M. 2009. Seagrass light acclimation: 2-DE protein analysis in Posidonia leaves grown in chronic low light conditions. *J. Expt. Marine Biol. Ecol.,* **374(2)**: 113- 122

Mengel, K. and Judel, G.K. 2006.Effect of light intensity on the activity of starch synthesizing enzymes and starch synthesis in developing wheat grains. *Physiologia Plantarum* **51**: 13-18

Mo, Z., Li, W., Pan, S., Fitzgerald, T.L., Xiao, F., Tang, Y., Wang, Y., Duan, M., Tian, H. and Tang, X. 2015. Shading during the grain filling period increases 2-acetyl-1-pyrroline content in fragrant rice. *Rice,* **8(1)**: 9

Murchie, E.H. and Horton, P. 1998. Contrasting patterns of photosynthetic acclimation to the light environment are dependent on the differential expression of the responses to altered irradiance and spectral quality. *Plant Cell Environ.,* **21(2)**: 139–148

Murchie, E.H., Hubbart, S., Chen, Y.Z., Peng, S.B. and Horton, P. 2005. Acclimation of photosynthesis to high irradiance in rice: gene expression and interactions with leaf development. *J. Exptl. Bot.,* **56**: 449 – 46

Murty, K.S., Nayak, S.K. and Sahu, G. 1975. Effect of low light stress on rice crop. In: Symposium on "Crop plant response to environmental stress", Almora. pp. 74 – 84

Murty, K.S. and Rao, C.N. 1982. Underutilized physiological potential for increasing productivity in rice. In: Plant Physiology - an underutilized potential for increasing crop productivity. Department of Atomic Energy, Bangalore, India, pp. 301-317.

Murty, K.S. and Sahu, G. 1987. Impact of low light stress on growth and yield of rice. In: Weather and rice. IRRI, Philippines, pp. 93- 101

Nayak, S.K., Janardhan, K.V. and Murty, K.S. 1978.Photosynthetic efficiency of rice as influenced by light intensity and quality. *Ind. J. Plant Physiol.,* **21**: 48-52

Nayak, S.K. and Murty, K.S. 1980. Effect of varying light intensities on yield and growth parameters in rice. *Agris,* **23(3)**: 309 - 316

Ögren, E. and Evans, J.R. 1993. Photosynthetic light-response curves. The influence of CO_2 partial pressure and leaf inversion. *Planta,* **189(2)**: 182-190

Paiva, E.A.S., Isaias, R.M.S., Vale, F.H.A. and Cristina Generosa de Senna Queiroz C.G.S. 2003. The influence of light intensity on anatomical structure and pigment contents of *Tradescantia pallida* (Rose) Hunt. cv. Purpurea Boom (Commelinaceae) leaves. *Braz. Arch. Biol. Tech.,* **46**: 617- 624

Panhwar, M.A., Memon, F.H., Kalhoro, M.A. and Somro, M.I. 2004. Performance of maize in intercropping system with soybean under different planting patterns and nitrogen levels. *J. Applied Sci.,* **4**: 201- 204

Pengelly, J.J., Sirault, X.R, Tazoe, Y., Evans, J.R., Furbank, R.T. and Caemmerer, S.V. 2010. Growth of the C4 dicot *Flaveria bidentis*: photosynthetic acclimation to low light through shifts in leaf anatomy and biochemistry. *J. Exptl. Bot.,* **61**: 4109– 4122

Pörs, Y. 1999. Anpassung von Tabakpflanzen (*Nicotiana tabacum* L.) an Licht-und Chlorophyllmangel. Humboldt-Universität zu Berlin, Germany, Ph.D thesis.

Powles, S.B. and Critchley, C. 1980. Effect of light intensity during growth on photoinhibition of intact attached bean leaflets. *Plant Physiol.,* **65(6)**: 1181-1187

Proietti, P. 2000. Effect of fruiting on leaf gas exchange in olive (*Olea europaea* L.). *Photosynthetica,* **38**: 397 - 402

Ren W.J., Yang, W.Y., Fan, G.Q., Zhu, X.M., Ma, Z.H. and Xu, J.W. 2003a. Effect of low light on dry matter accumulation and yield of rice. *J. Sichuan Agric. Univ.,* **29**(4): 292-296.

Ren, W.J., Yang, W.Y., Xu, J.W., Fan, G.Q. and Ma, Z.H. 2003b. Effect of low light on grains growth and quality in rice. *Acta Agron. Sin.,* **29(5)**: 785-790

Ren, W.J., Yang, W.Y., Zhang, G.Z., Zhu, X., Fan, G.Q. and Xu, J.W. 2003c. Effect of low-light stress on nitrogen accumulation, distribution and grains protein content of indica hybrid. *Plant Nutr. Fert. Sci.,* **9(3)**: 288-293

Restrepo, H. and Garces, G. 2013. Evaluation of low light intensity at three phenological stages in the agronomic and physiological responses of two rice (*Oryza sativa* L.) cultivars. *Agronom'a Colombiana,* **31(2)**:195- 200

Roacaas, G. and Scarano, F.R. 2001. Leaf anatomical variation in *Alchornea triplinervia* (Spreng) Müll. Arg. (Euphorbiaceae) under distinct light and soil water regimes. *Bot. J. Linn. Soc.,* **136**: 231– 238

Runkle, E. 2006. Technically speaking: Light it up! Greenhouse Product News 24 February 2009.

Salisbury, F.B. and Ross, C. 1978. Photomorphogenesis. In: Plant Physiology. 2nd Ed. Wadsworth Publ. Co., Belmont, pp. 438-463

Sato, K. and Kim, J.M. 1980. Relationship between environmental conditions and production and consumption activities of individual leaves in the population of rice plant in a paddy field: I. Changes in photosynthesis and dark respiration of individual leaves under field conditions. *Jap. J. Crop Sci.,* **49(2)**: 243– 250

Sims, D.A. and Pearcy, R.W. 1994. Scaling sun and shade photosynthetic acclimation to whole plant performance. I. Carbon balance and allocation at different daily photon flux densities. *Plant Cell Environ.,* **17**: 881–887

Singh, S. 2005. Effect of low-light stress at various growth phases on yield and yield components of two rice cultivars. *IRRN,* **30(2)**: 36-37

Sloger, C., Bezdicek, D., Milberg, R. and Boonkerd, N. 1975. Seasonal and diumal variations in N_2 (C_2H_2)-fixing activity in field soybeans. In: Nitrogen fixation by free-living micro-organisms, Stewart, W.D.P. (ed.). Cambridge University Press, New York, pp. 271-284

Sprent, J.I. and Bradford, A.M. 1977. Nitrogen fixation in field beans *(Vicia faba)* as affected by population density, shading and its relationship with soil moisture. *J. Agronc. Sci Camb.,* **88**: 303 - 310

Sprent, J.I. 1973. Growth and nitrogen fixation in *Lupinus arboreus* as affected by shading and water supply. *New Phytol.,* **72(5)**: 1005-1022.

Sun, Y.Y., Sun, Y.J., Chen, L., Xu, H. and Ma, J. 2012. Effects of different sowing dates and low-light stress at heading stage on the physiological characteristics and grain yield of hybrid rice. *Chin. J. Appl. Ecol.,* **23(10)**: 2737– 2744

Suzuki, Y., Makino, A. and Mae, T. 2001.Changes in the turnover of Rubisco and levels of mRNAs of *rbcL* and *rbcS* in rice leaves from emergence to senescence. *Plant Cell Environ.,* **24(12)**: 1353–1360

Syvertsen, J.P., Lloyd, J., McConchie, C., Kriedemann, P.E. and Farquhar, G.D. 1995. On the relationship between leaf anatomy andCO2 diffusion through the mesophyll of hypostomatous leaves. *Plant Cell Environ.,* **18**: 149- 157

Takahashi, H., and Sherman, C. 1993. The effect of zinc concentration in nutrient solution culture and the level of shade on growth, nitrogen content and nitrate redutase activity in tomato plants. *Jap. J. Trop. Agric.,* **37**: 12-27

Thomas, B. and Prue, D.V. 1984. Juvenility, Photoperiodism and Vernalisation. In: Advanced Plant Physiology, M.B. Wilkins (Ed.), Pitman, London, pp 408-439

Tombesi, A. and Cartechini, A. 1986. The effect of crown shading on the differentiation of flowering buds in olive. *Riv. Ortoflorofrutticolt Ital.,* **70**: 277 - 285

Umate, P. 2010. Genome-wide analysis of the family of light-harvesting chlorophyll a/b-binding proteins in Arabidopsis and rice. *Plant Signaling Behavior,* **5(12)**: 1537 – 1542

Valladares, F. and Niinemets, Ü. 2008. Shade tolerance, a key plant feature of complex nature and consequences. *Ann. Rev. Ecology Evolution Systematics,* **39**: 237-257

Vandenbussche, F., Vriezen, W.H., Smalle, J., Laarhoven, L.J.J., Harren, F.J.M. and Van Der Straeten, D. 2003. Ethylene and auxin control the Arabidopsis response to decreased light intensity. *Plant Physiol.,* **133**: 517-527

Venkateswarlu, B., Prasad, V.V.S.S. and Rao, A.V. 1977. Effect of low light intensity on different growth phases in rice (*Oryza sativa* L.). *Plant and Soil,* **47**: 37-47

Venkateswarlu, B. 1977. Influence of low light intensity on growth and productivity of rice *Oryza sativa* L. *Plant and Soil,* **47**: 713- 719

Venkateswarlu, B. and Visperas, R.M. 1987. Solar radiation and rice productivity. *IRRI Research Paper Series* (Philippines)

Wahua, T.A.T. and Miller, D.A. 1978. Effects of shading on the N_2-fixation, yield, and plant composition of field- grown soybeans. *Agron. J.,* **70(3)**: 387 – 392

Wang, L., Deng, F., Ren, W.J. and Yang, W.Y. 2013 Effects of shading on starch pasting characteristics of indica hybrid rice (*Oryza sativa* L.). *PLoS One,* **8(7)**: e68220

Wan-jun, R. Wen-yu, Y. Gao-qiong, F., Xue-mei, Z.H.E. Zhou-hua, M.A. and Jing-wen, X.U. 2003. Effect of low light on dry matter accumulation and yield of rice. *J. Sichuan Agril. Univ.,* 2003-2004

Wherley, B.G., Gardner, D.S. and Metzgerb, J.D. 2004. Tall fescue photo morphogenesis as influenced by changes in the spectral composition and light intensity. *Crop Sci.,* **45(2)**: 562-568

Yamamoto, Y., Kurokawa, H,. Nitta, Y. and Yoshida, T. 1995. Varietal differences of tillering response to shading and nitrogen levels in rice plant: Comparison between high tillering, semi dwarf indica and low tillering japonica. *Jap. J. Crop Sci.,* **64(2)**: 227– 234

Yamazaki, J.Y., Kamimura, Y., Okada, M., and Sugimara, Y. 1999. Changes in photosynthetic characteristics and photosystem stoichiometries in the lower leaves in rice seedlings. *Plant Sci.,* **148**: 155– 163

Yang, D., Duan, L.S., Xie, H.A., Li, Z.H. and Huang, T.X. 2011. Effect of pre-flowering light deficiency on biomass accumulation and physiological characteristics of rice. *Chin. J. Eco-Agric.,* **19(2)**: 347– 352

Yoshihashi, T., Kabaki, N., Nguyen, T.T.H. and Inatomi, H. 2002. Formation of flavor compound in aromatic rice and its fluctuations with drought stress. Research Highlights JIRCAS 2003(2004): 32–33

Zhang, J.W., Dong, S.T., Wang, K.J., Hu, C.H. and Liu, P. 2008. Effects of shading in field on key enzymes involved in starch synthesis of summer maize. *Acta Agron. Sin.*, **34**: 1470-1474.

■■■■

7

ABIOTIC DETERMINANTS OF TREE SEEDLING GROWTH IN TROPICAL DRY FORESTS

Sachchidanand Tripathi[1*], Rahul Bhadauria[2], Pratap Srivastava[2], Rishikesh Singh[3], and A. S. Raghubanshi[3]

[1] *Department of Botany, Deen Dayal Upadhyay College, University of Delhi, New Delhi-110017, India*
[2] *Ecosystems Analysis Laboratory, Department of Botany, Institute of Science, Banaras Hindu University, Varanasi-221005, India*
[3] *Institute of Environment & Sustainable Development (IESD), Banaras Hindu University, Varanasi-221005, India*
***Corresponding Author**: Email: snt191@gmail.com*

INTRODUCTION

Tropical dry forests (TDFs) account for about 42 percent of the total geographical area of tropical forests (Holdrige 1967). These forests are more threatened and less protected than moist and wet forests (Murphy and Lugo 1986). The extensive deforestation led habitat conversion in these forests has been recognized as the biggest cause of biodiversity crisis (Singh 2002). As a result, in India these forests appear as a mosaic of disturbed open canopied, and relatively undisturbed closed-canopied patches at several spatial scales (Jha and Singh 1990; Jha et al. 2005; Raghubanshi and Tripathi 2009). These patches show a marked heterogeneity in terms of irradiation, temperature, moisture and rate of nutrient release from decomposing litter (Raghubanshi and Tripathi 2009; Huante and Rincon 1993; Tripathi and Raghubanshi 2013).

Interactions among environmental drivers (i.e. light, water and nutrients) are a regular and inherent feature of any natural ecosystem, which have a great significance in shaping the plant community structure and dynamics (Portsmuth and Niinemets 2006). In TDFs, plants possess considerable variations in the life history strategies, which modulate the response of species to environmental drivers. Therefore, understanding the simultaneous effects of understory gradients of light, moisture and nutrients on the seedling performance is necessary to identify the crucial interactions controlling the regeneration dynamics in TDFs. The success of initial tree seedling establishment is related to the ability to capture and use of primary resources such as light, water and nutrients. So, information on the interactive effects of light, water and nitrogen on dry tropical tree seedlings is critical to formulating management practices to optimize the species and functional group specific resource availability. Furthermore, information about the strategies used by plants in response to the availability and interaction of these resources is indispensable to improve the management and sustainable use of different TDF species with increased economic potential.

It is necessary to select tree species having greater capacity to efficiently utilize nutrients and light, as well as greater C assimilation potential. This may facilitate the re-vegetation of degraded forests, particularly in a condition where irradiance is high and soil nutrient availability is low (Jr Santos et al. 2006). Therefore, eco-physiological studies to identify the optimal environmental requirements for growth and development of tree seedlings are very important for establishing the scientific basis of plantation and restoration programs in the dry tropical environment.

RESOURCE INTERACTION STUDIES IN TDFS

Most studies seeking the effect of key environmental drivers on plant growth in TDFs have been performed taking up single factor viz. irradiance (Khurana and Singh 2001, 2001; Rincón and Huante 1993; Huante et al. 1998) water or nutrient (Huante et al. 1998; Khurana and Singh 2004). However, interaction among key regulatory environmental variables across natural ecosystem is a regular and inherent feature (Portsmuth and Niinemets 2006; Khurana and Singh 2004). Therefore, it is necessary to design interaction studies to understand the combined effect of these factors on seedling establishment and growth.

Resource Acquisition in Relation to Life History Strategy

Plant's responses to the resource availability are strongly influenced by their life history strategies (Baker et al. 2003) Various life history traits provide plants with differential ability to match-up with the heterogeneous availability of resources (Westoby et al. 2002). The early successional pioneer species is thought to exhibit greater plasticity than late successional non-pioneers. It is attributed to the heterogeneous environments with high resource availability in early successional stage (Valladares et al. 2000; Rozendaal et al. 2006). In addition, the high resource availability enables the early successional species with sufficient C and nutrients to allocate rapidly in acclimatization (both morphological and physical) to changing light conditions (Rozendaal et al. 2006). They have high physiological capacities for light use in photosynthesis (Delagrange et al. 2004), which bestow them a higher growth rate and establishment at the early stages of succession. Pioneer species have short-lived leaves. Therefore, they can rapidly change tracks in the light environment by replacing old leaves at a high rate (Valladares et al. 2000).

Researches based on cost-benefit analyses emphasize that allocation to aboveground and belowground compartments plays a decisive role in differential adaptation of plants to resource gradients (Tilman 1988). A high biomass allocation to aboveground components may be beneficial in low light environments (such as understory with fertile soil), whereas high belowground allocation may be facilitating in nutrient or water-limited habitats (e.g. dry areas, infertile sites, and early stages of primary succession). It, therefore, involves a trade-off between the aboveground and belowground allocation (Tilman 1988). Further, most studies indicate that seedling growth rate and allocation pattern are generally very plastic in response to resource availability (Khurana and Singh 2004, 2006; Santiago et al. 2012). However, the tissue nutrient concentrations are largely fixed within genotypes (Hirose 1988). Furthermore, differences in the nutrient-use efficiency of species have been found to determine the allocation pattern and growth rate, which further drive the competition and succession in a given habitat (Bullock 1995; Alvarez-Añorve et al. 2012).

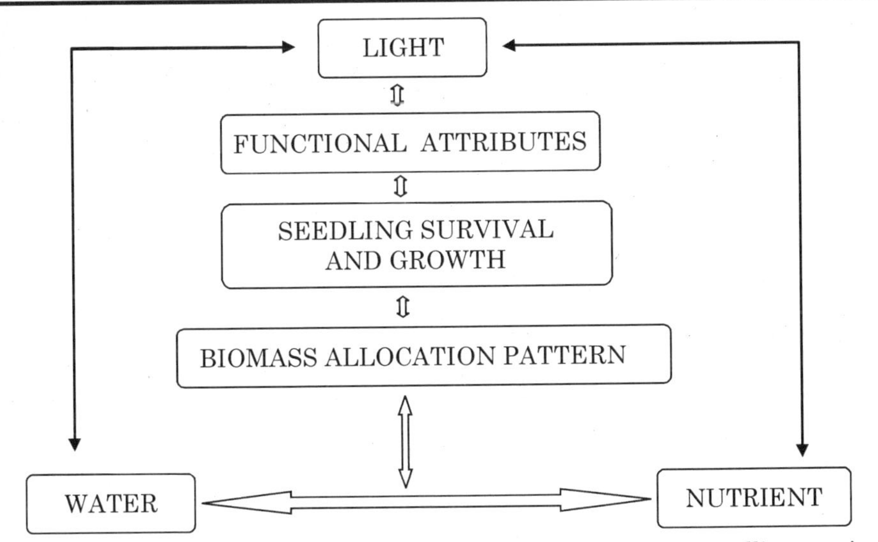

Fig. 1. A conceptual model of the effect of resource interaction on seedling survival and growth

The ecological significance of the interactive relationship among environmental factors on plant performance in communities is increasingly being acknowledged (Portsmuth and Niinemets 2006) emphasizing the importance of studies on interactions. Since, the variation in irradiance, water and nutrients availability can have complex interactive effects on plant's growth; inter-specific variation in potential growth rate may help to explain the limits to species distributions (Barbosa et al. 2014; Vadigi and Ward 2013). For example, the response of plant growth to nutrient availability increases at higher irradiance, because faster growth imposes an increased demand for nutrients. This positive interaction between irradiance and nutrient availability has been found in studies on pot-grown seedlings of tropical trees (Tripathi and Raghubanshi 2014).

SEEDLING GROWTH AND FUNCTIONAL TRAITS UNDER INTERACTION OF IRRADIANCE AND SOIL MOISTURE

Growth and survival of seedlings are strongly determined by irradiance (Rincón and Huante 1993) and water availability, which further determines the distribution of plants, globally. A thorough understanding of such effects is critical due to frequent occurrence of severe droughts in forests, worldwide (Peñuelas 2002). Climate change suggests an unpredictable pattern of rainfall in TDFs regions, which will further intensify the severity of drought in the coming decades (IPCC 2013). The heterogeneous supply of irradiance and water in natural ecosystems, even at the meter or centimeter scale (Villar Quero 2008), leads to a co-existence of different plant species (Baraloto 2006; Markesteijn 2011). The establishment pattern of woody seedlings in TDFs is driven by their responses to shade and drought, in interaction with the variation in response to nutrient supply and biotic factors (Tripathi and Raghubanshi 2014). Furthermore, level of irradiance and water availability may vary independently in different natural ecosystems (Lebrija-Trejos 2009; Markesteijn 2011). Also, they may be correlated positively or negatively. Therefore, there is a need to understand this complex interaction to understand, how tree seedlings in TDFs responsd to water availability across irradiance levels.

Seedlings respond functionally to the combination of shade and drought via exhibiting plasticity in biochemical, physiological, and structural components at the leaf and whole-plant level (Aranda 2005). Various hypotheses have been proposed to address the interactive impact of light and water availability on establishment of seedlings. According to intra-specific trade-off mechanism, deep shade will aggravate the stress imposed by drought. Consequently, the shaded plants allocate resources more to shoot, and to leaf area, than to root, which diminishes their ability to capture water (Smith and Huston 1989). However, a contrary facilitation hypothesis predicts that for leaf-level physiology, and for absolute plant growth and survival, the negative impact of drought is typically reduced in shade (Holmgren 1997). Further, a third hypothesis suggests that the effects of shade and water shortage are independent, and their impacts are orthogonal (Sack and Grubb 2002).

The manifestation of interactive effect of water and irradiance in plants is represented via trade-off in both morphological and physiological traits (Khurana and Singh 2001; Vadigi and Ward 2013). For example, it is generally found that net assimilation rate (NAR) decreases under low irradiance (Rincón and Huante 1993; Shipley and Meziane 2002). Moreover, an increase in specific leaf area (SLA) and leaf area ratio (LAR) may be a compensatory response, which is partly associated with a reduction in root mass fraction (RMF, root mass/plant mass (Paz 2003).

Large-seeded species strongly depend on seed reserves for their initial development, especially under deep shades (Rincón and Huante 1993; McLaren and McDonald 2003; Quero et al. 2006; Khurana and Singh 2006). Generally, deciduous species tend to grow faster than evergreen species, because, on average, they have a higher stomatal conductance, photosynthetic and respiration rates, leaf nitrogen concentration, and SLA (Wright et al. 2004). Plants grown in high light environments have high photosynthetic capacity and respiration rates. They require more light to saturate photosynthesis as compared to plants grown in low light conditions. Within and among species, with an increase in growth irradiance, leaves decrease the specific leaf area (SLA; $cm^2\,g^{-1}$) such that, a greater mass is displayed over a smaller area in sunny microsites. Therefore, species often show an increase in the maximum photosynthetic rate per unit leaf area (A_{area}) with increasing irradiance (Evans and Poorter 2001; Souza et al. 2010). From an economic perspective, with an increase in irradiance, C gained per unit of C invested (e.g. A_{mass}) should also increase. However, differences in the ability of plants to implement this strategy may exist (Sims et al. 1998). Plants in high light environments tend to allocate relatively more biomass to roots than leaves, reflecting the limitation of belowground resources relative to aboveground resources for light (Paz 2003).

Variation in the physiological response of seedlings in relation to interaction between light and water stress may be a compromise for survival. For instance, leaves of seedling grown under increasing shade conditions show lower ability of osmotic adjustment. Therefore, a complex condition arises under shade regimes during the dry periods between drought and shade tolerance. This acclimation is exhibited by the diminishing ability of seedlings to adopt the mechanisms of drought tolerance, such as osmotic adjustment under low-light environments (Khurana and Singh 2001; Amissah et al. 2015; Aranda et al. 2005) Furthermore, the ability of seedlings to adjust their physiological response under limiting light and water condition will act in concert to alleviate the worst situation, otherwise. This kind of conditions is usually occurring in the TDFs during the pre-monsoon summer months (Sack and Grubb

2002). In addition, soil dryness can vary across irradiances in different ways in different ecosystems under natural environment (Sack and Grubb 2002). It, therefore, emphasizes upon the need of controlled studies for analyzing the seedling responses to combinations of light and water availability.

SEEDLING GROWTH AND FUNCTIONAL TRAITS UNDER IRRADIANCE AND SOIL NUTRIENT INTERACTION

It has been well documented that plant growth is generally more strongly limited by nutrients in high than in low irradiance, even at a common nutrient supply (Poorter and Hayashida 2004; Tripathi and Raghubanshi 2013) This indicates that the interactive effects of light and nutrients could be an important factor in plant growth in forest environment. The interaction of N with high irradiance most probably reflects the central role it plays in photosynthetic machinery of the plants. It ranges from the major light-harvesting pigments and proteins to the critical enzymes such as Rubisco, that are involved at the rate limiting steps of C fixation (Evans 1989).

In natural ecosystems, it appears thatan inverse interaction between light and soil nutrient availability drives the growth of plants along natural gap-understorey continuum (Coomes 2000). The availability of limiting elements such as N increases temporarily in gaps from SOM than in understory conditions largely due to higher soil temperature, however, this inverse association otherwise, leads to faster immobilization of nitrogen and phosphorus in standing biomass under high light conditions.

On the basis of their growth requirements, forest tree species have been categorized into (1) pioneers, which regenerate only in the gap, and (2) non-pioneers, which can germinate and establish anywhere in the understory conditions (Costa et al. 2015). Pioneers typically exhibit marked differences in the growth characteristics. They show high relative growth rate (RGR), high leaf turnover rate and high photosynthetic rate (A_{max}) (Delagrange et al. 2004). It makes them more demanding of soil nutrients than non-pioneers and shade-bearers. Therefore, pioneers tend to show a greater response to nutrient limitations than shade-bearers. It was found that growth responses to irradiance can be amplified by mineral nutrient supply, which is evident strongly at the highest RGR (Peace and Grubb 1982; Khurana and Singh 2006; Tripathi and Raghubanshi 2013).

Nutrient availability is one of the important drivers associated with the distribution of tropical tree species (Alvarez-Aquino and Williams-Linera 2014). Most of the dry tropical forest of India occurs on nutrient poor soil, which also shows patchy distribution of nutrients (Singh et al. 1989; Raghubanshi 1990). Therefore, the species component of these ecosystems, which vary in life history traits such as successional status (e.g. pioneer and non-pioneer), seed mass and N_2 fixing ability, are expected to respond differently to N inputs at various level of irradiances provided in such heterogeneous environment. Moreover, variation in N availability in the patchy micro-sites in dry tropical forests may have a greater impact of the relative competitiveness of tropical tree seedlings during the regenerative phase (Royand and Singh 1994).

Furthermore, nutrient availability also affects plant biomass allocation. Increased nutrient supply often decreases the relative allocation for root production (van Noordwijk et al. 2004; Useche and Shipley 2010; Epron et al. 2012). Plants may acclimatize to lower resource availability by developing characteristics related to resource-use efficiency (Canham et al. 1996; Portsmuth and Niinemets 2006; Ferreira et al. 2015). For example, photosynthetic nitrogen-use efficiency (PNUE) may be higher at low N availability, depending on light environment and stomatal conductance. Furthermore, variability in light and N gradients may interact with water availability (Canham et al. 1996) lead to a trade-off, such that efficiency of using one resource goes down, if another is used more efficiently.

Most eco-physiological works focus on seedlings response to a change in single factor, while keeping the other factors at constant. Such single factor analysis with relative growth rate (RGR), leaf area ratio (LAR) and net assimilation rate (NAR; RGR= LAR x NAR) have provided the basic understanding of light and nutrient dependant variations in RGR. Increases in nutrient availability amplifies the leaf area ratio (LAR) due to increases in leaf mass fraction (LMF) at the expense of root mass fraction (RMF), and a decrease in leaf dry mass per unit area (LMA; LAR = LMF/LMA) (Portsmuth and Niinemets 2006). However, increase in irradiance result in higher NAR and LMA, but lower LAR (Poorter and Nagel 2000). Experiments performed at steady irradiance emphasize this variation in the morphological correlates (e.g. LMA and LAR) rather than the physiological correlates (such as, NAR drives RGR) (Wright and Westoby 2001). On the other hand, studies performed with irradiance gradients at a constant supply of nutrient imply that the variation in NAR determines the light-dependent changes in the RGR (Poorter and Hayashida 2000; Shipley and Meziane 2002). Therefore, it is difficult to predict the contrasting pattern of light and nutrient dependent variations in allocation and structural and physiological traits, from single factor analyses. It, further, emphasizes upon the need of multi-factorial studies to understand the interactive effect of light and nutrients on aforementioned growth traits. It is well documented that light and N in combination can affect plant growth and allocation (Kamaluddin and Grace 1993; Khurana and Singh 2001; Tripathi and Raghubanshi 2013; Amissah et al. 2014; Barbosa et al. 2014). The combined effects can also vary across the species (Portsmuth and Niinemets 2006), which ultimately modulates the species's competitive hierarchy at different resource levels.

Life history traits and successional status of the plant plays a pivotal role in their acclimatization to the marked heterogeneity of resource availability in natural environments (Tilman 1993). A number of studies demonstrate that shade tolerant plants perform poorly on nutrient deficient soils, due to extensive investment of nutrients to manufacture the broader leaf areas (Lusk and Contreras 1999). Furthermore, it has also been reported that nutrient availability most probably drives the light dependent plasticity in plant species, thus, giving an understanding that an improved nutrient availability is anticipated to amplify the light dependent plasticity strongly in shade tolerant species (Portsmuth and Niinemets 2007). Since fertile sites provide optimum conditions for faster increase in the leaf area, leading to a considerable decrease in light availability to the belowground plant communities, so a high plasticity to light modifications can bestow an important adaptive strategy to the plants. Plant growth is relatively strongly determined by the NAR than structural traits under optimal irradiance level. The shade-intolerant species show a greater range of variation in both NAR and LAR at a common RGR (Portsmuth and Niinemets 2007). Further, with an increase in irradiance, the plasticity in RGR, NAR and structural traits was found stronger in the shade intolerant species; however these adjustments in RGR and NAR were similar among all species at low irradiances (Shipley 2002). Both physiological plasticity and structural plasticity were reversely associated with shade tolerance in unfertilized plants. Moreover, the plasticity was found to be amplified by the nutrient addition, surprisingly (Meziane and Shipley 1999; Tripathi and Raghubanshi 2013). The nutrient dependent enhancement in plasticity was observed greater in more shade-tolerant plants. Therefore, differences in plasticity among species of varying shade tolerance at higher nutrient availability were non-significant. It has been suggested that both light and nutrient levels affected RGR and each of the growth components. With an increase in light, average NAR increases, while SLA and LWR decrease (Meziane and Shipley 1999). Increases in nutrient levels, however, increases all three growth parameters. Light and nutrient levels interacted in their effects on NAR and SLA, but not in their effects on LWR (Meziane and Shipley 1999). In TDFs, patchy distribution of soil nutrients along with heterogeneous light availability during the growing season have a significant impact on the growth determinants of tree seedlings (Tripathi and Raghubanshi 2014), therefore, it is important to gain more advanced insight into the combined effects of nutrients and light on plant performance in dry forests

The overall review of vital importance as described in the preceding pages are summarized herewith prior to coming to the precise conclusions. The present review emphasized upon the current scenario and future needs of the researches on resource interaction in relation to the tree species of TDFs. Pronounced seasonality in tropical dry forests (TDFs) provides a plethora of combinations of environmental conditions for the survival and growth of tree seedlings. Therefore, tree seedling faces a very heterogeneous availability of resources during their establishment phase in TDFs. Soil moisture availability in these forests is the key environmental regulator which determines the survival, growth and distribution of native tree species. However, the role of light conditions in these forests seems to be paradoxical. During the rainy season, low light conditions in the understories becomes limiting for the survival of tree seedlings. However, the shady conditions play a role of survival shelters in dry summer season, owing to the higher soil moisture availability. Moreover, the soils in dry tropics are nutrient poor with a patchy distribution of nutrients, and their availability is majorly regulated by climatic conditions. For example, available N and P in soil nutrient pool tend to be greater during the driest pre-rainy season as compared to rainy season. The major effects of this variation in resource availability can be demonstrated in the physiological and structural changes at the leaf and whole-plant level. It enables seedlings to acclimatize in different growth conditions and minimize the negative impacts of fluctuations in resources. It is proposed that a sound understanding of these relationships is critical for better interpretation of plant community dynamics under changing climate conditions. It may help in identifying a functional classification, which would enable prediction of species or functional groups responses to the environmental changes. Some of the related **key words are:** Tropical dry forests; resource interaction; seedlings; soil moisture; functional traits.

CONCLUSION

In TDFs, it appears that the interaction of light, water and nutrients resulting into a suite of soil physico-chemical and biochemical characteristics that have significant effects over the survival and growth of tree seedlings. Further, a considerable spatio-temporal variation in resource availability exists and species more plastic to this variability may be the dominant component of the forest plant community. Therefore, response of individual species and functional groups to environmental heterogeneity is critical for a better interpretation, restoration and management of plant communities in TDFs. Further, it may help in new functional classification of species on the basis of their responses to the resource combinations.

ACKNOWLEDGEMENT

Authors are highly thankful to the Prof. J.S. Singh, Emeritus professor, Department of Botany, Banaras Hindu University for guidance and University Grant Commission and Council of Scientific and Industrial Revolution for providing research fellowship.

REFERENCES

Alvarez-Añorve, M.Y., M. Quesada, G.A. Sánchez-Azofeifa, L.D. Avila-Cabadilla, J.A. Gamon (2012). Functional regeneration and spectral reflectance of trees during succession in a highly diverse tropical dry forest ecosystem. *American Journal of Botany*, 99, 816-826.

Alvarez-Aquino, C.L., G.U. Williams-Linera (2014). Seedling Survival and Growth of Tree Species: Site Condition and Seasonality in Tropical Dry Forest Restoration. *Botanical Sciences*, 90, 313-329.

Amissah, L., G.M. Mohren, B. Kyereh, L. Poorter (2015). The effects of drought and shade on the performance, morphology and physiology of Ghanaian tree species. *PloS One,* 2015 10:e0121004.

Aranda, I., L. Castro, M. Pardos *et al.* (2005). Effects of the interaction between drought and shade on water relations, gas exchange and morphological traits in cork oak *Quercus suber* L. seedlings. *Forest Ecology and Management,* 210, 117-129.

Baraloto, C., D. Bonal & D.E. Goldberg (2006). Differential seedling growth response to soil resource availability among nine neotropical tree species. *Journal of Tropical Ecology,* 22, 487-497.

Bullock, S.H. (1995). Plant reproduction in neotropical dry forest. In: Bullock SH, Mooney HA, editors. Seasonally Dry Tropical Forests. Cambridge: Cambridge University Press.

Canham, C.D., A.R. Berkowitz, V.R. Kelly, *et al.* (1996). Biomass allocation and multiple resource limitation in tree seedlings. *Canadian Journal of Forest Research,* 26, 1521-1530.

Chaturvedi, R.K., A.S. Raghubanshi, and J. S. Singh (2012). Effect of grazing and harvesting on diversity, recruitment and carbon accumulation of juvenile tree in tropical dry forest, *Forest Ecology and Management,* 284, 152-162.

Coomes, D.A. & P.J. Grubb (2000). Impacts of root competition in forests and woodlands: a theoretical framework and review of experiments. *Ecological Monographs,* 70, 171–207.

Costa, F.V., F.R. Costa, W.E. Magnusson, E. Franklin, J. Zuanon, R. Cintra, F. Luizão, J.L. Camargo, A. Andrade, W.F. Laurance, F. Baccaro (2015). Synthesis of the first 10 years of long-term ecological research in Amazonian Forest ecosystem–implications for conservation and management. *Natureza & Conservação,* 13, 3-14.

Delagrange, S., C. Messier, M.J. Lechowicz, *et al.* (2004). Physiological, morphological and allocational plasticity in understory deciduous trees: importance of plant size and light availability. *Tree Physiology,* 24, 775–784.

Easdale Tomás, A., Diego E. Gurvich, Alicia N Sersic & John R Healey (2007). Tree morphology in seasonally dry montane forest - shade tolerance and nutrient shortage. *Journal of Vegetation Science,* 18, 313-326.

Epron, D., J.P. Laclau, J.C. Almeida, J.L. Gonçalves, S. Ponton, C.R. Sette, J.S. Delgado-Rojas, J.P. Bouillet, Y. Nouvellon (2012). Do changes in carbon allocation account for the growth response to potassium and sodium applications in tropical Eucalyptus plantations? *Tree physiology,* 32, 667-79.

Evans, J. and H. Poorter (2001). Photosynthetic acclimation of plants to growth irradiance: the relative importance of specific leaf area and nitrogen partitioning in maximizing carbon gain. *Plant, Cell & Environment*, 24, 755-67.

Evans, J.R. (1989). Photosynthesis- the dependence on nitrogen partitioning, In *Causes and Consequences of Variation in Growth Rate and Productivity of Higher Plants* (eds. Lambers, H., Cambridge, M.L., Konings, H. & Pons, T. L.), pp. 159-174. SPB Publishing, The Hague.

Ferreira, W.N., C.F. Lacerda, R.C. Costa, S. Medeiros Filho (2015) Effect of water stress on seedling growth in two species with different abundances: the importance of Stress Resistance Syndrome in seasonally dry tropical forest. *Acta Botanica Brasilica,* 29, 375-82.

Gerhardt, K. (1993). Tree seedling development in tropical dry abandoned pasture and secondary forest in Costa Rica. *Journal of Vegetation Science,* 4, 95–102.

Gerhardt, K. (1996). Effect of root competition and canopy openness on survival and growth of tree seedling in tropical seasonal dry forest. *Forest Ecology and Management,* 82, 33-48.

Hirose T. (1988). Nitrogen availability, optimal root: shoot ratios and plant growth. In: M.J.A. Werger, P.J.M. Van der Aart, H.J. During, J.T.A. Verhoeven (Eds.). *Plant form and vegetation structure.* SPB Academic Publishing, The Hague, The Netherlands. pp 135–145.

Holmgren, M., M. Scheffer & M. Huston (1997). The interplay of facilitation and competition in plant communities. *Ecology,* 78, 1966–1975.

Huante, P., E. Rincón & F.S. Chapin (1998). Foraging for nutrients responses to changes in light and competition in tropical deciduous tree seedlings. *Oecologia,* 117, 209–216.

Jr Santos, U.M., J.F.C. Goncalves & T.R. Feldpausch (2006). Growth, leaf nutrient concentration and photosynthetic nutrient use efficiency in tropical tree species planted in degraded areas in central Amazonia. *Forest Ecology and Management,* 226, 299-309.

Kenzo, T., R. Yoneda, Y. Matsumoto, M.A. Azani, N.M. Majid (2008). Leaf photosynthetic and growth responses on four tropical tree species to different light conditions in degraded tropical secondary forest, Peninsular Malaysia. *Japan Agricultural Research Quarterly: JARQ,* 42, 299-306.

Khurana, E. & J. S. Singh (2001). Ecology of tree seed and seedlings: implications for tropical forest conservation and restoration. *Current Science,* 80, 748–757.

Khurana, E. & J.S. Singh (2001). Ecology of seed and seedling growth for conservation and restoration of tropical dry forest: a review. *Environmental Conservation,* 28, 39–52.

Khurana, E. & J.S. Singh (2004). Response of five dry tropical tree seedlings to elevated CO2: impact of seed size and successional status. *New Forests,* 27, 139–157.

Lusk, C.H. & O. Contreras. 1999. Foliage area and crown nitrogen turnover in temperate rain forest juvenile trees of differing shade tolerance. *Journal of Ecology,* 87, 973–983.

Markesteijn, L. & L. Poorter. 2009. Seedling root morphology and biomass allocation of 62 tropical tree species in relation to drought- and shade-tolerance. *Journal of Ecology*, 97, 311-325.

Markesteijn, L., L. Poorter, & F. Bongers (2007). Light-dependent leaf trait variation in 43 tropical dry forest tree species. *American Journal of Botany,* 94, 515–525.

May, R.M., J.H. Lawton & E.S. Nigel (1995). Assessing extinction rates. In: J.H. Lawton & R.M. May (Eds.). *Extinction Rates.* Oxford University Press, Oxford.

Meziane, D. & B. Shipley. 1999. Interacting components of interspecific relative growth rate: constancy and change under differing conditions of light and nutrient supply, *Functional Ecology*, 13, 611–622.

Murphy, P.G. & A.E. Lugo (1986). Ecology of Tropical Dry Forest, *Annual Review of Ecology and Systematics*, 17, 67-88.

Peace, W.H.J. & P.J. Grubb (1982). Interactions of light and mineral nutrient supply in the growth of Impatiens parviflora. *New Phytologist*, 90, 127-150.

Peñuelas, J., I. Filella, & P. Comas (2002). Changed plant and animal life cycles from 1952-2000 in the Mediterranean region. *Global Change Biology,* 8, 531-544.

Poorter, L. & O.Y. Hayashida (2000). Effects of seasonal drought on gap and understorey seedlings in a Bolivian moist forest. *Journal of Tropical Ecology,* 16, 481–498.

Portsmuth, A. & U. Niinemets (2006). Interacting controls by light availability and nutrient supply on biomass allocation and growth of *Betula pendula* and *B. pubescens* seedlings. *Forest Ecology and Management,* 227, 122-134.

Portsmuth, A. & U. Niinemets (2007). Structural and physiological plasticity in response to light and nutrients in five temperate deciduous woody species of contrasting shade tolerance. *Functional Ecology*, 21, 61-77.

Quero, J.L., R. Villar, T. Marañón, *et al.* (2006). Interactions of drought and shade effects on four Mediterranean *Quercus* species: physiological and structural leaf responses. *New Phytologist* 170, 819-834.

Raghubanshi, A.S., S.C. Srivastava, R.S. Singh, & J.S. Singh (1990). Nutrient release in leaf litter. *Nature,* 346, 227.

Rincón, E. & P. Huante (1993). Growth responses of tropical deciduous tree seedlings to contrasting light conditions. *Trees,* 7, 202-207.

Roy, S. & J.S. Singh. 1994. Consequences of habitat heterogeneity for availability of nutrients in a dry tropical forest. *Journal of Ecology,* 82, 503–509.

Rozendaal, D.M.A., V.H. Hurtado & L. Poorter (2006). Plasticity in leaf traits of 38 tropical tree species in response to light, relationships with light demand & adult stature, *Functional Ecology* 20, 207-216.

Sack, L. & P.J. Grubb (2002). The combined impacts of deep shade and drought on the growth and biomass allocation of shade-tolerant woody seedlings. *Oecologia,* 131, 175-185.

Sánchez-Gómez, D., M.A. Zavala, & F. Valladares (2006). Seedling survival responses to irradiance are differentially influenced by low-water availability in four tree species of the Iberian cool temperate–Mediterranean ecotone. *Acta- Oecologia,* 30, 322-332.

Santiago, L.S. (2015). Nutrient limitation of eco-physiological processes in tropical trees. *Trees,* 29, 1291-300.

Shipley, B. and D. Meziane (2002). The balanced-growth hypothesis and the allometry of leaf and root biomass allocation. *Functional Ecology,* 16, 326–331.

Sims, D.A., J.R. Seemann & Y. Luo (1998). The significance of differences in the mechanisms of photosynthetic acclimation to light, nitrogen and CO_2 for return on investment in leaves. *Functional Ecology,* 12, 185–194.

Singh, J.S., A.S. Raghubanshi, R.S. Singh, & S.C. Srivastava (1989). Microbial biomass acts as a source of plant nutrients in dry tropical forests and savanna. *Nature,* 238, 449–500.

Smith, T. & M. Huston (1989). A theory of the spatial and temporal dynamics of plant communities. *Vegetatio,* 83, 49-69.

Souza, G.M., A.M. Sato, R.V. Ribeiro, C.H. Prado (2010). Photosynthetic responses of four tropical tree species grown under gap and understorey conditions in a semi-deciduous forest. *Brazilian Journal of Botany,* 33, 529-38.

Tilman, D. (1988). Plant strategies and the structure and dynamics of plant communities. Princeton University Press, Princeton.

Useche, A. & B. Shipley (2010). Plasticity in relative growth rate after a reduction in nitrogen availability is related to root morphological and physiological responses. *Annals of Botany,* 106, 617-25.

Valladares, F., J.S. Wright, & E. Lasso (2000). Plastic phenotypic responses to light of 16 congeneric shrubs from a Panamanian rainforest. *Ecology,* 81, 1925–1936.

van Noordwijk, M., S. Rahayu, S.E. Williams, K. Hairiah, N.M. Khasanah, G. Schroth (2004). Crop and tree root-system dynamics. Below-ground interactions in Tropical Agroecosystems. Consepts and models with multiple component. CABI Publ., Wallingford, UK. ISBN 0-85199-673-6. pp:83-108.

Veenendaal, E.M., M.D. Swaine, R.T. Lecha *et al.* (1996). Responses of West African forest tree seedlings to irradiance and soil fertility. *Functional Ecology,* 10, 501–511.

Villar Quero, J.L. & R.T. Maráñon (2008). Relating leaf photosynthetic rate to whole-plant growth: drought and shade effects on seedlings of four *Quercus* species. *Functional Plant Biology,* 35, 725–737.

Wright, I.J. & M. Westoby. 2001. Understanding seedling growth relationships through specific leaf area and leaf nitrogen concentration: generalizations across growth forms and growth irradiance. *Oecologia,* 127, 21–29.

Wright, I.J., P.B. Reich, M. Westoby *et al.* (2004). The worldwide leaf economics spectrum. *Nature,* 428, 821–827.

■■■■

8

PHYSIOLOGICAL ADAPTATION OF RICE TO ANAEROBIC STRESS CONDITIONS CAUSED BY FLOODING DURING GERMINATION AND EARLY GROWTH AND ITS IMPLICATIONS IN CROP IMPROVEMENT

D. Vijayalakshmi[1*], R. Naveen Kumar[3], C. Vijayalakshmi[1] and M. Raveendran[2]

[1]Department of Crop Physiology, Tamil Nadu Agricultural University, Coimbatore
[2]Department of Plant Biotechnology, Tamil Nadu Agricultural University, Coimbatore
[3] Department of Sericulture, Tamil Nadu Agricultural University, Coimbatore

INTRODUCTION

Rice is the most important cereal crop worldwide. Rice is cultivated in a wide range of environments, from tropical to temperate climates, including in aerobic soils in uplands to wet lowlands with uncontrolled flooding. Despite being a semi-aquatic species, rice is sensitive to various types of flooding stress and this sensitivity varies with the genotype, stage of development, duration and depth of flooding and flood water conditions (Jackson and Ram, 2003; Sarkar *et al.*, 2006; Das *et al.*, 2009; Mackill *et al.*, 2012). In Tamil Nadu, rainfall being unpredictable and farmers depend on monsoon-season rice crops flooding of low-lying areas result in a severe reduction in rice establishment.

Rice is extremely sensitive to anaerobic conditions during germination (anaerobic germination) and early growth of the embryo (Yamauchi *et al.*, 1993; Ismail *et al.*, 2009; Angaji *et al.*, 2010). Rice seeds can germinate, and to some degree, extend their coleoptiles under hypoxic and even anoxic conditions, but fail to develop roots and leaves (Taylor, 1942; Biswas and Yamauchi, 1997; Ella and Setter, 1999). Due to increasing practices of direct seeding in rainfed and irrigated conditions, tolerance to hypoxic or anaerobic condition is becoming an important trait. If heavy rains happen to coincide during direct seeding, the entire crop may be lost. It significantly reduces the yield because of poor crop establishment. Hence, tolerance of anaerobic conditions at these early stages is a prerequisite for effective direct-seeded rice in rainfed and flood-affected areas. Also, the climate projections suggest that temperatures, precipitation and flooding, and sea level rise are likely to increase, with adverse impacts on crop yield and farm income (INCCA, 2010).

Considerable progress has been made in understanding the Genetic variation in the ability to germinate and establish in flooded soils has recently been observed in rice, and a few landraces were identified that are tolerant (Angaji *et al.,* 2010). Apart for that, advancing knowledge on the physiological mechanisms of tolerance to flooding during germination and early growth in rice could facilitate the development of improved rice varieties. Major traits that are unique to tolerant genotypes, include seed germination and seedling growth under low oxygen stress conditions facilitates an adaptive mechanism to withstand anaerobiosis (Magneschi and Perata, 2009), coleoptiles elongation, formation of roots and leaves under hypoxia, ability to catabolize starch into simple sugars for subsequent use in glycolysis and fermentative pathways to generate energy, role of enzymes, non-enzymatic proteins and hormones under anaerobic stresses. This review attempts to summarize the major physiological adaptive traits of rice for low-oxygen stress during seed germination and early growth. The review further discusses the prospects of using knowledge gained in plant adaptation to anaerobic conditions for application in crop improvement programmes.

PHYSIOLOGICAL TRAITS ASSOCIATED WITH TOLERANCE OF ANAEROBIC CONDITIONS DURING GERMINATION AND EARLY SEEDLING GROWTH IN RICE

Several growth, anatomical and physiological traits are identified that are associated with germination and survival of rice seedlings under anaerobic conditions.

Seedling Survival and Germination Percentage

The main trait used to study the tolerance of anaerobiosis is seedling survival after 21 days of submergence under 10 cm water head. The genotypes that recorded higher germination percentage under anaerobic condition are highly tolerant to low oxygen stress, as seed germination is a critical point in seedling establishment and subsequent plant vigor (Manigbas *et al.,* 2008; Miro and Ismail, 2013). Surviving seedlings are those that emerge from the water by fast germination and coleoptile elongation (Ismail *et al.,* 2009; Angaji *et al.,* 2010).

Coleoptile Elongation

Rice can germinate under hypoxic or anoxic conditions, but only tolerant genotypes have the ability of fast coleoptile elongation and root formation under submerged conditions in the field (Ismail *et al.,* 2009). Coleoptile elongation is usually targeted for selection of tolerant rice genotypes germinating under submergence because it is easy to phenotype. While screening for tolerance of early flooding, rapid coleoptiles growth is considered important. But tolerance of flooding at the later vegetative stages of growth, demands limited underwater elongation. Mechanisms such as rapid underwater elongation, as in deepwater rice varieties, or the dormancy strategy mediated by the SUB1 gene in submergence-tolerant varieties, become functional when the seedlings are flooded. Thus, it is very clear that different signaling mechanisms regulate the shifts in metabolic pathways and functioning of the genes involved in each type of flooding tolerance.

Coleoptile length increases mainly through cell elongation. Many genotypes are capable of initiating germination but cannot elongate further, as energy reserves and oxygen is limiting. Variation in the rate of coleoptile elongation under low oxygen (0–5%) was observed among 10 rice genotypes with maximum elongation observed at 0% oxygen (Turner *et al.,* 1981). Rapid elongation of coleoptiles could facilitate contact with air in waterlogged or flooded soils and subsequent aeration of the growing embryo.

Mobilization of Stored Carbohydrates

Generally, carbohydrate metabolism is strongly inhibited when oxygen is limiting. Hence, seeds with carbohydrate reserves, such as cereals, are known to be more tolerant of hypoxia during germination than seeds with fatty acid reserves (Al-Ani *et al.,* 1985; Raymond *et al.,* 1985). In rice, genotypes that are more tolerant of flooding during germination seem to have better capabilities for breaking starch into simple sugars, as demonstrated by faster depletion of starch in their germinating seeds compared with sensitive genotypes (Ismail *et al.,* 2009).

Amylases are believed to play a major role in starch breakdown in cereal seeds (Murata *et al.,* 1968), and rice seeds retain their ability to break down starch into readily fermentable carbohydrates when germinating under hypoxic or even anoxic conditions (Miro and Ismail., 2013). Tolerant genotypes also showed greater total amylase activity in germinating seeds, with a progressive increase in activity from sowing to reach several-fold at 3 days after seeding in flooded soil. This higher amylase activity was also associated with maintenance of higher soluble sugar concentrations in germinating seeds, greater starch depletion, better shoot and root growth, and higher seedling survival (Ismail *et al.,* 2009). In case of sensitive genotypes, most of the enzymes involved in the breakdown of starch into simple sugars are less active under low oxygen.

Among them are α-amylases, sucrose synthase and aldolase (Ismailetal, 2009). Another reason for energy crises under low oxygen stress is the break- down of pyruvate to generate energy through the TCA cycle. This cycle is inhibited by lack of oxygen, so plants move to activate alternative pathway called alcoholic fermentation pathway (Kato-Noguchi, 2006).

Alcoholic Fermentation Pathway

Under anoxia, carbohydrate catabolism shifts from aerobic to anaerobic pathways (alcoholic fermentation) for generating ATP for growth and maintenance processes. However, ATP production is 18-fold less efficient in this process than is aerobic respiration (Greenway and Setter, 1996).An increased rate of anaerobic respiration still remains one of the most important mechanisms to alleviate the adverse effects of reduced ATP supply during oxygen deficiency, and several reports demonstrated its importance in plants (Avadhani *et al.,* 1978; Jackson *et al.,* 1982; Waters *et al.,* 1991; Setter and Ella, 1994; Gibbs *et al.,* 2000). Activities of both pyruvate decarboxylase (PDC) and alcohol dehydrogenase (ADH) increased significantly within 12 h of imbibitions in both tolerant and sensitive genotypes, but remained progressively higher in the tolerant genotypes with time. Anaerobic respiration is probably one of the important attributes contributing to the superior performance of tolerant rice genotypes.

Alteration in Enzymes Activities under Anaerobic Stress

The enzyme alcohol dehydrogenase and pyruvate decarboxylase via anaerobic fermentation metabolism has been reported by several researchers to supply ATP for seedling germination and survival under anaerobic stress conditions (Gibbs and Greenway, 2003; Gibbs *et al.,* 2000; Jackson *et al.,* 1982). Another prominent enzyme that is found to be regulated under anaerobic condition is the cell wall peroxidases. Studies in some crop species like mung bean and peanut have reported that this enzyme acts antagonistically by inhibiting the cell wall extensibility thereby inhibiting the cell elongation (Goldberg *et al.,* 1987; Zheng and van Huystee, 1992; Ismail, 2009). These enzymes are responsible for the assembly of lignins and proteins in the cell wall (Whitmore, 1978) and for the binding of ferulic acid to the cell wall by the formation of diferuloyl cross links to matrix polysaccharides (Fry 1979), subsequently reducing wall extensibility and restricting cell elongation. Seeds of tolerant rice genotypes germinating under anoxic conditions showed substantially lower peroxidase activity than the intolerant genotypes (Ismail *et al.,* 2009). Furthermore, the activity of peroxidases in germinating embryos correlated negatively with coleoptile growth and seedling survival. Thus, the characterization of these biochemical changes under anaerobic germination in rice would reveal the adaptation mechanisms to withstand the anaerobic condition.

Non - Enzymatic Proteins under Anaerobic Stresses

Expansins are a group of non-enzymatic proteins known to regulate cell wall expansion. These proteins are encoded by a gene superfamily with at least two major groups: a-expansins and b-expansins. Their role in cell extensibility and coleoptiles elongation has been reported in many crops, including rice (Huang et al. 2000). Studies reveal that, *EXPA2* and *EXPA4* are specific to hypoxia (Lasanthi- Kudahettigeetal, 2007), while *EXPA7* and *EXP12* were expressed in coleoptiles under anoxia and *EXPA1* and *EXPB11* and *EXPB17* were found to be expressed in submerged coleoptiles (Takahashi *et al.,* 2011). However, further studies are required to ascertain the actual roles of these proteins in anaerobic stress tolerance.

ROLE OF PLANT HORMONES

Differential response to the content of the gaseous hormone ethylene is an important aspect of submergence tolerance, either through inhibiting elongation or through promoting active growth. ethylene promotes rice seedling growth under flooded conditions through effects mediated by GA (Fukao *et al.,* 2006), and this flooding induced elongation is one of the escape mechanisms that help submerged plants regain contact with air. This elongation is beneficial in early growth as elongation of the coleoptiles (Ku *et al.,* 1970), but a disadvantage for seedlings exposed to flash floods as it leads to exhaustion of carbohydrate reserves (Das *et al.,* 2009). Ethylene interactions with other plant hormones such as abscisic acid (ABA), auxins and gibberellins under low oxygen stress; however, the effects of these plant hormones are hard to predict

The phytohormones ABA and GA also seem to play important roles in submergence tolerance during germination. ABA acts as a positive regulator of rice germination under submergence and its plausible interaction with ethylene for coleoptile elongation. GA was related to inhibition of the coleoptile response to submergence during germination. A QTL for tolerance of anaerobic conditions during germination (*qAG12*, Septiningsih *et al.,* 2013) contains candidate genes that are possibly involved in promoting ABA synthesis during early stages of seed development. QTL for GA may regulate the QTL for anaerobic germination, since GA could promote coleoptile elongation under submergence according to Horton (1991).

USE OF PHYSIOLOGICAL TRAITS AND MOLECULAR TOOLS IN CROP IMPROVEMENT PROGRAMMES AIMED AT DEVELOPMENT OF SUITABLE VARIETIES FOR EXCESS WATER CONDITIONS

The enormous diversity and plasticity in adaptation to contrasting hydrological conditions, ranging from aerated soils in uplands to areas with water depths exceeding 5m in deep water areas, has made rice one of the most amenable crops for genetic manipulation to develop varieties suitable for excess water conditions (Ismail and Mackill, 2013; Kirk *et al.*, 2014). Several land races that can with stand different types of floods have been identified and characterized (Table 1)

Table 1. Physiological traits associated with tolerance for flash floods/submergence water logging during germination and early seedling growth in rice.

Tolerant varieties	**Main traits associated with tolerance**	**Major QTLs / Genes**
Khao Hlan On Ma –Zhan (Red) Khaiyan Kalonchi Kharsu Nanhi	– Fast germination and coleoptiles elongation – Ability to break down and use starch under low O2 – High anaerobic respiration to avoid energy crises when oxygen is low	qAG7. 1 qAG-9- 2

Genetic variation in the ability to germinate and establish in flooded soils has recently been observed in rice, and a few landraces were identified that are tolerant (Angaji *et al.*, 2010). Major QTLs (quantitative trait loci) associated with tolerance of flooding during germination have been recently identified and are being targeted for cloning and for use in marker-assisted breeding (Angaji *et al.*, 2010; Septiningsih *et al.*, 2013). Despite this progress, little is known about the molecular basis of tolerance of low oxygen stress in these contrasting rice genotypes.

CONCLUSION AND PERSPECTIVES

Little progress has been made in uncovering the major physiological traits associated with tolerance of flooding during germination and early seedling growth, but development of improved varieties for anaerobic stress tolerance has not yet been possible. New research in identifying new, rapid and repeatable protocols which will imitate the field conditions need to be developed. Characterization of environmental conditions and identification of target genotypes for genetic manipulation will help in speeding up breeding programmes to develop rice varieties where waterlogging is a serious problem during crop establishment stage.

REFERENCES

Angaji, SA; Septiningsih, EM; Mackill, DJ; Ismail AM. 2010. Identification of QTLs associated with tolerance of anaerobic conditions during germination in rice (*Oryza sativa* L.), *Euphytica* 172: 159–168

Avadhani, PN; Greenway, H; Lefroy, R; Prior, L. 1978. Alcoholic fermentation and malate metabolism in rice germinating at low oxygen concentrations. *Australian Journal of Plant Physiology,* 5: 15–25

Al-Ani ,A; Bruzau, F; Raymond, P; Saint-Ges, V.; Leblanc, JM; Pradet A. 1985. Germination, respiration, and adenylate energy charge of seeds at various oxygen partial pressures. *Plant Physiology,* 79: 885–890

Biswas, JK; Yamauchi, M. 1997. Mechanism of seedling establishment of direct-seeded rice (*Oryza sativa* L.) under lowland conditions. *Botanical Bulletin of the Academia Sinica.*, 38: 29–32

Das, KK; Panda, D; Sarkar, RK; Reddy, JN; Ismail, AM. 2009. Submergence tolerance in relation to variable floodwater conditions in rice. *Environmental and Experimental Botany,* 66: 425–434

Ella, ES; Setter, TL. 1999. Importance of carbohydrates in rice seedling establishment under anoxia. *Acta Horticulturae,* 504:209–216

Fry, SC. 1979. Phenolic components of the primary cell wall and their possible role in the hormonal regulation of growth. *Planta,* 146: 343–351

Fukao, T; Xu, K; Ronald, PC; Bailey-Serres, J. 2006. A variable cluster of ethylene response factor-like genes regulates metabolic and developmental acclimation responses to submergence in rice. *The Plant Cell,* 18: 2021–2034

Gibbs, J; Morrell, S; Valdez, A; Setter, TL; Greenway, T. 2000. Regulation of alcoholic fermentation in coleoptiles of two rice cultivars differing in tolerance to anoxia. *J Exp Bot.,* 51: 785–796.

Gibbs, J; Greenway, H. 2003. Mechanisms of anoxia tolerance in plants. I. Growth, survival and anaerobic catabolism. *Functional Plant Biol.,* 30: 1–47.

Goldberg, R; Liberman, M; Mathieu, C; Peirron, M; Catesson, AM. 1987. Development of epidermal cell wall peroxidase along the mung bean hypocotyls: possible involvement in the cell wall stiffening process. *J Exp Bot.,* 38: 1378–1390.

Greenway H, Setter TL. 1996. Is there anaerobic metabolism in submerged rice plants? A view point. In: Singh VP, Singh RK, Singh BB, Zeigler RS, eds. Physiology of stress tolerance in rice. Manila, Philippines: Narendra Deva University of Agriculture and Technology and International Rice Research Institute, 11–30.

Huang, J; Takano, T; Akita, S. 2000. Expression of a-expansion genes in young seedlings of rice. *Planta,* 211: 467–473.

Horton, RF.1991. The effect of ethylene and other regulators on coleoptiles growth of rice under anoxia. *Plant Sci.,* 79: 57–62. doi:10.1016/ 0168-9452(91)90069-K.

INCCA, (Indian Network for Climate Change Assessment). 2010. Climate change and India - A 4X4 Assessment, A Sectoral and Regional Analysis for 2030s: Ministry of Environment and Forests, Government of India.

Ismail, AM; Ella, ES; Vergara, GV; Mackill, DJ. 2009. Mechanisms associated with tolerance for flooding during germination and early seedling growth in rice (*Oryza sativa*). *Annals of Botany,* 103: 197–209.

Ismail, AM; Mackill, DJ. 2013. "Response to flooding: submergence tolerance in rice," In: *Plant Genetic Responses and Climate Change,Chapter15,* eds M. Jackson B. Ford-Lloyd and M. Parry (London: CABI International).

Jackson, M; Ram, PC. 2003. Physiological and molecular basis of susceptibility and tolerance of rice plants to complete submergence. *Annals of Botany,* 91: 227–241.

Jackson, MB; Herman, B; Goodenough A. 1982. An examination of the importance of ethanol in causing injury to flooded plants. *Plant Cell Environ.,*5: 163–172.

Kato-Noguchi, H. 2006. Pyruvate metabolism in rice coleoptiles under anaerobiosis. *Plant Growth Regul.* 50: 41–46.doi:10.1007/s10725-006- 9124-4.

Kirk, GJD; Greenway, BJ; Atwell, BJ; Ismail, AM; Colmer, D. (2014)."Adaptation of rice to flooded soils," in *Progress in Botany*, Vol. 75, eds U. Lüttge, W. Beyschlag, and J. Cushman (Berlin; Heidelberg : Springer-Verlag), doi:10.1007/978-3-642- 38797-5_8.

Ku, HS; Suge, H; Rappaport, L; Pratt, HK. 1970. Stimulation of rice coleoptile growth by ethylene. *Planta*, 90: 333–339.

Lasanthi-Kudahettige, R; Magneschi, L; Loret, E; Gonzali, S; Licausi, F; Novi, G; Beretta, O; Vitulli, F; Alpi, A; Perata, P. 2007. Transcript profiling of the anoxic rice coleoptile. *Plant Physiology,* 144:218–231.

Mackill, DJ; Ismail, AM; Singh, US; Labios, RV; Paris, TR. 2012. Development and rapid adoption of submergence-tolerant (Sub1) rice varieties. *Advances in Agronomy*, 115: 303–356.

Magneschi, L; Perata, P. 2009. Rice germination and seedling growth in the absence of oxygen.*Ann Bot.,* 103: 181-196.

Manigbas, NL; Renando, OS; Wilhelmina, VB; Angelo, JN; Emily, CA; Thelma, FP; Rolando, TC. 2008. Development of screening methods for anaerobic germination and seedling vigor in direct seeded wet seeded rice culture. *Philippine J Crop Science,* 33(3): 34-44.

Miro. B; Ismail, M. 2013. Tolerance of anaerobic conditions caused by flooding during germination and early growth in rice (*Oryza sativa* L.). *Frontiers Plant Sci.,* 4: 1-18.

Murata, T; Akazawa, T; Fukuchi, S. 1968. Enzymic mechanism of starch break-down in germinating rice seeds. I. An analytical study. *Plant Physiology*, 43: 1899–1905.

Raymond, P; Al-Ani, A; Pradet, A. 1985. ATP production by respiration and fermentation, and energy charge during aerobiosis and anaerobiosis in twelve fatty and starchy germinating seeds. *Plant Physiology*, 79: 9879–9884.

Sarkar,R.K; Reddy, J.N; Sharma, S. G; Ismail, A.M.(2006). Physiological basis of submergence tolerance in rice and implications for crop improvement. *Curr.Sci.,* 91: 899–906

Septiningsih, E.M; Ignacio, J.C.I; Sendon, P.M.D; Sanchez, D.L; Ismail, A.M; and Mackill, D.J. (2013). QTL mapping and confirmation for tolerance of aerobic conditions during germination derived from the rice landrace Ma-Zhan Red. *Theor. Appl. Gen.,* 126: 1357 1366.doi:10.1007/ s00122-013-2057-1.

Setter, TL; Ella, ES; Valdez, AP. 1994. Relationship between coleoptiles elongation and alcoholic fermentation in rice exposed to anoxia. II. Cultivar differences. *Annals of Botany*, 74: 273–279.

Takahashi, H; Saika, H; Matsumura, H; Nagamura, Y; Tsutsumi, N and Nishizawa, N.K. (2011). Cell division and cell elongation in the coleoptiles of rice alcohol dehydrogenase1 deficient mutant are reduced under complete sub-mergence. *Ann. Bot.,* 108: 253–261. doi:10.1093/aob/mcr137.

Taylor, DL. 1942. Influence of oxygen tension on respiration, fermentation and growth of wheat and rice. *American Journal of Botany,* 29: 721–738.

Turner, FT; Chen, CC; McCauley, GN. 1981. Morphological development of rice seedlings in water at controlled oxygen levels. *Agronomy Journal*, 73: 566–570.

Waters, I; Morrell, S; Greenway, H; Colmer, TD. 1991. Effects of anoxia on wheat seedlings. II. Influence of O_2 supplies prior to anoxia on tolerance to anoxia, alcoholic fermentation, and sugar levels. *Journal of Experimental Botany,* 42: 1437–1447.

Whitmore, FW. 1978. Lignin-protein complex catalyzed by peroxidase. *Plant Science Letters,* 13: 241–245.

Yamauchi, M; Aguilar, AM; Vaughan, DA; Seshu, DV. 1993. Rice (*Oryza sativa* L.) germplasm suitable for direct sowing under flooded soil surface. *Euphytica,* 67: 177–184.

Zheng X; van Huystee RB. 1992. Peroxidase-regulated elongation of segments from peanut hypocotyls. *Plant Science,* 81: 47–56.

■■■■

9

CADMIUM ENRICHMENT IN SOIL : PLANT PHYSIOLOGICAL MANIFESTATION

Tapas Kumar Pandit[1] and Dhiman Mukherjee[2]

[1]Regional Research Station (OAZ), U. B.-K. V, Majhian, Patiram, Balurghat, Dakshin Dinajpur-733133.West Bengal (India). Email: tapas_pandit@rediffmail.com.
2Dr. Dhiman Mukherjee, AICRP on Wheat and Barley Improvement, Directorate of Research, B.C.K.V, Kalyani -741235, West Bengal (India). Email: dhiman_mukherjee@yahoo.co.in

INTRODUCTION

Of all non-essential heavy metals, cadmium (Cd) is perhaps the metal that has attracted the most attention in soil science and plant nutrition due to its potential toxicity to humans, and also its relative mobility in the soil–plant system. Cadmium (Cd) loading in soil and the environment has been accelerated worldwide due to enhanced industrialization and intensified agricultural production, particularly in the developing countries. Soil Cd pollution, resulting from both anthropogenic and geogenic sources, has posed an increasing challenge to soil quality and food security as well as to human health. Compared with other heavy metals such as copper (Cu) and lead (Pb), Cd demonstrates greater mobility along food chain and in the environment. Cd is harmful to human health and potentially toxic to biota at lower concentrations. Therefore, it is imperative to develop management strategies for control of pollution sources and remediation of contaminated soils. This chapter emphasises Cd movement in plant system and its toxicity effect on plants and soil ecosystems and its remedial measures. Extensive studies have been conducted in recent years to understand biogeochemical processes of Cd in soils, its cycle in agroecosystems, impacts on soil quality and food security, and the remediation of Cd-contaminated soils. The uptake of cadmium may take place in competition with essential elements. In the field, crops are routinely subjected to a combination of different abiotic stresses. Progress has been made in the understanding of mechanisms that govern Cd accumulation and detoxification in accumulating plants.

It is estimated that around 30,000 tones cadmium (Cd) are released into the environment annually, of which 13,000 tones resulted from human activity (Zawoznik *et al.,* 2012). Cd is a relatively rare element and is not found in a pure state in nature. In the air, Cd is rapidly oxidised into cadmium oxide. It easily reacts with carbon dioxide, water vapour, sulphur dioxide, sulphur trioxide, or hydrogen chloride and produces cadmium carbonate, hydroxide, sulphide, or chloride. Cd can undergo weak bonding to carbon and other more electronegative atoms. Vast areas of agricultural soils are contaminated by Cd. The Cd concentration in soil solution of uncontaminated soils is in the range of 0.004-0.32 μM, whereas moderately polluted soils contained 0.32-1.00 μMCd. *Thalaspi caerulescens,* a plant species with Cd hyperaccumulation, is capable of surviving in soils containing more than 35μM Cd in the soil solution (Brown *et al.,* 1994). Cd is one of the most readily uptake and rapidly transported heavy metals in higher plants. The level of Cd uptake in higher plants is determined by the Cd concentration of the soil. It was assumed that for Cd being an essential element, there would be no specific uptake mechanisms. Instead, Cd would enter plant cells via uptake systems for essential elements. In other words, the uptake of Cd ions takes place in competition with that of essential elements such as K, Mg, Ca and Fe, across the same transmembrane transporter (Pál *et al.,* 2006). Such competition results in decrease in the uptake of essential elements. Thus, the possibility that toxic effect of Cd may be attributable to the decline of essential elements cannot be excluded. Rice (*Oryza sativa* L.) is the world's most important food crop and a primary source of food for more than half of the population (Mukherjee, 2002). Rice is also the model system for monocot species regarding Cd accumulation. Nutrient deprivation and Cd stress are becoming important abiotic stresses limiting the productivity of rice. The most general symptoms of Cd stress are chlorosis and growth inhibition. It has long been recognized that analysis of chlorotic young leaves showed low leaf Fe when soils or culture solutions contained high Cd, suggesting that toxic effect is attributable to the decrease in Fe concentration. Zinc andCd have many physical and chemical similarities as they both belong to group II of periodic table. They usually compete with each other for various ligands. Thus, interaction between those two elements in rice is becoming important and lot of scope for future study.

CADMIUM ENRICHMENT IN SOIL

Cadmium (Cd) in soils is derived from both natural and anthropogenic sources. Natural sources include underlying bedrock or transported parent material such as glacial till and alluvium. The major natural sources for mobilisations of Cd from the earth's crust are volcanoes and weathering of rocks. Within the biosphere the Cd is translocated by different processes. Naturally a very large amount of Cd is released into the environment, about 25,000 t a year. The main anthropogenic input of Cd to soils occurs by industrial waste from processes such as electroplating, manufacturing of plastics, mining, paint pigments, alloy preparation, and batteries that contain Cd, composts, or fertilisers. Even domestic sewage sludge, which originated from the strictest control sources, contains Cd and adds it to pollution. From the sewage systems, Cd enters rivers and streams and therefore contaminates other places or accumulates in the sludge. The addition of Cd in metal-rich sewage sludge may also result in contamination of groundwater (Moradi *et al.,* 2005). Environmental pollution by metals became extensive as mining and industrial activities increased in the last 19th and 20th century. These pollutants ultimately derived from a growing number of diverse anthropogenic sources (industrial effluents and waste, urban runoff, domestic garbage, dumps and mining operation), have progressively affected more and more different ecosystem (Macfarlane and Burchett, 2001). Metal toxicity and tolerance in plants is a subject that has been broadly reviewed on several occasions over the last 30 years (Brown and Jones, 1975; Foy *et al.,* 1978; Ernst *et al.,* 1992; Das *et al.,* 1973; Sanita di Toppi and Gabrielli, 1999; Hall, 2003; Clemens *et al.,* 2002).

As a cheap alternative means of treatment and disposal, a major part of industrial wastes have found their way into farm lands in the peri-urban areas during recent years (Chhonkar *et al.* 2004, Mukherjee, 2013). As an alternative for chemical fertilizers, a large amount of sewage and sludge also go the farm lands, mainly for vegetable crop production, providing plant nutrients (Bhaise *et al.,* 2007). Organic matter to soil (Smith *et al.* 1992) and irrigation water (Patel *et al.,* 2004; Sur *et al.,* 2006) reported that Pb and Zn contend were much greater in Khardah and Titagarh areas of West Bengal, where soils were amended with greater amount of sewage water. Vegetables cultivation in the peri-urban areas plays a major role in supporting the livelihood (primarily through food provision), income generation and employment of the family members of small and marginal farmers (Marshall *et al.,* 2001). Most vegetables are prone to heavy metal pollution and the content of heavy metals in sewage-irrigated vegetables could be 2 to 40 times higher than the non-sewage irrigated vegetables due to higher availability of Cd and other heavy metals like Pb, Cr and Ni (Alam *et al.,* 2003). Crops raised on toxic-metal contaminated soils accumulate heavy metals in quantities excessive enough to cause clinical problems and health hazards both to animal and human beings (Chow, 1970. Pandit *et al.* (2006) reported that absorption of heavy metal by leafy vegetables (spinach, cabbage, cauliflower) was higher followed by root crops (radish). Cadmium belongs to the group12 elements with a d^{10} s^2 electronic arrangement. The average Cd content of the lithosphere is estimated at 0.2 mg / kg and the range for soil is given as 0.01-0.70 mg / kg (Lindsay, 1979). Its normal concentration in soil ranges from 0-1 mg / kg, while 1-3 mg / kg indicates slight contamination (Rodriguez-Flores and Rodriguer-Castellon, 1982). The Cd polluted soil may comprise 3-10 mg / kg of soil Cd (Rodriguez-Flores and Rodriguer-Castellon, 1982) to a lethal level of < 100 mg / kg in sewage sludge treated soils as shown in the study by NEERI (1999-2002). Cd over-accumulated in plant (Wuana and Okieimen, 2011) with an enrichment ratio (ER) ranging from 1-10. FAO/WHO recommended maximum tolerable intake of Cd 400-500 μg/week or equal to 70 μg/ day. Soil contaminated with the heavy metals above the permissible limit lead to declines in agricultural yields (Nellessen and Fletcher, 1993, Salt and Rauser, 1995; Akinola and Eki yoyo, 2006). Soil can act as a physical filter by its sieving action, a chemical filter absorption and precipitation of chemical substances and biological filter by decomposing organic materials. The food production in the Cd contaminated soil in an eco-friendly manner is a serious challenge.

SOIL FACTORS AND CADMIUM UPTAKE IN PLANTS

Among various soil parameters known to affect the availability of Cd, soil pH was considered the most important. Many investigations showed that there was a linear trend between soil pH and Cd uptake: the decreasing of soil pH leads to increasing concentration of Cd in plants, provided that other soil properties remain unchanged (Kirkham, 2006). Soil pH affects the availability of Cd present in soil solution but increasing of soil pH does not always reduce Cd uptake by plants. Soil pH is also one of the important factors regulating Cd extractability in soils. The increase in soil pH increases the adsorption of Cd by soils and thus reduces its extractability. Under field conditions, the uptake of Cd by plants may be affected by many variable soil and climatic parameters. Adams et al. (2004) investigated the relationship between soil properties and the concentration of Cd in wheat and barley grain by analysing 162 wheat and 215 barley grain samples collected from paired soil and crop surveys in Britain, and wheat and barley samples from 2 long-term sewage sludge experiments. The results showed that soil total Cd and pH were the significant factors influencing the Cd concentration in grain. Li et al. (2005) conducted field experiments with rice plants grown on the acidic red soil in China, and showed that, at soil pH of 4.95, Cd content in grain was 0.36 mg / kg, while, at pH of 6.54, Cd content in grain was 0.43 mg / kg. According to Chaudri et al. (2007), among the main factors responsible for Cd accumulation in wheat grain are Cd, pH, and organic carbon in soil. In barley grains, Singh and Myhr (1998) did not observe any significant correlation between the extractable Cd in soil, soil pH, and Cd accumulation. It is not easy to extrapolate results from greenhouse studies to field conditions. In greenhouse experiments application of NPK-fertilisers containing Cd increase its concentration in crops (He and Singh, 1994), but no such increase in Cd concentration in crops grown under field conditions was found as a result of long-term application of phosphate fertilisers. Many greenhouse and pot experiments also have yielded results that showed that the uptake of Cd in plants was affected by soil pH. The extent of Cd accumulation also depends on plant genotype (Li *et al.,* 2005). The uptake of ions takes place in competition with that of elements such as Zn, P, Cl^-, Ca and Cu (Kudo *et al.,* 2011). Soil, environmental, and management factors impact on the amount of Cd accumulated in plants. Much of the Cd taken up by plants is retained in the roots, but a portion is translocated to the aerial portions of the plant and into the seed. The amount of Cd accumulated and translocated in plants varies with species and with cultivars within species. Shen et al. (2006) reported that mycorrhizal inoculation increased plant growth with enhancement of P nutrition, and thus may increase plant tolerance to Zn and Cd by a dilution effect. Cadmium induced oxidative signals at root membranes get intensified at toxic Cd level to elicit PCD. Leakage of energy pool as reactive oxygen species (ROS) alters the growth metabolism toward defence metabolism constraining the primary metabolism, particularly the enzymes of Krebs cycle and photosynthesis (Fig. 1).

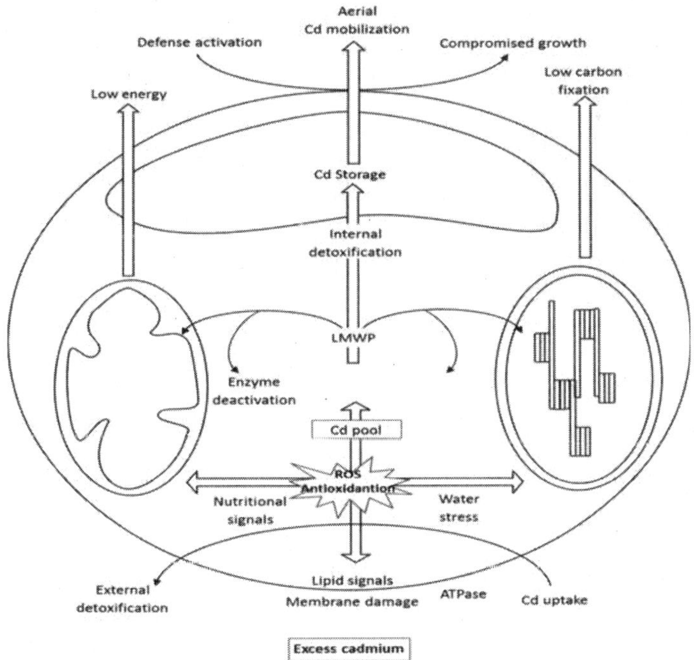

Fig. 1: Cadmium uptake and processing in plant (Source: Mohd*et al.,* 2013).

TOXICITY OF CADMIUM

Cadmium enters the soil through various anthropogenic sources including application of phosphate fertilizers, waste water and anthropogenic emissions from power stations, metal industries, urban traffic and cement industries (Alloway and Steinnes, 1999; Sanita di Toppi and Gabrielli, 1999; Wu *et al.,* 2004; Yang *et al.,* 2004). In plants, Cd toxicity can disturb photosynthesis rate, chlorophyll content, stomatal conductance, transpiration rate, relative leaf water content, and other physiological processes, which ultimately reduces plant growth and development (Seregin, 2001). Cd taken up by plants transferred to particular organs (Kabata-Pendias and Pendias, 2001) induces chlorosis, necrosis, vein reddening, and root and shoot growth retardation, besides reduced nutrient uptake (Sanita di Toppi and Gabbrielli, 1999; Mohamed *et al.,* 2012). Heavy metals, including Cd, can induce nutrient deficiency and even decrease concentrations of several macronutrients in plants (Siedleska, 1995; Mukherjee, 2004). Thus, it seems possible to mitigate some of the metal-induced negative effects by optimization of plant mineral nutrition. Some data supporting this point of view have been recently accumulated. It is known that plant nutrients affect the activity and bioavailability of Cd in the soil-plant environment. Several nutrients have many direct as well as indirect effects on Cd availability and toxicity. Indirect effects include dilution of Cd concentration by increased plant biomass and alleviation of physiological stress (Sarwar *et al.* 2010). Direct impacts result in decreased Cd solubility in soil by promoted precipitation in and adsorption, competition between Cd and plant nutrients for the same membrane transporters as well as Cd sequestration in vegetative parts avoiding its accumulation in grain/edible parts. For example, phosphate promotes the precipitation of Cd^{2+} (Hong *et al.,* 2010), while ferrous iron (Fe^{2+}) competes with Cd^{2+} for the same membrane transporters in plant cells (Vert *et al.,* 2002; Kovacs *et al.* 2010). Efficiency of nitrate supply recently was shown for phytoextraction of Cd and Zn by the hyperaccumulator *Thlaspi caerulescens,* which showed special relevance for phytoremediation (Xie *et al.* 2009). Cd also reduces the absorption of nitrate and its transport from roots to shoots, by inhibiting the nitrate reductase activity in the shoots (Hernandez *et al.* 1996). Nitrogen (N) is an essential macronutrient deficient in most soils (Tisdale, 1975). It is an important component of many structural, genetic and metabolic compounds in plants (Hassan *et al.* 2005; Mukherjee, 2008).This becomes significantly influenced with high concentration of Cd in soil. N as a component of certain molecules can perform a signal function in the defense reactions of plants against various stresses (Durner, 1999). Previously, the effect of N deficiency, or application of either nitrate or ammonium on plants and subsequent defense responses have mostly been studied (Polesskaya *et al.* 2004; Zhao *et al.,* 2005; Chaffei *et al.* 2008). Chlorosis, leaf rolls and stunting are the main and easily visible symptoms of cadmium toxicity in plants. Chlorosis may appear due to Fe deficiency (Haghiri, 1973), P deficiency or reduced Mn transport (Godbold and Hutterman, 1985). In general, Cd has been shown to interfere with the uptake, transport and use of several elements (Ca, Mg, P and K) and water by plants (Das *et al.,* 1997).

EFFECT OF CADMIUM ON ANTIOXIDANT SYSTEM

Plants possess a number of antioxidant systems that protect them from oxidative damage (Smeets *et al.,* 2005; Pal *et al.,* 2006). Superoxidase (SOD) is the first enzyme in the detoxifying process that converts O_2^- radicals to H_2O_2 at a very rapid rate (Polle and Rennenberg, *et al.,* 1992, Somashekaraiah *et al.,* 1992) by either inducing oxygen free radical production (Balaknina *et al.,* 2005; Demirevska-Kepava *et al.,* 2006) or by decreasing concentrations of enzymatic and non-enzymatic antioxidants (Somashekaraiah *et al.,* 1992) by either inducing oxygen free radical production (Balaknina *et al.,* 2005; Demirevska-Kepava *et al.,* 2006) or by decreasing concentrations of enzymatic and non-enzymatic antioxidants (Somashekaraiah *et al.,* 1992; Stohs and Bagchi, 1995; Shaw, 1995; Gallego *et al.,* 1996; Sandalio *et al.,* 2001; Balestrasse *et al.,* 2001; Fornazier *et al.,* 2002; Cho and Seo, 2004; Mohan and Hosetti, 2006). These defense systems are composed of metabolites such as ascorbate, glutathione, tocopherol, etc., and enzymatic scavengers of activated oxygen such as peroxidases, catalases and superoxide dismutases (Noctor and Fayer, 1998; Asada, 1999; Sandalio *et al.,* 2001; Khan *et al.,* 2002; Borctal *et al.,* 2003; Panda and Khan, 2003; Chaoui *et al.,* 2004; Demiral and Turkan, 2005; Mandhania *et al.,* 2006).Peroxidase induction is a general response of higher plants after uptake of toxic quantities of metals (Van Assche and Clijsters, 1990). Cd ions can inhibit or sometimes stimulate the activity of several antioxidative enzymes. In *Halianthus annuus* leaves, Cd enhanced lipid peroxidation, increased lipoxygenase activity and decreased the activity of superoxide dismutase (Sandalio *et al.,* 2001; Khan *et al.,* 2002; Panda and Khan, 2003), glutathione reductase, catalase, ascorbate peroxidase, glutathione reductase and dehydroascorbate reductase (Gallego *et al.,* 1996). Cadmium induced the activity of peroxidase (POX) in soybean (Fuhrer, 1982), bean leaves (Lee *et al.,* 1996) and in roots and leaves of *Oryza sativa* (Reddy and Prasad, 1993), *Brassica juncea* (Singh and Tiwari, 2003; Hayat *et al.,* 2007), *Cicer arietinum* (Hasan *et al.,* 2007b), *Bacopa monniera* (Mishra *et al.,* 2006) and in the leaves of *Calamus tenuis* (Khan and Patra, 2007). In *Phaseolus aureus*, Cd ions decreased catalase activity and increased the activities of gluaicol peroxidase and ascorbate peroxidase (Shaw, 1995; John *et al.,* 2007). Membrane damage due to lipid peroxidation caused by metals is mediated by activated oxygen radicals (hydrogen peroxide, hydroxyl and superoxide radicals) and could be quenched by the induction of specific enzymes like peroxidase, superoxide dismutase and catalase (De Vos and Schat, 1981). *Phaseolus vulgaris* roots exposed to 5 mM Cd had higher activities of guaiacol and ascorbate peroxidase and elevated levels of lipid peroxidation (Chaomi *et al.,* 1997). Cadmium treatment also increased lipid peroxidation (Lazono-Rodriguez *et al.,* 1997; Sandalio *et al.,* 2001; Astolfi *et al.,* 2004; Chaoui *et al.,* 2004; Srivastava *et al.,* 2004), whereas no effect on lipid peroxidation was noticed in the roots of *Daucus carota* plants exposed to Cd (Sanita di Toppi *et al.,* 1998). In addition to these antioxidant molecules, thiols also possess strong antioxidative properties and are able to counteract oxidative stress imposed by Cd (Pichorner *et al.,* 1993; Shanthala *et al.,* 2006). Germinating pigeon pea seedlings exposed to Cd^{2+} had altered enzyme activity and mobilizations of food reserves (Bishnoi *et al.,* 1993).

MOBILIZATION AND TRANSPORT IN PLANT SYSTEM

The bioavailability of some metals is limited because of low solubility in oxygenated water and strong binding to soil particles. Both the acidification of the rhizosphere and the exudation of carboxylates are considered potential targets for enhancing metal accumulation (Clemens *et al.,* 2002). The degree to which higher plants are able to take up Cd depends on its concentration in the soil and its bioavailability, modulated by the presence of organic matter, pH, redox potential, temperature and concentrations of other elements. With the exception of Fe, which is solubilized by either reduction to Fe (II) or extrusion of Fe (III)-chelating phytosiderophores (Hirsch, 1998), little is known about active mobilization of trace elements by plant roots. In particular, the uptake of Cd ions seems to be in competition for the same transmembrane carrier with nutrients, such as K, Ca, Mg, Fe, Mn, Cu, Zn, Ni (Clarkson and Luttge, 1989; Rivetta *et al.,* 1997). The cell membrane plays a role in metal homeostasis, preventing or reducing entry into the cell. However, examples of exclusion or reduce uptake mechanisms in higher plants are limited. Cadmium is one of the most dangerous metals due to its high mobility and the small concentration at which its effects on plants begin to appear (Barcelo and Poschenrieder, 1990). The apoplast continuum of the root epidermis and cortex is readily permeable for solutes. The cell walls of the endodermal cell layer act as a barrier for apoplastic diffusion diffusion into the vascular system. In general, solutes have to be taken up into the root symplasm before they can enter the xylem (Tester and Leigh, 2001). Subsequent to metal uptake into the root symplasm, three processes govern the movement of metals from the root into the xylem: sequestration of metals inside root cells, symplastic transport into the stele and release into the xylem (Clemens *et al.,* 2002). The membrane potential, which is negative on the inside of the plasma membrane and might exceed-200mV in root epidermal cells, provides a strong driving force for the uptake of cations through secondary transporters (Hirsch *et al.,* 1998).

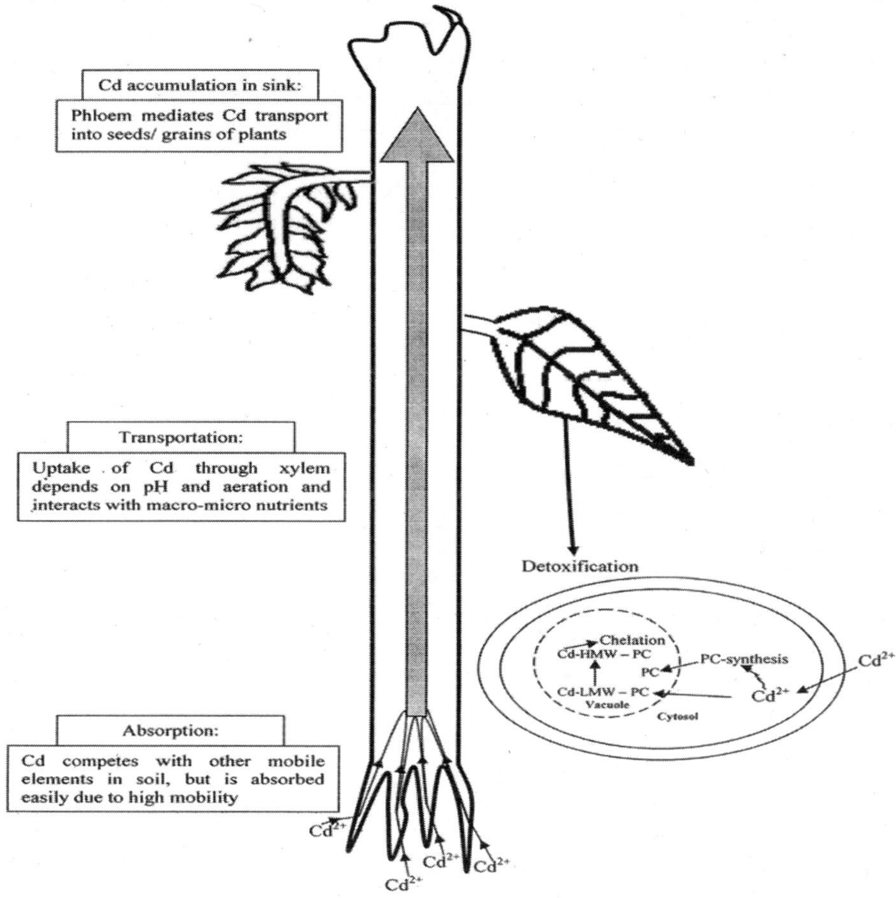

Fig. 2. An overview of Cd absorption, transportation, accumulation and detoxification in plants (Nazar *et al.,* 2012).

Despite the different mobility of metal ions in plants, the metal content is generally greater in roots than in the above ground tissues (Ramos *et al.,* 2002). In most environmental conditions, Cd enters first the roots, and consequently they are likely to experience Cd damage first (Sanita di Toppi and Gabrielli, 1999). Cd easily penetrate the root through the cortical tissue and is translocated to the above- ground tissues (Yang *et al.,* 1998). As soon as Cd enters the roots, it can reach the xylem through an apoplastic and/ or symplastic pathway (Salt *et al,* 1995a), complexed by several ligands, such as organic acids and/or phytochelatins (Senden *et al.,* 1992, 1994; Salt *et al.,* 1995b). Normally, Cd oins are mainly retained in the roots, and only small amounts are transported to the shoots (Cataldo *et al.,* 1983). In general, the content of Cd in plants decreases in the order: roots> stems >leaves >fruits >seeds (Blum, 1997). In one of study, a Cd^{2+}-selective vibrating microelectrode was constructed using a neutral carrier-based Cd ionophore to investigate ion-transport processes along the roots of wheat (*Triticum aestivum* L.) and two species of Thlaspi, one a Zn/Cd hyperaccumulator and the other a related nonaccumulator. In simple Cd (NO3)2 solutions, the electrode exhibited a Nernstian response in solutions with Cd2+ activities as low as 50 nM. Addition of Ca2+ to the calibration solutions did not influence the slope of the calibration curve but reduced the detection limit to a solution activity of 1 μM Cd^{2+} Addition of high concentrations of K^+ and Mg^{2+} to the calibration solution to mimic the ionic composition of the cytoplasm affected neither the slope nor the sensitivity of the electrode, demonstrating the pH insensitive electrode's potential for intracellular investigations. The electrode was assayed for selectivity and was shown to be at least. 1000 times more selective for Cd^{2+} than for any of those potentially interfering ions tested. Flux measurements along the roots of the two Thlaspi species showed no differences in the pattern or the magnitude of Cd^{2+} uptake within the time frame considered. The Cd^{2+} selective microelectrode will permit detailed investigations of heavy-metal ion transport in plant roots, especially in the area of phytoremediation (Pineros *et al.,* 1998). Naik *et al.* (2013) reported that DTPA Cd was significantly negatively correlated with leaf area and total chlorophyll content while positively correlated with root and shoot Cd concentrations of spinach and cabbage. Moral *et al.* (1994) reported that Cd was easily transported to the aerial parts of tomato plants but was not detected in fruits. It has recently been hypothesized that Cd accumulation in developing fruits could occur via phloem-mediated transport (Hart *et al.,* 1998). Hinesly *et al.* (1984) reported that soil pH greatly influences Cd uptake and transportation in corn while street *et al.* (1977) reported that uptake of Cd by corn was lower in acid soils with higher organic matter content. Cadmium concentrations in maize and ryegrass were negatively related to the product of cadmium in soil and pH (Tudoreanu and Phillips, 2004). A positive coefficient between soil pH and cadmium in rye-grass may drive from ionic competition, for example, sodium has been demonstrated to increase plant cadmium (Chiy and Phillips, 1999). With the exception of the recently described Cd-carbonic anhydrase of marine diatoms (Lane and Morel, 2000), no biological function has been reported to date for the metals Pb and Cd. Thus, it is unlikely that metal transporters with specificities for the respective metal cations exist (Clemens, 2001). However, these cations without specific metal trans-porters are likely to enter cells through cation transporters with broad substrate specificity (Clemens, 2001). Cd and Zn have been found to be co-accumulated in aerial parts of *Arabidopsis halleri* (Bert *et al.,* 2003) plants. This shows that Cd and Zn uptake are genetically correlated, suggesting that the metals are taken up (at least partly) by the same transporters or that their transporters, when different, are controlled by common regulators. *Arabidopsis halleri* L., previously known as *Cardaminopsi shalleri* L. Hayek, is one of the two species known to hyperaccumulate Cd (Brooks,

1998; Kupper *et al.,* 2000), is also a zinc (Zn) hyperaccumulator and usually occurs on Zn, Cd and Pb contaminated sites (Bert *et al.,* 2003). Differences in grain Cd accumulation between two wheat species (*Triticum aestivum* and *Triticum turgidum* var *duram*) may not only result from differences in root Cd influx, but seem to be associated with differences in plant-internal Cd allocation (Hart *et al.,* 1998). Pandit *et al.* (2012) reported that application of lime and organic matter to soil decrease the avability of Cd to the soil and plant throughout the crop growth. Several cation transporters have been identified in recent years with the use of molecular techniques, largely owing to the complementation of *Saccharomyces cerevisiae* mutants (Clemens, 2001). Most of the transporters through to be involved in the uptake of micronutrients are in the ZIP (ZRT, IRT-like protein) and the Nramp (Natural resistance –associated macrophase protein) family (Guerinot, 2000; Williams *et al*; 2000). Recently, several plant transporters have been identified that show affinity for both Zn and Cd. By complementation of a yeast Zn-transport defective mutant with a *T. caerulescens*c DNA library, Lasat *et al.* (2000) cloned the ZNT1 cDNA, which encodes a high affinity Zn transporter. However, ZNT1 can also mediate low affinity Cd transport (Lasat *et al.,* 2000; Pence *et al.,* 2000). Based on the study of two *T. caerulescens* ecotypes, Lombi *et al.* (2000) suggested that Cd may be transported in the low Cd accumulation ecotype via ZNT1 but, conversely, that Cd may be mediated in the high accumulation ecotype via a high affinity Cd transporter. Additional studies in yeast showed that IRT1, an iron transporter belonging to the ZIP family, has a broad substrate range and also transports Zn and possibly Cd (Korshunova *et al.,* 1999; Clemens, 2001). Furthermore, At Nramp3, an Arabidopsis metal transporter involved in iron metal uptake, showed Cd^{2+} transport activity (Thomine *et al.,* 2000).

1. Cell wall and plasma membrane

The binding property of the cell wall and its role in the mechanism of metal tolerance has been controversial (Thurman and Collins, 1983; Verkleij and Schat, 1990). The walls of roots cells are directly exposed to the metals in soil solution. The interaction of the metals with the cell wall has been reported in several articles reviewed by Ernst *et al.* (1992) but since then, only a few more papers appeared covering the topic. Most of the cell wall-associated heavy metals are bound to polygalacturonic acids, to which the affinity of metal ions vary according to the metal (Ernst *et al.,* 1992). The plasma membrane is the first "living" structure that is target for heavy metal toxicity and consequently, could also be involved in tolerance. Such toxicity could result from various mechanisms including the oxidation and cross-linking of protein thiols, inhibition of key membrane proteins such as H^+-ATPase, or changes in the composition and fluidity of membrane lipids (Meharg, 1993). A direct effect of Cd and Cu has been reported on the lipid composition of membranes (Fodor *et al.,*

1995; Hernandez and Cooke, 1997; Quartacci *et al.,* 2001). Moreover, Cd treatment has been shown to reduce ATPase activity of the plasma membrane fraction of wheat and sunflower roots (Fodor *et al.,* 1995).

2. Xylem transport

The chemical composition of xylem sap is very different from phloem sap, xylem sap having a pH ranging from 5.0– 6.0, a more oxidizing redox potential and a much lower concentration of organic compounds, such as sugars, peptides and proteins. As described earlier Cd not only prefers to form bonds with sulphydryl ligand groups, but also binds to N and O ligand groups. Thus, cysteine and other sulphydryl containing compounds (phytochelatins, glutathione etc.) and various organic acids (citrate) and other amino acids in xylem sap could be important in transporting Cd from roots to shoots. Although there are no definitive studies on the forms of Cd in xylem sap, Senden *et al.* (1994) reported that treating xylem sap with citric acid increased Cd transport through xylem vessels of excised tomato stem/leaf systems. Furthermore, when citric acid was supplied to the roots of tomato plants, Cd (II) uptake increased two- fold and Cd transport from roots to shoots increased 6- to 8- fold. Thus, citrate can stimulate Cd uptake in the root and transports it in xylem sap. Mori *et al.* (1991) reported the occurrence of deoxymugineic and epihydroxymugineic acid phytometallophores in xylem sap collected from rice seedlings and Cataldo *et al.* (1983) reported finding Cd (II) associated primarily with components of the amino acid/peptide fraction of xylem sap collected from xylem exudates of soybean, while polyvalent ion such as Fe (III) were found associated with organic acid complexes.

3. Phloem transport

The abundance of organic ligands (organic acids, amino acids, sugars, peptides and proteins) and the alkaline pH (pH 7.0-8.0) of phloem sap ensures that virtually all Cd^{2+} carried in the phloem is present in complexed conditions that favour the stability of sulphydryl-containing ligands, which are likely to be carriers of Cd. However, there is no direct evidence for any specific Cd complexes in phloem sap, but they could include phytometallophores, such as nicotinamine, metallothioneins, the phytochelatins as well as glutathione, cysteine and other sulphydryl containing molecules. Deoxymugineic acid a phytometallophone was identified in phloem sap of rice plants (Mori *et al.,* 1991). It is presumed that phytochelatins and phytometallophores play a role in Cd movement in phloem sap and in loading Cd into seeds and grains. Moreover, nicotinamine could also function as an iron transporter in the phloem of all higher plants as nicotianamine was essential for iron mobilization in plants from phloem sources (like mature leaves) to phloem sinks (reproductive organs, newly forming roots leaves and growing points)

(Scholz, 1989). Zinc- binding substances similar to the phytochelatins have been reported in phloem sap of citrus (Taylor *et al.,* 1988).

EFFECT OF CADMIUM ON VARIOUS PARAMETERS

Plant Growth

Cadmium is not an essential nutrient and at high concentration inhibits plant growth (Anita *et al.,* 1990; Aery and Rana, 2003). It has also been reported that even at relatively low concentrations it alters plant metabolism (Van Assche and Clijsters, 1990). The presence of cadmium in the soil decreases the growth of soybean (Dewdy and Ham, 1997; Cataldo *et al.,* 1983) and chickpea plants (Hasan *et al.,* 2007 b). High concentrations of Cd decreased cell growth as well as whole plant growth (Prasad, 1995). Cd toxicity causes inhibition and abnormalities of general growth in many plant species. After long-term exposure to Cd, roots are mucilaginous, browning, and decomposing; reduction of shoots and root elongation, rolling of leaves, and chlorosis can occur. Under high altitude range, water flow from top hill has huge source of Cd in soil and water debris, this leads to loss of natural vegetation on hill slope (Mukherjee, 2012). Cd was found to inhibit lateral root formation while the main root became brown, rigid, and twisted (Rascio and Navari-Izzo, 2011). The main reason indicated is disordered division and abnormal enlargement of epiderma and cortical cell layers in the apical region. The changes in the leaf included alterations in chloroplast ultrastructure, low contents of chlorophylls, which caused chlorosis, and restricted activity of photosynthesis (Miyadate *et al.,* 2011). Rascio *et al.* (2008) reported that treatment of rice seedlings with Cd led to inhibition of root growth and alterations in their morphogenesis. In pea plants, the Cd stress also caused disorders in root elongation and the mitotic process and caused chromosomal aberrations of root tips. The observation showed that in these abnormalities as lagands, bridges, stickiness, precocious separation, and fragments were most common. At high Cd concentration (250 μM), the disorder of mitosis of roots in pea happens rapidly, even after 24 h of treatment. An unusual number of nucleus populations in the differentiated roots were found (Fusconi *et al.,* 2007). The effect of growing pea (*Pisum sativum* L.) plants with $CdCl_2$ (0–50 μM) on different plant physiological parameters and antioxidative enzymes of leaves was studied in order to know the possible involvement of this metal in the generation of oxidative stress. In roots and leaves of pea plants Cd produced a significant inhibition of growth as well as a reduction in the transpiration and photosynthesis rate, chlorophyll content of leaves, and an alteration in the nutrient status in both roots and leaves. The ultrastructural analysis of leaves from plants grown with 50 μM $CdCl_2$, showed cell disturbances characterized by an increase of mesophyll cell size, and a reduction of intercellular spaces, as well as severe disturbances in chloroplast structure. Alterations in the activated oxygen metabolism of pea plants were also detected, as evidenced by an increase in lipid peroxidation and carbonyl-groups content, as well as a decrease in catalase, SOD and, to a lesser extent, guaiacol peroxidase activities. Glutathione reductase activity did not show significant changes as a result of Cd treatment. A strong reduction of chloroplastic and cytosolic Cu, Zn-SODs by Cd was found, and to a lesser extent of Fe-SOD, while Mn-SOD was only affected by the highest Cd concentrations. Catalase isoenzymes responded differentially, the most acidic isoforms being the most sensitive to Cd treatment. Results obtained suggest that growth of pea plants with $CdCl_2$ can induce a concentration-dependent oxidative stress situation in leaves, characterized by an accumulation of lipid peroxides and oxidized proteins as a result of the inhibition of the antioxidant systems. These results, together with the ultrastructural data, point to a possible induction of leaf senescence by cadmium (Sandalio *et al.,* 2001).The interactions between sulphur nutrition and Cd exposure were investigated in maize (*Zea mays* L.) plants by Astolfi *et al.* (2004) and his group. Plants were grown for 12 days in nutrient solution with or without sulphate. Half of the plants of each treatment were then supplied with 100 microM Cd. Leaves were collected 0, 1, 2, 3, 4 and 5 days from the beginning of Cd application and used for chemical analysis and enzyme assays. Cd exposure produced symptoms of toxicity (leaf chlorosis, growth reduction) and induced a noticeable accumulation of non-protein SH compounds. As phytochelatins are glutamate- and cysteine-rich peptides, the effect of cadmium on some enzyme activities involved in N and S metabolism of maize leaves was studied in relation to the plant sulphur supply. In vivo Cd application to S-sufficient plants resulted in a drop of all measured enzyme activities. On the other hand, S-deficient plants showed a decrease in nitrate reductase (NR; EC 1.6.6.1) and glutamine synthetase (GS; EC 6.3.1.2) activity, and an increase in NAD-dependent glutamate dehydrogenase (GDH; EC 1.4.1.2) and phosphoenolpyruvate carboxylase (PEPc; EC 4.1.1.31) activity as a result of the Cd treatment. Furthermore, in the same plants ATP sulphurylase (ATPs; EC 2.7.7.4) and O-acetylserine sulphydrylase (OASs; EC 4.2.99.8) showed a particular pattern as both enzymes exhibited a transient maximum value of activity after 4 days from the beginning of Cd exposure. Results provide evidence that the increase of ATPs, OASs, GDH and PEPc activities, observed exclusively in S-deficient Cd-treated plants, may be part of the defence mechanism based on the production of phytochelatins.

Fresh and Dry Biomass Production

Cadmium (Cd) is a nonessential element, which strongly inhibits plant growth, fresh and dry biomass production, and causes plant death even at very low concentrations. The effect of cadmium treatment on the growth, Cd accumulation and macronutrient content of durum wheat (Dena CV.) growing in soil supplemented with 0, 15 and 30 mg/kg Cd were studied at

Iran under the supervision of Mojtaba and Barya (2013). Results showed that Cd treatment had significant effect. Cadmium treatment declined the shoot length; shoot dry weight and N, P, K, Ca content in shoot of durum wheat. The maximum decrease was observed at 30 mg/ kg Cd treatment. On the other hand, Cd accumulation in root and shoot of plant increased with increasing Cd concentration in soil. The results suggest that Cd toxicity in durum wheat suppresses the growth development by inhibiting cell division and some enzymes related to growth. Besides, cadmium may have antagonistic interactions especially at high concentrations with macronutrient content in plant which this led to decrease in macronutrient content in durum wheat. Treatment of cowpea (*Vigna unguiculata* L.) with different concentrations of cadmium decreased the germination percentage, growth parameters and biochemical contents. The experiment was conducted at Tamil Nadu, during the period of January to March 2011. In the pot culture experiment, cowpea plants were analysed on three different sampling (viz., 15th, 30th and 45th) days, in soil amended with various levels of cadmium (viz., 10, 30, and 50 mg /kg). The inner surfaces of pots were lined with a polythene sheet. Each pot containing 3 kg of air dried soil. Six seeds were sown in each pot. All pots were watered to field capacity daily. Plants were thinned to a maximum of three per pots, after a week of germination. Cadmium at all levels (10,30 and 50 mg/kg) tested, decreased the growth parameters such as, root and shoot length, fresh and dry weight of root and shoot, total leaf area and biochemical constituents such as, chlorophyll-a, chlorophyll-b, total chlorophyll and carotenoid contents of cowpea plants compared to untreated plants (Vijayaragavan *et al.,* 2011). An experiment was conducted during 2006–08 to find out the effect of cadmium on plant growth and its accumulation in different amaranthus (*Amaranthus tricolor* L.) varieties. Three amaranthus cultivars, namely Local, 'Arka Arunima' and 'Arka Suguna' were evaluated for their Cd tolerance and accumulating capacity in greenhouse on sandy loam soil. Variable levels of Cd were created through the application of $CdCl_2$ @ 0 (control), 50, 100 and 200 mg Cd/kg soil. Fresh and dry biomass yield of shoots and roots in all the cultivars decreased with increased Cd levels. The relative reduction in fresh biomass yield of cultivar followed the order: Local> 'Arka Suguna'> 'Arka Arunima'. Significant reduction in the biomass of Local was recorded at applied Cd level of 50 mg/kg where as percentage reduction in biomass was lower in 'Arka Arunima'. The Cd content of foliage increased with increase in soil Cd levels in all the cultivars. However, the concentration of Cd in 'Arka Arunima' was highest at all levels of applied Cd. This cultivar accumulated 43.5% more Cd than that in Local and 25.8% more Cd than that in 'Arka Suguna' at soil Cd level of 50.0 mg/kg. At highest soil applied Cd level of 200 mg/kg the Cd concentration in 'Arka Arunima' was 16.3 and 13.7% more than that in Local and 'Arka Suguna' respectively. Local cultivar accumulated relatively lower levels of Cd and this cultivar can be preferred for cultivation in high Cd soils.

Nodule System

The presence of heavy metals in the soil decreased the yield of symbiotic nitrogen- fixing organisms and the number of nodules per plant (Vigue *et al.,* 1981). The presence of Cd decreased nodulation and nitrogenase activity in *Phaseolus vulgaris* (Dewdy and Ham, 1997; Vigue *et al.,* 1981), Trifoliumrepens (McGranth *et al.,* 1988), soybean (Vespa *et al.,* 1978), Alnusrubra (Wickliff *et al.,* 1980) and in *Pisum sativum* (Dhingra and Priefer, 2006). The interaction of *Rhizobium* in the nodules of chickpea was found to be very sensitive to heavy metals resulting in a decrease in dry mass of chickpea and greengram (Woolhouse, 1983; Rana and Ahmad, 2002). An increase in Cd concentration decreased the fresh mass in mungbean (Shen *et al.,* 1990). Moreover, a marked decrease in root and shoot mass of *Vignaam bacensis* was observed when treated with low concentration of Cd (Al-Yemens, 2001). Naik *et al.* (2009) reported that the lead and nickel concentrations in soil and shoot dry matter of sesame gradually decrease with the progress of crop growth with Cd treatment. Nitrogen assimilation in pea plants was severely affected on exposure to Cd (Hernandez *et al.,* 1995; Dhingra and Priefer, 2006). A positive correlation was observed between leghemoglobin content and nitrogenase activity (Darkaro, 1995; Comba *et al.,* 1998) and both these parameters exhibited a parallel decrease in the presence of Cd (Farnandez *et al.,* 1996). The oxidation stress generated by Cd^{2+} accelerated senescence of nodules in soybean plants (Balestrasse *et al.,* 2001). Competitive nodulation abilities of *Bradyrhizobium japonicum* strains USDA110ET and A1017ET were significantly depressed by growing the strains in yeast extract-mannitol broth supplemented with 2 μM $CdCl_2$. Soybean seedlings were co-inoculated with each test strain and its competitor strain 138NR, and the bacteria in nodules formed on 21 days old plant roots were identified by the antibiotic resistant markers. Cell surface hydrophobicity of each test strain increased by growing the strains with 2 μM $CdCl_2$, though definite increase in bacterial attachment to plant root surface was hindered by excessive secretion of exopolysaccharides of the strains. Polyacrylamide gel electrophoresis revealed that growing the test strains in the medium with 2 μM $CdCl_2$ induced the production of lipopolysaccharides of small molecular sizes. The results in this study suggest that cadmium of low concentration causes the weakening of competitive nodulation ability of rhizobia through inhibition of the lipopolysaccharide synthesis (Ozawa and Kenji, 2015). Heavy metals have deleterious the effects on nodulation and N_2 fixation of *Rhizobium*-Legume symbiosis, due to their inhibitory effects on the growth and activity of both symbionts. This research has been undertaken to evaluate the effect of Cd tolerance of native rhizobial strains on diminution of the Cd detrimental effects on *Sinorhizobium meliloti* alfalfa symbiosis by

Sepehri *et al.* (2006). For this purpose, a greenhouse experiment was conducted based on Randomized Complete Block Design. The treatments in this experiment included: plants inoculation with 6 bacterial strains (sensitive, partially tolerant and tolerant to Cd), 5 levels of Cadmium (0, 2, 5, 10, 20 mg/kg soil) and non-inoculated control. In different levels of Cd, the effects of bacterial inoculation on root nodule number and total amount of nitrogen in plant shoot were compared. The results indicated that soil pollution by Cd even at 2 mg/kg had significant effect on symbiotic properties of rhizobial strains, and according to Cd tolerance of various strains, the mentioned effect was different. Decreasing effect of Cd concentration on root nodules and nitrogen concentration in plants that were inoculated with sensitive strains in comparison with plants inoculated with tolerant strains was 68.31% and 40.8%, respectively. In this research, R95m was introduced as the best strain because of its ability for nodulation and nitrogen fixation. The effects of cadmium stress on nodulation, N2 fixation capabilities of the root nodule, the change in ultrastructure of the root nodule, soybean growth, and the distribution of cadmium in plants were studied by Chen *et al.* (2013). The results obtained show that the nodulation of soybean roots was greatly inhibited by the addition of Cd, especially at the addition level of 10 and 20 mg / kg soil. The inhibition of plant growth, especially the root growth, increased as the cadmium concentration increased, with deleterious effects observed for the roots. The weight ratio of soybean root/leaf decreased as the Cd concentration increased, which might explain the reason for nodulation decreases. The results also indicate that N_2 fixation of root nodule was stimulated to some extent at the low levels of Cd addition, but decreased sharply with further increase of the Cd concentration. High Cd levels were also associated with changes in the ultrastructure of root nodule, in which the effective N_2 fixing area was reduced and the N_2 fixing cells in the area also reduced. In addition, the results also reveal that the content of Cd in different parts of the plants was as follows: roots>>stems>seeds, indicating that the accumulation of Cd by roots is much larger than that by any other part of the soybean plant, and might cause deleterious effects to root systems. 'The effect of cadmium (Cd) on nitrogen (N_2) a fixing pea plant was studied Hernandez *et al.* (1995). Perlite and vermiculite were compared as substrates in a hydroponic system where the plants were grown. In both experiments, there was no significant effect on nitrogenase activity when referred to fresh weight of nodules, but there was a reduction in the nodulation ratio (number of nodules developed per plant), even at low concentrations of available Cd. In plants grown in vermiculite, the nitrogenase activity expressed in terms of activity per nodule increased as Cd application increased; while in plants cultivated in perlite that activity decreased when the applied Cd increased at the highest levels of Cd in the nutrient solution. Vermiculite had a significant interaction with the Cd in the nutrient solution, and therefore, it assumes there was less Cd available to the plants than when perlite was used as the substrate. The loss of N_2 fixation by the presence of Cd in plants grown in perlite has been related to a change in the distribution of manganese (Mn) in the plant tissues. There was a reduction of Mn content both in the root and nodules. The reduction was greater in the nodules, while the concentration of Mn increased in the shoot.

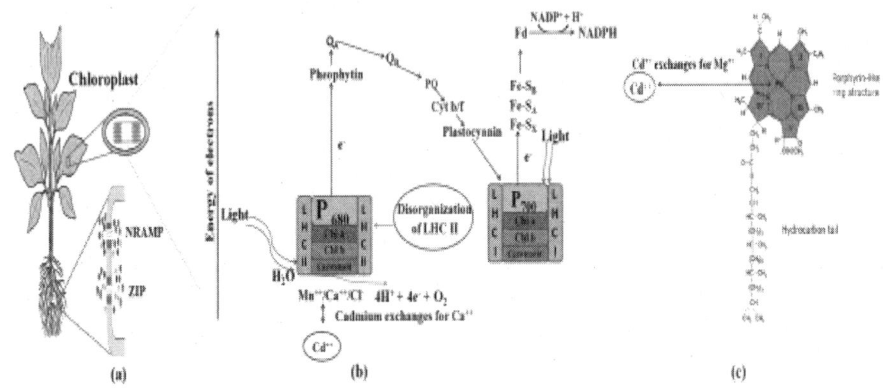

Fig.3 : Illustrates the effects of Cd as a potent inhibitor of photosynthesis (An overview of effects of Cd exposure to plants at different levels in photosynthetic machinery. (a) Cd uptake in cells through plasma membrane transporters. (b) Alteration in organisation of oxygen evolving and light harvesting complexes, Cd also binds with Q_B pocket thus slows down electron flow from Q_A to Q_B. (c) Incorporation of Cd in chlorophyll molecule. Stomatal closure due to entry of Cd into the guard cells in competition to Ca^{+2} (Perfus-Barbeoch *et al.,* 2002) and reduction in stomata count per unit area are also characteristic symptoms of Cd stress resulting in lesser conductance to CO_2 (Pietrini *et al.,* 2010) which consequently leads to overall inhibition of photosynthesis). (Source: Parmer *et al.,* 2013).

Photosynthesis

In many species, such as oilseed rape (*Brassica napus*), sunflower (*Helianthus annuus*), Thlaspi caerulescens, maize, pea, barley, mungbean (*Vigna radiata*), and wheat (Moussa & El-Gamal, 2010), the evidence showed that photosynthesis was inhibited after both long-term and short-term Cd exposure. A large number of studies have demonstrated that the primary sites of action of Cd are photosynthetic pigments, especially the biosynthesis of chlorophyll and carotenoids (Prasad, 1995). According to Baryla *et al.* (2001), the observed chlorosis in oilseed rape was not due to a direct interaction of Cd with the chlorophyll biosynthesis pathway and most probably it was caused by decreasing of chloroplast density the Cd-induced decrease in pigment content was more powerful at the leaf surface (stomatal guard cells) than it was in the mesophyll. In addition, the change of cell size, and the reducing of stomata density in the epidermis in Cd-treated leaves were observed. Thus, Cd might interfere directly with chloroplast replication and cell division in the leaf. This research also revealed that stomatal conductance was strongly reduced by Cd. Cd stress affects photosynthesis in various ways. It inhibits the synthesis of chlorophylls (Chl) (Azevedo *et al.*, 2005, Shukla *et al.*, 2008) and their stable binding to proteins (Horváth *et al.*, 1996), thereby decreasing the accumulation of pigment– lipoprotein complexes, particularly photosystem (PS) I (Wang *et al.*, 2009). The primary targets of Cd toxicity are also PSII and an enzymatic phase of photosynthesis, particularly ribulose-1, 5 bisphosphate carboxylase/oxygenase (Krantev *et al.*, 2008; Popova *et al.*, 2008).

Cadmium is an effective inhibitor of photosynthesis (Greger *et al.*, 1994; Chugh and Sawhney, 1999; Vassilev *et al.*, 2005). Alinear relationship between photosynthesis and inhibition of transpiration was observed in clover, lucerne, and soybean that suggest Cd inhibited stomatal opening (Barcelo *et al.*, 1986). Cadmium damages the photo-synthetic apparatus, in particular the light harvesting complex II (Krupa, 1988) and photosystems I and II (Siedlecka and Baszynsky, 1993; Siedlecka and Krupa, 1996). The inhibition of root Fe (III) reductase induced by Cd leads to Fe (II) deficiency which seriously affects photosynthesis (Alkantara *et al.*, 1994). Cadmium also causes stomatal closure in higher plants (Poschenreder *et al.*, 1989) and an overall inhibition of photosynthesis (Sheoran *et al.*, 1990; Krupa *et al.*, 1993; Chaugh and Sawhny, 1999). Cd stress affects photosynthesis in various ways. It inhibits the synthesis of chlorophylls (Azevedo *et al.*, 2005; Shukla *et al.*, 2008) and their stable binding to proteins (Horvath *et al.*, 1996), thereby decreasing the accumulation of pigment lipoprotein complexes, particularly photosystem (PS) I (Wang *et al.*, 2009). A large number of studies have demonstrated that the primary sites of action of Cd are photosynthetic pigments, especially the biosynthesis of chlorophyll (Baszynski *et al.*, 1980) and carotenoids (Prasad, 1995). According to Baryla *et al.* (2001), the observed chlorosis in oilseed rape was not due to a direct interaction of Cd with the chlorophyll biosynthesis pathway and most probably it was caused by decreasing of chloroplast density. The Cd-induced decrease in pigment content was more powerful at the leaf surface (stomatal guard cells) than it was in the mesophyll. In addition, the change of cell size, and the reducing of stomata density in the epidermis in Cd-treated leaves were observed. Thus, Cd might interfere directly with chloroplast replication and cell division in the leaf. This research also revealed that stomatal conductance was strongly reduced by Cd.

Chlorophyll and Protein Content

The presence of Cd decreased the content of chlorophyll and carotenoids, and increased non-photochemical quenching in *Brassica napus* (Larsen *et al.*, 1998). Similarly, the synthesis and level of chlorophyll decreased in other plant species under the influence of the cadmium (Czuba and Ormond, 1973; Griffiths *et al.*, 1995; Stiborova *et al.*, 1986; Bishnoi *et al.*, 1993; Ferretti *et al.*, 1993; Imai *et al.*, 1996; Phetsombat *et al.*, 2006;

Pandey *et al.,* 2007). Growth reduction associated with cadmium treatment was probably caused by inhibition of protein synthesis (Foy *et al.,* 1978). Phytotoxicity of the metal in other crop plants has been observed in the form of a loss in protein levels (Dubey and Dwivedi, 1987; Tamas *et al.,* 1997). Moreover, the grains developed on the plants grown under Cd stress had lower protein content (Salgare and Achareke, 1992).

OXIDATIVE STRESS AND ROS GENERATION BY CADMIUM

Generally, heavy metals cause oxidative damage to plants, either directly or indirectly through reactive oxygen species (ROS) formation. Plants are organisms exposed to different kinds of stresses, such as air pollution, drought, temperature, light, heavy metals, salinity, freezing, UV radiation and nutritional limitation. The intoxication with pollutant metals induces oxidative stress because they are involved in several different types of ROS-generating mechanisms (Fig. 4) (Pinto *et al.,* 2003). ROS intermediates are partially reduced forms of atmospheric oxygen (O_2) (Halliwell and Gutteridge, 1999). The typically result from the transfer of one, two or three electrons to O_2 to from, respectively, a superoxide radical (O_2^-), hydrogen peroxide (H_2O_2) or a hydroxyl radical (HO^-) (Dat *et al.,* 2000; Asada, 1999). These radicals occur transiently in aerobic organisms because they are also generated in plant cells during normal metabolic processes, such as respiration and photosynthesis (Asada and Takahashi, 1987). Although some of them may function as important signaling molecules that alter gene expression and modulate the activity of specific defense proteins, all ROS can be extremely harmful to organisms as high concentrations. ROS can oxidize proteins, lipids, and nucleic acids, often leading to alterations in cell structure and mutagenesis (Halliwell and Gutteridge, 1999). There are many others potential sources of ROS in plants, in addition to those that come from reactions involved in normal metabolism, such as photosynthesis and respiration. The balance between the steady state levels of different ROS are determined by the interplay between different ROS producing and ROS scavenging mechanisms, and can change drastically depending upon the physiological condition of the plant and the integration of different environmental, developmental and biochemical stimuli (Asada and Takahashi, 1987; Asada, 1999; Polle, 2001). A variety of proteins function as scavengers of superoxide and hydrogen peroxide. These include, among others, superoxide dismutase (SOD), catalase (CAT), ascorbate peroxidase (APOX), glutathione reductase (GR), thioredoxin, and the peroxiredoxin family of proteins (Asada, 1999; Mittler, 2002; Bowler *et al.,* 1992). These protein antioxidants are supplemented with a host of nonprotein scavengers, including, but not limited to, intracellular ascorbate and glutathione (Noctor and Foyer, 1998). The intoxication with pollutant metals induces oxidative stress because they are involved in several different types of ROS-generating mechanisms (Stohs and Bagchi, 1995). For example, transition metals (such as Fe^{3+} and Cu^{2+}) participate in the well-known Haber-Weiss cycle, producing HO^{-} from O_2^{-} and H_2O_2 (Winterbourn, 1982). Metals without redix capacity (such as Cd^{2+}, Pb^{2+}, and Hg^{2+}) can enhance the prooxidant status by reducing the antioxidant glutathione (GSH) pool, activating calcium-dependent systems and affecting iron-mediated processes (Pinto *et al.,* 2003). These heavy metals can also disrupt the photosynthetic electron chain, leading to O_2^{-} and $^1O_2$ production (Asada and Takahashi, 1987).

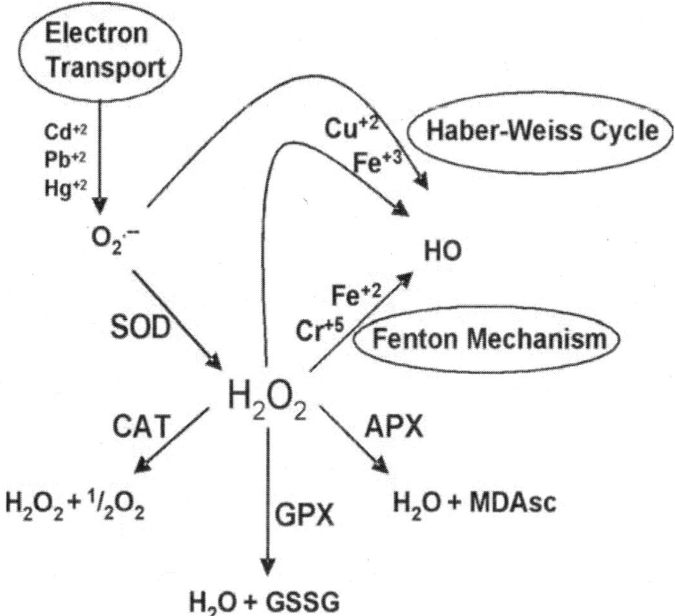

Fig. 4: ROS generation by heavy metals. Adapted from Pinto *et al.* (2003)

Cd was found to produce oxidative stress (Hendry *et al*; 1992; Somashekaraiah *et al*, 1992), but, in contrast with other heavy metals such as Cu, it does not seem to act directly on the production of oxygen reactive species (via Fenton and: or Haber Weiss reactions) (Salin, 1988). As it was previously observed for other stresses, activation or inhibition of antioxidative enzymes depends not only on stress intensity and duration but also on the tissue type and the age of the plant (Sgherri *et al.,* 2001). Table 1 shows several examples of the oxidative response to cadmium reported in higher plants. Treatment with Cd stimulated accumulation of lipid peroxides in *Pisum sativum* (Lozano-Rodriguez *et al.,* 1997; Dixit *et al.,* 2001; Metwally *et al.,* 2004), different barley genotypes (Wua *et al.,* 2003), sunflower seedlings and leaf discs (Groppa *et al.,* 2001; Gallego *et al.,* 1996), *Arabidopsis thaliana* seedlings (Cho and Seo, 2004) and soybean nodules (Balestrasse *et al.,* 2004), although decreased the rate of lipid peroxidation in peroxisomes of pea plants (Romero-Puertas *et al.,*

1999). No peroxidation was noticed in Cd-exposed plants and hairy roots of *Daucus carota* (Sanita di Toppi *et al.*, 1998). SOD, GR, APOX, POD and CAT showed variations in their activities that depend on the Cd concentrations and plant species used (Wua *et al.*, 2003; Sandalio *et al.*, 2001; Metwally, 2003, 2004; Gallego *et al.*, 1996; Balestrasse *et al.*, 2003). In *Helianthus annuus* leaves, Cd decreased GSH content, and decreased or increased the activity of the antioxidative enzymes SOD, CAT, APOX, GR and DHAR, deepening on Cd concentration, the organ used and the age of the plants (Gallego *et al.*, 1996, 1999; Groppa *et al.*, 2001). In *Phaseolus vulgaris* roots and leaves, 5mM Cd enhanced activities of the peroxidases GPOX and APOX, and raised lipid peroxidation (Chaoui *et al.*,, 1997). In two species of *Alyssum*, GR activity increased at 0.02mM Cd but decreased at 0.05mM Cd (Schickler and Caspi, 1999).

Table 1: Antioxident enzymes modified in different species exposed to variable cadmium concentrations

Cadmium concentration (μM)	**Exposure time**	**Plant species**	**Antioxidant enzymes modified**	**References**
5	10d	*Pisum sativum*	CAT, APOX, GPOX	Metwally *et al.*, 2003
1 and 10	10d	*Triticum duram*	CAT, SOD, APOX, GPOX	Milone *et al.*, 2003
4 and 40	7d	*Pisum sativum*	CAT, SOD, APOX, GPOX	Dixit *et al.*, 2001
5 and 50	48h	*Populus canescens*	CAT, SOD, APOX, GR, MDAR	Schutzendubel and Polle, 2002
50	21d	*Phragmites australis*	CAT, SOD, APOX, GR	Ianelli *et al.*, 2002
50, 100 and 200	48h	*Glycine max*	CAT, SOD, APOX	Balestrasse *et al.*, 2001
500	12h	*Helianthus annuus*	CAT, SOD, APOX, GR, DHAR	Gallego *et al.*, 1996
100 and 500	20d	*Oryza sativa*	CAT, SOD, GPOX	Shah *et al.*, 2001
300 and 500	21d	*Arabidopsis thaliana*	CAT, SOD, APOX, GPOX, GR	Cho and Seo, 2004
2000 and 5000	0-96h	*Saccharum officinarum*	CAT, SOD, GR	Fornazicr *et al.*, 2002
5000	0-24h	*Oryza sativa*	CAT, SOD, APOX, GPOX, GR	Hsu and Kao, 2004

Cadmium treatment induced lipo-oxygenase, with the simultaneous inhibition of the antioxidative enzymes, SOD and CAT (Somashekaraiah *et al.,* 1992). In particular, CAT actively often decreased following exposure to elevated Cd concentrations (Shim *et al.,* 2003; Dalurzo *et al.,* 1997; Sandalio *et al.,* 2001; Fornazier *et al.,* 2002). However, Vitoria *et al.* (2001) reported that the activities of CAT, GR and specific isoenzymes of SOD increased in the leaves and roots of a resistant variety of radish, following exposure to increasing (between 0.25 and 1 mM) concentrations of Cd. Although the concentrations of Cd used were high when compared to those found in contaminated soils, the same concentrations have also been used in assays with species of *Thlaspi* that are able to accumulate high concentrations of Cd (Lombi *et al.,* 2000). In soybean nodules and root's low Cd concentrations (50 uM) induced the antioxidant defenses, which in turn protect nodules and roots against oxidative stress, whereas moderate (100 uM) and high (200 uM) concentrations of the metal produced oxidative stress, in roots, or roots and nodules, respectively (Balestrasse *et al.,* 2001). In several reports, Cd concentrations studied are frequently not realistic when compared with the levels of contamination in the natural environment (Sanita di Toppi and Gabrielli, 1999). Taking this into account, Milone *et al.* (2003) presented Cd-induced antioxidative responses in wheat treated with realistic concentrations of Cd (up to 10 uM Cd). In this work, SOD, APOX and CAT were inhibited in the roots of the most sensitive cultivar, Adamello, while SOD was scarcely affected in both roots and leaves of Adamello and the most tolerant cultivar, Ofanto. Lipid peroxidation and H_2O_2 levels, SOD, CAT, APOX and GR activites increased in pea roots and leaves under Cd stress (Dixit *et al.,* 2001). While APOX and CAT decreaded at high Cd concentrations (Sandalio *et al.,* 2001).A severe suppression of SOD and CAT, and almost complete loss of APOX activities after 48 h of exposure to 50 μM Cd was observed in pine roots (Schutzendubel and Polle, 2001). Cd induced inhibition of APX and CAT was also associated with H_2 O_2 accumulation and growth retardation in the poplar roots (Schutzendubel and Polle, 2002b). Cho and Seo (2004) reported that Cd- induced oxidative stress in Arabidopsis is due to H_2 O_2 accumulation. Romero-Puertas *et al.* (2004) studied the involvement of H_2 O_2 and O_2 in the signaling event that lead to the variation of the transcript levels of CAT, Grand Cu Zn-SOD in pea plants under Cd strees. Thiols possess strong antioxidative properties, and they are consequently able to counteract oxidative stress (pichorner *et al.,* 1993). Groppa *et al.* (2001) and Gallego *et al.* (1996) reported a Cd induced decay in GSH content in sunflower plants or leaf discs, respectively. There are numerous papers reporting Cd induced depletion of GSH in several plant species (Rauser *et al.,* 1995; Dixit *et al.,* 2001; Balestrasse *et al.,* 2001). However, GSH increased in Phragmitesaustralis roots and leaves, treated either with 50 uM Cd (Ianelli *et al.,* 2002), or 50 and 100μM Cd (Pietrini *et al.,* 2003). Cadmium induced depletion of GSH has been mainly attributed to phytochelatin synthesis (Grill *et al.,* 1985). In tobacco leaves, PC-heavy metal complexes have been reported to be accumulated in the vacuole (Vogeli-Lange and wagner, 1990). Also in *Avena sativa,* PC-Cd complexes have been shown to be transported across the tono-plast (Salt and Rauser, 1995). In two varieties of two wheat species, a significant increase in total PC accumulation with increasing Cd exposure has been reported (Stolt *et al.,* 2003). The decline in the levels of GSH might alco be attributed to a decreased GR activity (Dixit *et al.,* 2001), an increased utilization for ascorbate synthesis or for a direct interaction with metals lime Cd (Pietrini *et al.,* 2003) or xenobiotics. The variety of response to Cd-induced oxidative stress is probably related not only to the levels of Cd supplied, but also to the plant species, the age of the plant and duration of treatment. Table 1 show antioxidant enzyme variation reported by several authors using different Cd-concentrations, exposure time and plant species.

DEFENSE MECHANISMS AGAINST CADMIUM IN PLANTS

There is a growing interest in problems concerning heavy metal contamination of cultivated lands and little is known regarding metal homeostasis and tolerance at the organismic level. For that reason, it is essential to clarify those problems related to metal transport, accumulation, detoxification and tolerance, and in this sense, it would be of enormous interest to use of model systems other than plants, such as *Saccharomyces cereviciae,* as well as the molecular analysis of hyperaccumulators like *Arabidopsis halleri* and some of the *Thlaspi* species, very little is known about the biochemistry of metal homeostasis factors. Physical interaction of transporters, chelators and chaperones is likely to play an important role. An improved knowledge in these crucial areas will help to further elucidate the molecular mechanism that lie beyond plant metal tolerance and homeostasis. The mechanisms leading to heavy metal tolerance can be divided into avoidance strategies and tolerance strategies. Avoidance leads to limitation of Cd uptake. Plant tolerance mechanisms include accumulation and storing of Cd by binding it to amino acids, proteins, and peptides. Other mechanisms that plants have developed to cope with damage caused by Cd are related to some stress signalling molecules, such as salicylic acid, jasmonic acid, nitric oxide, and ethylene. All these compounds were induced by Cd treatment, which suggests that they are involved in cell response to Cd toxicity (Popova *et al.,* 2012). An important pathway by which plants detoxify heavy metals is through sequestration with heavy-metal-binding peptides called phytochelatins or their precursor, glutathione. To identify limiting factors for heavy-metal accumulation and tolerance, and to develop transgenic plants with an increased capacity to accumulate and/or tolerate heavy metals, the *Escherichia coli* gshII gene encoding glutathione synthetase (GS) was overexpressed in the cytosol of Indian mustard (*Brassica juncea*). The transgenic GS plants accumulated significantly more Cd than the wild type: shoot Cd concentrations were up to 25% higher and total Cd accumulation per shoot was up to 3-fold higher. Moreover, the GS plants showed enhanced tolerance to Cd at both the seedling and mature-plant stages. Cd accumulation and tolerance were correlated with the gshII expression level. Cd treated GS plants had higher concentrations of glutathione, phytochelatin, thiol, S, and Ca than wild-type plants. This may be concluded that in the presence of Cd, the GS enzyme is rate limiting for the biosynthesis of glutathione and phytochelatins, and that overexpression of GS offers a promising strategy for the production of plants with superior heavy-metal phytoremediation capacity (Zhu *et al.*, 1999). Many plants survive, grow, and develop in Cd-polluted soils even in high concentrations of Cd. Investigations showed that some of these plants exhibit a hypertolerant capacity of their organelles and tissues. Strategies to cope with Cd toxicity involve the uptake and the distribution of Cd, defined as "hyperaccumulation". On the other hand, some plants increased cleaning up of the ROS by antioxidants to protect cells and tissues from destruction. Thus, the mechanism of Cd tolerance in plants can include both antioxidant defence and/or hyperaccumulation defence (Rascio and Navari-Izzo, 2011). "Hyperaccumulator" is the term used for plants that actively take up exceedingly large amounts of one or more heavy metals from the soil. Moreover, the heavy metals are not retained in the roots but are translocated to the shoot and accumulated in aboveground organs, especially leaves, at concentrations 100-1000 folds higher than the accumulate in non-hyper accumulating species. Hyperaccumulating plants show no symptoms of phytotoxicity. In hyper accumulating plants, the toxic effects of heavy metal at high accumulation are minimised, under the influence of detoxification mechanisms. Such mechanisms may be mainly based on chelation and sub-cellular compartmentalisation (Yadav, 2010). An important role in heavy metal hyperaccumulation seems to be played by free amino acids, such as histidine and nicotinamine, which form stable complexes with bivalent cations (Hassan and Aarts, 2011). Enhanced metal xylem loading and translocation to the shoots is the next key physiological step in the metal hyperaccumulation trait that accounts for the increased metal flow towards the shoot. Storage and detoxification/ sequestration of heavy metals, including Cd, in the shoot are key strategies of hyperaccumulating plants. The heavy metal detoxification/ sequestration occurs in locations such as the epidermis, trichomes, and even cuticle, where they do least damage to the photosynthetic machinery. In many cases, heavy metals are also excluded from both subsidiary and guard cells of the stomata. This may protect the functional stomatal cells from metal phytotoxic effects. The detoxifying/sequestering mechanisms in the aerial organs of hyperaccumulators consist mainly of heavy metal complexation with ligands and/or in their removal from metabolically active cytoplasm by moving them into inactive compartments, mainly vacuoles and cell walls (Rascio and Navari-Izzo, 2011). The small ligands, such as organic acids, have a major role as detoxifying factors. These ligands may be instrumental to prevent the persistence of heavy metals as free ions in the cytoplasm and even more in enabling their entrapment in vacuoles where the metal–organicacid chelates are primarily located. For example, in leaves of Thlaspi goesingense, citrate is the main ligand of Ni; in leaves of *Solanum nigrum*, citrate and acetate bind Cd, while most Zn in *Arabidopsis halleri* and Cd in Thlaspi caerulescens are complexed with malate (Rascio and Navari- Izzo, 2011).

Phytorededitiation has emerged as an economical, ecofriendly and aesthetically acceptable technology in the recent years (Huang *et al.*, 2011). Amongst the various strategies adopted for removal of toxic heavy metal from the contaminated soils, phytoremeditiation has emerged as an economical, eco-friendly acceptable technology in the recent years. *Brassica juncea* is one of the most studied plants found tolerant to most of the toxic heavy metals and suggested to be a potential phytoremediator. The contaminated lands bearing industrial wastes and waste drought and excessive heavy metals simultaneously. *B. juncea* have been found to be sensitive to most of the stresses and its green leaves are used as vegetable and fodder in many part of the Indian subcontinent, therefore, other alternatives are required to be investigated for removal of the metals from the contaminated agricultural ecosystems. We have recently reported *Ricinus communis*. (Bauddh and Singh, 2012), as better alternative to *B. juncea*. A considerable portion of the agricultural ecosystem supporting cultivation of *B. juncea* in Indian subcontinent is affected by cadmium; therefore it will be very significant to investigate comparative tolerance and phytoremediation efficiency of *B. juncea* and *R. communis* from the metal contaminated soil. Study conducted by Gharehbaghli (2015), to investigated comparative tolerance and phytoremediation efficiency of Indian mustard and *R. communis* from the metal contaminated soil. The results show that Castor was found more tolerant to Cd and removed more Cd in a given time than Indian mustard. Castor produced 23 and twelve folds higher biomass in terms of fresh weight and dry weight. The amont of Cd accumulated in the roots and shoots of Indian mustard was higher as per unit biomass than that in caster. Total removal of metal from soil was much higher in caster on per plant basis in the same period in presence of stresses. *R. communis* accumulated about seventeen and 1.5 fold higher Cd in their roots and shoots.

FACTORS ALLEVIATING CADMIUM TOXICITY IN PLANTS

Survival under stressful conditions depends on the plant's ability to perceive the stimulus, generate and transmit signals, and induce biochemical changes that adjust the metabolism accordingly. Therefore, the search for signal molecules that mediate stress tolerance is an important step in our better understanding of how plants acclimate to the adverse environment. In general, Cd has been shown to interfere with the uptake, transport, and use of several elements (Ca, Mg, P, and K). The application of 10 mM Mg in the nutrient solution of Japanese mustard spinach (*Brassica rapa* L. var. *pervirdis*) can alleviate Cd toxicity (Kashem and Kawai, 2007). The results showed that additional Mg in the nutrient solution enhanced the growth of plants suffering from Cd toxicity, resulting in a reduced Cd concentration in the plant. The authors suggested that additional Mg counteracted and detoxified physiological Cd toxicity in plants, especially in shoots. In a transcriptomic study of Mg starvation in Arabidopsis, Hermans *et al.* (2011) showed that a Mg pretreatment of 7 days alleviated the bleaching of young leaves caused by Cd. No or little difference in Cd tissue concentration between the +Mg and −Mg plants was observed. The authors suggested that lower Cd toxicity was probably not attributable to modified root to shoot translocation. A protective effect of Mg pretreatment was also observed on Fe starvation. However, Fe foliar spray partially alleviated Cd-induced chloroses, while it almost completely restored chlorophyll content in Fe-deficient leaves. The author's conclusion was that the protective effect of Mg against Cd toxicity could be attributable partly to the maintenance of Fe status but also to the increase in antioxidative capacity, detoxification, and/or protection of the photosynthetic apparatus. A hydroponic experiment with 2 rice cultivars differing in Cd tolerance was conducted by Hassan *et al.* (2005) to investigate the alleviating effect of Zn on growth inhibition and oxidative stress caused by Cd. The data showed that the addition of Zn to the medium solution alleviated Cd toxicity, which was reflected in a significant increase in plant height, biomass, chlorophyll concentration, and photosynthetic rate, and a marked decrease in MDA and activity of anti-oxidative enzymes. It was also noted that Zn increased shoot Cd concentration at higher Cd supply, probably due to the enhancement of Cd translocation from roots to shoots. Similar results have been shown for maize plants (Adilogllu*et al.,* 2005). Plant resistance can be induced by adopting various strategies. One of these, exogenou use of various growth regulators and other chemicals, has been proven worthwhile in producing resistance to many stresses in a number of plants. Salicylic acid (SA) response against heavy metal stress is a new study subject in the field of crop physiology. Results indicated that seed imbibitions with SA affected physiological processes related to growth and development and photosynthesis in maize plants. The beneficial effect of SA during the earlier growth period may help plants to avoid cumulative damage upon exposure to Cd. Alternatively SA could be involved in the expression of specific proteins or defence-related enzymes. These results may provide a good background for strategies aimed at manipulating plants for decreased Cd content in order to develop crops capable of tolerating environmental changes with as little damage as possible (Popova *et al.,* 2012). A protective effect of abscisic acid (ABA) against Cd toxicity has been suggested by Hsu and Kao (2003). The authors showed that exogenous application of ABA reduced the transpiration rate, decreased Cd content, and enhanced Cd tolerance of rice seedlings. There are data that another phytohormone, gibberellin, is also involved in plant adaptation to Cd stress. Ghorbanli *et al.* (2000) showed that the addition of 10 mg m^{-3} gibberellin reduced the negative effects of Cd^{2+} in shoot and root growth of soybean plants. The addition of gibberellin caused a partial elimination of the Cd effects on the roots and shoots and increased leaf area and length of stem.

Future Prospects

There is a growing interest in problems concerning heavy metal contamination of cultivated lands and little is known regarding plant tolerance at the organism level. It is necessary to minimize the entry of Cd into the food chain because of the number of associated health risks. Many strategies have been devised to minimize Cd toxicity. Proper plant nutrition is a good strategy to alleviate the damaging effects of Cd on plants and to avoid its entry into the food chain. Use of plant nutrients to alleviate Cd toxicity in plants is a relatively inexpensive, time saving, and effective approach to avoid Cd contamination of food. Growers are already applying nutrients to obtain good crop yield, and so to alleviate Cd toxicity the proper management of these plant nutrients is needed, keeping in mind the interactions between Cd and plant nutrients. Crop rotation, and the use of other organic and inorganic amendments are some other approaches being used to remediate Cd contaminated soils, but these approaches are time consuming and require extra resources. Selection and breeding of crop plants/cultivars that accumulate low Cd in the grain and other edible plant parts is one of these approaches (Chaney *et al.,* 1999; Liu *et al.,* 2007). It seems an attractive approach to change the Cd profile of crop plants and the benefit continues generation after generation in plants through the seed. Cadmium is a potentially toxic heavy metal that hampers plant productivity by interfering with their photochemistry. Cd causes disturbances in a range of physiological processes of plants such as photosynthesis, water relations, ion metabolism and mineral uptake. Cd pronouncedly affects photosynthesis by alteration of its vital machinery in all aspects. Photosynthesis is a well organised and sequential process fundamental to all green plants and microorganisms which involves various components, including photosynthetic pigments and photo systems, the electron transport system and CO_2 reduction pathways. Any damage at any level caused by Cd, critically affects overall photosynthetic capacity. Future research should be focus on key effects of Cd on photosynthetic apparatus including chloroplast structure, photosynthetic pigments, Chl-protein complexes and photo systems resulting in overall decrease in efficiency of carbon assimilation pathway.

CONCLUSION

Mineral nutrition of crops is an essential component of our agricultural system. However, higher levels of Cd in soil alter the uptake and translocation of mineral nutrients elements, resulting in nutrient deficiencies, oxidative stress, and reduction in plant growth and development of agricultural crops. Sufficient availability of mineral nutrients competes with Cd for transporters and reduces Cd accumulation in plants. Also, the involvement of mineral nutrients in physiological and biochemical processes and molecular mechanisms regulating photosynthesis and antioxidants system accounts for the alleviation of Cd stress. Thus, the management of mineral nutrients may be a useful tool to alleviate Cd stress, and improved mineral nutrition may provide a strategy to reduce Cd toxicity and alleviate Cd stress. The mechanisms of nutrient uptake and accumulation associated with the growth responses in Cd treated plants are not well characterized. Further research is required for a better understanding of the interactions between Cd and plant nutrients in soil-plant system, since heavy metals are important environmental pollutants and their toxicity is a problem of increasing significance for ecological, evolutionary, nutritional, and environmental reasons.

References

Adams, C.A., Kar, S. R., Hopper, J. E. (2004) Self-association of the amino-terminal domain of the yeast TATA-binding protein. *Journal of Biological Chemistry*, 279(2):1376-82

Adiloglu, A.I., Adiloglu, S., Gonulsuz, E. and Oner, N. (2005). Effect of zinc application on cadmium uptake of maize grown in zinc deficient soil. *Pakistan Journal of Biological Sciences,* 8: 10–12.

Aery, N.C. and D.K. Rana (2003).Growth and cadmium uptake in barley under cadmium stress. *Journal of Environment Biology,* 24, 1 17-123.

Akinola, M.O. and T.A. Ekiyoyo (2006). Accumulation of lead, cadmium and chromium in some plants cultivated along the bank of river Ribilia atOdo-nla area of Ikirodu, Lagos State, Nigeria. *Journal of Environment Biology,* 27: 597-599.

Al Yemens, M.N. (2001).Effect of cadmium, mercury and lead on seed germination and early seedling growth of *Vigna ambacensis* L. *Indian Journal of Plant Physiology*, 6 : 147-151.

Alam, M. G. M., Snow, E. T., and Tanaka, A (2003). Arsenic and heavy metal contamination of rice, pulses and vegetables grown in Samta village, Bangladesh. *Science Total Environment,* 308 : 83–96.

Alkantara, E., Romerza, M. and LaGuardia, M.D. (1994). Effect of heavy metals on both induction and function of root Fe(III) reductase in Fe-deficient cucumber (*Cucumis sativus* L.) *plants. Journal of Experiment Botany,* 45: 1893-1898.

Alloway, B.J. and Steinnes, E. (1999). Anthropogenic additions of cadmium to soils. In: Laughlin M.J., Singh B.R. (eds) Cadmium in Soil and Plants. Kluwer Academic Publishers, Dordrecht, pp. 97–123.

Anita, R., Chugh, L.K., Sawhney, V. and Sawhney, S.K. (1990). Proceedings of environmental pollution (Eds.: S.K, Arora, M. Singh and R.P. Aggarwal). *Haryana Agricultural University Press, Hisar, India.*pp.66-77.

Asada, K and Takahashi, M. (1987). Production and scavenging of active oxygen in photosynthesis. In: Kyle DJ, Osmond C, Arntzen CJ (eds.), Photoinhibition, pp.227-297. *Elsevier,* New York.

Asada, K. (1999). The water-water cycle in chloroplast: scavenging of active oxygen and dessipation of excess photons. *Annual Review of Plant Molecular Biology,* 50: 601-639.

Astolfi, S., Zuchi, S. and Passera, C. (2004). Effects of cadmium on the metabolic activity of *Avena sativaplants* grown in soil or hydroponic culture. *Biology of Planta.*, 48:413-418.

Astolfi, S., Zuchi, S. and Prassera, C. (2004). Role of sulphur availability on cadmium-induced changes of nitrogen and sulphur metabolism in maize (*Zea mays* L.) leaves. *Journal of Plant Physiology*, 161 (7) : 795-802.

Azevedo, H., Pinto, C. G. G., Fernandes, J., Loureiro, S. and Santos, C (2005). Cadmium effects on sunflower growth and photosynthesis. *Journal of Plant Nutrition,* 28: 2211-2220.

Balaknina, T., Kosobryukhov, J., Ivanov, A. and Kreslauskii, V. (2005). The effect of cadmium on CO_2 exchange, variable fluorescence of chlorophyll and the level of antioxidant enzymes in pea leaves. *Russian Journal of Plant Physiology,* 52: 15-20.

Balestrasse, K. B., Benavides, M. P., Gallego, S. M. and Tomaro, M. L. (2003). Effect on cadmium stress on nitrogen metabolism in nodules and roots of soybean plants. *Functional Plant Biol*ogy, 30:57-64.

Balestrasse, K. B., Gallego, S. M. and Tomaro, M. L (2004). Cadmium induced senescence in nodules of soybean (*Glycine max* L.) plants. *Plant Soil,* 262:373-381.

Balestrasse, K.B., Gardey, L., Gallego, S.M and Tomaro, M.L. (2001). Response of antioxidant defense system in soybean nodules and root subjected to cadmium stress. *Australian Journal of Plant Physiology*, 28:497-504.

Barcelo, J. and Poschenrieder, C. (1990). Plant water relations as affected by heavy metal stress: a review. *Journal of Plant Nutrition*, 13:1-37.

Barcelo, J., Poschenrieder, C., Andreu, I. and Gunse, B. (1986). Cadmium induced decrease of water stress resistance in bush bean plants (*Phaseolus vulgaris* L. cv. V *ontendev*). I. Effect of Cd on water potential, relations water content and cell wall elasticity. *Journal of Plant Physiology*, 125 : 17-25.

Baryla, A., Carrier, P., Franck, F., Coulomb, C., Sahut, C. and Havaux, M. (2001). Leaf chlorosis in oilseed rape plants (*Brassica napus*) grown on cadmium-polluted soil: causes and consequences for photosynthesis and growth. *Planta,* 212:696–709

Baszynski.T., Wajda. L., Krol, M., Wolinska, D., Krupa, Z. and Tuken-Dorf, A. (1980). Photosynthetic activities of cadmium-treated tomato plants. *Physiol Plant.* 48:365–370

Bert, V., Meerts, P., Saumitou-Laprade, P., Salis, P., Gruber, W. and Verbruggen, N. (2003). Genetic basis of Cd tolerance and hyper accumulation in Arabidopsis halleri. *Plant Soil,* 249:9-18.

Bhise, P. M., Challa, O., and Venegopalan, M. V. (2007). Effect of waste water irrigation on soil heavy metal toxicity in plants.*Journal of Plant and Soil,* .6: 421-438.

Bishnoi, N.R., Sheoran, I.S. and Singh, R. (1993). Influence of cadmium and nickelon photosynthesis and water relations in wheat leaves of differentialinsertion level. *Photosynthetica,* 28 : 473-479.

Blowler, C., Van Montagu, M. and Inze, D. (1992). Superoxide dismutage & stress tolerance.*Annu. Rev. Plant Physiol. Plant Mol. Biol.* 43:83-116.

Blum,W, H. (1997). Cadmium uptake by higher plants. In: proceedings of extended abstracts from the Fourth International Conference on the Biochemistry of Trace Elements, pp.109-110, *Berkeley,* UAS. University of California.

Borctal, M., Ozdemir, F. and Turkan, I. (2003). The effect of salt stress on lipid peroxidation and antioxidants in leaves of sugarbeet *Beta vulgaris* L. and wild beet *Beta maritime* L. *Plant Science,* 164:77-84.

Brooks, R. R (1998). Plants that hyperaccumulate heavy metals. *CAB International,* Wallingford, UK.

Brown, J. C. and Jones, W.E (1975). Heavy metal toxicity in plants 1. A crisis in embryo. *Communication Soil Science and Plant Annals,* 6: 421-438.

Brown, S.L., Chaney, R.L., Angle, J.S. and Baker, AJM. (1994). Phytoremediation potential of *Thlaspi caelulescens* and bladder campion for zinc-and cadmium-contaminated soils. *Journal of Environment.* 23:1151-1157.

Cataldo , D.A., Garland, T. R. and Wildung, R E. (1983). Cadmium uptake kinetics in intact soybean plants. *Plant Physiology,* 73:844-848.

Chaffei, C., Gouia, H., Debouba, M. and Ghorbel, M.H. (2008). Differential toxicological response to cadmium stress of bean seedlings grown with $NO3^-$ or NH^{4+}as nitrogen source. *International Journal of Botany,* 4: 14–23.

Chaney, R.L., Ryan, J.A., Li, Y.M. and Brown, S.L. (1999). Soil cadmium as a threat to human health. In: McLaughlin MJ & Singh BR (eds.). Cadmium in Soil and Plants,. Dordrecht: *Kluwer Academic Publisher.* pp. 219–256.

Chaomi, A., Maznoudi, S., Gharbal, M.H. and Ferjani, E. El. (1997). Cadmium and zinc induction of lipid peroxidation and effect on antioxidant enzyme activities in bean (*Phaseolus vulgaris* L.). *Plant Science,* 127:139-147.

Chaoui, A., B. Jarrar and Ferjani, E.E.L. (2004). Effects of cadmium and copper on peroxidase, NADH oxidase and IAA oxidase activities in cell wall, soluble and microsomal membrane fractions of pea roots. *Journal of Plant Physiology,* 161:1225-1234.

Chaoui, A., Mazhoudi, S., Ghorbal, M. H and Ferjani, E. (1997). Cadmium and zinc induction of lipid peroxidation and effects on antioxidant enzyme activities in bean (*Phasrolus vulgaris* L.). *Plant Science.* 127:139-147.

Chaudri, A., McGrath, S., Gibbs, P., Chambers, B., Carlton-Smith, C., Godley, A., Bacon, J., Campbell, C. and Aitken, M. (2007). Cadmium availability to wheat grain in soils treated with sewage sludge or metal salts. *Chemosphere,* 66: 1415–1423.

Chen, Y.X., Yang, Y., Zheng, S.J., Tian, G.M., Luo, Y.M. and Wong, M.H. (2003). Effect of cadmium on nodulation and N^2 fixation of soybean in contaminated soils. *Chemosphere,* 50 (6) : 781-787.

Chhonkar, P. K., Datta, S. P., Joshi, S. C., and Pathak, H. (2004). Impact of industrial effluents on soil health and agriculture: Distillery and paper mill effluents. *Journal of Scientific Industrial Research,* 59 : 350–361.

Chiy, P.C; and Phillips, C.J.C. (1999). The effects of mild water stress and soil type on the responses of grass and clover to sodium fertilizer. *Journal of Food Agriculture,* 79: 1399-1405.

Cho, U. and Seo, N. (2004). Oxidative stress in *Arabidopsis thaliana* exposed to cadmium is due to hydrogen peroxide accumulation. *Plant Science,* 168:113-120.

Chow, T. J. (1970). Lead accumulation in road side soil and grass. *Nature,* 225 : 295–296.

Chugh, L.K. and Sawhney, S.K. (1999). Photosynthetic activities of *Pisum sativum* seedlings grown in presence of cadmium. *Plant Physiology Biochemistry,* 37: 297-303.

Clarkson, D.T. and Luttge, U. (1989).Mineral nutrition. Divalent cations, transport and compartmentalization. *Progressive Botany,* 51: 93-112.

Clemens, S. (2001). Molecular mechanisms of plant metal tolerance and homeostasis. *Planta.* 212 : 475-486.

Clemens, S., Palmgreen, M. G. and Kramer, U. (2002). A long way ahead: understanding and engineering plant metal accumulation. *Trends Plant Science,* 7: 309-315.

Comba, M.E., Benavioes, M.P., Gallego, S.M. and Tonaro, M.L. (1998). Relationship between nitrogen fixation and oxidative stress induction in nodules of salt-treated soybean plants. *Journal of Experimental Botany.* 60 : 115-126.

Czuba, M. and Ormond, D. (1973).Effect of cadmium and zinc on ozone induced phytotoxicity in cress and lettuce. *Canadian Journal of Botany,* .52: 645-649.

Dakora, F .D. (1995). A functional relationship between legheamoglobin and nitrogenase based on novel measurement of the two proteins in legume root nodules. *Annals of Botan,* 75 : 4-54.

Dalurzo, H. C., Sandalio, L. M., Gomez, M. and Del Rio, L.A. (1997). Cadmium infiltration of detached pea leaves: effect on its activated oxygen metabolism. *Annals of Botany,* 37:59-64.

Das, P, Samantaray, S. and Rout, G.R (1973). Studies on cadmium toxicity in plants: a review. *Environment Pollution,* 98 : 29-36.

Dat, J. F., Van Breusegem., Vandenabeele, S., Vranova, E., Van Montagu, M. and Inze, D. (2000). Dual action of active oxygen species during plant stress responses. *Celleluar and Molecular Life Science,* 57 : 779-795.

De Vos, C.H.R. and Schat, H. (1981). Free radicals and heavy metal tolerance. In: Ecological response to environmental stress (Eds.: J. Rozema and J. Verkleij). *A.C. Kluwer Publishers, The Netherlands.* pp. 22-30.

Demiral, T. and Turkan, I. (2005). Comparative lipid peroxidation, antioxidant defense systems and proline content in roots of two rice cultivars differing in salt tolerance. *Environment of Experimental Botany,* 53 : 247-57.

Demirevska, K., Simova-Stoilova, L., Stoyanova, Z. and Feller, U. (2006). Cadmium stress in barley: Growth, leaf pigment and protein composition and detoxification of reactive oxygen species. *Journal of Plant Nutrition,* 29 : 451-468.

Dewdy, R.H. and Ham, G. E. (1997). Soybean growth and elemental content as influenced by soil amendments of sewage sludge and heavy metals: Seedling studies. *Agronomy Journal .,* 69 : 300-303.

Dhingra, H.R. and Priefer, U. R. (2006). Impact of cadmium on structural and functional aspect of pea (*Pisum sativum* L.) root nodules. *Journal of Plant Biology,* 33 : 207-207.

Di, L.S. and Gabrielli R. (1999). Response to cadmium in higher plants. *Environmental Experimental Botany,* 41: 105–130.

Dixit, V., Pandey, V. and Shyam, R. (2001). Differential oxidative responses to cadmium in roots and leaves of pea (*Pisum sativum* L cv Azad). *Journal of Experimental Botany.* 52:1101-1109.

Dubey, R.S. and R.S. Dwivedi. (1987). Effect of heavy metal on seed germination and seedling growth of soybean. *National Academy of Science Letters,* 10 : 121-124.

Durner J. and Klessig D.F. (1999). Nitric oxide as a signal in plants. Current Opinion. *Plant Biology,* .2: 369–374.

Ernst, W.H.O, Verkleij, J.A.C, Schat, H (1992). Metal tolerance in plants. *Acta Botanical. Neerl,* 41: 229-248.

Ernst, W.H.O., Verkleiji, J.A.C. and Schat, H. (1992). Metal tolerance in plants. *Acta Botany, Neeri.,* 411: 229-248.

Fernandez, P., de Lorenozo, M.C., de Felipe, M.R., Rajalakshmi, S., Gordon, A.J., Thomas, B.J. and Minchin, F. R. (1996). Possible reasons for relative salt stress tolerance in nodules of white lupine cv. Multolypa. *Journal of Experimental Botany.,* 47 : 1709-1716.

Ferretti, M., Ghisi, L. Merlo, F., Dallavecchia and Passera, C. (1993). Effect of cadmium on photosynthesis and enzymes of photosynthesis sulfate and nitrate assimilation pathways in maize (*Zea mays* L.). *Photosynthetica,* 29 : 49-54.

Fodor, A., Szabo-Nagy, A. and Erdei, L. (1995). The effects of cadmium on the fluidity and H^+-ATPase activity of plasma membrane from sunflower and wheat roots. *Journal of Plant Physiology,* 14:787-792.

Fornazier, R.F., Ferreira, R.R., Vitoria, A.P. Molina, S.M.G., and Azevedo, R.A. (2002). Effects of cadmium on antioxidative enzyme activities in sugarcane. *Biology of Plant,* 45: 91-97.

Foy, C.D., Chaney, R.L. and White, M.C. (1978).The physiology of metal toxicity in plants. *Annual Review of Plant Physiology,* 29:11-566.

Fuhrer, J. (1982). Ethylene biosynthesis and cadmium toxicity in leaf tissue of beans *Phaseolus vulgaris* L. *Plant Physiology,* 70: 162-167.

Fusconi, A., Repetto, O., Bona, E., Massa, N., Gallo, C., Dumas-Gaudot, E. and Berta, G. (2006). Effects of cadmium on meristem activity and nucleus ploidy in roots of *Pisum sativum* L. cv. Frisson seedlings. *Environmental and Experimental Botany,* 58: 253– 260.

Gallego, S.M., Benavides, M. P. and Tomaro, M. L. (1996). Effect of heavy metal ion excess on sunflower leaves: Evidence of involvement of oxidative stress. *Plant Science,* 121: 151-159.

Gallego, S.M., Benavides, M.L. and Tomaro, M.L. (1999). Effect of cadmium ions on antioxidant defense system in sunflower cotyledons.*Biology of Plant,* 43: 49-55.

Gharehbaghli, N. (2015). Investigation of growth and phytoremediation potential of *Ricinus cimmunis* (L.) and *Brassica juncea* (L.) in cadmium contaminated soil Nesa Gharehbaghli. *Biological Forum – An International Journal,* 7(2): 35-38.

Ghorbanli, M., Kaveh, S.H. and Sepehr, M.F. (2000). Effects of cadmium and gibberellin on growth and photosynthesis of Glycine max. *Photosynthetica,* 37: 627–631.

Godbold, D.L and Hutterman, A (1985). Effect of zinc, cadmium and mercury on root elongation on *Picea abies* (Karst.) seedlings and the significance of these metals to forest die-back. *Environment Pollution,* 230:674-676.

Greger, M., Johansson, M., Stihi, D.and K. Humza. (1994).Foliar uptake of Cd bypea (*Pisum sativum*) and sugar beet (*Beta vulgaris*). *Physiology of Plant,* 88 : 563-570.

Griffiths, P .G., Sasse, J.M., Yokota, T. and Comeron, D.W. (1995). 6-deoxoty-phasteroland 3-dehydro-6-deoxoteasterore, possible precursors to brassino steroids in pollen of *Cupressus urizonica. Bios. Biotech. Biochem,* 59 : 956-959.

Grill, E., Winnaacker, E.L. and Zenk, M. H (1985). Photosynthesis: the principal heavy metal complexing peptides of higher plants. *Science,* 230 : 674-676.

Groppa, M. D., Tomaro, M. L. and Benavides, M. P. (2001). Polyamines as protectors against cadmium or cropper-induced oxidative damage in sunflower leaf discs. *Plant. Science,* 161 : 481-488.

Guerinot, M, L. (2000). The ZIP family of metal transporters. *Biophysics Acta.* 1465:190-198.

Haghiri, F. (1973). Plant uptake of cadmium as influenced by cation exchange capacity, organic matter, zinc and soil temperature. *.Journal of Environment,* 2 : 93-93.

Hall, J.L. (2003). Cellular mechanisms for heavy metal detoxification and tolerance. *Journal of Experimental Botany,* 53 : 1 -11.

Halliwell, B., Gutteridge, J.M.C. (1999). Free radicles in Biology and Medecine.3rd ed. Oxford University Press, New York.

Hart, J. J., Welch, R. M., Norvell, W.A., Sullivan, L.A. and Kochian, L. V. (1998). Characterization of cadmium building, uptake and translocation in intact seedlings of bread and duram wheat cultivars.*Plant Physiology,* 116 : 1413-1420.

Hasan, S.A., Ali, B., Hayat, S.and Ahmad, A. (2007b). Cadmium induced changes in the growth and carbonic anhydrase activity of chickpea. *Turkish Journal of Biology,* 31: 137-140.

Hassan M.J., Wang F., Ali, S. and Zhang G. 2005. Toxic effects of cadmium on rice as affected by nitrogen fertilizer form. *Plant Soil,* 277: 359–365

Hassan, M.J., Zhang, G., Wu, F., Wei, K. and Chen, Z. (2005). Zinc alleviates growth inhibition and oxidative stress caused by cadmium in rice. *Journal of Plant Nutrition and Soil Science,* 168: 255–261.

Hassan, Z. and Aarts, M.G.M. (2011). Opportunities and feasibilities for biotechnological improvement of Zn, Cd or Ni tolerance and accumulation in plants. *Environmental and Experimental Botany,* 72: 53–63.

Hayat, S., Ali, B., Hasan, S.A. and Ahmad, A. (2007. Brassino steroid enhanced the level of antioxidants under cadmium stress in *Brassica juncea. Environment of Experimental Botany,* 60 : 33-41.

He, QB. And Singh, B.R. (1994). Crop uptake of cadmium from phosphorus fertilizers: I. Yield and cadmium content. *Water, Air & Soil Pollution,* 74: 251–265.

Hendry, G.A.F., Baker, A.J.M. and Ewart, C.F .(1992). Cadmium tolerance and toxicity, oxygen radical processes molecular damage in cadmium-tolerant and cadmium-sensitive clones of *Holcus lanatus. Acta Botanical, Neerl.* 41:271-281.

Hermans, C., Chen, J., Coppens, F., Inzé, D. and Verbruggen, N. (2011). Low magnesium status in plants enhances tolerance to cadmium exposure. *New Phytologist,* 192: 428–436.

Hernandez, L.E., Garate, A.R. and Carpena, Ru. 1995. Effect of cadmium on nitrogen fixing pea plants grown in perlite and vermiculite. *Journal of Plant Nutrition,* 18 (2): 287-303.

Hernandez, L. E., Carpena-Ruiz, R. and Garate, A. (1996). Alterations in the mineral nutrition of pea seedlings exposed to cadmium. *Journal of Plant Nutrition,* 19 : 1581-1598.

Hinesly, T. D., Redberg, K. E., Pites, R.I. and Ziegler, E. L (1984). Cadmium and Zinc uptake by corn (*Zea mays* L) with repeated applications of sewage sludge. *Journal of Agriculture Food Chemistry,* 32:155-163.

Hirsch, R. E., Lewis, B., Spalding, E. P. and Sussman, M. R. (1998). A role for the AKT1 potassium channel in *plant nutrition. Science,* 280 : 918-921.

Hong C.O., Chung D.Y., Lee do K. and Kim, P.J. (2010). Comparison of phosphate materials for immobilizing cadmium in soil. *Arch. Environment Contamination Toxicology,* 58 : 268–274.

Horvath, G., Droppa, M., Oravecz, A., Raskin, V.I. and Marder, J.B. (1996). Formation of the photosynthetic apparatus during greening of cadmium-poisoned barley leaves. *Planta,* 199:238-244.

Hsu, Y.T. and Kao, C.H. (2003). Role of abscisic acid in cadmium tolerance of rice (*Oryza sativa* L.) seedlings. *Plant, Cell and Environment,* 26: 867–874.

Hsu, Y.T. and Kao, C.H. (2008). Distinct roles of abscisic acid in rice seedlings during cadmium stress at high temperature. *Botanical Study,* 49:335–342.

Hsu, Y.T. and Kao, C.H. (2003).Role of abscisic acid in cadmium tolerance of rice (*Oryza sativa* L.) seedlings. *Plant, Cell and Environment,* 26: 867–874.

Iannelli, M. A., Pietrini, F., Fiore, L., Petrilli, L. and Massacci, A. (2002). Antioxidant response to cadmium in *Phragmites australis* plants. *Plant Physiology and Biochemisytry,* 40 : 977-982.

Imai, E., H. Obata, K. Shimizu and Komiya, T. (1996).Conservation of glutathione into cadystins and their analogs catalyzed by carboxypeptides. *Bioscience Biotechnology and Biochemistry,* 60 : 1193-1194.

John, R., Ahmad, P., Gagdil, K. and Sharma, S. (2007). Antioxidative response of *Lemna polyrrhiza* L.to cadmium stress. *Journal of Environment Biology,* 28:583-589.

K. Bauddh, Rana, P., Singh. (2012) Growth, tolerance efficiency and phyto-remediation potential of *Ricinus communis* (L.) and *Brassica juncea* (L.) in salinity and drought affected cadmium contaminated soil. *Ecotoxicology and Environmental safety,* 85:13-22.

Kabata-Pendias, A. and Pendias, H. (2001) Trace elements in soils and plants. CRC, London

Kashem, M.D.A. and Kawai, S. (2007). Alleviation of cadmium phytotoxicity by magnesium in Japanese mustard spinach. *Soil Science & Plant Nutrition,* 53:246–251.

Khan, M.H. and Patra, H. K. (2007). Sodium chloride and cadmium induced oxidative stress and antioxidant response in *Calamus tenuisleaves. Indian Journal of Plant Physiology,* 12 : 34-40.

Khan, M.H., Singha, L.B. and Panda, S.K. (2002). Changes in antioxidant levels in *Oryza sativa* L. roots subjected to NaCl-salinity stress. *Acta Physiology of Plant,* 24 : 145-148.

Kirkham, M., Morgan, R. K. and Davies, C. (2006). Why do midwives stay. Department of Health. London. (www. Nhsemployens.org & www. Rcm.org)

Korshunova, Y. O., Eide, D., Clark, W. G., Guerinot, M. L and Prakasi, H. B. (1999). The IRT1 protein from *Arabidopsis thaliana* is a metal transporter with a broad substrate range. *Plant Molecular Biology,* 40 : 37-44.

Kovacs K., Kuzmann E., Vertes, A., Levai, L., Cseh, E. and Fodor, F (2010). Effect of cadmium on iron uptake in cucumber roots: a Mössbauer-spectroscopic study. *Plant Soil*, 327 : 49–56

Krantev, A., Yordanova, R., Janda, T., Szalai, G., Popova, L (2008). Treatment with salicilyc acid decreases the effect of cadmium on photosynthesis in maize plants. *Journal Plant Physiology,* 165 : 920–931.

Krupa, Z. (1988). Cadmium induced changes in the composition and structure of the light-harvesting complex II in radish cotyledons. *Physiology of Plant,* 73 : 518-524.

Krupa, Z., Quist, G. and Hurner, G. (1993).The effect of cadmium on photosynthesis of *Phaseolus vulgaris*– A fluorescence analysis. *Physiology of Plant,* 88 : 626-630.

Kudo, H., Kudo, K., Ambo, H., Uemura, M. and Kawai, S. (2011). Cadmium sorption to plasma membrane isolated from barley roots is impeded by copper association onto membranes. *Plant Science,* 180: 300–305.

Kupper, H., Lombi, E., Zhao, F. J. and McGrath, S. P. (2000). Cellular comparation of cadmium and zinc in relation to other elements in the hyper accumulator *Arabidopsis halleri. Planta,* 212 : 75-84.

Lane, T. W. and Morel, F. M. (2000).A biological function for cadmium in marine diatomeas. *Proceeding of National Academy of Science,* USA. 97:4627-4631.

Larsen, P .B., Degenhart, L.M., Stenzler, S.H., Howell and Kochian, L.V. (1998). Aluminium-resistant Arabidopsis mutant that exhibit altered pattern of aluminium accumulation and organic acid release from roots. *Plant Physiology,* 117 : 9-18.

Lasat, M. M., Pence, N. S., Garvin, D. F., Ebbs, S. D. and Kochian, L. V. (2000). Molecular physiology of zinc transport in the Zn hyper accumulator *Thlaspi caerulescens. Journal of Experimental Botany,* 51 : 71-79.

Lazono-Rodriguez, E., Hernandez, L.E. Bonay, P. and Carpena, R. O.R. (1997). Distribution of Cd in shoot and root tissue of maize and pea plants: Physiological distribution. *Journal of Experimental Botany,* 48:123-128.

Lee, J.G., Adner, B.A. and Morel, F.M.M. (1996). Export of cadmium phytochelatinby the marine diatom *Thalassiosira weissflogii. Environment Science and Technology,* 39:1814-1821.

Li, Z., Li, L. and Chen, G.P.J. (2005). Bioavailability of Cd in a soil–rice system in China: soil type versus genotype effects. *Plant and Soil,* 271: 165–173.

Lindsay, W. L. (1979). Chemical Equlibria in Soils. John Wiley and Sons. New York, p. 449.

Liu, J., Qian, M., Cai, G., Yang, J. and Zhu, Q. (2007). Uptake and translocation of Cd in different rice cultivars and the relation with Cd accumulation in rice grain. *Journal of Hazardous Materials*, 143: 443–447

Lombi, E., Zhao, F. J., Dunham, S. J., McGrath, S. P (2000). Cadmium accumulation in populations of *Thlaspi caerulescens* and *Thlaspi goesingense. New Phytologist*, 145:11-20.

Lozano-Rodriguez, E., Hernandez, L. E., Bonay, P. and Carpena-Ruiz, R.O. (1997). Distributions of Cd in shoot and root tissues of maize and pea plants: physiological distribution. *Journal of Experiment Botany*, .48:123-128.

Lu, L.L., Tian, S.K., Yang, X.E. and He, ZL (2009). Cadmium uptake and xylem loading are active processes in the hyper accumulator Sedum alfredii. *Journal of Plant Physiology*, 166: 579–587.

Macfarlane, G.R, Burchett, M.D. (2001). Photosynthesis pigments and peroxidase activity as indicators of heavy metal stress in the grey mangrove. *Avicennia marina* (Forsk.) Vierh. Mar. Pollution Bulletin, 42:233-240.

Marshall, W. E., Seo, C. W. and Chamarthy, S. (2001). Adsorption of selected toxic metals by modified peanut shell. *Journal of Chemical Technology and Biotechnology*, 76, 113–118

McGranth, S.P., Brookers, P.C. and Giller, K.E. (1988). Effects of potentially toxic metals in soil derived from past applications of sewage sludge on nitrogen fixation by *Trifoliumrepens* L. *Soil Biology and Biochemistry*, 20 : 415-425.

Meharg, A.A., Cumbes ,Q.J., Macnair, M. R. (1993). Pre-adaptation of Yorkshire Fog, *Holcus lanatus* L. (Poaceae) to arsenate tolerance. *Evolution*, 47: 313–316

Metwally, A., Safronova, V. I., Belimov, A.A. and Dietz, K. J. (2004). Genotypic variation of the response to cadmium toxicity in *Pisum sativum* L. *Journal of Experimental Botany*, . 56:167-178.

Metwally, A., Finkemeier, I., Georgi, M. and Dietz, K. J (2003). Salicylic acid alleviates the cadmium toxicity in barley seedlings. *Plant Physiology*, 132:272-281.

Milone, M, T., Sgherri, C., Clijters, H. and Navari-Izzo, F. (2003). Antioxidative responses of wheat treated with realistic concentrations of cadmium. *Environment of Experimental Botany*, 50 : 265-273.

Mishra, S., Srivastava, S., Tripathi, R.D., Govindarajan, R., Kuriakaseand, S.V. and Prasad, M.N.V. (2006). Phytochelatin synthesis and response of antioxidants during cadmium stress in *Bacopa monnieri* L. *Plant Physiology and Biochemistry*, 44: 25-37.

Mittler, R (2002). Oxidative stress, antioxidants and stress tolerance. *Trends Plant Sciences*, 7: 405 - 410.

Miyadate, H., Adachi, S., Hiraizumi, A., Tezuka, K., Nakazawa, N., Kawamoto, T., Katou, K., Kodama, I., Sakurai, K., Takahashi, H., Satoh-Nagasawa, N., Watanabe, A., Fujimura, T. and Akagi, H. (2011). OsHMA3, a P18-type of ATPase affects root-to-shoot cadmium translocation in rice by mediating efflux into vacuoles. *New Phytologist*, 189: 190–199.

Mohamed, A.A., Castagna, A., Ranieri, A. and Sanita di Toppi, L (2012).Cadmium tolerance in*Brassica juncea* roots and shoots is affected by antioxidant status and phytochelatin biosynthesis. *Plant Physiology and Biochemistry*, 57 : 15–22.

Mohan, B.S. and Hosetti, B.B. (2006). Phytotoxicity of cadmium on the physiological dynamics of *Salvinia natans* L. grown in macrophyte ponds. *Journal of Environment Biology*, 27:701-704.

Mohd, I. Hayat, S. Ahmad, A. and Alyemani, M, N. (2013). Soil cadmium enrichment: allocation and plant physiological manifastations. *Saudi Journal of Biological Sciences*. 20:1-10.

Mojtaba, S.Y. and Baya, H.R. (2013). Effect of cadmium toxicity on growth, cadmium accumulation and macronutrient content of durum wheat (Dena CV.).International *Journal of Agriculture and Crop Sciences*. 6 (15): 1099-1103 (www.ijagcs.com IJACS/2013/6-15/1099-1103).

Moradi, A, Abbaspour, K.C. and Afyuni, M. (2005). Modelling field-scale cadmium transport below the root zone of a sewage sludge amended soil in an arid region in Central Iran. *Journal of Contaminant Hydrology,* 79: 187–206.

Moral, R., Palacios, G., Gomez, I., Navarro-Pedreno, J. and Mataix, J. (1994). Distribution and accumulation of heavy metals (Cd, Ni and Cr) in tomato plant. *Fresenius Environ. Bull.* 3: 395-399.

Mori, S., Nishizawa, N., Hayashi, H., Chino, H., Yoshimura, E. and Ishihara, J. (1991). Why are young rice plants highly susceptible to iron deficiency. In: Iron nutrition and interactions in plants (Eds.: Y . Chen and Y . Hadav). *Kluwer Academic Publishing, Dordrecht.* pp. 175-188.

Moussa, H. and El-Gamal, S. (2010). Effect of salicylic acid pretreatment on cadmium toxicity in wheat. *Biologia Plantarum,* 54: 315–320.

Mukherjee, D. 2013. Studies on resource management for sustainable ecosystem in Eastern Himalaya. *Asian Journal of Agriculture and Food Science,* 1 (5): 222-35.

Mukherjee, D. (2002). Food security coping with nutritional challenge ahead. *Agriculture Today,* 8 : 30-32.

Mukherjee, D. (2008). Potential of herbals (medicinal plant) in context of India. *MAPs Dew*, 4 (3) : 1-5.

Mukherjee, D. (2012). Resource conservation through indigenous farming system in hills of West Bengal. *Journal of Crop and Weed,* 8 (1): 160-64.

Mukherjee, D.(2004). Influence of stresses on crop efficiency. *Kisan world,* 31 (10) : 40.

Naik, S. K., Pandit, T. K., Sur, P., Patra, P. K. and Das, D. K (2013). Effects of graded levels of cadmium on spinach and cabbage grown in an Inceptisol. *Soil Science & Plant Analysis*, 44:1629-1642.

Naik, S. K., Sasmal, S., Sur, P., Pandit, T. K., Sur, P., Mondal, M. and Das, D. K (2009). Mobility of lead and nickel in soil in relation to nutrition of sesame. *Indian Society of Soil Sci 4:* 1-14.

Nazar, R., Iqbal, N., Masood, A., Iqbal, M., Khan, R., Syeed, S. and Khan, N.A. (2012). Cadmium toxicity in plants and role of mineral nutrients in its alleviation. *American Journal of Plant Sciences*, 3 : 1476-1489.

Nellessen, J.E. and J.S. Fletcher (1993). Assessment of published literature on the uptake, accumulation and translocation of heavy metals by vascular plants. Chemosphere, 9:1669-1680.

Noctor, G. and Foyer, C.H. (1998). Ascorbate and glutathione: Keeping active oxygen under control. *Ann. Rev. Plant Physiol. Plant Molecular Biology,* 49 : 249-279.

Ozawa, T. and Kenji, I. (2015). Inhibitory effect of cadmium on competitive nodulation ability of *Bradyrhizobium japonicum. Bulletin Hiroshima Institute of Technical Research,* 49 : 27-33.

Pal, M., E. Horvath, T., Janda, E., Paldi, E. and Szalai, G. (2006). Physiological changes and defence mechanisms induce by cadmium stress in maize. *Journal of Plant Nutrition,* 169:239-246.

Pál M.E, Horvá, T., Janda, E., Páldi, D and Szalai, C. (2006) Physiological changes and defense mechanisms induced by cadmium stress in maize. *Journal of Plant Nutrition,* 169 : 239-246.

Panda, S.K. and Khan, M.G. (2003).Salt stress influences lipid peroxidation and antioxidants in the leaf of an indica rice (*Oryza sativa* L.). *Physiology of Molecular Biology of Plantd,* 9:273-278.

Pandey, S., Gupta, K. and Mukherjee, A.K. (2007). Impact of cadmium and lead on *Catharanthus roseus*– A phyto remediation study. *Journal of Environment and Biology,* 28: 655-662.

Pandit, T. K., Naik, S. K., Patra, P. K. and Das, D. K (2012). Influence of lime and organic matter on the mobility of cadmium in cadmium-contaminated soils in relation to nutrition of spinach. *Soil Sediment Contamination,* 21:419-433.

Pandit, T. K., Sur, P., Sarkar, S., Naik, S. and Das, D. K. (2006). Distribution of some heavy metals in soils & plants of some urban & peri urban areas of West Bengal. *Journal of Inter academicia,* 10 (4) : 54-58.

Parmar, P., Kumari, N. and Sharma, V. (2013). Structural and functional alterations in photosynthetic apparatus of plants under cadmium stress. Botanical Studies :*An international Journal,* 54 : 45-47.

Patel, P. K., Pandaya, R. R., Maliwal, G. L., Patel, K. C., Ramani, V. P., and George, V (2004). Heavy metal content of different effluents and their relative availability in soils irrigated with waters around major industrial cities of Gujarat. *Journal of Soil Science,* 52 : 89–94.

Pence, N. S., Larsen, P. B., Ebbs, S. D., Letham, D. L. D., Lasat, M. M. and Garvin, D. F. (2000). The molecular physiology of heavy metal transport in the Zn/Cd hyper accumulator *Thlaspi caerulescens. Proceeding of National Academy of Soil Science, . USA.* 97 : 4956-4960.

Perfus-Barbeoch, L., Leonhardt, N, Vavasseur, A., Forestier, C. (2002). Heavy metal toxicity: cadmium permeates through calcium channels and disturbs the plant water status. *Plant Journal,* 32 :539–548.

Phetsombat, S., Kruatrachue, M. Pokethitiyook, P.and Upatham, S. (2006). Toxicity and bioaccumulation of cadmium and lead in *Salvinia cucullata. Journal of Environment and Biology,* 27:647-652.

Pichorner, H., Korori, S.A.A., Thur, A. and Ebermann, R. (1993). The two and the fourelectron transfer to molecular oxygen-mediated by plant peroxidase in the presence of thiols. In : Plant peroxidases: Biochemistry and Physiology (Eds.: K.G. Welinder, S.K. Rasmussen, C. Penel and H. Greppin). *University of Geneva.* pp. 131-136.

Pietrini , F., Iannelli, M. A., Pasqualini, S. and Massacci, A (2003). Interaction of cadmium with glutathione and photosynthesis in developing leaves and chloroplast of Phragmites australis (Cav.) Trin. Ex. Steudel. *Plant Physiology,* 133 : 829-837.

Pietrini, F., Zacchini, M., Iori , V., Pietrosanti, L., Ferretti, M. and Massacci, A. (2010). Spatial distribution of cadmium in leaves and on photosynthesis: examples of different strategies in willow and poplar clones. *Plant Biology,* 12:355–363.

Pinto, E; Sigaud-Kutner, T. C. S; Leitao, M.A S; Okamoto, O. K; Morse, D; Colepicolo, P; (2003). Heavy metal-induced oxidative stress in algae. *Journal of Phycology,* 39 : 1008-1018.

Polesskaya O.G., Kashirina E.I. and Alekhina N.D. (2004). Changes in the activity of antioxidant enzymes in wheat leaves and roots as a function of nitrogen source and supply. *Russsian Journal of Plant Physiology,* 51: 615–620.

Polle, A (2001). Dissecting the superoxide didmutase-ascorbate- peroxidase-glutathione pathway in chloroplasts by metabolic modeling. Computer simulations as a step towards flux analysis. *Plant Physiology,* 126:445-462.

Polle, A. and Rennenberg, H. (1994). Photooxidative stress in trees. In: Causes of photoxidative stress and amelioration of defense systems in plants.(Ed.: C.H. Foyer and P .M. Mullineauv). *CRC Press, Boca Raton.*pp.199-218.

Popova, .LP., Maslenkova, L.T. Ivanova, A.P. and Stoinova, Z. (2012). Role of Salicylic Acid in Alleviating Heavy Metal Stress. In: Ahmad P & Prasad MNV (eds.) *Environmental Adaptations and Stress Tolerance of Plants in the Era of Climate Change,* pp. 441–466. New York, Dordrecht, Heidelberg, London: Springer.

Popova, L. P., Maslenkova, L.T., Yordanova, R.Y., Ivanova, A.P., Krantev, A.P., Szalai, G., Janda, T. (2008).Exogenous treatment with salicylic acid attenuates cadmium toxicity in pea seedlings. *Plant Physiology and Biochemistry.* 47(3):224–231.

Poschenrieder, C., Gunse, B. and Barcelo, J. (1989). Influence of cadmium on waterelation, stomatal resistance, and absicisic acid content in expanding bean leaves. *Plant Physiology,* 90, 1365-1371.

Prasad, M.N.(1995). Cadmium toxicity and tolerance in vascular plants. *Environ. Experimental Botany,* 35 : 525-545.

Quartacci, M. F., Cosi, E., Navari-Izzo, F. (2001). Lipids and NADPH-dependent superoxide production in plasma membranes vesicles from roots of wheat grown under copper deficiency or excess. *J. Exp. Bot.* 52:77-84.

Ramos, I., Esteban, E., Lucena, J. J. and Garate, A. (2002).Cadmium uptake and subcellular distribution in plants of Lactuca sp. Cd-Mn interaction. *Plant Science,* 162:761-767.

Rana, A. and Ahmad, M. (2002).Heavy metal toxicity in legume micro symbiont system. *J. Plant Nutrition,* 25, 369-386.

Rascio, N. and Navari-Izzo, F. (2011). Heavy metal hyper accumulating plants: How and why do they do it? And what makes them so interesting? *Plant Science,* 180: 169–181.

Rascio, N. and Navari-Izzo, F. (2011). Heavy metal hyper accumulating plants: How and why do they do it? And what makes them so interesting? *Plant Science,* 180: 169–181.

Rascio, N., Dalla Vecchia, F., La Rocca, N., Barbato R., Pagliano, C., Raviolo, M., Gonnelli, C. and Gabbrielli, R. (2008). Metal accumulation and damage in rice (cv. Vialone nano) seedlings exposed to cadmium. *Environmental and Experimental Botany,* 62: 267–278.

Rauser, W. E (1995). Phytochelations and related peptides: structure, biosynthesis, and function. *Plant Physiology,* 109:1141-1149.

Reddy, G.N. and Prasad, M.N.V. (1993): Tyrosine is not phosphorylated in cadmium induced hsp70 cognate in maize (*Zea mays* L.) seedlings: Role in chaperone function. *Biochem. Arch.,* 9: 25-32.

Rivetta , A., Negrini, N. and Cocucci, M (1997). Involvement of Cd^{2+}-calmodulin in Cd^{2+}toxicity during the phases of radish (*Raphanus sativus* L.) seed germination. *Plant Cell Environment,* 20:600-608.

Rodriguez-Flores, M., Rodriguez-Castellon, E. (1982). Lead and cadmium levels in soil and plants near highways and their correlation with traffic density. *Environment Pollution Series,* 4: 281–290.

Romero-Puertas, M. C., McCarthy, I., Sandalio, L. M., Palma, J. M., Cropas, F. J., Gomez, M. and del Rio, R. A (1999). Cadmium toxicity and oxidative metabolism of pea leaf peroxisomes. *Free Radical Research,* 31:25-31.

Salgare, S.A. and Acharekar, C. (1992).Effect of industrial pollutant on growth and content of certain weeds. *Journal of Nature Conservation,* .4: 1-6.

Salin, M. L (1988). Toxic oxygen Species and protective systems of the chloroplasts. *Physiology of Plant,* 72:681 689.

Salt, D. E., Prince, R. C., Pickering, I. J and Raskin, I (1995a).Mechanism of cadmium mobility and accumulation in Indian mustard. *Plant Physiology,* 109:427-433.

Salt, D. E; Blaylock, M; Kumar, NBPA, Dushenkov, V, Ensley, B, D; Chet, Raskin, I (1995b). Phytoremediation: a novel strategy for the removal of toxic metals from the environment using plants. *Biotechnol,* 13:468-474.

Salt, D.E. and Rauser, W.E. (1995). MgATP-dependent transport of oat roots. *Plant Physiology,* 107:1293-1301.

Sandalio, L. M., Dalurzo, H. C., Gomez, M., Romero-Puertas, M.. C., del Rio, L. A. (2001). Cadmium-induced changes in the growth and oxidative metabolism of pea plants. *Journal of Experimental Botany,* 52:2115-2126.

Sandalio, L.M., Dalurzo, H.C., Gomez, M.C., Romero Puertas, M.C. and del Rio, L.A. (2001).Cadmium induced changes in the growth and oxidative metabolism of pea plants. *Journal of Experimental Botany,* 52:2115-2126.

Sandalio, L.M., Dalurzo, H.C., Gomez, M., Puertas, M.C.R. and Del rio, L.A. (2001). Cadmium-induced changes in the growth and oxidative metabolism of pea plants. *Journal of Experimental Botany*, 52 (364) : 2115-2126.

Sanita di Toppi, L. and Gabbrielli, R. (1999). Response of cadmium in higher plant. *Environment of Experimental Botany,* 41:105-130.

Sanita di T oppi, L., Lambardi, L., Pazzagli, G., Cappugi, M., Durante, L. and Gabbrielli, R. (1998). Response to cadmium in carrot in vitro plants and cell suspension cultures. *Plant Science,* 137: 1- 19.

Sarwar N., Saifullah, Malhi S.S., Zia, M.H., Naeem, A., Bibi, S. and Farid, G. 2010. Role of mineral nutrition in minimizing cadmium accumulation by plants. *Journal of Science and Food Agriculture,* 90: 925–937

Schickler , H. and Caspi, H (1999). Response of antioxidant enzymes to nickel and cadmium stress in hyperaccumulator plants of the genus *Alyssum. Physiology of Planta.* 105:39-44.

Scholz, G. (1989). Effect of nicotianamine on iron-mobilization in de-rooted tomato seedlings. *Biologica Metals,* 2 : 89-91.

Schutzendubel, A. and Polle, A. (2002b). Cadmium and H_2O_2–induced oxidative stressin *Populus x canescens* roots. *Plant Physiology and Biochemistry,* 40 : 577-584.

Schutzendubel, A., Schwanz, P., Teichmann, T. and Polle, A. (2001). Cadmium induced changes in antioxidative systems, H_2O_2 content and differentiation in pine (*Pinus sylvestris*) roots. *Plant Physiology,* 127 : 887-898.

Schutzendubel, A.and Polle, A. (2002a). Plant responses to abiotic stress: heavy-metal induced oxidative stress and protection by mycorrhization. *Journal of Experimental Botany,* .53 : 1351-1365.

Senden, M.H.M.H., Van der, A.J.G.M., Verburg, T.G. and Wolterbeek, H.T. (1994). Effect of cadmium on the behaviour of citric acid in isolated to matoxylem cell walls. *Journal of Experimental Botany,* 45 : 597-606.

Senden, M.H.M.N., Van der Meer, A.J.G.M. and Wolterbeek, H. T. (1994). Effects of cadmium on the behavior of citric acid in isolated tomato xylem cell walls. *Journal of Experimental Botany,* 45 : 597-606.

Senden, M.H.M.N., Van der Meer, A.J.G.M., Wolterbeek, H. T. (1992). Cadmium-citric acid xylem cell wall interactions in tomato plants. *Plant Cell Environment,* 15 : 71-79.

Sepehri, N .M.., Saleh Rastin, H., Asadi, R. and Alikhani, H. (2006). Effects of soil pollution by Cadmium on nodulation and nitrogen fixation ability of native strains of *Sinorhizobium meliloti. Journal of Science and Technology of Agriculture and Natural Resources,*64:11681175. (http://agris.fao.org/agris-search/search.do?recordID=DJ2012041982).

Seregin I.V. and Ivanov V.B. (2001). Physiological aspects of cadmium and lead toxic effects on higher plants. *Russian Journal of Plant Physiology,* 48: 523–544.

Sgherri, C.L.M., Milone, A.M.T., Clijsters, H. and Navari-Izzo, F. (2001). Anti-oxidative enzymes in two wheat cultivars, differently sensitive to drought and subjected to sub-symptomatic cropper doses. *Journal of Plant Physiology,* 158 : 1439-1447.

Shah, K., Kumar, R. G., Verma, S. and Dubey, R. S. (2001). Effect of cadmium on lipid perooxidation, superoxide anion generation and activities of antioxidant enzymes in growing rice seedlings. *Plant Sciences,* 161 : 1135-1144.

Shanthala, L., Venkatesh, B., Lokesha, A.N., Prasad, T.G. and Sashidhar, V.R. (2006). Glutathione depletion due to heavy metal-induced phytochelatin synthesis caused oxidative stress damage: Beneficial adaptation to one abiotic stress in linked to vulnerability to a second abiotic stress. *.Journal of Plant Biology,* 33 : 209-214.

Shaw, B.P. (1995). Effect of mercury and cadmium on the activities of antioxidative enzymes in seedlings of *Phaseolus aureaus. Biology of Plantarum,* 37 : 587-596.

Shen, H., Christie, P. and Li, X. (2006). Uptake of zinc, cadmium and phosphorus by arbuscular mycorrhizal maize (*Zea mays* L.) from a low available phosphorus calcareous soil spiked with zinc and cadmium. *Environmental Geochemistry and Health*, 28: 111–119.

Shen, Y .Y ., Dai, J.Y., Hu, D.C., Gu, H.L., He, R.Y. and Zheng, B. (1990). Studies on physiological effects of brassinolide on drought resistant in maize. *Journal of . Shenyang Agricultural University,* 21 : 191-195.

Sheoran, I.S., Signal, H.R. and Singh, R. (1990). Effect of cadmium and nickel on photosynthesis and the enzymes of photosynthetic carbon reduction cycle in pigeon pea (*Cajanus cajan* L.). *Photosynthesis. Research,* 23: 345-351.

Shim, I. S., Momose, Y., Yamamoto, A., Kim, D. W. and Usui, K. (2003). Inhibition of catalase activity by oxidative stress and its relationship to salicylic acid accumulation in plants. *Plant Growth Regulator,* 39 : 285-292.

Shukla, U.C., Murthy, R.C. and Kakkar, P. (2008).Combined effect of ultraviolet–B radiation and cadmium contamination on nutrient uptake and photosynthetic pigments in *Brassica campestris* L. seedlings. *Environment. Toxicology,* 23 : 712-719.

Siedlecka, A. and Baszynsky, T. (1993). Inhibition of electron flow around photo system I in chloroplasts of cadmium treated maize plants is due to cadmium induced iron deficiency. *Physiology of Plantarum,* 87 : 199-202.

Siedlecka, A. and Krupa, Z. (1996). Interaction between cadmium and iron and itseffect on photosynthetic capacity of primary leaves of *Phaseolus vulgaris. Plant Physiology and Biochemistry,* 34 : 833-841.

Siedleska A. (1995). Some aspects of interactions between heavy metals and plant mineral nutrients. *Acta Sociceyt of Botanical* 64: 265–272.

Singh, B.R. and Myhr, K. (1998). Cadmium uptake by barley as affected by Cd sources and pH levels. *Geoderma,* 84: 185–194.

Singh, P .K. and Tewari, R.K. (2003).Cadmium toxicity induced changes in plant water relations and oxidative metabolism of *Brassica juncea* L. plants. *Journal of Environment and Biology,* 24 : 107-112.

Smeets, K., Cypers, A., Lamrechts, A., Semane, B., Hoet, B., Laere, A.V. and Vangronsveld, J. (2005). Induction of oxidative stress and antioxidative mechanism in Phaseolus vulgaris after Cd application. *Plant Physiology and Biochemistry,* 43 : 437-444.

Smith, S. R., Hall, J. E., and Hadley, P. 1992. Composting sewage sludge wastes in relation to their sustainability for use of fertilizer materials for vegetables crop production. *Acta Horticulture,* 302 : 203–215.

Somashekaraiah, B. V., Padmaja, K. and Prasad, A. R. K. (1992). Phytotoxicity of cadmium ions on germinating seedlings of mung bean (*Phaseolus vulgaris*): involment of lipid peroxides in chlorophyll degradation. *Physiology of Plant,* 85 : 85-89.

Srivastava, S., Tripathi, R.D. and Dwivedi, U.N. (2004). Synthesis of phytochelatinsand modulation of antioxidants in response to cadmium stress in *Cuscuta reflexa*– An angiospermic parasite. *Journal of Plant Physiology,* 161 : 665-674.

Stiborova, M., Doubrovava, M., Brezninova, A. and Frederick, A. (1986). Effect of heavy metal ion on growth and biochemical characteristics of photosynthesis of barley Hordeum vulgareL. *Phytosynthetica,* 20 : 418-425.

Stohs, S.J. and Bagchi, D. (1995).Oxidative mechanism in the toxicity of metal ions.*Free Radical Biology Med.,* 18 : 321-336.

Stolt , J. P., Sneller, F. E. C., Bryngelsson, T., Lundborg, T. and Schat, H. (2003). Phytochelation and cadmium accumulation in wheat. *Environment of Experimental Botany,* .49 : 21-28.

Street, J. J., Lindsay, W. L. and Sabey, B. R. (1977). Solubility and plant uptake by cadmium in soils amended with cadmium and sewage sludge. *Journal of Environment Quality,* 6 : 72-77.

Sur, P., Pandit, T. K., Naik, S. K., Mandal, M., Chakroborty, A., Das, D. K. (2006). Occurrence of some heavy metals in some contaminated soils and crops of West Bengal. *Indian Agriculturist,* 50 (1&2) : 79-84.

Tamas, L., Huttova, J. and Zigova, Z. (1997).Accumulation of stress protein inintracellular spaces of barley leaves induced by biotic and abiotic factors. *Biology of Planta,* 39:387-394.

Taylor, K., Albrigo, L.G. and Chase, C.D. (1988). Zinc complexation in the phloem ofblight affected citrus. *Journal of Horticultural Sciences,* American Society,113 : 407-411

Tester, M. and Leigh, R. A (2001). Partitioning of nutrient transport processes in roots. *Journal of Experiment Botany,* 52 : 445-457.

Thomine, S., Wang, R., Ward, J. M., Crawford, N., Schroeder, J.I. (2000). Cadmium and iron transport by members of plant metal transporter family in Arabidopsis with homology to Nramp genes. *Proceeding of National Academy of Sciences,* USA. 97:4991-4996.

Thurman, D.A. and Collins, J.C.L. (1983).Metal tolerance mechanism in higher plants review. In: Proceedings of International Conference on Heavy Metals in the Environmental. Heidelberg, CEP *Consultan's Edimburg. pp.* 298-300.

Tisdale S.L. and Nelson W.L (1975). Elements required in plant nutrition. In: Tisdale S.L., Nelson W.L. (eds) *Soil Fertility and Fertilizers. Macmillan,* New York, pp. 66–104

Tudoreanu, L. and Phillips, C.J.C. (2004). Modeling cadmium uptake and accumulation plants.*Advances in Agronomy,* 84 : 121-157.

Van Asshe, F and Clijsters, M. (1990). Effects of metals on enzyme activity in plant. *Plant Cell Environment,* 13: 195-206.

Varalakshmi, L.R. and Ganeshamurthy.(2009). Effect of cadmium on plant biomass and cadmium accumulation in amaranthus (*Amaranthus tricolor*) cultivars. *Journal of Agricultural Sciences,* 79 (11) : 21-26.

Vassilev, A., Perez-Sanz, A., Semane, B., Carteer, R. and Vangronsveld, J. (2005). Cadmium accumulation and tolerance of two salix genotypes hydrophonically grown in presence of cadmium. *Journal of Plant Nutrition,* 28 : 2159-2177.

Verkleiji, J.A.C. and Schat, H. (1990). Mechanisms of metal tolerance in higher plants. In: Shaw J. (eds), Heavy metal tolerance in plants: evolutionary aspects, pp. 179-193. CRC Press, Boca Raton.

Vert G., Grotz N., Dedaldechamp, F., Gaymard, F., Guerinot, M.L., Briat, J.F. and Curie, C. (2002). IRT1, an Arabidopsis transporter essential for iron uptake from the soil and plant growth. *Plant Cell,* 14 : 1223–1233.

Vesper, S.J. and Weindensaul, T.C. (1978). Effect of cadmium nickel, copper and zinc on nitrogen fixation by soybean. *Water Air Soil Pollution,* 9 : 413-422.

Vigue, G.T., Pepper, I.L. and Bezdicek, D.F. (1981). The effect of cadmium on nodulation and nitrogen-acetylene fixation by dry mass (*Phaseolus vulgaris* L.). *Journal of . Environment Quality,* 10: 87-90.

Vitoria, A. P., Lea, P. J and Azevedo, R.A. (2001). Antioxidant enzymes responses to cadmium in radish tissues. *Phytochemistry,* 57 : 701-710.

Vogeli-Lange, F. and Wagner, G. (1990). Sub cellular localization of cadmium and cadmium-binding peptides in tobacco leaves. Implication of a transport function for cadmium binding peptides. *Plant physiology,* 92:1086-1093.

Wang, H., Zhao, S.C., Liu, R.L., Zhou, W. and Jin, J.Y. (2009). Changes of photosynthetic activities of maize (*Zea mays L.*) seedlings in response to cadmium stress.*Photosynthetica,* 47 : 277-283.

Wilcklif, C., Evans, H.J., Carter, K.R. and Russell, S.A. (1980). Cadmium effects on the nitrogen fixation system of red alder. *Journal of Environment Quality,* 9: 180-184.

Williams, L. E., Pittman, J. K. and Hall, J. K. (2000). Emerging mechanisms for heavy metal transport in plants. *Biochmical Biophysis Acta,* 1465 : 104-126.

Winterbourn, C. C. (1982). Superoxide-dependent formation of hydroxyl radicals in the presence of iron salts is a feasible source of hydroxyl radicals in vivo. *Biochemical Journal Letter,* 205 : 461-463.

Wool House, H.W. (1983). Encyclopaedia of plant physiology. In: Toxicity and tolerancein the responses of plants of metals (Eds.: O.L. Lange, P.S. Nobel,C.B. Osmond and H. Ziiegler). New Series, Vol. 12C, *Springer-Verlag,Berlin.* pp. 245-300.

Wu F.B., Chen F.K., Wei K.G. and Zhang, P. (2004). Effects of cadmium on free amino acids, glutathione, and ascorbic acid concentration in two barley genotypes (*Hordeum vulgare* L.) differing in cadmium tolerance. *Chemosphere.* 57: 447–45.

Wua, F, Zhang, G. and Dominy, P. (2003). Four barley genotypes respond differently to cadmium: lipid peroxidation and activities of antioxidant capacity. *Environment of Experimental Botany,* 50 : 63-78.

Wuana, R.A. and Okieimen, F.E (2011). Heavy metals in contaminated soils: a review of sources, chemistry, risk and best available strategies for remediation. (ISRN. http://dx.doi.org/10.5402/2011/402647)

Xie H.L., Jiang R.F., Zhang F.S. and McGrath S.P. (2009). Effect of nitrogen form on the rhizosphere dynamics and uptake of cadmium and zinc by hyper accumulator *Thlaspi caerulescens. Plant Soil.* 318: 205–215.

Yadav, S.K. (2010). Heavy metals toxicity in plants: An overview on the role of glutathione and phytochelatins in heavy metal stress tolerance of plants. *South African Journal of Botany,* 76: 167–179.

Yang X.E., Long X.X., Ye H.B., He Z.L., Calvert, D.V. and Stoffella, P.J. (2004). Cadmium tolerance and hyper accumulation in a new Zn-hyper accumulating plant species (*Sedum alfredii* H.). *Plant Soil,* 259 : 181–189.

Yang, M.G., Lin, X. Y. and Yang, X. E. (1998). Impact of Cd on growth and nutrient accumulation of different plant species. *China Journal of Applied Ecology,* 19 : 89-94.

Zawoznik, M.D., Gropppa, M.P. and Benavides, L. (2012). Unraveling cadmium toxicity and tolerance in plants: insight into regulatory mechanisms. *Environment Experimental Botany*, 83 : 33-46.

Zhao, D.L., Reddy, K.R., Kakani, V.G. and Reddy, V.R. (2005). Nitrogen deficiency effects on plant growth, leaf photosynthesis, and hyper spectral reflectance properties of sorghum. *European Journal of Agronomy,* 22: 391–403.

Zhu, Y.L., Elizabeth, A.H., Smits, A.H.P., Jouanin, L. and Terry, N. (1999). Over expression of glutathione synthetase in Indian mustard enhance cadmium accumulation and tolerance. *PlantPhysiology,*119 : 73–79.

■■■■

10

LEGUME SYMBIOSIS UNDER ABIOTIC STRESSES-A REVIEW

Ranju Gulati

Department of Botany, DAV College, Chandigarh, India
E mail: ranjugulatidav@gmail.com

INTRODUCTION

Nitrogen is the most vital nutrient for plant growth both quantitatively as well as qualitatively. It is the main plant mineral nutrient needed for chlorophyll production and functioning of other plant cell components, i.e., proteins, nucleic acids, amino acids. Though nitrogen is present in abundance in the earth's atmosphere (approximately 78%) but due to its inert nature it is metabolically unavailable directly to the higher plants. Nitrogen must be reduced into a plant accessible form and then converted to other forms like ammonia and nitrate. The process is referred to as nitrogen fixation. The availability of fixed nitrogen is often a limiting factor for crop productivity, creating pressure on global agriculture to provide food security as world's population is continuously growing in the 21st century. The abundance of nitrogen, contrasted with the difficulty of chemically utilizing this source, creates a paradox which nature has ingeniously solved by the process of biological nitrogen fixation i.e. the reduction of nitrogen to the metabolically usable form of ammonia with the help of some restricted prokaryotes. (Kim and Rees, 1994). These prokaryotes break the strong triple bond within the nitrogen molecule, with the key participation of the enzyme nitrogenase, to produce ammonia and this phenomenon is termed as biological nitrogen fixation (Franche *et al.*, 2009). Biological nitrogen fixation is a phenomenon occurring in all known ecosystems and is, undoubtedly, of greatest agricultural importance. Biological nitrogen fixation represents the most significant contributor to the global nitrogen cycle (Christiansen *et al.,* 2001). It can be a major source of nitrogen in agriculture where symbiotic N_2 fixing systems are used. Only a relatively small number of microorganisms, referred to as diazotrophs, restricted to the Archaebacteria, are able to catalyze nitrogen fixation. The symbiotic relationship between diazotrophs called rhizobia and legumes can provide large amounts of nitrogen to the plant and can have a significant impact on agriculture. Biological nitrogen fixation (BNF) is the distinguishing feature of a legume in a cropping system. Legume –rhizobia association is critical for plant productivity in the view that soil nitrogen often limits plant growth (Graham and Vance, 2003). BNF introduces approximately 40 million tonnes of nitrogen into agricultural soils each year (Herridge *et al.,* 2008) but this symbiosis is affected by environmental stress and in light of this the chapter will discuss the process of biological nitrogen fixation and effects of abiotic stresses on legume production and biological nitrogen fixation.

PROCESS OF BIOLOGICAL NITROGEN FIXATION

In most undisturbed non-agricultural ecosystems - as well as some agricultural ones - soluble nitrogen (ammonium) is made available to plants via a process of biological nitrogen fixation (BNF), in which bacteria (termed "diazotrophs") that contain the enzyme complex called nitrogenase can fix atmospheric N_2 into ammonia (NH_3) using energy from ATP and reductant (electrons) supplied by the metabolism of carbohydrates, such as sugars.

$$atmN_2 + 8e^- + 8H^+ + 16ATP = 2NH_3 + H_2 + 16ADP + 16Pi.$$

This ammonia, which is potentially toxic to the organism, is usually then immediately converted into amino acids or amides for use by the diazotrophic bacterium in the production of proteins and peptides to facilitate its growth. The fixed nitrogen incorporated into diazotrophic bacteria is then released into the environment when they die, usually in the form of amino acids that then become mineralized and available for uptake by other bacteria and by plants. Biological nitrogen fixation contributes about 100 million tons of nitrogen for terrestrial ecosystems, 30 to 300 million tons for marine ecosystems and 20 million tons from chemical fixation due to atmospheric phenomena (Rakash and Rana, 2013).

Symbiosis

Many plants, particularly legumes, but also some other higher plants (e.g. *Gunnera,* and "actinorhizal" plants, such as *Alnus, Casuarina* and *Myrica*), as well as some cycads and the fern *Azolla,* form mutualistic symbiotic relationships with nitrogen-fixing bacteria. In these systems, soil bacteria located in specialized organs take nitrogen from the atmosphere and convert it to soluble nitrogen which is then available for plant growth. Nitrogen-fixing legumes are one of the main sources of nitrogen in both natural environments and low-input agricultural systems.

N-fixing Legumes and their Bacterial Symbionts

Legumes (*Fabaceae* or *Leguminosae*) are the third largest angiosperm family, with more than 700 genera and 18,000 species with an exceptionally wide range of habitats (Lewis *et al.,* 2005). It has been divided taxonomically into three sub-families, the Papilionoideae (468 genera), Mimosoideae (78 genera) and Caesalpinioideae (157 genera), with only the Papilionoideae being represented in temperate regions. The vast majority of legume species of all three sub-families are confined to the tropics and sub-tropics, where they are particularly abundant and diverse in forests, wetlands and savannahs. Although it is often assumed that all legumes can nodulate and fix N but research indicates that only 20% of legume genera have been tested for their ability to nodulate and fix. Of those tested, few genera (5%) in the Caesalpinioideae are known to nodulate, whereas most genera nodulate in the Papilionoideae (93%) and Mimosoideae (88%).

In addition, legumes have been estimated to contribute 20% of the nitrogen needed for global grain and oilseed production and can potentially fix about 80% of their own nitrogen need and in addition can contribute to the yield of subsequent crops and this outstanding contribution is due to ability of most legumes to establish a symbiotic relationship with soil nitrogen-fixing (N_2-fixing) belong to diverse groups of α and β proteobacteria, known collectively as rhizobia, e.g., *Rhizobium, Mesorhizobium, Bradyrhizobium, Azorhizobium, Allorhizobium,* or *Sinorhizobim* (Chen *et al.,* 2003; MacLean *et al.,* 2007; Velázquez *et al.,* 2010). Rhizobia are very important for crop production because they form symbiotic relationship with legume the process that converts atmospheric elemental Nitrogen (N_2) into ammonia (NH_3) accounting for 65% of the nitrogen currently utilized in agriculture (Matiru and Dakora, 2004).

Legumes are categorized into four groups as per there to their ability to fix nitrogen in experiments. Common bean has the lowest capacity for nitrogen fixation and is in a group by itself, with an average Nitrogen derived from atmosphere (Ndfa) of 40%. The second group consists of most of the winter and summer pulses, with an average Ndfa of 63%. The third group consisting of soybean and groundnut has Ndfa of 68% and the final group includes faba bean and lupin with Ndfa of 75%. The ranges of values within the four groups are large and reflect variations in legume growth, set by genetic, agronomic, environmental and experimental factors, the availability of soil mineral N and numbers and effectiveness of rhizobia in the vicinity of the growing root system (Herridge *et al.,* 2008).

Infection and Nodule Development

Development of the symbiosis results in the production of a new plant organ, the root nodule, where symbiotic nitrogen fixation occurs. The induction of nodule involves the attachment of free-living soil bacteria rhizobia to the legume roots and prepares the legume for infection (Schultze and Kondorosi, 1998). Rhizobia attached to a root hair result in deformation and curling of root hairs and then are transported toward root cortical cells using an infection thread. Before being released into root cortical cells they are surrounded by a plant-derived membrane, the symbiosome membrane (SM) that separates them from the plant cell cytoplasm (Whitehead and Day, 1997). Rhizobia once differentiated into their symbiotic nitrogen fixing form are called as bacteroids which can fix the atmospheric nitrogen (Mylona *et al.,* 1995, Chanway *et al.,* 2014). The bacteroid, SM and the space surrounding the bacteroid, the symbiosome space, together comprise the symbiosome (Catalano *et al.,* 2004).

Biochemistry of Symbiotic Nitrogen Fixation

Symbiotic nitrogen fixation in legumes takes place in specialized organs called nodules that result from rhizobial infection (Krusell *et al.,* 2005) and the enzyme complex nitrogenase comprising the Fe and Mo subunits is responsible for the fixation of nitrogen in root nodules. The nitrogen molecule consists of two nitrogen atoms joined by a triple covalent bond which renders the molecule almost chemically inert and nonreactive (Wagner, 2012). It does not combine with other elements easily. The breaking of this bond is catalyzed by the microbial enzyme nitrogenase. Nitrogenase is the complex metalloenzyme that plays a crucial role in the reduction of biological nitrogen as it catalyses the conversion of atmospheric dinitrogen to ammonia. Nitrogenase consist of two component proteins called as component I and component II and iron is essential in the two components that make up nitrogenase. The iron protein is the smaller component, which is reduced and provides electrons to the molybdenum-iron protein, a larger, heterotetrameric component that contains the catalytic site (Dixon and Kahn, 2004). At the catalytic site, N_2 binds and is reduced (Peters and Szilagyi, 2006). Both the iron protein and the molybdenum-iron protein are sensitive to oxygen. These proteins are soluble in nature. Component I is also known as Mo Fe protein or dinitrogenase and component II is referred as Fe protein or dinitrogenase reductase (Burris, 1991). The Mo Fe protein contains the active site for substrate reduction having a molecular weight of approximately 250kDa (Christiansen *et al.,* 2001). According to Rees *et al.,* 2005 the Mo-Fe protein is associated with two Mo atoms, 30 non-heme Fe atoms and 32 acid labile sulfides. The Mo, Fe and S proteins of component I are organized into two unique metalloclusters called P (or [8Fe-7S]) cluster and iron-molybdenum cofactor (Fe-Mo cofactor) (Dos Santos *et al.,* 2004). The actual reduction occurs in the Fe-Mo centre (Rees *et al.,* 2005). Component II has a molecular weight of 55000-65000Da (Christiansen *et al.,* 2001). The Fe protein contains four non-heme Fe atoms and four γ homodimer consisting of acid labile sulfides referred as [4Fe-4S] cluster which is bridged between the dimer (Howard *et al.,* 1989).

The enzyme nitrogenase possesses a high degree of sensitivity towards oxygen and thus the nodule must maintain low oxygen concentrations while maintaining oxygen supply for bacterial metabolism (Appleby, 1984; Downie, 2005). An oxygen binding protein leghaemoglobin, LHb (red pigment of legume root nodule) acts as an oxygen buffer that cycle between oxidized Fe^{3+} and reduced Fe^{2+} and helps to maintain oxygen level sufficiently low inside the nodule (Arredondo Peter *et al.,* 1998; White *et al.,* 2007). Leghemoglobin buffers the concentration of free oxygen in the cytoplasm of infected plant cells to ensure the proper function of root nodules. Leghemoglobin is present within the cytoplasm of infected cells, at a concentration of approximately 3 mM (Bergersen and Appleby, 1981). Whether leghemoglobin is present within the symbiosome space is controversial (Appleby, 1984). However, if present it is found at low concentrations, approximately 200–500 μM (Bergersen and Appleby, 1981). The apoprotein and heme moiety, both components of leghemoglobin are synthesized by the plant (O' Brian, 1996).

Nitrate Assimilation

Nitrogen derived from soil generally in the form of nitrate is reduced to ammonia by a two-step process called nitrate assimilation. The first step i.e. reduction of nitrate to nitrite, is catalyzed by the plant nitrate reductase (NR, EC 1.6.6.4) and involves reduction of two electrons and the second step is a six electron reduction of nitrite to ammonia cataylsed by nitrite reductase (NiR , EC 1.7.7.1) (Rosales *et al.* 2011). Ammonia is then assimilated into glutamine and glutamate and the major enzymes involved are glutamate synthase (GS) or glutamine-2-oxoglutarate amino transferase (GOGAT), and glutamate dehydrogenase (GDH) (Esposito *et al.,* 2005; Wickert *et al.,* 2007). Although the GS/GOGAT metabolic pathway is the main route of N assimilation in higher plants, these plants possess the ability to use alternative routes, such as the reversible amination of 2-oxoglutarate to produce glutamate by GDH (Lasa *et al.* 2002). For the GS/GOGAT cycle to work, N metabolism must interact with C metabolism, since GS activity requires energy in the form of ATP and the GOGAT uses C skeletons and reductant in the form of 2-oxoglutarate and reduced ferredoxin or NADH, respectively. This pathway is of crucial importance, since the gultamine and glutamate produced are donors for the biosynthesis of major N-containing compounds, including amino acids, nucleotides, chlorophylls, polyamines, and alkaloids (Lea and Ireland, 1999).

ABIOTIC STRESSES AFFECTING BIOLOGICAL NITROGEN FIXATION

In the entire world plant productivity is limited by some environmental factors most common being: salinity drought, pH imbalances, metal toxicities and extremes temperatures. The various abiotic stresses are discussed as follows.

Drought

Due to changing climatic condition affecting pattern of rainfall, drought is inevitable and food demand for increasing world population will aggravate effects of drought (Somerville and Briscoe, 2001). Scarcity of water is a severe environmental constraint leading to loss of crop productivity due to reduced canopy absorption of photosynthetically active radiation, decreased radiation-use efficiency and reduced harvest index (Earl and Davis, 2003, Farooq *et al.,* 2009). The first and foremost effect of drought on legumes is impaired germination, seedling establishment and early seedling growth (Harris *et al.,* 2002, Okcu *et al.,* 2005, Kaya *et al.,* 2006 and Zeid and Shedeed, 2006). Drought manifests itself in the rhizobia via osmotic stress, which alters rhizobia morphology which eventually leads to reduction in infection and nodulation of legumes (Zahran, 1999, Elboutahiri *et al.,* 2010, Mhadhebi *et al.,* 2011). Drought reduces the rhizobia population in soils, inhibits nodulation and nitrogen fixation. Nitrogen fixation process is highly sensitive to the deficiency of soil moisture and reduction in nitrogen fixing ability is linked to reduced permeability of the diffusion barrier for oxygen flux in the nodule and thus prevents nitrogenase damage (Hunt and Layzell, 1993). Streeter (2003) in soyabean found that, under drought conditions, nitrogen fixation activity in nodules was depressed because demand for fixed nitrogen to support growth was lower.

Drought problems for legumes are likely to worsen with the projected rapid expansion of water-stressed areas of the world from 28 to 30 countries today to 50 countries encompassing 3 billion people by 2030 (Postel, 2000). Hence, there is a crucial need to increase drought tolerance in legumes. Several mechanisms have been suggested to explain the varied physiological responses of several legumes to water stress. The legumes with a high tolerance to water stress usually exhibit osmotic adjustment like proline, the free amino acids and low-molecular-weight solutes such as pinitol, production of chaperones, stress enzymes synthesis (Hussain *et al.,* 2014). Potassium can also alleviate the effects of water shortage on symbiotic N_2 fixation in legumes (Marschner 1995, Kurdali *et al.,* 2002).

Salinity

Salt stress is one of the major environmental stresses conditions that adversely affect legume production in arid and semiarid regions, particularly because these plants take most of their nitrogen demands from symbiotic N_2 fixation (El-Sheikh *et al.,* 1995). Legumes and the process of nodule formation and functioning are sensitive to salt stress (Bruning and Rozema, 2013). The main cause of salinity is the nutrient imbalance in the soil which is considered as a constraint influencing the nitrogen fixing symbiosis and the survival of both the partners (Mohammadi *et al.,* 2012) and can reduce crop yield (Jouyban, 2012). Dhanapackiam and Muhammad Ilyas 2010, Fahmi *et al.,* 2011 reported the declension in nodulation, nitrogenase activity and ultimately nitrogen content under salt stress in *Sesbania grandiflora* and faba bean. The legume-*Rhizobium* symbioses and nodule formation on legumes are more sensitive to salt or osmotic stress than are the rhizobia. Hasan (2009) and Kenenil *et al.* (2010) examined the effects of salt stress on nitrogen fixation. The reduction of N_2-fixing activity by salt stress is usually attributed to a reduction in respiration of the nodules and a reduction in cytosolic protein production, specifically leghemoglobin, by nodules. Reduction in photosynthetic activity might also affect N_2 fixation by legumes under salt stress. Salt stress also causes the formation of reactive oxygen species (ROS) which causes oxidative damage to various cellular components (Pallavi *et al.,* 2012)

One of the adaptive salinity-tolerance strategies, that maintain the ionic and intracellular osmotic homeostasis, is the accumulation of osmolytes (or osmoprotectant), mainly amino-acids (proline, glycine betaine) and sugars (Farissi *et al.,* 2011) and to overcome oxidative stress and damages caused by salinity, plants have developed antioxidant non-enzymatic and enzymatic systems (Mittler, 2002; Wang, 2009).

Heavy Metals

Heavy metal toxicity can limit growth and development of plants (Qadir *et al.,* 2004; Agehi *et al.,* 2009) and excessive metal concentrations on the other hand cause Metal enriched soils can adversely affect the legume growth by causing undeniable damage to rhizobia, legumes and their symbiosis due to their toxicity on their physiological processes (Bibi and Hussain, 2005; Wani and Khan, 2010). Formation of ineffective nodules under heavy metal toxicity in the soils has been reported by Brady *et al.,* 1990. .Singh *et al.* (1994) and Reichman (2007) reported the declension in the length of root and number and size of root hair which are essential for nodulation under heavy metals. Heavy metals disrupt nodule structure, inhibit activity of nitrogenase which will lead to decline in nitrogen fixation and will ultimately affects biomass production in legumes (Balestrasse *et al.* 2003; Mumtaz *et al.* 2006; Younis, 2007)

Temperature

Heat stress is a complex function of intensity, duration, and the rate of the increase in temperature beyond a threshold level for a period sufficient to cause permanent damage to plant growth and development (Wahid *et al.*, 2007). High soil temperature influences the survival and persistence of rhizobia. temperature based delay in infection process, nodulation and nitrogen fixing is reported in legumes (Zahran, 1999; Aranjuelo *et al.*, 2007, Alexandre and Oliveria, 2013) and the temperature optimum for the growth of rhizobia in culture is from 28-31°C though rhizobia from Savannah environment were reported to grow well at 40°C (Zahran, 2001). Every legume and *Rhizobium* combination has an optimum temperature relationship, which is around 30°C for clover and pea, between 35 to 40°C for soybean, peanut and cowpea, and between 25 to 30°C for common bean (Evans *et al.,* 1988, Evans *et al.,* 1990; Long, 2001). Tsonev *et al.* (2003) studied the effect of low temperature on *Phaseolus vulgaris* and found that it enhanced photosynthetic down regulation with protective acclimatization to cold, highlighting the role of photorespiration in acclimatization to temperature. They supported the hypothesis for the provision of a safety valve for excess energy, which cannot be replaced by compensating increased CO_2 fixation.

Soil Acidity

Worldwide, more than 1.5 Giga hectares of acid soils limit agriculture production (Graham and Vance 2000) and as much as 25% of the agriculture land is impacted by problems associated with soil acidity (Munns 1986).The soil acidity can affects seriously affect nodulation, N_2 fixation and plant growth (Lie, 1981). Nitrogen fixation by rhizobia to most leguminous plants is effective at neutral or slightly acidic soils (Francis, 1982; Bordeleau and Prévost, 1994). Researchers have reported that most legume species fail to nodulate at pH less than 5.0 because cannot withstand acidic condition but few strains can grow below pH 4.5 (Ayanba *et al.,* 1983, Chen *et al.,* 1993) The optimum pH for rhizobial growth is considered to be between 6.0-7.0 (Hungaria and Vargas, 2000). Rhizobia can be more sensitive to acidic condition than their legume host and according to Munns and Keyser (1981) slow growing *Rhizobium* species are generally more acid tolerant than fast growing species Foster (2000) recorded that *Sinorhizobium meliloti* was viable only below to pH 5.5. Some rhizobia have wide range of pH such as *Sinorhizobium fredii* can grow well between pH 4 – 9.5 but *Bradyrhizobium japonicum* cannot grow at the extremes of that range (Fujihara and Yoneyama, 1993). Cheng *et al.* (2002) reported that soil acidity had different effects on the nodulation of *Medicago* species. *Medicago murex* produced fewer nodules in the soil pH 4.3 than the plants grown in soil of pH 7.0 but these nodules developed at the similar rate to the plants grown in soil of neutral pH. Rao *et al.* (2002) observed pH of 8.9 as critical limit for nodulaton in *Cicer arietinum*. Indeed, it is in many cases the inability of the rhizobia to persist and survive under acidic conditions that reduces the effectiveness of the symbiosis. Therefore, the selection of rhizobial strains tolerant to acidic conditions may improve the acid tolerance of the legume through an efficient symbiotic nitrogen fixation under acidity conditions.

FUTURE SCOPE

There is an urgent need to improve the inputs of organics and BNF in agriculture. Development of effective and competitive rhizobial strains tolerant to high temperature, drought, nitrate, acidity and other abiotic stresses is of high priority. Soil management practices such as soil reclamation, correcting nutrient deficiencies, application of organics and screening of segregating material in legumes in low nitrogen soils are some of the quickest means to increase the contribution of BNF in agriculture. Where inoculation is not feasible, the selection/breeding for high nodulating cultivars could be an option. Developing transgenic inoculants, nitrogen-independent cereals, and endophytic nitrogen fixation requires a very careful re-look and stronger justification for funding.

CONCLUSIONS

Environmental factors affect symbiotic process and nitrogen fixation by *Rhizobium* in soil. There have been discovered some symbiotic systems that are tolerant to certain stress factors like: salinity, drought, extreme temperature, metal toxicity. This relationship can ensure successful molecular nitrogen fixation under less favorable factors. Under poor conditions *Rhizobium*/legumes symbiosis is very important because it may be the only way to fix nitrogen. In the past century the tremendous progress have been made in understanding the process of symbiosis and nitrogen fixation, and using genetic engineering will lead to improve the plant resistance to environmental factors.

REFERENCES

Aghaei, K; Ehsanpour, AA; Shah, AH and Komatsu, S 2009. Proteome analysis of soybean hypocotyl and root under salt stress. *Amino Acids.*, 36: 91-98.

Alexandre, A and Oliveira, S 2013. Response to temperature Stress in Rhizobia. *Critical Reviews in Microbio.,* 39: 219-228.

Appleby, CA 1984. Leghemoglobin and rhizobium respiration. *Annu. Rev. Plant Physiol. Plant Mol. Biol.,* 35: 443–478.

Aranjuelo, I; Irigoyen, J and S´anchez-D´ıaz, M 2007. Effect of elevated temperature and water availability on CO2 exchange and nitrogen fixation of nodulated alfalfa plants. *Environ. Exp. Bot.*, 59: 99-108.

Arredondo-Peter, R; Hargrove, MS; Moran, JF; Sarath, G and Klucas, RV 1998. Plant hemoglobins. *Plant Physiol.*, 118:1121–1125.

Ayanaba, A; Asanuma, S and Munns, DN 1983. An agar plate method for rapid screening of *Rhizobium* for tolerance to aluminum stress. *Soil Sci. Soc. Am. J.*, **47**: 256-258.

Balestrasse, KB; Benavides, MP; Gallego, SM and Tomaro, M L 2003. Effect of cadmium stress on nitrogen metabolism in nodule and roots of soybean plants. *Funct. Plant. Biol.*, 30: 57-64.

Bergersen, FJ and Appleby, CA 1981. Leghemoglobin within bacteroid enclosing membrane envelopes from soybean root nodules. *Planta.*, 152: 534–543.

Bibi, M and Hussain, M 2005. Effect of copper and lead on photosynthesis and plant pigments in black gram (*Vigna mungo* L.). *Bull Environ Contam Toxicol.,* 74:1126–1133.

Bordeleau, LM and Prevost, D 1994. Nodulation and nitrogen fixation in extreme environments. *Plant Soil.*, 161: 115-125.

Brady, DJ; Hetch-Buchholz, CH; Asher, CJ and Edwards, DG 1990. Effect of low activities of aluminium on Soybean (*Glycine max* L.) early growth and nodulation. In: M. L. Van Blusichem (Ed.), Plant Nutrient Physiology and Application. Kluwer, Dordrecht. 329-334.

Bruning, B and Rozema, J 2013. Symbiotic nitrogen fixation in legumes: perspectives for saline agriculture. *Environ. Exp. Bot.*, 92:134–143.

Burris, RH 1991. Nitrogenase. *J. Biol. Chem.*, 266: 9339-9342.

Catalano, CM; Lane, WS and Sherrier, DJ 2004. Biochemical characterization of symbiosome membrane proteins from *Medicago truncatula* root nodules. *Electrophoresis.,* 25:519–531.

Chanway, CP; Anand, R and Yang, H 2014. Nitrogen fixation outside and inside plant tissues. In: Ohyama T, ed. Advances in biology and ecology of nitrogen fixation. In Tech, 3–21.

Chen, H; Richardson, AE and Rolfe, BG 1993. Studies of the Physiology and Genetic Basis of Acid Tolerance in *Rhizobium leguminosarum* biovar *trifolii. Applied Env. Microbio.*, 59: 1798 – 1804.

Chen, WM; Moulin, L; Bontemps, C; Vandamme, P; Béna, G and Boivin-Masson, C 2003. Legume symbiotic nitrogen fixation by ß-proteobacteria is widespread in nature. *J. Bacteriol.*, 185: *7266–7272.*

Cheng, WH; Endo, A; Zhou, L; Penney, J; Chen, HC; Arroyo, A; Leon, P; Nambara, E; Asami, T; Seo, M; Koshiba, T and Sheen, J 2002. A unique short-chain dehydrogenase/reductase in *Arabidopsis* glucose signaling and abscisic acid biosynthesis and functions. *Plant Cell.,* 14: 2723–2743.

Christiansen, J; Dean, DR and Seefeldt, LC 2001 Mechanistic features of the Mo-containing nitrogenase. *Annu. Rev. Plant physiol. Plant Mol. Biol.*, 52: 269-95.

Dhanapackiam, S and Muhammad Ilyas, MH 2010. Effect of NaCl salinity on growth, nodulation and total nitrogen in *Sesbania grandiflora. Ind. J. Sci. Tech.* 3(1):87-89.

Dixon, R and Kahn, D 2004. Genetic regulation of biological nitrogen fixation. *Nature Rev. Microbiol.*, 2: 621-631.

Dos Santos, PC; Dean, DR; Hu, YL and Ribbe, MW 2004. Formation and insertion of nitrogenase iron-molybdenum cofactor. *Chem. Rev.*,104: 1159-1173.

Downie, JA 2005. Legume Haemoglobins: Symbiotic Nitrogen fixation needs bloody nodules. *Curr. Biol.*, 15(6): 196-198.

Earl, H and Davis, RF 2003. Effect of drought stress on leaf and whole canopy radiation use efficiency and yield of maize. *Agron. J.*, 95: 688– 696.

Elboutahiri, N; Thami-Alami, I and Udupa, SM 2010. Phenotypic and genetic diversity in *Sinorhizobium meliloti* and *S. medicae* from drought and salt affected regions of Morocco. *BMC Microbiol* ., 10:15.

El-Sheikh, EAE and Wood, M 1995. Nodulation and N_2 fixation by soybean inoculated with salt-tolerant rhizobia or salt-sensitive bradyrhizobia in saline soil. *Soil. Biol. Biochem.*, 27:657–661.

Esposito, S; Guerriero, G; Vona, V; Rigano, VDM; Carfagna, S and Rigano, C 2005. Glutamate synthase activities and protein changes in relation to nitrogen nutrition in barley: The dependence on different plastidic glucose-6P dehydrogenase isoforms. *J. Exp. Bot.*, 56:55-64.

Evans, J; Dear, B and O'Connor, GE 1990. Influence of an acid soil on the herbage yield and nodulation of five annual pasture legumes. *Aust J Exp Agric.*, 30:55–60.

Evans, J; Hochman, Z; O'Connor, GE and Osborne, GJ 1988. Soil acidity and *Rhizobium*: their effect on nodulation of subterranean clover on the slopes of southern New South Wales. *Aust J Agric Res.,* 38:605–618.

Fahmi, AI; Agate, HH; Eissa, RA and Hassan, MM 2011. Effect of salt stress on some nitrogen fixation parameters in Faba Bean. *Pakistan J Biol Sci.,* 14: 385-391.

Farissi, M; Bouizgaren, A; Faghire, M; Bargaz, A and Ghoulam, C 2011. Agro-physiological responses of *Moroccan alfalfa* (*Medicago sativa* L.) populations to salt stress during germination and early seedling stages. *Seed Sci. Technol.*, 39:389–401.

Farooq, M; Wahid, A; Kobayashi, N; Fujita, D and Basra, SMA 2009. Plant drought stress: effects, mechanisms and management Review Article. *Agro for Sustainable Dev.*, 29:185-212.

Foster, JW 2000. Microbial Responses to Acid Stress. In G. Storz & R. Hengge-Aronis (ed.), Bacterial Stress Response. ASM Press, Washington, D.C, 99-115.

Franche, C; Lindström, K and Elmerich, C 2009.Nitrogen-fixing bacteria associated with leguminous and non-leguminous plants. *Plant Soil.*, 321:35-59.

Francis, AJ 1982. Effects of Acidic Precipitation and Acidity on Soil Microbial Processes. *Water, Air, Soil Pollution.*, 18(1): 375-394.

Fujihara, S and Yoneyama, T 1993. Effects of pH and osmotic stress on cellular polyamine contents in the soybean rhizobia *Rhizobium fredii* p220 and *Bradyrhizobium japonicum* A 1007. *Applied Environ, Microbio.,* 59: 1104–1109.

Graham, PH and Vance, CP 2000. Nitrogen fixation in perspective: an overview of research and extension needs. *Field Crops Res.,* 65: 93-106

Graham PH and Vance CP 2003. Legumes: importance and constraint to greater use - Update on legume utilization. *Plant Physiol.,* 131.

Hassan, MM 2009. Genetic studies on *Rhizobium* and legumes symbiosis relationship. Ph.D. Thesis, Department, Faculty Agriculture, Min. University, Egypt.

Harris, D; Tripathi, RS and Joshi, A 2002. On-farm seed priming to improve crop establishment and yield in dry direct-seeded rice, in: Pandey S., Mortimer M., Wade L., Tuong T.P., Lopes K., Hardy B. (Eds.), Direct seeding: Research Strategies and Opportunities, International Research Institute, Manila, Philippines, 231-240.

Herridge, DF; Peoples, MB and Boddey, RM 2008. Global inputs of biological nitrogen fixation in agricultural systems. *Plant Soil* 311 1–18s to greater use. *Plant Physiol.,*131: 872–877.

Howard, JB; Davis R; Moldenhauer, B; Cash, VL and Dean, D 1989. FeS cluster ligands are the only cysteines required for nitrogenase Fe-protein activities. *J. Biol. Chem.*, 264: 11270- 11274.

Hungria, M and Vargas, MA 2000. Environmental Factors Affecting N2 Fixation in Grain Legumes in the Tropics, with an Emphasis on Brazil. *Field Crops Res.*, 65:151-164.

Hunt, S and Layzell, DB 1993. Gas exchange of legume nodules and the regulation of nitrogenase activity. *Ann. Rev. Plant Physiol. Plant Mol. Biol.,* 44: 483-511.

Hussain, Z; Ali, S: Hayat, Z; Zia, MA: Iqbal, A and Ali, GM 2014. Agrobacterium mediated transformation of DREB1A gene for improved drought tolerance in rice cultivars (*Oryza sativa* L.) *Aust. J. Crop Sci.,* 8 (7): 1114–1123.

Jouyban, Z 2012. The effects of salt stress on plant growth. *Tech. J. Engg. App. Sci.*, 2(1):7-10.

Kaya, MD; Okçu, G; Atak, M; Çıkılı, Y and Kolsarıcı, O 2006. Seed treatments to overcome salt and drought stress during germination in sunflower (*Helianthus annuus* L.). *Eur. J. Agron.*, 24: 291-295.

Kenenil, A; Assefa, F and Prabu, PC 2010. Characterization of acid and salt tolerant rhizobial strains isolated from faba bean fields of Wollo, Northern Ethiopia. *J. Agric. Sci. Tech.,* 12, 365-376

Kim, J and Rees, DC 1994. Nitrogenase and biological nitrogen fixation. *Biochem.,* 33: 389-397.

Krusell, L; Krause, K; Ott, T; Desbrosses, G; Krämer, U; Sato, S; Nakamura, Y; Tabata, S; James, EK; Sandal, N; Stougaard, J; Kawaguchi, M; Miyamoto, A; Suganuma, N and Udvardi, MK 2005. The Sulfate Transporter SST1 Is Crucial for Symbiotic Nitrogen Fixation in *Lotus japonicus* Root Nodules. *The Plant Cell.*, 17(5): 1625-1636.

Kurdali,, F; Al-Ain, F and Al-Shamma, M 2002. Nodulation, dry matter production and N fixation by faba bean and chickpea as affected by soil moisture and potassium fertilizer. *J. Plant Nut.*, 25: 355-368.

Lasa, B; Frechilla, S; Aparicio-Tejo, PM and Lamsfus, C 2002. Role of glutamate dehydrogenase and phosphoenolpyruvate carboxylase activity in ammonium nutrition tolerance in roots. *Plant Physiol. Biochem.*, 40: 969-976.

Lea, PJ and Ireland, RJ 1999. Nitrogen metabolism in higher plants. In B Singh (ed.), *Plant Amino Acids: Biochemistry and Biotechnology.* Marcel Decker Inc, New York, 1-47

Lewis, GP; Schrire, B; Mackinder, B and Lock, M 2005. *Legumes of the World.* Richmond, VA: Kew Publishing.

Lie, TA 1981. Environmental physiology of the legume *Rhizobium* symbiosis. In Nitrogen fixation Vol. 1. Ecology Ed. WJ Broughton, Clarendon Press. Oxford. 104-134.

Long, SR 2001. Genes and signals in the Rhizobium-legume symbiosis. *Plant Physiol.*, 125: 69- 72.

MacLean, AM.; Finan, TM and Sadowsky, MJ 2007. Genomes of the symbiotic nitrogen-fixing bacteria of legumes. *Plant Physiol.,*144: 615–622.

Marschner, H 1995. Mineral nutrition of Higher Plant. Academic Pres, New York, pp: 890.

Matiru, VN and Dakora, FD 2004. Potential Use of Rhizobial Bacteria as Promoters of Plant Growth for Increased Yield in Landraces of African Cereal Crops. *African J. Biotech.*, 3:1-7.

Mhadhbi, H; Fotopoulos, V; Mylona, PV; Jebara, M; Aouani, ME and Polidoros, AN 2011. Antioxidant gene-enzyme responses in *Medicago truncatula* genotypes with different degree of sensitivity to salinity. *Physiol. Plant.* 141:201–214.

Mittler, R 2002 Oxidative stress, antioxidants and stress tolerance. *Trends Plant Sci.*, 7:405–410.

Mohammadi, S; Kollias, G and Grama, A 2012. Role of synthetic genetic interactions in understanding functional interactions among pathways. *Pac Symp Biocomput*: 43-54

Mumtaz, H; Sajid, M and Ahmad, A 2006. Effect of lead and chromium on growth, photosynthetic pigments and yield components in mash bean [*Vigna mungo* (L.) Hepper]. *Pak. J. Bot.*, 38:1389–1396.

Munns, DN 1986. Acid soils tolerance in legumes and rhizobia. *Adv. Plant Nutr.*, 2: 63-91.

Munns, DN and Keyser, HH 1981.Tolerance of rhizobia to acidity, aluminium and phosphate. *Soil Sci. Soc. Am. J.,* 34: 519-523.

Mylona, P; Pawlowski, K and Bisseling, T 1995. Symbiotic nitrogen fixation. *Plant Cell.*, 7: 869-85.

O'Brian, MR 1996. Heme synthesis in the rhizobium–legume symbiosis: a palette for bacterial and eukaryotic pigments. *J. Bacteriol.,* 178: 2471–2478.

Okçu G ; Kaya, MD and Atak, M 2005. Effects of salt and drought stresses on germination and seedling growth of pea (*Pisum sativum* L.). *Turk. J. Agric. For.*, 29: 237-242.

Pallavi, S; Ambuj Bhushan, J; Dubey, RS and Pessarakli, M 2012. Reactive Oxygen Species, Oxidative Damage, and Antioxidative Defense Mechanism in Plants under Stressful Conditions. *J. Bot.*, 217027–217052.

Peters, JW and Szilagyi, RK 2006. Exploring new frontiers of nitrogenase structure and mechanism. *Curr. Opin. Chem. Biol.,* 10: 101–108.

Postel, SL 2000. Entering an era of water scarcity. *Ecol Appl.,* 10:941–948.

Qadir, S; Qureshi, MI; Javed, S and Abdin, MZ 2004. Genotypic variation in phytoremediation potential of *Brassica juncea* cultivars exposed to Cd stress. *Plant Sci.,* 167, 1171-1181 .

Rakash, N and Rana, K 2013. Food Legumes for Livelihood and Nutritional Security in North Eastern Himalayan Region: Prospects and Constraints. *Ind. J. Agri. Sci.,* 83: 899-906.

Rao, SSR; Vardhini, BVV; Sujatha, E and Anuradha, S 2002. Brassinosteroids - a new class of phytohormones. *Curr. Sci.*, 82: 1239-45.

Rees, DC; Tezcan, FA; Haynes, CA; Walton, MY; Andrade, S; Einsle, O and Howard, JB 2005. Structural basis of biological nitrogen fixation. *Royal Soc.*, 1829: 971-984.

Reichman, SMA 2007. The potential use of the legume-*Rhizobium* symbiosis for the remediation of arsenic contaminated sites. *Soil Biol.,* 39: 2587–2593.

Rosales, EP; Iannone, MF; Groppa, MD and Benavides, MP 2011. Nitric oxide inhibits nitrate reductase activity in wheat leaves. *Plant Physiol. Biochem.,* 49; 124–130.

Schultze, M and Kondorosi, A 1998. Regulation of symbiotic root nodule development. *Annu. Rev. Genet.,* 32 : 33–57.

Singh, RP., Maheshwari, R and Sinha, SK 1994. Recovery of lead caused decrease in biomass accumulation of mungbean (*Vigna radiata* L.) seedlings by K_2HPO_4 and $CaCl_2$. *Ind. J. Exp. Biol.*, 32: 507–510.

Somerville, C and Briscoe, J 2001. Genetic engineering and water. *Sci.*, 292 : 2217.

Streeter, JG 2003. Effects of drought on nitrogen fixation in soybean root nodules. *Plant Cell Envir.,* 26(8): 1199–1204.

Tsonev, T; Velikova, V; Georgieva, K; Hyde, PF and Jones, HG 2003. Low temperature enhances photosynthetic down-regulation in French bean (*Phaseolus vulgaris* L.) plants. *Ann. Bot.*, 91: 343-352

Velázquez, E; García-Fraile, P; Ramírez-Bahena, MH; Rivas, R and Martínez-Molina, E 2010. Bacteria involved in nitrogen-fixing legume symbiosis: current taxonomic perspective. In: Khan MS, Zaide A, Musarrat J (eds) Microbes for legume improvement, Springer, Wien, 1-25.

Wagner, SC 2012. Biological Nitrogen Fixation. *Nature Edu. Knowl.,* 3(10): 15.

Wahid, A; Gelani, S; Ashraf, M and Foolad, MR 2007. Heat tolerance in plants: an overview. *Environ. Exp. Bot.,* 61:199-223.

Wang, WB; Kim, YH; Lee, HS; Kim, KY; Deng, XP and Kwak, SS 2009. Analysis of antioxidant enzyme activity during germination of alfalfa under salt and drought stresses. *Plant Physiol. Biochem.*,4: 570–577.

Wani, PA and Khan, MS 2010. *Bacillus* species enhance growth parameters of chickpea (*Cicer arietinum* L.) in chromium stressed soils. *Food Chem. Toxicol.,* 48:3262–3267.

White, J; Prell, J; James, EK and Poole, P 2007. Nutrient sharing between symbionts. *Plant Physiol.*, 144: 604-614.

Whitehead, LF and Day, DA 1997. The peribacteroid membrane. *Physiol. Plant.,* 100: 30–44.

Wickert, E; Marcondes, J; Lemos MV and Lemos, EG 2007. Nitrogen assimilation in citrus based on CitEST data mining. *Genet. Mol. Biol.*, 30:810-818.

Younis, M 2007. Responses of *Lablab purpureus-Rhizobium* symbiosis to heavy metals in pot and field experiments. *World J Agric Sci.*, 3:111–122.

Zahran, HH 1999. *Rhizobium*-legume symbiosis and nitrogen fixation under severe conditions and in an arid climate. *Microbiol. Mol. Biol. Rev.*, 63: 4, 968 – 989.

Zahran, HH 2001. Rhizobia from Wild Legumes: Diversity, Taxonomy, Ecology, Nitrogen Fixation and Biotechnology. *J. of Biotech.*, 91: 143-153.

Zeid, IM and Shedeed, ZA 2006. Response of alfalfa to putrescine treatment under drought stress. *Biol. Plant.,* 50: 635-640.

■■■■

11

INTRODUCTION TO PHENOMICS AND ITS APPLICATION IN PHYSIOLOGICAL BREEDING

A. Nishant Bhanu*, H.R. Bhandari, M.N. Singh and K. Srivastava

Department of Genetics and Plant Breeding, Institute of Agricultural Sciences, Banaras Hindu University, Varanasi 221 005, India
**Corresponding Author, e-mail: nishant.bhanu@gmail.com*

INTRODUCTION

To feed the world population in the future propounds a tremendous challenge in front of the researchers of agriculture and the allied plant science as rapid growth in global population from ~6 billion in 2000 to >13 billion in 2100 is expected. (Tester and Langridge, 2010). In addition, crop production is itself facing the consequences of global climate change. Compounding the pressure of exposure to extreme weather, limited availability of nutrient and water resources, decrease of the arable lands restricting it in the tropics and subtropics is overall leading to a reduction in the available area for cultivation. In particular, a 50% loss of land may take place by the year 2050 in certain regions because of drought and increased salinization of the arable land (Wang *et al.,* 2003). Hence, a second green revolution with more sustainable management of renewable resources and more efficient utilization of genetic diversity is needed to facilitate the breeding for climate resilient crops that could assure the rising food demand.

Genetic diversity is the basis for all the plant improvement. The plant breeders have obtained valuable insights regarding heritability of traits, breeding value of the parents, influence of environment and selection approaches for genetically superior offspring using conventional tools like replication and sophisticated experimental designs. The advent of the genomic era emphasised the evaluation of genetic diversity unswervingly at the DNA level. The rapid advancement of high-throughput marker genotyping, next generation sequencing and resequencing, collection of high-density genetic variation information viz. SNPs (single-nucleotide polymorphisms) and CNVs (copy-number variants) necessitates for its correlation with specific functions. Over the past few years, a multitude of genome wide association studies (GWAS) and related strategies have provided valuable insights related to genetic variants associated with complex diseases or other traits in plants. Findings like this are definitely elevating our knowledge about the genetic basis of phenotypic variation thus providing a prospect for genetic testing. Hitherto, only a small proportion of the causal genetic factors have been explained from the variants identified so far, leaving unfolding of the remaining "missing" heritability (Manolio *et al.,* 2009). Moreover, prediction of phenotypic variation from individual genetic codes is still a challenge.

A key goal of biology is to understand phenotypic characteristics and link it with genetic variants. A complex web of interactions between genotype and environment produces the phenotypic variation. Understanding and accessibility of such a 'genotype–phenotype' map cannot be established without the detailed phenotypic data that allow these interactions to be studied (Houle *et al.,* 2010). However, we are lagging a lot in characterizing phenomes (the full set of phenotypes of an individual) with respect to characterizing genomes. Consequently, a suit of new promising technologies-high throughput and high-dimensional phenotyping - phenomics to accelerate progress in our perception of the relationship between genotype and phenotype is emerging (Houle *et al.,* 2010, Fubrank and Tester, 2011)

Geneticists are concerned in comprehending the nature and origin of allelic alterations and mutations of functional significance in the context of both qualitative and quantitative traits. Genomics is embraced by the plant breeders as

a. an approach to deed and protect the genetic conformation of plant varieties,
b. sketching evolutionary relationships,
c. identification of diversity among the genotypes,
d. identification and selection of valuable mutations, and
e. gaining insight regarding the nature of genotype by genotype (GxG) and genotype by environment (GxE) interactions.

Hence, the ultimate goal of genomics research in plant breeding is to pay rich dividends in improving the competence, efficacy and cost-benefit of cultivar improvement.

Further, to understand the effect of genomic variants on phenotypes, pleiotropy and to work out the causes related with complex phenomenon like crop yield, diseases etc. phenomic-level data are indispensable. With the limited understanding of many important biological phenomenons we are unable to measure all the important factors which can be of fringe benefits.

Several reasons corroborating the fact that the moment has arrived to consider the significance of phenomic level attributes are

i. availability of high throughput phenotyping technologies;
ii. brisk advancement of conceptual, analytical and bioinformatics approaches enabling use of high dimensional data;
iii. accessibility of dynamic models for genomic, cellular as well as organism level and
iv. availability of phenotypic information.

Consequently, judicious reviewing of relevant literature on phenomics, this topic was thought worthwhile. Hence, this chapter on phenomics and its application in physiological breeding will cover from simple to complex, focusing upon and discussing the genotyping and phenotyping; concept of a genotype–phenotype map; phenotyping system; plant phenotyping to phenomics; forward and reverse phenomics; phenomic tools; application of plant phenomics to trait-based physiological breeding; plant phenotyping platforms and phenomic projects; plant phenomics and data analysis; and plant phenomics in closing the 'gene to genotype' loop, which could be of interest to readers.

GENOTYPING AND PHENOTYPING

The set of programs and protocols used for elucidating the genetic makeup i.e. genotype of an individual refers to genotyping technology. For deciphering the genetic causes of complex phenomena, including vigour, disease, crop yields and evolutionary fitness genotyping of the character/ individual is essential. The genotyping was performed initially through molecular markers like RFLP (Restriction Fragment Length Polymorphism), RAPD (Random Amplified Polymorphic DNA), AFLP (Amplified Fragment Linked Polymorphism) and SSRs (Simple Sequence Repeats) or microsatellites. In the recent times SNPs (Single nucleotide polymorphism) have become the choice due to their high abundance, accountability to high throughput analysis and low mutation rate. Nowadays, with the advent of next generation sequencing many new technologies have developed which has made possible to discover the genome wide markers.

Phenotype (derived from Greek word *phainein* meaning to show and *typos* meaning type) is the resultant of the complex interactions between genotype (morphological, developmental, physiological, pathological or biochemical properties, phenology and behaviour) and the environmental factors that can be monitored, quantified and/or visualised by some technical procedure. Phenotype is the physical expression of the allelic combinations (genotype) which an organism possesses. Phenotyping is needed for

i. Functional analysis of specific genes
ii. Forward and reverse genetic analysis
iii. Production of new plants with valuable characteristics (Parthsarathii *et al.*, 2015) (vol. 15 advances in pp)

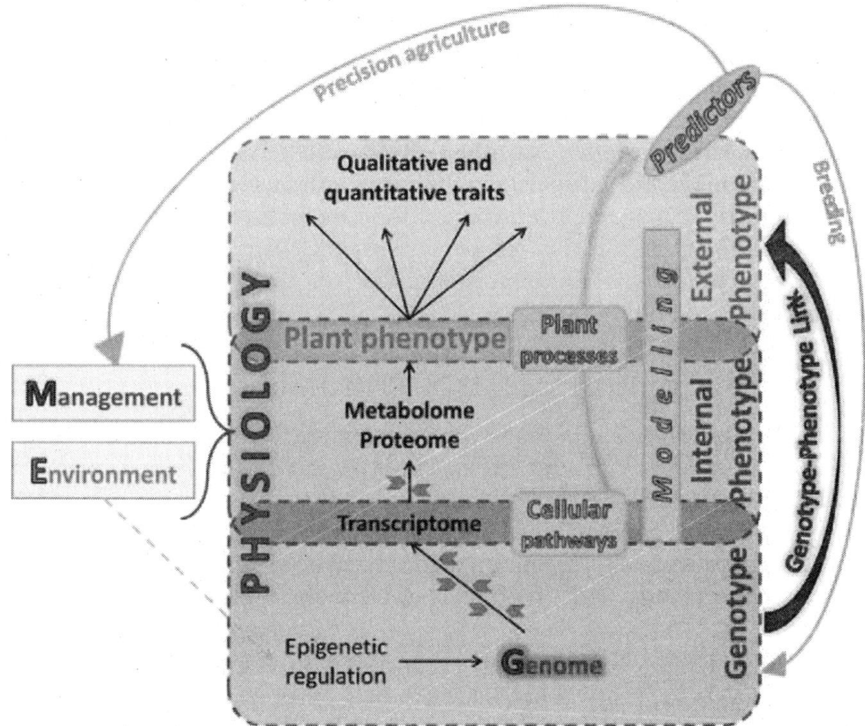

Fig. 1. Physiology is the key interface between the genome and the plant phenotype which consists of multiple levels of information processing via the transcriptome, proteome, and metabolome, as well as regulatory events. Physiology amalgamates genotype, environment, management, and their intense interactions (G×E×M), which determines the internal phenotype as a link between the genotype and external phenotype. Modelling approaches of cellular and whole plant level can be framed by linking the internal phenotype with the genetic background and external phenotypes which can allow the establishment of robust, external predictors. These predictors can be directly applied in precision agriculture as markers in breeding. (Source: Großkinsky *et al.* 2015)

CONCEPT OF A GENOTYPE–PHENOTYPE MAP

Genotype–phenotype map (Burns, 1970; Gjuvsland *et al.,* 2013) is a metaphor used mainly for the multiple ways in which genomics information influences the phenotype of an organism (Houle *et al.,* 2010). The interaction between the genomic information (G) with external parameters (E, M) creates the internal phenotypic state of the organism throughout its life which comprises cellular, tissue, and physiological properties (Fig. 1). These internal phenotypes in turn shape the external phenotypes and a complex feedback relationship is created between gene, phenotype and environment. This external phenotype can be measured in much more simplified way using the high-throughput non-invasive techniques. A detailed phenotypic data is required to study the interactions occurring in a genotype- phenotype map is inaccessible without the detailed phenotypic data that allow these interactions to be studied. Thus, to study a genotype-phenotype map phenomics has enabled us to trace causal links between genotypes, phenotypes and environmental factors. Various approaches for studying a genotype-phenotype map has been listed in table 1.

Table 1. Various approaches towards genotype–phenotype map

S.No.	Approach	Level	Mapping	Information	Related Technology	Potential Application
1	Genomics	DNA	GWAS, linkage Mapping	Sequence, SNPs, CNVs, indels	Genome sequencing, array-based CGH, SNP typing (NGS), exon sequencing	High-throughput genotyping and resequencing reference panels; Identification of rare alleles; High-density genetic markers for QTL detection
2	Epigenomics	Chromatin	methQTL, EWAS	DNA methylation, histone modifications, siRNA	Bisulfite sequencing, ChIP-chip, ChIP-seq	High-throughput examination of epigenome; High-density epigenetic markers for QTL mapping
3	Transcriptomics	RNA	eQTLs, TWAS	Transcripts, miRNAs	EST, gene chip, micro-array, RNA-seq	High-throughput; unbiased and cost-effective examination of transcriptome
4	Proteomics	Proteins	pQTLs, PWAS	Proteins	LC-MS/MS, 2D-PAGE	Detecting quantitative and qualitative variation in proteins

S.No.	Approach	Level	Mapping	Information	Related Technology	Potential Application
5	Metabolomics	Metabolites	mQTLs, MWAS	Metabolites	MS based, NMR, HPLC	Detecting quantitative and qualitative variation in cellular metabolites
6	Phenomics	Cell, tissue, organism, development, physiology, morphology, behaviour	phQTLs, PheWAS	Biomass, morphology	Automated imaging, Non-invasive pheno typing	Phenotyping large samples required for systems genetic analyses Automated noninvasive phenotyping for growth modelling Dissecting quantitative traits for QTL Mapping
7	Systems biology	System	Network mapping	Network	Integrated analysis	Combinational use of genetic and multi-omic data into an integrative analysis

QTLs: quantitative trait loci; methQTLs: QTLs; eQTLs: expression QTLs; pQTLs: protein QTLs; mQTLs: metabolic QTLs; phQTLs: phenotypic QTLs; GWAS: genome-wide association studies; EWAS: epigenome-wide association studies; TWAS: transcriptome-wide association studies; PWAS: proteome-wide association studies; MWAS: metabolome-wide association studies; PheWAS: phenome-wide association studies; SNPs: single-nucleotide polymorphisms; CNVs: copy-number variations; indels: insertion–deletion polymorphisms; siRNAs: small interfering RNAs; miRNAs: microRNAs; NGS: next-generation sequencing; CGH: comparative genome hybridisation; ChIP-chip: chromatin immunoprecipitation (ChIP) followed by microarray technology; ChIP-seq: ChIP followed by high-throughput DNA sequencing; EST: expressed sequence tag; RNA-seq: RNA sequencing; LC-MS/MS: liquid chromatography–tandem mass spectrometry; 2D-PAGE: two-dimensional polyacrylamide gel electrophoresis; MS-based technologies: such as gas chromatography–mass spectrometry (GCMS) and liquid chromatography–mass spectrometry (LC-MS); NMR: nuclear magnetic resonance; HPLC: high-performance liquid-phase chromatography [***Source:** Chen *et al.* 2014]

PHENOTYPING SYSTEM

Dhont *et al.* (2013) described plant phenotyping as the set of structural, physiological, and performance related traits of a genotype in a given environment. Plant phenotyping is a complex affair comprises of a plethora of system and tools. The plant phenotyping system therefore, not only includes the tools which predicts the phenotype rather, also includes the measures related to plant growth in a certain environment which may range from strictly controlled conditions in a growth chamber to the natural field conditions. Dhont *et al.* (2013) also suggested that the phenotyping system can be described by means of throughput, resolution, and dimensionality.

Throughput

Throughput refers to the number of individual units at particular organizational levels within plants, or at the plant or canopy level, that can be analyzed for a particular set of traits at a given time. High-throughput refers to automatized, non-invasive phenotyping systems that have a high capacity with respect to standardized data acquisition, processing, visualization, and feeding into integrated databases.

Resolution

Resolution is the process or act of separating something into its constituent parts or elements. Spatial resolution refers to the level of separation of plants or plant organs, tissues, and cells into their elementary or organizational units used for the measurement of plant traits, straddling the range from cell wall and plastids to the field. whereas temporal resolution indicates the level of separation into elementary time periods significant for plant processes and used for the measurement of plant traits i.e. spanning seconds (e.g. stomatal conductance) or minutes (e.g. cell expansion) to days (e.g. cell division) or weeks and months (e.g. yield).

Dimensionality

Dimensionality in plant phenotyping refers to the diversity of phenotypic traits measured at different spatial and temporal resolutions and in different categories, such as plant structure, physiology, and performance. In phenomics, dimensionality additionally includes the number of genotypes and the diversity of environmental conditions and treatments taken into account upon phenotyping.

Hence, in plant phenotyping system throughput, resolution, and dimensionality are pivotal aspects with consequences for data quality.

PLANT PHENOTYPING TO PHENOMICS

According to Houle *et al.* (2010); Chen *et al.* (2014), the genome (G), environmental conditions (E), and their complex dynamic interactions (G×E) determines the plant phenome. An additional determinant of the phenotype is the crop management (M). Thus, in relevance to the specific agricultural system, Porter and Christensen (2013) suggested that plant phenotype has to be considered as the result of the complex G×E×M interactions. The term 'Phenomics' introduced in 1997 (Schork, 1997) is in analogy to genomics and the many other 'omics' technologies which refers to the systematic characterization of phenotypes through acquisition of high-dimensional phenotypic data on an organism-wide scale (Houle *et al.*, 2010). Moreover, plant phenomics can also be considered as the study of phenomes of multiple genomic expression states with countless possibilities of environmental conditions which can be measured at specific time during development, or alternatively in a dynamic manner.

With the establishment and advancement of the non invasive techniques with automated facilities and their application on large scale, the high-throughput phenotyping techniques which were formerly developed for controlled environments has gained attention since the last decade. In addition, this has been progressively applied from then onwards in the field of basic plant research and crop breeding (Araus and Cairns, 2014; Langridge and Reynolds, 2015). Thus, Plant phenomics may enable to receive the full assistance of the available genomic information, and to relieve the phenotyping bottleneck in breeding (Furbank and Tester, 2011). The term phenomics is defined in analogy to genomics to proceed for the overall characterization of the complete set of a phenotype of a genome. The analogy is somewhat misleading as this ultimate goal of characterization of complete set will remain hypothetical and elusive since phenotypes differ from cell to cell and from time to time, and hence entire characterization can never be achieved. This information applies in particular to plants with a phenome that is greatly influenced by external factors through the G×E×M interactions. Therefore, in phenomics there is a need to prioritize what to measure and when to measure along with meeting solidity between probing and descriptive goals.

FORWARD AND REVERSE PHENOMICS

The phenomics involves two concepts: Forward phenomics and Reverse phenomics

i. Forward phenomics uses phenotyping tools to "sieve" collections of the information of visible and valuable traits of germplasm (Furbank and Tester, 2011). It is used to determine the best suitable germaplasm for a particular trait. High-throughput and fully automated and low resolution followed by higher-resolution, lower-throughput measurements speeds up the phenotyping process and leads to identification of desirable traits fitting to a particular situation (Furbank and Tester, 2011).

ii. Reverse phenomics is the one in which exhaustive dissection of traits is performed to reveal mechanistic basis. In this, the considered necessary traits present in particular germplasm is already known. It attempts to find out the reasons behind why the particular genotype is performing superior. Large scale biochemical or biophysical processes are used to identify the candidate gene linked to the particular biochemical or physiological trait (Furbank and Tester, 2011). The expression of the identified candidate gene is then compared with other genotypes. Thus, reverse genetics provides detail dissection of system underlying specific trait. It also allows exploitation of the candidate gene related with the trait by introducing it into new varieties or transferring to other plant species using transformation techniques approaches (Furbank and Tester, 2011).

PHENOMIC TOOLS

In Table 2 a broad classification of phenotypes which can be characterized is depicted along with the capabilities to do extensive and intensive sampling. As the combined measurement of all the phenotypic classes in same individual is tough, a few promising tools for phenomic scale measurement has been highlighted (Houle *et al.* 2010).

Transcriptomics and Epigenomics

The **transcriptome** comprises the set of all messenger RNA molecules (transcripts) in one cell or a population of cells (genome) under specific circumstances and the study of transcriptome using high throughput methods is transcriptomics. Similarly, the study of regulatory mechanisms that are not accompanied by the action of genetic elements is known as **epigenetics** and the study of the complete set of epigenetic modifications on the genetic material of a cell is epigenomics. For extensive phenomic data, nucleic acid-based measurements of transcriptomes and epigenomes are the most commonly known source. As the expression of gene varies with the cell type and developmental stage, acquiring intensive data is exigent. To draw the inference regarding the causal relationship RNA data sets have been used extensively (Rockman, 2008; Kliebenstein, 2009). However, mixed samples violate the assumptions behind such inferences (Chu *et al.,* 2003). Suitable techniques essential for intracellular or single-cell assays are being developed (Sindelka *et al.,* 2010).

Proteomics and Metabolomics

Proteomics is the large scale comprehensive study of genetic information within a cell system at the protein level (**proteome**), particularly their structure and function. **Metabolomics** refers to the systematic study of the unique chemical fingerprints that specific cellular processes leave behind under a given set of conditions. Both proteomics and metabolomics owe a similar set of procedures i.e. separation of the sample followed by mass spectrometry to identify the existing compounds. Both approaches integrate diverse post-transcriptional and regulatory events. On one side, where metabolites provide important information regarding the environment that different individuals experience, the analysis of proteome detects most peptides in a sample (de Godoy *et al.,* 2008). Even though the metabolomes are substantially less complex than proteomes, the characterization of the entire metabolites is not feasible (Sawada *et al.,* 2009; Mayr, 2008).

Table 2. Phenomic scale measurements across biological hierarchy

S.N.	Level	Extensive	Intensive
1	DNA and RNA	Bioinformatics capabilities is the limitation; Cost declining	Costly; explicit sampling is exigent
2	Chromatin	Possible through chromatin immune-precipitation; low precision; costly	Costly; explicit sampling is exigent owing to instability
	Proteins and metabolites	Mostly solved; possibilities for rare constituents are still improving	Costly; explicit sampling is exigent
	Cells	Huge diversity of assays is possible; most assays are of low throughput; Image-based techniques for high throughput are encouraging	Many temporally incessant assays. Combination of extensive and intensive measurement through Image-based approaches
	Development	Abundant low-throughput measures; high-throughput image-based methods possible for small living individuals and sectioned tissues	Temporal depth by recurring sampling or image-based approaches
	Physiology	Huge diversity of assays is possible; most assays are of low throughput; Image-based techniques for high throughput are encouraging; high throughput proteomic or metabolomic data required	Many short-term continuous assays. Spatial sampling is often possible
	Morphology	Solved in principle with the use of imaging, but assays often require extensive sample preparation. Post-processing to extract and measure features requires specialized informatics capability	Extensive sampling gives intensive coverage. Specific aspects of behaviour can be intensively measured with data loggers
	Behaviour	Possible with continuous observation of video of confined or local populations, or for humans, self-reporting. Data extraction from video using specialized software or human labour	Extensive sampling gives intensive coverage. Specific aspects of behaviour can be intensively measured with data loggers

***Source:** Houle *et al.* (2010)

Behaviour

The temporal nature and context dependence resulting in high variance of behavior propounds special challenges for phenomics. Miniaturized and sophisticated automated data loggers, global positioning system locators, accelerometers, electroencephalography or neural activity recorders are in use for studying the behaviour. However, these technologies are expensive, unable in providing information for large samples and used mainly for studying behaviour of the free living animals. Web-based tools may prove to be effective in recovering the behavioural data from volunteer subjects (Blider *et al.,* 2009; Carlbring *et al.,* 2007).

Imaging

With the availability of number of imaging technologies that span molecular to organismal spatial scales is paying dominancy in phenomic studies (Walter *et al.,* 2007). Automatic image analysis offers an additional advantage of increased objectivity and precision in capturing spatial and temporal data of many phenotypic groups such as morphology, behaviour, physiological state, and locations of proteins and metabolites. Two-dimensional imaging system such as photography and traditional microscopy are cheap and quick to acquire. However, Mass spectrometry is the most common method used for the detection of analytes. The three-dimensional imaging combines the information from different real or virtual slices and combines to reconstruct a three dimensional structure. Similarly, MRI or computerized tomography scans can be used for volumetric imaging. However, the scanners are time consuming and expensive. To recover data in multiple modalities combination of positron emission tomography and optical imaging (Nagel *et al.,* 2009) or multiple fluorescent features can be utilized. But for all these techniques, sample preparation is expensive and time consuming.

HIGH THROUGHPUT PHENOTYPING

Study of Below Ground Parts

Below ground parts (roots) constitute an important part of plant life. The environment near roots called rhizosphere is very complex consisting of different living and non-living constituents. Roots are the basis for plant growth but different physiological activities being performed by roots are still unidentified. Most of the observations carried out on roots are based on controlled conditions or are based on destructive sampling. Non-destructive techniques of studying roots have the advantage of repeated observations. One such technique is Minirhizotron Technique. This technique is capable of studying root distribution, root demography and interaction between roots and root-soil and soil organisms. This method permits calculation of fine-root length production, mortality and turn-over. The same fine-root segments can be monitored over their lifetime and pictures are stored in a database for processing (Maeght *et al.,* 2013). Indirect method of root studies involves tracer technique, electrical capacitance and electrical resistivity tomography. Tracers (radio-active or stable isotopes) are injected at different depths and later recovered in biomass. The presumption that the equivalent parallel resistance-capacitance of the electrical circuit formed by the inter-face between soil water and plant root surfaces is proportional to the overall amount of active roots present has led to the development of electrical capacitance method for measuring root biomass. Electrical resistivity tomography can be used to study soil water movement, root water uptake and drainage. RGB imaging can also be used to study roots which use light waves in visible range (400-700 nm).

Study of Above Ground Parts

Foliage is the important parts of plants responsible for photosynthesis. Chlorophyll is the major active component of leaf tissues participating in photosynthesis. Chlorophyll upon receipt of the light energy gets excited and emits fluorescence. The measurement of chlorophyll fluorescence is the basic to study different activities going inside the chlorophyll. Chlorophyll fluorescence measurements are highly attractive because they allow fast and non-invasive determination of relevant photosynthetic parameters such as quantum yield, electron transport rate, or non-photochemical quenching (Maxwell and Johnson 2000; Baker 2008). GROWSCREEN FLUORO is an automated screening device that combines a fluorescence imager with computer controlled moving stages and automated data analyses routines for high-throughput measurement of relevant photosynthetic parameters. GROWSCREEN FLUORO analyses can reveal influences of mutations, transgenes or environmental factors on diverse phenotypic properties of the plants. Such high-throughput phenotyping facilitates functional genomics by linking phenotypes to given genotypes to help select plants with enhanced resource use efficiency or stress tolerance (Rascher *et al.,* 2011). Thermal imaging has potential for studying canopy or leaf surface temperatures. This can be used to study plant stress and for stress responses induced by any biotic or abiotic stresses. Tomography, such as X-ray and magnetic resonance imaging (MRI) is suitable to study internal structures of plants, roots, and soils (Li *et al.,* 2014). MRI also has great potential for studying dynamics of metabolites (Borisjuk *et al.,* 2012). MRI detects signals from protons (hydrogen nuclei) in a given specimen. The signals are measured in a spatial array and plotted as an image, where the brightness of the pixels is roughly proportional to the proton density. Metzner *et al.,* (2015) compared CT and MRI for the study of root characteristics. They found that both the techniques performed equally well. CT was advantageous to investigate specific root details or finely graduated root diameters of thin roots, as it provided the higher spatial resolution. MRI delivered higher fractions of the root systems than CT, most likely because of the strong root-to-soil contrast achievable by MRI. These two modalities were complementary to each other. However, MRI is less superior to CT as soil elements interfere with MRI signal and radiation effects making it less superior to CT. Despite the potential especially for root studies, tomography has to undergo a great deal of development to improve the signal-to-noise ratio due to non root elements affecting measurements. Radiation from X-rays may also have a negative effect on the root itself. Stereo image analysis, can be used for characterizing plant height, leaf shape and leaf angle distribution (Biskup *et al.,* 2007; Yu *et al.,* 2007).

Thermal imaging is based on the principal of cooling by transpiration. This imaging can be used to identify water stress and temperature stress. It uses cameras to measure spatial heterogeneity of heat emissions, usually from leaves; the heat is electromagnetic radiation in the infrared region, usually between 8–13 μm. The technique can be applied to different scales, e.g., from single seedlings/leaves through whole trees or field crops to regions. Hyperspectral imaging is based on accumulation of certain compounds in plant parts which have certain spectral properties. RGB imaging is also used in shoot studies. Hyperspectral imaging involves the acquisition of electromagnetic spectra at every pixel in an image, thus combining spatial and spectral information (Bock *et al.,* 2010). A major advantage of hyperspectral imaging is the wide range of measurements that can be derived from the data collected. Different plant diseases can cause distinct spectral reflectance patterns, so hyperspectral imaging offers the potential to not only detect disease but also identify specific diseases (Mukta and Bart 2015).Thermal imaging can be used to assess the diease symptoms in plant because pathogens initiate changes in stomatal behavior includind closure of stomata which in turn, affects leaf temperature. Multispectral imaging includes the near-infrared range of the light spectrum and enables a more detailed analysis of plant and leaf biochemistry such as the composition of important pigments (e.g. chlorophyll and carotenoids) and water content in leaves, and to calculate a range of vegetation indices (VIs).

Study of Plant-Insect Interaction

Breeding for resistance to insect-pest has been the major area of agricultural research. The application of plant phenomics to pathology is at its inaugural stage as the genetics of resistance are comparatively simpler for most of the pathogens such as rust. However, in addition to genotypic challenges, a greater challenge lies in the phenotypic scoring of disease at multiple time points in case of adult plant resistance. Non-destructive methods such as fluorescence and hyperspectral imaging can provide a great assurance in quantitative scoring.

In many cases, plant-insect interaction is visible in terms of leaf symptoms. In other cases, no visible symptoms appear. Assessment of actual degree of damages done by insect-pests many times is faulty inside the crop canopy. Goggin *et al.* (2015) has reviewed the use of high throughput techniques in assessment of plant-insect interaction. As a result of insect-feeding on plant parts especially leaves, chlorophyll loss is observed in leaf tissues. This can be measured using hand-held spectrophotometer (SPAD meter). Screening for viral infection in population can be done by using high-throughput antibody-based assay (ELISA). Many a times, different chemicals are stimulated, over produced or under produced under different biotic stresses. These chemicals may have certain spectral characteristics of the plant that varies with resistance/tolerance. In such cases, near-infrared (NIR) reflectance spectroscopy can be used as a screening tool to select/reject cultivars with resistance/tolerance. Near-infrared and far-infrared sensors are used to estimate water content, leaf temperature, and stomatal conductance. This can differentiate normal plants with stressed plants. NIR can be used in remote sensing by measuring plant reflectance and predict degree of biotic stresses. Visible cameras acquire high-resolution images that allow the characterization of plant pigments and the quantification of plant size, architecture, chlorosis and necrosis. Fluorescence cameras especially multicolour fluorescence camera can be utilized in identification of resistant sources by differential measures of chlorophyll fluorescence of different lines.

Assessment of Plant Disease Symptoms

Mukta and Bart (2015) reviewed the use of different techniques to assess the disease symptom in plants. Visible light imaging can be used to assess the leaf area damaged and degree of damage caused by pathogen. Image-based phenotyping offers the potential to improve reproducibility and sensitivity of disease quantification. Further, image analysis enhances the ability to distinguish between genotypes with different levels of disease severity. Hyperspectral images provide a wealth of information for investigating plant disease phenotypes. Typical wavelengths observed for plant imaging experiments are the visible (400–700 nm), near-infrared (NIR; 700–1100 nm), and short-wave infrared (1100–2500 nm) regions.

In addition, Grobkinsky *et al.*, 2015 gave a brief overview of currently applied high throughput technologies for the determination of crop phenotype as given in Table 3.

APPLICATION OF PLANT PHENOMICS TO TRAIT-BASED PHYSIOLOGICAL BREEDING

In plant breeding, integration of physiological traits at all level is required to speed up the yield improvement (Chapman *et al.*, 2012; Mir *et al.*, 2012; Long *et al.*, 2015). Physiological breeding adds-on our knowledge to exemplify genetic resources in planning crossing strategies. Large range of traits including genetically complex physiological characteristics is exercised in physiological breeding which is not so in the case of conventional breeding (Reynolds *et al.*, 2012; Cossani *et al.*, 2012; Richards *et al.*, 2010). It also enriches our command over suitable traits taken into consideration for breeding purposes using phenomic as well as genomic information whereas the molecular breeding utilizes only genomic information.

Table 3. Current High throughput technologies for crop phenotyping

Dimensonality	Equipment	Measurements	Traits	Species	Cost	Limitations	References
Morphology, outer parts of organism, limited overall physiology	RGB imaging	Projected structures, surfaces, colours of shoot, root, seeds, leaf spots	Growth dynamics, chlorosis and necrosis, simple pigmentation, senescence, root architecture	Rice, wheat, barley, cotton , Arabidopsis	Low	Limited output on physiological processes, 2D, illumination	Garnier *et al.*, 2006; Skiryex *et al.*, 2011; Tisne *et al.*, 2013; Vasseur *et al.*, 2014; Bresson *et al.*, 2014
	Laser, stereo imaging, time of flight imaging	Deep 3D structure, leaf angle, leaf area, plant height, coverage	Biomass quantity, structural composition, changes	Rice, wheat, barley	Medium	Limited information on physiological processes, occlusion	Tilley *et al.*, 2012; Eitel *et al.*, 2014; Metzner *et al.*, 2015
Morphology and physiology	Multispectral imaging	Discrete spectral reflectance including NIR, simple pigmentation, simple VI	Photosynthetic apparatus (indirect), biomass development, nitrogen content, senescence	Barley	Medium	Plant structure and development and natural illumination influence signal, discrete spectra give non-detailed information on tissue biochemistry	Cseri *et al.*, 2013; Honsdorf *et al.*, 2014; Chen *et al.*, 2014; Neumann *et al.*, 2015
	Thermal imaging	Leaf or canopy temperature	Coupling to stomatal conductance, stress response from biotic and abiotic factors	Rice, wheat, barley	Low	Plant structure, illumination, timing, RH, wind, air temperature, background (soil) temperature, calibration to specific reference	Jones *et al.*, 2009; Rajendran *et al.*, 2009; Harris *et al.*, 2014; Richard *et al.*, 2015

Dimensonality	Equipment	Measurements	Traits	Species	Cost	Limitations	References
Morphology and more advanced physiology	Hyperspectral imaging	Continuous spectral reflectance including NIR, snapshot of physiology, advanced pigmentation and specific VI	Specific pigments related to biochemical composition and photosynthesis e.g. xanthophyll, senescence, leaf water status, chlorosis/ necrosis	Rice, wheat, barley, sorghum, tomato	Medium	Plant structure and development and natural illumination influence signal, data load, complex analysis	Golzarian *et al.*, 2011; Schilling *et al.*, 2014
	Fluorescence imaging	Quantum yield, nonphotochemical quenching, photosynthetic apparatus and its associated metabolism, projected leaf area	Study leaf stress from biotic and abiotic factors, snapshots of photosynthetic performance, senescence	Rice, wheat, barley, Arabdiopsis, pea	Medium	Requires intensive illumination, potential dark adaption takes time, difficult to image deep complex canopies, preferably rosette plants	Woo *et al.*, 2008; Jansen *et al.*, 2009; Munns *et al.*, 2010;
	MRI, CT (X-ray)	Internal structures, insoil structures, Volumetric data	Root development, Metabolites. Tillering, grain quality	Rice, Wheat	Medium	MRI is less superior to CT; soil elements interfere with MRI signal, radiation effects	Gregory *et al.*, 2003; Buhler *et al.*, 2011; Hargreves *et al.*, 2009
Physiological processes	PAM-fluorescence, wet chemistry, destructtive sampling	Photosystem II activity, metabolite and phytohormone profiles, enzyme activities	Advanced studies of plant physiological functions and mechanisms	Barley	Medium	Hand on time, throughput	Meroni *et al.*, 2009; Gorbe and Calatayud, 2013)

CT: Computed Tomography; MRI: Magnetic Resonance Imaging; NIR: Near infrared; PAM: Pulse amplitude modulation, RGB: Red-Green-Blue; RH: Relative Humidity; VI: Vegetative Index

With the use the molecular marker approach alleles associated with genes of major effect can be identified which can be directly used in the genomic selection and marker assisted selection. Since, there is a lack of comprehensive knowledge regarding the genetic basis of the differences existing in cultivar, the role of phenomics become pivotal in the physiological breeding. Hence, in physiological breeding approach both genomics and phenomics go conjointly. Non invasive techniques are now applied to directly measure the traits for physiological breeding. This has solved the challenge of computing the thorough and quantitative analysis of traits. As the significant traits for developing a superior variety is identified, the genomic region associated with those particular traits is needed to be identified using the marker assisted selection approach. If the trait is a quantitative trait i.e. governed by multigenes, quantitative physiological breeding should be followed to identify the robust phenotypic marker. For obtaining the values related to biomass and growth digital imaging technique can be exercised. The imaging technique can be used for specific traits of interest like tillering, early vigor, coleoptile length and biomass at anthesis (Richards, *et al.,* 2010). Use of phenomic tools can increase the efficiency and accuracy of evaluation of these traits. 3D models prepared through imaging techniques will assist in drawing more accurate and precise information. These pieces of information can be quantified using automated techniques which can be utilized by the plant breeders in improving the traits under consideration.

PLANT PHENOTYPING PLATFORMS AND PHENOMIC PROJECTS

With availability of the new imaging sensors like high-resolution imaging spectrometers and the advanced software for image analysis and feature extraction, it has become easy to address the "phenotyping bottleneck". PHENOPSIS (Granier *et al.,* 2006), one of a pioneer platform was developed that assisted in dissecting the effect of genotype x environment interaction in different physiological and biochemical processes in *Arabdiopsis thaliana.* Similarly, a number of platforms developed viz., TraitMill (Reuzeau *et al.,* 2005; Reuzeau *et al.,* 2006), GROWSCREEN (Walter *et al.,* 2007; Jansen *et al.,* 2009; Meyer *et al.,* 2010), LIMINA (Bylesjo *et al.,* 2008), HYPOTrace (Wang *et al.,* 2009), HTPheno (Hartmann *et al.,* 2011) and LeafAnalyser (Weight *et al.,* 2008) which provides solution for general image-processing related to morphological traits such as plant height, leaf length and width etc.) and colorimetric analysis. The recently developed imaging sensors and software for image analysis to evaluate the performance and plant function under controlled condition and some of the phenomics project has been enlisted in Table 4 and 5 respectively.

Table 4. Plant Phenotyping Platforms

Name	**Description**	**URL**
PHENOPSIS	Automated platform developed by Optimalog; reproducible; used for phenotyping of plant responses to soil water deficit in Arabidopsis	http://bioweb.supagro.inra.fr/phenopsis/
LemnaTec	Robotic greenhouse system; uses non-destructive imaging; monitors plant growth under fully controlled conditions	http://www.lemnatec.com/
Phenodyn	Measures growth rate and transpiration rate	http://bioweb.supagro.inra.fr/phenodyn/
GROWSCREEN	Studies leaf growth and root architecture	http://www2.fz-juelich.de/icg/icg-3/jppc/growscreen/
TraitMill	A gene engineering system; developed by CropDesign; used in large-scale plant transformation and high-resolution phenotyping e.g. rice	http://www.cropdesign.com/tech traitmill.php
RootReader3D	high-throughput; 3D analysis of plant roots	http://www.plantmineralnutrition.net/rootreader.htm
PlantScan 3D meshes	Automated; mesh based; high-throughput; 3D plant analysis of aerial parts	http://www.plantphenomics.org.au/node/157
HTPheno	Phenotypic measurements viz., height, width, shoot area	http://htpheno.ipkgatersleben.de/
WinRHIZO	Root measurement	http://www.regent.qc.ca/index.html
WinFOLIA and WinSEEDLE	Leaf analysis	http://www.regent.qc.ca/index.html
WinCELL	Wood cell analysis	http://www.regent.qc.ca/index.html

Table 5 Phenomics-related projects

Name of the project	**Description**	**URLs**
IPPN (International Plant Phenomics Network)	Assists in developing concepts and technologies to assess the plant phenome; Its goal is to increase the visibility and impact of plant phenotyping	http://www.plantphenomics.com; http://www.plantphenomics.org.au; http://www.fz-juelich.de/icg/icg-3/jppc

Name of the project	Description	URLs
EPPN (European Plant Phenotyping Network)	Its goal is to establish the link between genotype and phenotype and their interaction with the environment	http://www.plantphenotyping network.eu/
DPPN (German Plant Phenotyping Network/ (Deutsches Pflanzen Phänotypisierungs netzwerk)	Its goal is to establish new concepts and technologies for phenotyping	http://www.dppn.de/
PHENOME	Its aim is to provide versatile, high-throughput infrastructure and methods for characterization of genotypes	http://urgi.versailles.inra.fr/ Projects/PHENOME/
APPF (Australian Plant Phenomics Facility)	It aims to explore and improve plant function and performance through high-throughput plant phenotyping and "reverse phenomics" approaches	http://www.plantphenomics. org.au/

This extensive research utilizing these above mentioned resources will help in exploring the genetic basis of yield and yield related traits in physiological breeding to develop resilient plants.

PLANT PHENOMICS AND DATA ANALYSIS

Final step in phenotyping is the interpretion of data obtained through high throughput techniques. The major challenge of phenomics data is their great number and heterogeneity of data obtained with diverse methods. Thus, handling and integration of data is of that need Image data is pre-processed for determining quantitative or qualitative values of phenotypic traits of a plant/genotype in a given environment. Pre-processing steps adjust for differences in lighting and alignment of the image. Algorithms can be used for classification or quantification of image features. A commonly used example for hyperspectral imaging applications is the Spectral Angle Mapper (SAM) classification algorithm (Bauriegel *et al.,* 2011; Mahlein *et al.,* 2012). In this step, quality of image data is taken into consideration. Some noisy image data may affect the dataset negatively. Likewise, post-processing of phenotypic traits is done for prediction of plant behavior using statistical analysis. Different statistical methods and algorithms are used for analyzing the processed phenotype image data set, so as to move from the raw data to final results. Data normality tests and outlier detection is done to improve the data quality. Accordingly, data transformation is carried out. Phenotypic data obtained through high throughput techniques are multi-dimensional which may be correlated with each other. Hence, to reduce the dimensionality of dataset different data reducing techniques like Principal Component Analysis (PCA), Factor Analysis (FA), Artificial Neural Network (ANN) are used. This algorithm compares vectors representing experimentally determined spectra and reference spectra, and calculates an angle to represent the degree of difference between the two at each pixel.

PLANT PHENOMICS IN CLOSING THE 'GENE TO GENOTYPE' LOOP

The prime aim of the plant breeders seeks improvements in yield and yield per unit of input. With the advent of high throughput phenotyping and the approaches described above can be used in many different ways to facilitate the process of identification of gene and trait and genotype development necessary to produce a new crop variety. Furbank and Tester (2011) demonstrated a "gene to genotype' scheme to use phenomics in developing develop a crop genotype or variety with tolerance to a particular stress drought (Figure 2).

In this scheme phenomics is utilized at a number of levels. Forward phenomics can be used to collect the information of valuable traits of germplasm (1 in Figure 2), or reverse phenomics for identifying important traits validated by physiological studies with differing drought tolerance (2 in Figure 2).

Genetic association is the study of genetic variation to identify the candidate gene that contributes to a specific trait. For analysis of the genetic association, bi-parental or multi-parental genetic approach can be taken up. The resequencing technologies are in use to assess the genetic variation, cost effective, high throughput precision phenotyping is pivotal for fine mapping. In addition to this, to test the hypotheses regarding role of a particular gene through high quality reverse genetics (4 in figure 2) and to test the effect of different alleles or varying levels of target genes on the traits (5 in Figure 2), and tolerance against drought of the resultant crop (6 in Figure 2), the phenomics has an essential role to play.

Reverse phenomics (2 and 3 in Figure 2) dissects trait to elucidate mechanisms to find out candidate genes associated with the trait. Reverse phenomics supports in elucidating the common strategies responsible for stress tolerance by utilizing a suite of new tools applied. Detail studies of mechanisms through these methods may permit to elucidate the next generation of traits for the breeding toolbox (Furbank and Tester, 2011). Moreover, crop modelling approaches using APSIM (Agricultural Production Systems Simulator; http://www.apsim.info/Products/Downloads.aspx) (Hammer *et al.* 2006) are equally vital (7 in Figure 2) in estimating the value of trait for constructing the plant ideotype. The schematic representation in Figure 2 is circular because phenomics facilitates both the discovery of genes and assists to enhance tolerance by alteration of the genes that contribute to stress tolerance (Furbank and Tester, 2011).

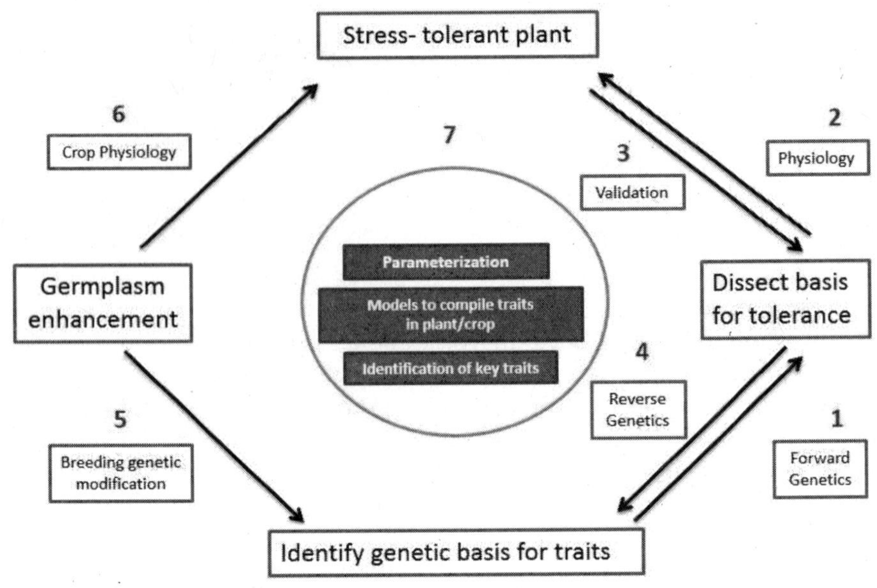

Fig. 2. Closing the gene to genotype loop with phenomics (Furbank and Tester, 2011)

CONCLUSION

The major challenge ahead in plant-based agriculture is to have security of global food, fiber and fuel. The target becomes more challenging with the face of a shift in the climatic scenario. For this, the scale of quantum advances is required in the field of technology. To be precise, there is no single approach to develop a crop variety which can withstand the repercussion of changing climatic variables. It can be said that with the embrace of coordinated approach of conventional and modern plant breeding and plant physiology one can find ways to develop crop varieties and agricultural techniques that can augment crop production and provide sustainability.

Plant phenomics, an analogy to "omics" can be considered simply as "manifestation of plant physiology" to bring physiology up to speed with genomics. By instigating the implausible recent advances made in photonics, robotics and computers to the wider field of plant biology this has become achievable. The combination of high-resolution high-throughput phenotyping offers new prospects of gathering detailed complementary information on important physiological plant traits which are of agricultural importance. With the establishment of robust physiological traits (predictors), they can be directly applied in precision agriculture as markers in breeding programmes. Phenomics is an emerging transdiscipline which endeavours to leverage revolution in genome-wide genotyping. However, phenomics lags predominantly behind genomics and therefore, there is need to elevate the existing laboratory-based phenotyping methods, and web-based ascertainment. In addition, considerable work is indispensable to decipher the available set of information into pragmatic deployment. Moreover, the major challenge exists in the multitudes of phenotypic traits and the influence of the environment on these traits as well as extremely high cost of the current running phenome projects. Inspite of this, with the readiness of the tools and the available knowledge it can now become a lot more possible to meet the agricultural challenges of the 21st Century to feed the world at present and in the future. Hence, the emerging science of phenomics will be significant in discerning this vision once the assorted challenges are vanquished.

REFERENCES

Araus JL and Cairns JE 2014 Field high-throughput phenotyping: the new crop breeding frontier. *Trends in Plant Science* **19** 52–61

Baker NR 2008 Chlorophyll fluorescence: a probe of photosynthesis in vivo. *Annual Review of Plant Biology* **59** 89–113

Bauriegel E, Giebel, A and Herppich, WB 2011 Hyperspectral and chlorophyll fluorescence imaging to analyse the impact of *Fusarium culmorum* on the photosynthetic cintegrity of infected wheat ears. *Sensors (Basel)* **11** 3765–3779

Bilder RM *et al* 2009 Phenomics: the systematic study of phenotypes on a genome-wide scale. *Neuroscience* **164** 30–42 (2009)

Biskup B, Scharr H, Schurr U, and Rascher U 2007 A stereo imaging system for measuring structural parameters of plant canopies. *Plant Cell Environment* **30** 1299–1308

Bock CH, Poole GH, Parker PE and Gottwald TR 2010 Plant disease severity estimated visually by digital photography and image analysis, and by hyperspectral imaging. *Critical Review Plant Science* **29** 59–107

Borisjuk L, Rolletschek H and Neuberger T 2012 Surveying the plant's world by magnetic resonance imaging. *The Plant Journal* **70** 129–146

Bresson J, Vasseur F, Dauzat M, Labadie M, Varoquaux F, Touraine B *et al* 2014 Interact to survive: *Phyllobacterium brassicacearum* improves Arabidopsis tolerance to severe water deficit and growth recovery. *PLoS One* **9** e107607

Bühler J, Huber G, Schmid F and Blümer P 2011 Analytical model for long dostance tracer-transport in plants. *Journal of Theoretical Biology* **270** 70-79

Burns J 1970 The synthetic problem and the genotype–phenotype relation in cellular metabolism. In: Waddington CH, ed. *Towards a theoretical biology. 3. Drafts. An IUBS Symposium* . Chicago: Aldine Publishing Company, 47–51

Bylesjo M, Segura V, Soolanayakanahally RY, Rae AM, Trygg J, Gustafsson P, Jansson S and Street NR 2008 LAMINA: a tool for rapid quantification of leaf size and shape parameters. *BMC Plant Biology* **8** 82

Carlbring P, Brunt S, Bohman S, Austin D, Richards J *et al* 2007 Internet vs. paper and pencil administration of questionnaires commonly used in panic/agoraphobia research. *Computer in Human Behaviour* **23** 1421–1434

Chapman SC, Chakraborty S, Dreccer MF, Howden SM 2012 Plant adaptation to climate change — opportunities and priorities in breeding. *Crop Pasture Science* **63** 251-268

Chen D, Chen M, Altmann T and Klukas C 2014 Bridging genomics and phenomics. In: Chen M, Hofestädt R, eds. *Approaches in integrative bioinformatics: towards the virtual cell.* Berlin: Springer-Verlag, pp 299–333

Chu TJ, Glymour C, Scheines R and Spirtes P 2003 A statistical problem for inference to regulatory structure from associations of gene expression measurements with microarrays. *Bioinformatics* **19** 1147–1152

Cossani CM, Reynolds MP 2012 Physiological traits for improving heat tolerance in wheat. *Plant Physiology* **160** 1710-1718

Cseri A, Sass L, Törjék O, Pauk J, Vass I and Dudits D 2103 Monitoring drought responses of barley genotypes with semi-robotic phenotyping platform and association analysis between recorded traits and allelic variants of some stress genes. *Australian Journal of Crop Science* **7** 1560–157

de Godoy LMF, Oslen JV, Cox J, Nielsen ML, Hubner NC, Frolich F, Walther TC and Mann M 2008 Comprehensive mass-spectrometry-based proteome quantification of haploid versus diploid yeast. *Nature* **455** 1251–1254 (2008)

Dhont S, Wuyts N and Inzé D 2013 Cell to whole-plant phenotyping: the best is yet to come. *Trends in Plant Science* **18** 428–439

Eitel JUH, Gessler PE, Smith AMS and Robberechet 2006 Suitability if exsting and novel spectral indices to remotely detect water stress in *Populus* spp. *Forest Ecology and Management* **229** 170-182

Furbank RT and Tester M 2011 Phenomics–technologies to relieve the phenotyping bottleneck. *Trends in Plant Sciences* **16(12)** 635–644

Gjuvsland AB, Vik JO, Beard DA, Hunter PJ and Omholt SW. 2013 Bridging the genotype–phenotype gap: what does it take? *Journal of Physiology* **591** 2055–2066

Goggin FL, Lorence A and Topp CN 2015 Applying high-throughput phenotyping to plant–insect interactions: Picturing more resistant crops. *Current Opinion in Insect Science* **9** 69–76

Golzarian MR, Frick RA, Rajendran K, Berger B, Roy S, Tester M and Lun DS 2011 Accurate inference of shoot biomass from high throughput images of cereal plants. *Plant Methods* **7** 2

Gorbe E and Calatayud A 2013 Applications of chlorophyll fluorescence imaging technique in horticultural research: a review. *Scientia Horticulturae* **138** 24–35

Granier C, Aguirrezabal L, Chenu K, Cookson SJ, Dauzat M, Hamard P, Thioux JJ, Rolland G, Bouchier-Combaud S, Lebaudy A *et al* 2006 PHENOPSIS, an automated platform for reproducible phenotyping of plant responses to soil water deficit in Arabidopsis thaliana permitted the identification of an accession with low sensitivity to soil water deficit. *New Phytology* **169(3)** 623–635

Gregory PJ, Hutchinson DJ, Read DB, Jenneson PM and Gilboy WB *et al* 2003 Non-invasive imaging of roots with higher resolution X-ray micro-tomography. *Plant and Soil* **255** 251-59

Großkinsky DK, Svensgaard J, Christensen S and Roitsch T 2015 Plant phenomics and the need for physiological phenotyping across scales to narrow the genotype-to-phenotype knowledge gap. *Journal of Experimental Botany* **66** 5429–5440

Hammer G, Cooper M, Tardieu F, Welch S and Walsh B *et al.* 2006 Models for navigating biological complexity in breeding improved crop plants. *Trends in Plant Sciences* **11** 587–593

Hargreaves C, Gregory P and Bengough AG 2009 measuring root traits in barley (*Hordeum vulgare* ssp. Vulgare and ssp. Spontaneum) seedlings using gel chambers, soil sacs and X-ray microtomography *Plant and Soil* **316** 285-97

Harris B, Sadras V and Tester M 2010 A water-centred framework to assess the effects of salinity on the growth and yield of wheat and barley. Plant Soil **336(1–2)** 377–389

Hartmann A, Czauderna T, Hoffmann R, Stein N and Schreiber F 2011 HTPheno: an image analysis pipeline for high-throughput plant phenotyping. BMC Bioinformatics **12**148

Honsdorf N, March TJ, Berger B, Tester M and Pillen K 2014 High-throughput phenotyping to detect drought tolerance QTL in wild barley introgression lines. *PLoS One* **9(5)** e97047

Houle D, Govindaraju DR and Omholt S 2010 Phenomics: the next challenge. *Natural Reviews Genetics* **11(12)** 855–866

Jansen M, Gilmer F, Biskup B, Nagel KA, Rascher U, Fischbach A, Briem S, Dreissen G, Tittmann S, Braun S *et al* 2009 Simultaneous phenotyping of leaf growth and chlorophyll fluorescence via GROWSCREEN FLUORO allows detection of stress tolerance in Arabidopsis thaliana and other rosette plants. *Functional Plant Biology* **36(10–11)** 902–914

Jones AR, Kramer EM, Knox K, Swarup R, Bennett MJ, Lazarus CM, Leyser HM and Grierson CS 2009 Auxin transport through non-hair cells sustains root-hair development. *Nature Cell Biology* **11** 78–84

Kliebenstein, D 2009 Quantitative genomics: analyzing intraspecific variation using global gene expression polymorphisms or eQTLs. *Annual Review of Plant Biology* **60** 93–114

Langridge P and Reynolds MP 2015 Genomic tools to assist breeding for drought tolerance. *Current Opinion in Biotechnology* **32** 130–135

Li L, Zhang Q and Huang D 2014 A review of imaging techniques for plant phenotyping. *Sensors* **14** 20078–20111

Long SP, Marshall-Colon A, Zhu X-G 2015 Meeting the global food demand of the future by engineering crop photosynthesis and yield potential. *Cell* **161** 56-66

Mahlein AK, Steiner U, Hillnhütter C, Dehne HW and Oerke, EC 2012 Hyperspectral imaging for small-scale analysis of symptoms caused by different sugarbeet diseases. *Plant Methods* **8** 3

Manolio TA, Collins FS, Cox NJ, Goldstein DB, Hindorff LA, Hunter DJ, McCarthy MI, Ramos EM, Cardon LR, Chakravarti A *et al* 2009 Finding the missing heritability of complex diseases. *Nature* **461(7265)** 747–753

Maxwell K, Johnson GN 2000 Chlorophyll fluorescence – a practical guide. *Journal of Experimental Botany* **51** 659–668

Mayr M 2008 Metabolomics ready for the prime time? *Circulation: Cardiovascular Genetics* **1** 58–65

Maeght JL, Rewald B and Pierret A 2013 How to study deep roots-and why it matters. *Frontiers in Plant Science* **4 (299)** 1-14

Meroni M, Rossini M, Guanter L, Alonso L, Rascher U, Colombo R and Moreno J 2009 Remote sensing of solar-induced chlorophyll fluorescence: review of methods and applications. *Remote Sensing of Environment* **113** 2037–2051

Metzner R, Eggert A, Van Dusschoten D, Pflugfelder D, Gerth S, Schurr U *et al* 2015 Direct comparison of MRI and X-ray CT technologies for 3D imaging of root systems in soil: potential and challenges for root trait quantification. *Plant Methods* **11** 1–11

Meyer RC, Kusterer B, Lisec J, Steinfath M, Becher M, Scharr H, Melchinger AE, Selbig J, Schurr U, Willmitzer L *et al* 2010 QTL analysis of early stage heterosis for biomass in Arabidopsis. *Theoretical and Applied Genetics* (Theoretische und angewandte Genetik) **120(2)** 227–237

Mir RR, Zaman-Allah M, Sreenivasulu N, Trethowan R, Varshney RK 2012 Integrated genomics, physiology and breeding approaches for improving drought tolerance in crops. *Theoretical and Applied Genetics* **125** 625-645

Munns R 2010 Approaches to identifying genes for salinity tolerance and the importance of timescale. In *Plant Stress Tolerance.* Edited by: Sunkar R. Methods in Molecular Biology. Humana Press, New York, USA pp 25–38

Mutka AM and Bart R S 2015 Image-based phenotyping of plant disease symptoms. *Frontiers in Plant Science* **5(734)** 1-8

Nagel KA, Kastenholz B, Jahnke S, van Dusschoten D, Aach T, Mühlich M, Truhn D, Scharr H, Terjung S, Walter A and Schurr U 2009 Temperature responses of roots: impact on growth, root system architecture and implications for phenotyping. *Functional Plant Biology* **36** 947–959

Neumann K, Klukas C, Friedel S, Rischbeck P, Chen D *et al* 2015 Dissecting spatio-temporal biomass accumulation in barley under different water regimes using high throughput analysis *Plant Cell and Environment* **38(10)** 1980-96 doi: 10.1111/pce.12516

Parthsarathi T, Vanitha K and ravichandran V 2015 Plant phenotyping under drought. In: *Advances in Plant Physiology,* Ed. A. Hemantaranjan, Scientific Publishers (India), Jodhpur pp 41-89

Rajendran K, Tester M and Roy SJ 2009 Quantifying the three main components of salinity tolerance in cereals. Plant, Cell Environ. **32(3)** 237–249

Rascher U, Blossfeld S, Fiorani F, Jahnke S, Jansen M, Kuhn AJ *et al* 2011 Non-invasive approaches for phenotyping of enhanced performance traits in bean. *Functional Plant Biology* **38** 968–983

Reuzeau C, Frankard V, Hatzfeld Y, Sanz A, Van Camp W, Lejeune P, De Wilde C, Lievens K, de Wolf J, Vranken E *et al* 2006 Traitmill™: a functional genomics platform for the phenotypic analysis of cereals. Plant Genetics Resources **4(1)** 20–24

Reuzeau C, Pen J, Frankard V, de Wolf J, Peerbolte R, Broekaert W 2005 TraitMill: a discovery engine for identifying yield-enhancement genes in cereals. Fenzi Zhiwu Yuzhong *Molecular Plant Breeding* **3** 7534

Reynolds M, Foulkes J, Furbank R, Griffiths S, King J, Murchie E, Parry M, Slafer G 2012 Achieving yield gains in wheat. *Plant Cell Environment* **35** 1799-1823

Richard CA, Hickey LT, Fletcher S, Jenning R, Chenu K *et al* 2015 High-throughput phenotyping of seminal root traits in wheat. *Plant Methods* **11** 13

Richards RA, Rebetzke GJ, Watt, Condon AG, Spielmeyer W and Dolferus R 2010 Breeding for improved water productivity in temperate cereals: phenotyping, quantitative trait loci, markers and the selection environment. *Functional Plant Biology* **37** 85–97

Rockman MV 2008 Reverse engineering the genotype–phenotype map with natural genetic variation. *Nature* **456** 738–744

Sawada Y, Akiyama K, Sakata A, Kuwahara A, Otsuki H, Sakurai T *et al* 2009 Widely targeted metabolomics based on large-scale MS/MS data for elucidating metabolite accumulation patterns in plants. *Plant Cell Physiology* **50** 37–47

Schilling RK, Marschner P, Shavrukov Y, Berger B, Tester M, Roy SJ and Plett DC 2014 Expression of the Arabidopsis vacuolar H^+-pyrophosphatase gene (AVP1) improves the shoot biomass of transgenic barley and increases grain yield in a saline field. Plant *Biotechnology Journal* **12** 378 38

Schork NJ 1997 Genetics of complex disease—approaches, problems, and solutions. *American Journal of Respiratory and Critical Care Medicine* **156** S103–S109

Sindelka R, Sidova M, Svec D and Kubista M 2010 Spatial expression profiles in the *Xenopus laevis* oocytes measured with qPCR tomography. *Methods* **51** 87–91

Skirycz A, Vandenbroucke K, Clauw P, Maleux K, de Meyer B *et al* 2011 Survival and growth of *Arabdiopsis* plants given limited water are not equal. *Nature Biotechnology* **29** 212-14

Tester M and Langridge P 2010 Breeding technologies to increase crop production in a changing world. *Science* **327** 818-822

Tisné S, Serrand Y, Bach L, Gilbault E, Ben Ameur R *et al* 2013 Phenoscope: An automated large-scale phenotyping platform offering high spatial homogeneity. *The Plant Journal* **74** 534–544

Tilly N, Hoffmeister D, Liang H, Cao Q, LenzWiedemann V, Miao Y and Bareth G 2012 Evaluation of terrestrial laser scanning for rice growth monitoring. In: *The International Archives of the Photogrammetry, Remote Sensing and Spatial Information Sciences*, Melbourne, Australia, Vol. XXXIX, Part B7 pp 351-356

Vasseur F, Bontpart T, Dauzat M, Garnier C and Viely D 2014 Multivariate genetic analysis of plant responses to water deficit and high temperature revealed contrasting adaptive strategies. *Journal of Experimental Biology.* **65** 6457-69

Walter A, Scharr H, Gilmer F, Zierer R, Nagel KA, Ernst M, Wiese A, Virnich O, Christ MM, Uhlig B *et al* 2007 Dynamics of seedling growth acclimation towards altered light conditions can be quantified via GROWSCREEN: a setup and procedure designed for rapid optical phenotyping of different plant species. *New Phytology* **174(2)** 447–455

Wang W, Vinocur B and Altman A 2003 Plant responses to drought, salinity and extreme temperatures: towards genetic engineering for stress tolerance. *Planta* **218** 1–14

Wang L, Uilecan IV, Assadi AH, Kozmik CA and Spalding EP 2009 HYPOTrace: image analysis software for measuring hypocotyl growth and shape demonstrated on Arabidopsis seedlings undergoing photomorphogenesis. Plant Physiology **149(4)** 1632–1637

Weight C, Parnham D and Waites R 2008 Leaf Analyser: a computational method for rapid and large-scale analyses of leaf shape variation. *The Plant Journal* **53(3)** 578–586

Woo NS, Badger MR and Pogson BJ 2008 A rapid, non-invasive procedure for quantitative assessment of drought survival using chlorophyll fluorescence. *Plant Methods* **4** 27

Yu S, Wilson R, Edmondson R and Parsons N 2007 Surface modelling of plants from stereo images. In: *Proceeding of 6th International Conference on 3-D Digital Imaging and Modeling*, 3DIM'07, Montreal, QC pp 312–319.

■■■■

12

IMAGE-BASED PLANT PHENOTYPING

Ruchi Bansal[1*], Jyoti Kumari[1], Rashmi Yadav[1] and Sundeep Kumar[2]

[1] Germplasm Evaluation Division,
National Bureau of Plant Genetic Resources, New Delhi 110012, India
[2]Division of Genomic Resources,
National Bureau of Plant Genetic Resources, New Delhi 110012, India
**Corresponding author: ruchibansal06@gmail.com*

INTRODUCTION

To feed the increasing population of India, we need to enhance our food grain production from 260 Mt now to 300 Mt by 2020. The challenge is to increase the crop yield at the rate of 2.4% per year with limited land, water and nutrient resources (Ray *et al.,* 2013). It can be achieved by developing input use efficient crop varieties, which can perform well under low input conditions and are able to tolerate the extreme conditions of climate change. Extensive breeding and agronomic efforts led to yield increase during past five decades. Now the focus is to achieve the yield target with climate resilient crops. To integrate the resilience in to our crops, we require new sources of resistance against different abiotic and biotic stresses. During the past decade, omics has emerged as the most powerful technology to understand plant's response to stress conditions. Recent developments in the molecular biology techniques offer the potential to increase the rate of genetic improvement. Annotation of gene has predicted the structure and function of different genes. However, functional validation is possible only by precise phenotyping (Arvidsson *et al.,* 2011). Although a large collection of germplasm of different crop species are available worldwide, phenotypic descriptions of these collections are still limited. Identification of a candidate germplasm which carries genes for targeted traits requires precise and accurate phenotyping of the germplasm (Miyao *et al.,* 2007). Replicated trials are conducted across multiple environments and seasons for identification of an elite germplasm. Precise phenotyping is widely recognized as the most laborious, time consuming and technically challenging task. Hence, efforts are being made to develop new high throughput phenotyping techniques by using imaging, spectroscopy, robotics and high-performance computing.

TRADITIONAL PHENOTYPING vs HIGH-THROUGHPUT PHENOTYPING

The phenotype is the resultant of the complex interactions between genotype and the environmental factors. It has been a part of variety selection since the beginning of human civilization and crop domestication. Plant phenotyping is the comprehensive evaluation of complex traits as well as individual parameters, which form the basis of complex traits. Traditionally crops have been characterized by using different qualitative and quantitative traits (Rao, 2004). Qualitative traits are the heritable traits which are not influenced by the environment. These traits are easy to score and provide fast discrimination among the crop accession. These quantitative traits are highly affected by genotype and environment interactions. For qualitative traits, degree of expression varies from nominal, ordinal, continuous and binary. They may be defined by number, order or presence/absence of a trait (e.g. colour, shape). Quantitative traits are recorded by weighing, measuring or counting (e.g. seed weight, yield) (Biodiversity, 2007). Agro-morphological traits used to characterize the plant genetic resources are called descriptors. These traits have been catalogued by International Plant Genetic Resources Institute and different international plant research organizations. It contains passport data as well as descriptive features for identification of a particular crop germplasm (Summerfield *et al.,* 1996). Conventional methods of phenotyping involve destructive analysis for measuring different physiological parameters, but now the recent developments in imaging technologies have enabled the comprehensive and non destructive analysis of plants. Realising the potential of phenomics, different research organizations took over the establishment of automated high throughput phenomics facilities to bridge the phynotype-genotype gap and for enhancing the breeding efficiency, resource management and basic research (Furbank and Tester, 2011).

High-throughput phenotyping involves the combined application of robotics and image analysis and has led to a revolutionary change in the phenotyping (Arvidsson *et al.,* 2011). Bioinformatics approaches enable us in analyzing the very high dimensional data rapidly. The application of image analysis has been known for decades in the field of remote sensing in agriculture and forestry; however, its recent application in plant phenotyping for stress tolerance in a high-throughput manner has opened new opportunities for plant biologists.

Tremendous growth in technological interventions has reduced the cost of genotyping, but we lag behind in our ability of phenotyping precisely. It is especially true in the case of very complex traits such as abiotic stress tolerance. Several physiological traits are being used for the phenotyping of population under stress, but inaccuracy in the measurement of a trait at a specific growth stage has been a problem since a very long time. Due to bottlenecks in the present phenotyping techniques, high-throughput phenotyping has emerged as a powerful tool to characterize large number of lines/individual plants (Furbank, 2009). Now the plant phenotyping is equipped with the sensors, system technologies and algorithms. Non destructive measurements, precision, quick operation and direct data storage are the advantages of digital phenotyping. Nowadays high-throughput phenotyping platforms are being used in the controlled conditions as well as in the field (White *et al.,* 2012; Petrozza *et al.,* 2014).

IMAGING TECHNIQUES IN PHENOTYPING

Use of image analysis algorithm is the base of phenotyping studies. Different image analysis software tools have been developed. These input driven softwares are manual, semi automated and automated. Imaging basically relies on the interaction between light and plants reflecting in the form of wavelength specific absorbance, reflectance and transmittance (Li *et al.,* 2014). Plant phenotyping involves imaging at different wavelengths. Visible cameras detect light in the visible range (400-700 nm). Infrared camera use near infra red light and near infrared cameras detect near infrared and short wave infra red light. Thermal cameras sense long wave infra red radiations, while, hyper spectral cameras detect spectral bands ranging from 350-2500 nm.

Development in automation, robotics, sensor and imaging technologies has provided lot of choice for lab scale screening systems. These imaging techniques are widely used for phenotyping in different crops (Table 1.) In the last decade, automated phenotyping systems have been deployed in the controlled environments. These systems are designed for large scale phenotyping of model plant Arabidopsis to cereals (Jansen *et al.,* 2009; Golzarian *et al.,* 2011). There are two types of imaging systems depending upon the movement of the object. They may be categorized as Sensor to plant: systems, in which the sensor moves at different positions to reach the plant (Granier *et al.,* 2006; Arvidsson *et al.,* 2011) and Plant to sensor, when the position of sensor is fixed and the plant moves to reach the sensor (Golzarian *et al.,* 2011). Most of the high-throughput phenotyping platforms are suitable for the water deficit environments, where the target environment is to be defined in terms of moisture content (Granier *et al.,* 2006; Berger *et al.,* 2010). High-throughput phenotyping has also been done for tolerance to salt stress, cold stress and nutrient deficiency (Rajendran *et al.,* 2009; Humplik *et al.,* 2015; Neilson *et al.,* 2015).

Table 1. Application of imaging techniques in non invasive phenotyping in plants

S.No.	Imaging Technique	Crop	Parameters	References
1.	Visible light imaging	Arabidopsis	Projected area, growth dynamics, yield traits, root architecture, Germination rate, early embryonic axis growth, height, size, morphology, flowering time	Granier *et al.,* 2006; Skiryex *et al.,* 2011 ; Tisne *et al.,* 2013 ; Bresson *et al.,* 2014
2.	Fluorescence imaging	Arabidopsis, wheat, barley	Photosynthetic status, Quantum yield, non photochemical quenching, shoot architecture	Jansen *et al.,* 2009; Munns *et al.,* 2010; Feher-Juhasez *et al.,* 2014 ; Richard *et al.,* 2015
3.	Thermal imaging	Wheat, barley	Canopy temperature	Jones *et al.,* 2009; Rajendran *et al.,* 2009 ; Harris *et al.,* 2010 ; Munns *et al.,* 2010
4.	Near infra red imaging	Barley	Water content composition parameters for seeds, leaf area index	Cseri *et al.,*2013; Honsdorf *et al.,* 2014 ; Chen *et al.,* 2014 ; Neumann *et al.,* 2015
5.	Hyperspectral imaging	Arabidopsis	Leaf water status, leaf growth, coverage density	Golzarian *et al.,* 2011 ; Schilling *et al.,* 2014
6.	3D imaging	Maize, Wheat	Shoot structure, leaf angle, root architecture	Ivanov *et al.,* 1995; Omasa *et al.,* 2007; Eietel *et al.,* 2014; Metzner *et al.,* 2015
7.	Laser imaging	Rice	Shoot structure, biomass, leaf angle, canopy structure, root architecture	Tilly *et al.,* 2012
8.	Magnetic resonance images	Berkheya	Morphometry, water content	Moradi *et al.,* 2010
10	Computed tomography	Barley	Tillers, morphometric parameters, grain quality	Gregory *et al.,* 2003; Hargreaves *et al.,* 2009

PLANT TRAITS IN PHENOTYPING

There are different morphological and physiological traits used for non destructive phenotyping. The most widely used traits based on imaging are summarized here.

Chlorophyll Flourescence

Chlorophyll fluorescence is an important parameter, which is being extensively used in the phenotyping under controlled as well as field conditions. Fluorescence is the emission of light of shorter wavelength after absorption of light. Chlorophyll molecules emit the fluorescence in red and far red spectral bands. Chlorophyll fluorescence provides an estimate of quantum efficiency of photosystem II (PSII) (Fv/Fm) for early detection of abiotic stress. Fv is variable fluorescence from leaves adapted to darkness and indicates PSII ability to present photochemistry. Fm is maximal emission of electromagnetic radiation from dark- and light-adapted leaf and demonstrates level of fluorescence when primary quinone electron acceptor of PSII is maximally reduced. Fv/Fm generally decreases when plants are exposed to stress in the light, and this event provides an easy and fast tool for observing stress (Baker, 2008). Fluorescence imaging has been widely used in the measurement of chlorophyll fluorescence. Pulse amplitude modulation based instruments measure quantum efficiency, electron transport rate, non-photochemical quenching (Maxwell and Johnson, 2000). Chlorophyll fluorescence has been used as a screening criterion in many cereal and horticultural crops. Fluorescence imaging was used to study the chlorophyll fluorescence as well as leaf area estimation. In Arabidopsis, measurement of chlorophyll fluorescence has been used for detection of stress tolerance under drought and chilling conditions (Jansen *et al.,* 2009) and in wheat under saline conditions (James *et al.,* 2002). Chlorophyll fluorescence was more relevant in case of heat, cold and UV stress. Under these stress conditions, light harvesting complex in the photosystems are more prone to damage due to changes in photosynthesis and membrane properties. Semi automated systems have used fluorescence cameras and validation of the traits during drought and chilling stress was carried out (Jansen *et al.,* 2009). In wheat, chlorophyll fluorescence has been used as early stress detection parameter. QTL mapping was done under heat stress at seedling stage. A very large population of mutants was characterized by using the fluorescence meters and fluorescence cameras (Lu *et al.,* 2011).

Automated imaging of mature plants is complex because it requires the construction of a three dimensional model. Therefore, most of the studies have been carried out at the seedling stage. Recent developments for measuring photosynthesis at the canopy level are laser induced fluorescence and sun induced fluorescence, but they need to be validated for reproducibility.

Canopy Temperature

Canopy temperature depression (canopy temperature differences with respect to the surrounding air) is currently being used as one of the most important selection trait in plant phenotyping for abiotic stress tolerance. Canopy temperature depends on the evaporation/transpiration from leaf surface. Crop temperature may rise under stress conditions due to reduced transpiration. Estimation of canopy temperature is used to detect the changes in plant water relations in response to stress. Stomatal conductance, transpiration rate and canopy temperature are directly correlated to each other (Jones *et al.,* 2002; 2009). Thermal imaging is used to estimate the leaf surface temperature. Measurement of canopy temperature by infra red thermometer is generally carried out in the breeding programs for drought prone environment. Hand-held thermometers and guns are now cheaply available and can be used conveniently. But manual phenotyping of very large populations by hand-held gadgets require time and labour for walking in the plots to record canopy temperature. Changes in the environmental conditions during data recording over long distance also affect the data. But phenotyping by the imaging sensors can acquire data in a large number of plots at the same time. Furthermore, thermal cameras allow the more precise measurements in a single image in a fraction of time (Jones *et al.,* 2009). Phenotyping by automated imaging set ups in the controlled environment provides us an opportunity to evaluate large population in a target environment. Thermal imaging was used at individual plant level in wheat seedlings to capture the differences in canopy temperature in response to salt stress with a differential of 1°C in temperature (Sirault *et al.,* 2009).

Precise measurement of the trait is challenged by the background noise due to soil and effect of wind and cloudiness (Jones *et al.,* 2009, Matsuda *et al.,* 2012). Though, the use of thermal cameras provide a fast screening strategy, but the understanding of the basic principles of heat fluxes in the natural environment is necessary before phenotyping. Surface temperature measurements require extensive calibrations. Orientation of leaf towards the incident radiation and camera angle used to capture the images should be taken into consideration during data analysis.

Chlorophyll Content

It is the most commonly used trait in phenotyping. It can be easily recorded using a portable optical meter, which convert the ratio of absorbance into a digital signal and then into SPAD (Soil plant analytical development) value or CCI (chlorophyll content index) (Mullan and Mullan, 2012). SPAD is calculated on the basis of transmission values. Transmission of red light at 650 nm, which is absorbed by chlorophyll and transmission of infra red light at 940 nm at which no chlorophyll absorption takes place. CCI is the ratio of transmittance at 940 nm to 650 nm from specified leaf area of clamped leaf portion. Such values are positively correlated with the destructive chlorophyll measurements in the crops (Marquad and Tipton, 1987). Leaf cholorophyll content was found to be strongly correlated with dry root biomass (Songsri *et al.,* 2009), seed yield (Puangbut *et al.,* 2009) and nitrogen fixation status (Voltmann, 2011). SPAD measurements and NDVI (Normalized difference vegetation index) are the most commonly used indices for phenotyping in the field (Ruan *et al.,* 2001; Richardson *et al.,* 2002). Hyperspectral imaging has been used in pigment determination in cucumber in response to biotic stress (Zhao *et al.,* 2016), Arabidopsis (Harshvardhan *et al.,* 2014) and barley (Chen *et al.,* 2014) under drought stress.

Use of multispectral and hyperspectral cameras generate very large volume of data for analysis. Therefore, reduction in cost and data analysis is the major issue, which need to be addressed for large scale phenotyping.

Growth Pattern

Plant growth analysis is a key descriptive parameter among physiological traits. Since it involves destruction of plant providing end point analysis and is time consuming, it is not suitable for large scale phenotyping.

Nowadays, the imaging techniques have been developed for the three dimensional reconstruction of canopy to estimate plant area density, plant height, leaf area and angle, surface texture to study plant growth and development. The most commonly used techniques are RGB imaging, light detection and ranging (LIDAR), stereovision, photon mixer devices, time of flight cameras and Microsoft kinect. LIDAR is a remote sensing technology, which uses sensors for measurement of three dimensional distribution of plant canopy. It uses distance by illuminating an object and analyzing the reflected light. It is a robust and accurate technique with high resolution. In wheat, laser scanners have been used to map the leaf area index and plant area density profile (Hosoi and Omasa, 2009). LIDAR was used to estimate canopy structure, biomass and nitrogen in rice, populus and tomato (Omasa *et al.,* 2007; Tilly *et al.,* 2012; Eitel *et al.,*2014). Stereovision has also been used to provide the estimates of plant height, leaf area to estimate growth in maize (Ivanov *et al.,*1995). At present, three dimensional imaging has been used in the study of growth and development in the plant species, but its potential for abiotic stress is yet to be explored.

Root Phenotyping

Phenotyping for root traits is a difficult task due to technical difficulty in accessing the roots under soil. Therefore, in past, majority of the phenotyping studies were focused on the shoot phenotyping. With the development in technology, root phenotyping has been done for different traits like root length density, root size, root surface area, root length, diameter, root growth response, and number of lateral roots, root hairs (Weitz, 2009). Root architecture depends on the plant species and environmental conditions. Destructive sampling for root phenotyping includes root excavation or washed soil cores for measurement of root traits to understand the root architecture. But these measurements show differences due to difficulties in precise excavations (Hund, 2010). Non destructive sampling includes the use of soil filled rhizotrons with clear panels and soil free techniques like hydroponics, aeroponics and use of gel chamber or agar plate systems. With the advancement of non-destructive imaging techniques, automated phenotyping of root architecture has been performed (Iyer-Pascuzzi *et al.* 2010). High-throughput automated phenotyping of root growth with gel method was done in arabidopsis to quantify root length and curvature in response to gravitropism (French *et al.,* 2009). But the restriction of the method was chamber size restricting the root growth and generation of two dimension data (Hargreaves *et al.,* 2009). Shovelomics, a recently coined term for high-throughput phenotyping method using root excavations was used to study the root architectural traits (number of whorls occupied by brace roots, number of brace roots and their branching density, number, angle and branching density of crown roots) in maize (Trachsel *et al.,* 2011; 2013). Different imaging techniques like stereoscopy, neutron radiography, magnetic resonance imaging, ground penetrating radar and X-ray computed tomography (CT), laser capture microdissection (LCM) has increased our visibility of root architecture and its relation to the growing environment (Chen *et al.,* 2015). LCM based method has been used to identify root tissue specific genes involved in nutrient transport or homeostasis. Several genes involved in accumulation/exclusion of nutrients or toxic metals in crop grains have been identified and characterized in different crops such as maize, rice, and wheat using this approach (Sarkar, 2016).

AUTOMATED PHENOTYPING PLATFORMS

Recent developments in sensing technologies, automation, aeronautics and computing are being used to develop controlled environment based and field based phenotyping platforms. Controlled environment based phenotyping platforms are developed in public domain and are also being sold commercially (Lemna Tec, Plantscreen, PHENOPSIS, WIWAM, GROWSCREEN). These platforms are special structures designed for large scale phenotyping of selected species ranging from arabidopsis to cereals. These controlled environment systems are deployed in growth chambers and green houses. Studies in the controlled environment are generally performed to minimize the variability between plants in statistically significant way. Accurate measurement of the environmental conditions in controlled environment can help in finding out the principal environmental factor affecting the phenotype. But controlled environment has its own limitations and can not truly represent the field conditions.

Space availability in the growth chambers affect the pattern of growth and development of plants and therefore, it is not feasible to assess the effect of stress on reproductive growth of plant. Real characterization of plant's response to stress is problematic in controlled environment systems. Therefore, field based phenotyping platforms are getting more attention as compared to controlled platforms (White *et al.,* 2012). Field based phenotyping platforms may be ground or aerial based platforms depending on the vehicle used to carry the sensor. Ground based options include high clearance tractors, linear move or central pivot irrigation systems, while, aerial based platforms use manned fixed and rotatory wing aircraft, unmanned aircraft and tethered aerostats.

PHENOMICS FACILITIES

Australian Plant Phenomics Facility (APPF) has two facilities, one the high resolution plant phenomics centre at Canberra and another plant accelerator at Adelaide. This facility was developed for high-throughput phenotyping of plants across a range of growth conditions (glass houses and growth chambers). Others high-throughput phenotyping facilities are CropDesign in Gatersleben, Germany and phenoArch in France. In Germany, the Institute for Phytosphere Research (IPR) in Jülich established the Jülich Phenomics Centre in 2007. Recently, Indian Council of Agricultural Research, New Delhi under National Initiative on Climate Resilient Agriculture (NICRA) project has established managed stress environment phenomics facilities with visual, IR thermal and NIR root and shoot imaging facilities at Indian Agricultural Research Institute (IARI), New Delhi and Central Research Institute for Dryland Agriculture, Hyderabad, while these facilities are being established at National Institute for Abiotic Stress Management, Baramati and Indian Institute of Horticultural Research, Bengaluru. Further, Indian Council of Agriculture Research is establishing a National Phenomics Facility for high-throughput non-destructive automated controlled environment phenotyping facility at IARI, New Delhi.

CONCLUSIONS

High-throughput phenotyping is an emerging research area. It is a combined application of plant biology, sensor technology and automated engineering. It has significant potential to systematically address the challenges in the phenotypic studies. Use of imaging techniques in the high-throughput phenotyping has opened the new horizons for large scale screening of crop genotypes. Application of non-invasive phenotyping is increasing day by day for screening in controlled as well as field conditions for abiotic stress tolerance. Canopy structure, growth analysis and photosynthetic performance can be measured non-destructively in model plant and cereal crops. Study of root architecture in the field is still a problem to be addressed. Mechanistic understanding and precise phenotyping play a key role in identification and evaluation of useful parameters. Success of these technologies depends on precise environmental monitoring and use of information technology for data management. Achieving high-throughput screening requires comprehensive facilities for different target environments. For large scale phenotyping, reduction in cost and gain in time are the future targets. Successful application of high-throughput phenotyping can be achieved only through the collective efforts of plant biologists, physicists, mathematicians and engineers.

FUTURE PROSPECTS

Translating our research under controlled conditions into field conditions is the future target. We need to develop more understanding of the heterogenous conditions prevalent in the field, so that we can link the knowledge generated through the phenotyping studies in controlled and semi-controlled environment with the natural environment. Cheaper phenotyping solutions within acceptable limits are needed to be developed for field as well as controlled conditions. Quick and robust root phenotyping techniques are required to characterize the root architecture.

REFERENCES

Arvidsson S, Perez-Rodriguez P and Mueller-Roeber B 2011 A growth phenotyping pipeline for *Arabidopsis thaliana* integrating image analysis and rosette area modeling for robust quantification of genotype effects. *New Phytologist* **191** 895–907

Baker NR 2008 Chlorophyll florescence: a probe of photosynthesis in vivo. *Annual Review of Plant Biology* **113** 59–89

Berger B, Parent B and Tester M 2010 High-throughput shoot imaging to study drought responses. *Journal of Experimental Botany* **61** 3519-28

Bioversity International 2007 Guidelines for the development of crop descriptor lists. Bioversity technical bulletin series. Rome, Italy

Bühler J, Huber G, Schmid F and Blümler P 2011 Analytical model for long-distance tracer-transport in plants. *Journal of Theoretical Biology* **270** 70–79

Chen D, Neumann K, Friedel S, Kilian B, Chen M, Altmann T and Klukas C 2014 Dissecting the phenotypic components of crop plant growth and drought responses based on High-throughput image analysis. *Plant and Cell* **26** 4636-55

Chen YL, Djalovic I and Rengel Z 2015 Phenotyping for root traits. In Phenomics in crop plants. Trends, options and limitations. Kumar J., Pratap A. and Kumar S, Springer, New Delhi Heidelberg, New York, Dodrecht, London 101-128.

Cseri A, Sass L, Torjek O, Pauk J, Vass I and Dudits D 2013 Monitoring drought responses of barley genotypes with semi robotic phenotyping platform and association analysis between recorded traits and allelic variants of some stress genes. *Australian Journal of Crop Sciences* **7** 1560-70

Eitel JU, Magney TS, Vierling LA, Brown TT and Huggins DR 2014 Lidar based biomass and crop nitrogen estimates for rapid, non-destructive assessment of wheat nitrogen status. *Field Crops Research* **159** 21–32

Feher-Juhasez E, Majer P, Sass L, Lantos C, Csiszar J, Turoczy Z, Milhaly R, Mai A, Horvath GV, Vass I, Dudits D and Pauk J 2014 Phenotyping shows improved physiological traits and seed yield of transgenic wheat plants expressing the alfalfa aldose reductase under permanent drought stress. *Acta Physiologiae Plantarum* **36** 663-73

Furbank RT 2009 Plant phenomics: from gene to form and function. *Functional Plant Biology* **36** 5

Furbank RT, Tester M 2011 Phenomics-techniques to relieve the phenotyping bottleneck. *Trends in Plant Sciences* **16** 635-644

Golzarian M, Frick R, Rajendran K, Berger B, Roy S, Tester M, Lun DS 2011 Accurate inference of shoot biomass from high-throughput images of cereal plants. *Plant Methods* **7** 2

Granier C, Aguirrezabal L, Chenu K, Cookson SJ, Dauzat M, Hamdard P, Thioux JJ, Rolland G, Bouchier-Combaud S, Lebaudy A, Muller B and Simonneau T, Tardieu F 2006 PHENOPSIS, an automated platform for reproducible phenotyping of plant responses to soil water deficit in *Arabidopsis thaliana* permitted the identification of an accession with low sensitivity to soil water deficit. *New Phytologist* **169** 623–635

Gregory PJ, Hutchison DJ, Read DB, Jenneson PM, Gilboy WB and Morton EJ 2003 Non-invasive imaging of roots with high resolution X-ray micro-tomography. *Plant and Soil* **255** 251–59

Hargreaves C, Gregory P and Bengough AG 2009 Measuring root traits in barley (Hordeum vulgare ssp. Vulgare and ssp. spontaneum) seedlings using gel chambers, soil sacs and X-ray microtomography. *Plant and Soil* **316** 285-97

Harris BN, Sadras VO and Tester M 2010 A water-centred framework to assess the effects of salinity on the growth and yield of wheat and barley. *Plant and Soil* **336** 377-89

Harshavardhan VT, Vanson L, Seiler C, Junker A, Weigelt-Fischer K, Klukas C, Altmann T, Sreenivasulu, Baumlein H and Kuhlmann N 2014 AtRD22 and AtUSPL1, members of the plant specific BURP domain family involved in *Arabidopsis thaliana. PLos One* **9** e110065

Honsdorf N, March TJ, Berger B, Tester M and Pillen K 2014 High-throughput phenotyping to detect drought tolerance QTL in wild barley introgression lines. *PLoS One* **9** e97047

Hosoi F, Omasa K 2009 Estimating vertical plant area density profile and growth parameters of a wheat canopy at different growth stages using three-dimensional portable lidar imaging. *ISPRS Journal of Photogrammetry and Remote Sensing* **64** 151–158

Humplik JF, Lazar D, Furst T, Husickova A, Hybl M and Spichal L 2015 Automated integrative high-throughput phenotyping of plant shoots: a case study of the cold-tolerance of pea (*Pisum sativum* L.). *Plant methods* **11** 1-11

Ivanov N, Boissard P, Chapron M, Andrieu B 1995 Computer stereo plotting for 3-D reconstruction of a maize canopy. *Agricultural and. Forest Meteorology* **75** 85–102

Iyer–Pascuzzi AS, Symonova O, Mileyko Y, Hao Y, Belcher H, Harer J, Weitz JS, Benfey PN 2010 Imaging and analysis platform for automatic phenotyping and trait ranking of plant root systems. *Plant Physiology* **152** 1148–1157

James RA, Rivelli AR, Munns R and von Caemmerer S 2002 Factors affecting CO_2 assimilation, leaf injury and growth in salt-stressed durum wheat. *Functional Plant Biology* **29** 1393-1403

Jansen M, Gilmer F, Biskup B, Nagel KA, Rascher U, Fischbach A, Briem S, Dreissen G, Tittmann S, Braun S, Jaeger ID, Metzlaff M, Schurr U, Scharr H, Walter A 2009 Simultaneous phenotyping of leaf growth and chlorophyll fluorescence via GROWSCREEN FLUORO allows detection of stress tolerance in *Arabidopsis thaliana* and other rosette plants. *Functional Plant Biology* **36** 902–914

Jones HG, Serraj R, Loveys BR, Xiong L, Wheaton A, Price AH 2009 Thermal infrared imaging of crop canopies for the remote diagnosis and quantification of plant responses to water stress in the field. *Functional Plant Biology* **36** 978–89

Jones HG, Stoll M, Santos T, de Sousa C, Chaves MM and Grant OM 2002 Use of infrared thermography for monitoring stamatal closure in the field: application to grapevine. *Journal of Experimental Botany* **53** 2249-2260

Li L, Zhang Q and Huang D 2014 A review of imaging techniques for plant phenotyping. *Sensors* **14** 20078-20111

Lu Y, Savage LJ and Last RL 2011 Chloroplast phenomics: systematic phenotypic screening of chloroplast protein mutants in *Arabidopsis*. *Methods in Molecular Biology* **775** 161–85

Marquad RD and Tipton JL 1987 Relationship between extractable chlorophyll and an in situ method to estimate leaf greenness. *Horticultural Sciences* **22(6)** 1327.

Matsuda O, Tanaka A, FujitaT, Iba K 2012 Hyperspectral imaging techniques for rapid identification of Arabidopsis mutants with altered leaf pigment status. *Plant and Cell Physiology* **53** 1154-70

Maxwell K and Johnson GN 2000 Chlorophyll fluorescence-a practical guide. *Journal of Experimental Botany* **51** 659-68

Metzner R, Eggert A, Dusschoten D, Pflugfelder D, Gerth S, Schurr, Uhlmann N and Jahnke S 2015 Direct comparison of 3D imaging of root systems in soil: potential and challenges for root trait quantification. *Plant methods* **11** 17

Miyao A, Iwasaki Y, Kitano H, Itoh J, Maekawa M, Murata K, Yatou O, Nagato Y, Hirochika H 2007 A large-scale collection of phenotypic data describing an insertional mutant population to facilitate functional analysis of rice genes. *Plant Molecular Biology* **63** 625–635

Moradi A, Oswald S, Nordmeyer-Massner J, Pruessmann K, Robinson B and Schulin, R 2010 Analysis of nickel concentration profiles around the roots of the hyperaccumulator plant *Berkheya coddii* using MRI and numerical simulations. *Plant and Soil* **328** 291–302

Mullan D, Mullan D 2012 Chlorophyll content. In: Pask A, Pietragalla J, Mullan D, Reynolds M (eds) Physiological breeding II: a field guide to wheat phenotyping. CIMMYT, Mexico, pp 41–43

Munns R, James RA, Sirault XRR, Furbank RT and Jones HG 2010 New phenotyping methods for screening wheat and barley for beneficial responses to water deficit. *Journal of Experimental Botany* **61** 3499–507

Neilson EH, Edwards AM, Blomstedt CK, Berger CK, Berger B, Moller BL and Gleadow RM 2015 Utilization of a high-throughput shoot imaging system to examine the dynamic phenotypic responses of a C4 cereal crop plant to nitrogen and water deficiency over time. *Journal of Experimental Botany* **66** 1817-32

Neumann K, Klukas C, Friedel S, Rischbeck P, Chen D, Entzian A, Stein N, Granar A and Kilian B 2015 Dissecting spatio-temporal biomass accumulation in barley under different water regimes using high-throughput image analysis. *Plant Cell and Environment* Doi: 10.1111/pce.12516. In press.

Omasa K, Hosoi F and Konishi A 2007 3D lidar imaging for detecting and understanding plant responses and canopy structure. *Journal of Experimental Botany* **58** 881–98

Petrozza A, Santaniello A, Summerer S, Di Tommaso G, Di Tommaso D, Paparelli E, Piaggesi A, Perata P, Cellini F 2014 Physiological responses to Megafol treatments in tomato plants under drought stress: A phenomic and molecular approach. *Scientia Horticulturae* **174** 185-92

Puangbut D, Jogloy S, Vorasoot N, Akkasaeng C, Kesmalac T, Patanothai A 2009 Variability in yield responses of peanut (*Arachis hypogaea* L.) genotypes under early season drought. *Asian Journal of Plant Sciences* **8** 254–264

Qin F, Shinozaki K and Yamaguchi-Shinozaki K 2011 Achievements and challenges in understanding plant abiotic stress responses and tolerance. *Plant and Cell Physiology* **52(9)** 1569-1582

Rajendran K, Tester M and Roy SJ 2009 Quantifying the three main components of salinity tolerance in cereals. *Plant Cell and Environment* **32** 237–49

Rao NK 2004 Plant genetic resources: advancing conservation and use through biotechnology. *African Journal of Biotechnology* **3** 136–145

Ray DK, Ramankutty N, Mueller ND, West PC and Foley JA 2012 Recent patterns of crop yield growth and stagnation. *Nature Communication* **3** 1293

Ray DK, Muller ND, West PC and Foley JA 2015 Yield trends are insufficient to double global crop production by 2050. *PLoS ONE* **8** e66428

Richard CA, Hickey LT, Fletcher S, Jennings R, Chenu K and Christopher 2015 High-throughput phenotyping of seminal root traits in wheat. *Plant Methods* **11** 13

Richardson AD, Duigan SP and Berlyn GP 2002 An evaluation of noninvasive methods to estimate foliar chlorophyll content. *New Phytologist* **153** 185-194

Ruan WR, Solie JB, Johnson GV, Stone GV, Stone ML, Lukina EV, Thomason WE and Schepers JS 2001 In-season prediction of potential grain yield in winter wheat using canopy reflectance. *Agronomy Journal* **93** 131-138

Sarkar AK (2016). Laser capture microdissection (LCM) based approach: technique, application, and potential in plant functional genomics and nutrient studies. In: Training Manual on "Genomics and phenomics for enhancement of crop nutrient use efficiency at NRCPB, New Delhi. 61-63

Schilling RK, Marschner P, Shavrukov Y, Berger B, Tester M, Roy SJ, Plett DC (2014) Schilling RK, Marschner P, Shavrukov Y, Berger B, Tester M, Roj SJ et al. 2014 Expression of the Arabidopsis vacuolar H+ -pyrophosphatase gene (AVP1) improves the shoot biomass of transgenic barley and inceased\s grain yield in a saline field. *Plant Biotechnology Journal* **12** 378-86.

Sirault XRR, James RA and Furbank RT 2009 A new screening method for osmotic component of salinity tolerance in cereals using infrared thermography. *Functional Plant Biology* **36** 970-977

Skirycz A, Vandenbroucke K, Clauw P, Maleux K, De Meyer B, Dhondt S, Pucci A, Gonzalez N, Hoeberichts F, Tognetti VB, Galbiati M, Tonelli C, Van Breusegem F, Vuylsteke M, Inzé D 2011 Survival and growth of *Arabidopsis* plants given limited water are not equal. *Nature Biotechnology* **29** 212–14

Songsri P, Jogloy S, Holbrook CC, Kesmala T, Vorasoot N, Akkasaeng C, Patanothai A 2009 Association of root, specific leaf area and SPAD chlorophyll meter reading to water use efficiency of peanut under different available soil water. *Agricultural Water Management* **96** 790–798

Summerfield RJ, Ellis RH, Craufurd PQ 1996 Phenological adaptation to cropping environment. From evaluation descriptors of times to flowering to the genetic characterisation of flowering responses to photoperiod and temperature. *Euphytica* **92** 281–286

Tilly N, Hoffmeister D, Liang H, Cao Q, Liu Y, Lenz-Wiedemann V, Miao Y and Bareth G 2012 Evaluation of terrestrial laser scanning for rice growth monitoring. *The International Archives of Photogrammetry,. Remote Sensing and Spatial Information Sciences.* **39** B7

Tisné S, Serrand Y, Bach L, Gilbault E, Ben Ameur R, Balasse H, Voisin R, Bouchez D, Durand-Tardif M, Guerche P, Chareyron G, Da Rugna J, Camilleri C, Loudet O 2013 Phenoscope: an automated large scale phenotyping platform offering high spatial homogeneity. *The Plant Journal* **74** 534-44

Trachsel S, Kaeppler S, Brown K, Lynch J 2011 Shovelomics: high throughput phenotyping of maize (*Zea mays* L.) root architecture in the field. *Plant and Soil* **341**75–87

Trachsel S, Kaeppler SM, Brown KM, Lynch JP 2013 Maize root growth angles become steeper under low N conditions. *Field Crops Research* **140** 18–31

Uddling J, Gelang-Alfredsson J, Piikki K, and Pleijel H 2007 Evaluating the relationship between leaf chlorophyll concentration and SPAD-502 chlorophyll meter readings. *Photosynthesis Research* **91** 37–46

Vollmann J, Walter H, Sato T, Schweiger P 2011 Digital image analysis and chlorophyll metering for phenotyping the effects of nodulation in soybean. *Computers and Electronics in Agriculture* **75** 190–195

Walter A, Scharr H, Gilmer F, Zierer R, Nagel KA, Ernst M, Wiese A, Virnich O, Christ MM, Uhlig B, Jünger S, Schurr U 2007 Dynamics of seedling growth acclimation towards altered light conditions can be quantified via GROWSCREEN: a set up and procedure designed for rapid optical phenotyping of different plant species. *New Phytologist* **174** 447-55

Weitz J 2009 Automated phenotyping of plant root systems. EPSO workshop on plant phenotyping, Forschungszentrum Ju ¨lich, Germany, 02 Nov 2009

White JW, Andrade-Sanchez P, Gore MA, Bronson KF, Coffelt TA, Conley MM, Feldmann KA, French AN, Heun JT and Hunsaker DJ 2012 Field-based phenomics for plant genetics research. *Field Crops Research* **133** 101–112

Xie XD, Wang YB, Williamson L, Holroyd GH, Taglivia C, Murchie E, Theobald J, Knight MR, Davies WJ, Leyser HMO, Hetherington AM 2006 The identification of genes involved in the stomatal response to reduced atmospheric relative humidity. *Current Biology* **16** 882-887.

■■■■

13

AN INSIGHT INTO SIGNAL TRANSDUCTION PATHWAYS AND THE ROLES OF PHYTOHORMONES: AN EMERGING CHALLENGE IN PLANT PHYSIOLOGICAL RESEARCH

Md. Afjal Ahmad, Savita Jangde and Pravin Prakash*

Department of Plant Physiology, Institute of Agricultural Sciences, Banaras Hindu University,BHU, Varanasi - 221 005, India
**E-mail: prakashpbhu@gmail.com*

INTRODUCTION

The plant form is attained through the action of specific "organ forming substances" or chemical messengers were the opinion of German plant physiologist Julius Von Sachs (1832-1897). His observations were confirmed by Charles Darwin through his work on Phototropism and Gravitropism. Now it is understood that hormones initiate biochemical reactions and cause changes in chemical composition within the plant. Accompanying the changes in chemical composition are changes in growth patterns that lead to the formation of roots, shoots, leaves, flowers and other structural entities characteristic of the plant. Environmental factors, such as light, temperature etc. interact with the phytohormones and biochemical processes during growth and differentiation.

Hormones are organic compounds synthesized in one part of plant and translocated to another part, wherein, very low concentrations, it cause a physiological response. Hormones are usually effective at concentrations of μM or less. The hormones involved in various physiological responses are auxins, gibberellins, cytokinins, abscissic acid and ethylene. Some new regulator of growth and organic compounds involved in signal transduction are plant steroidal hormone brassinosteroids, jasmonic acid, salicylic acid and polypeptide, systemin. The sequence of events initiated by hormones in signal transduction pathways are resolved into three sequential stages, viz., initial signal perception, amplification of the signal or signal transduction pathway and the final response.

The signal perception involves the binding of the hormone with a receptor site. Cells that are destined to respond to the hormone (target cells) possess a cellular receptor, specific to the hormone molecule. Receptors are glycoproteins that bind reversibly with hormones. Change in conformation of the receptor, due to reversible binding of hormones assumes an activated state. Formation of active hormone –receptor complex completes the signal perception stage. The hormone receptor should follow certain important characteristics such as the receptor should bind only one type of hormone and its structural analogue (specificity), receptor should have high affinity for the hormone, receptors can be saturated by increasing the concentration of hormone molecules, hormones must bind reversibly with the putative receptor and molecules of similar structure should be able to compete with the hormone for binding site. Hormone receptors are also called as hormone binding proteins.

The second stage of hormone action is the signal transduction and amplification stage. In this stage, the activated hormone receptor complex sets into motion a cascade of biochemical events that ultimately leads to the final, characteristic response. The hormone receptor complex activates a membrane protein called the "G protein" which in turn binds to the cytoplasmic surface of the membrane. Binding of the G-protein to the adenylate cyclase activates the enzyme, thus stimulating the formation of cyclic adenosine monophosphate (cAMP) in the cytoplasm. Alternatively, as with certain hormone receptor complexes, the G protein may interact with an ion channel that controls the flow of calcium into the cell. Once in the cytoplasm, the calcium will bind with one of a number of cytosolic calcium binding proteins, such as calmodulin. The effect of either cAMP or the Ca^{2+} – calmodulin complex is to activate specific protein kinases. Protein kinases are enzymes that modify other proteins by chemically adding phosphate groups to them (phosphorylation). i.e., protein kinase phosphorylate other proteins by transferring a phosphate group from ATP. Phosphorylation usually results in a functional change of the target protein (substrate), thereby altering the metabolism of the cell and triggering the cell's ultimate response. The substrate (target protein) may be a transcriptional activator of a gene of interest or enzymes. Reversible protein phosphorylation is a key mechanism for switching proteins on and off in plants.

The hormone is considered first or primary messenger because it brings the original message to the cell surface. Diacyl glycerol (DAG), cAMP, calcium, inositol triphosphate (IP_3), cGMP, changes in intracellular pH etc. serves as secondary messengers. The function of a secondary messenger is to relay information from the plasma membrane to the biochemical machinery inside the cell. Secondary messengers also do the amplification of the original signal. For example, a single hormone molecule will activate only one molecule of adenylate cyclase may produce hundreds of cAMP molecules, thereby initiating a cascade of biochemical events.

An attempt has been made below to give a brief insight into signal transduction pathways of some of the phytohormones based on plant physiological research done till now.

AUXIN SIGNALING

The term auxin is generally given to IAA and other natural and synthetic compounds having similar structure and growth regulating properties. Generally, auxins are produced in the growing apices of stem and root where from they migrate to the other parts of the plant. Auxins such as IAA and phenyl acetic acid (PAA) are natural auxin. They regulate apical-basal polarity in embryonic development; apical dominance in shoots; induction of lateral and adventitious roots; vascular tissue differentiation; and cell growth in both stems and coleoptiles, including the asymmetric growth associated with phototropic and gravitropic responses (Davies, 1995). Auxin is transported from the shoot downward. However, the prevailing model of the initiation of auxin gradients in the apical meristem has been questioned by the demonstration that all parts of young plants can synthesize IAA, thus potentially diminishing the importance of polar auxin transport (Ljung *et al.,* 2001). Cholodny and Went in the 1920s independently suggested the chemiosmotic hypothesis of auxin transport, which was later refined by Rubery and Sheldrake (1974). The theory postulates the existence of an auxin efflux carrier that actively and asymmetrically redistributes auxin in root and stem tissue on gravitropic or phototropic stimulation.

Auxin movement both into and out of cells requires specialized carriers (Friml and Palme, 2002a). Several Arabidopsis genes encoding putative auxin carriers have been identified during the past decade. The amino acid permease-like gene *AUX1* and the family of bacterial transporter-like *PIN* genes encode putative auxin influx (Bennett *et al.,* 1996) and efflux carriers, respectively. Characterization of the first putative auxin efflux carrier PIN1 (Gälweiler *et al.,* 1998) gave context to auxin's asymmetric localization. PIN1 encodes a 622- amino-acid protein with 12 predicted transmembrane spanning segments. It shares similarity with a group of transporters from bacteria of the major facilitator class, evidence supporting a transport function (Gälweiler *et al.,* 1998; Pao *et al.,* 1998).

A search of the Arabidopsis genome for genes with homology to PIN1 revealed another seven genes belonging to the same family. Similar sequences have been found in all other plants now sequenced, but not in animals, indicating that PIN proteins have evolved exclusively in plants. Based on genetic evidence, PIN proteins are strong candidates for either the auxin efflux carrier itself or an important regulatory component of the efflux machinery (Palme and Gälweiler, 1999). More important, the distributions of PIN1 and other PIN proteins in the plasma membrane of auxin-transporting cells of stems and roots were found to be dynamic and asymmetric according to the direction of auxin flux (Friml *et al.,* 2002b; Gälweiler *et al.,* 1998; Geldner *et al.,* 2001; Müller *et al.,* 1998; Steinmann *et al.,* 1999). Auxin gradients in plant tissue appear to be sink driven; gradient formation seems to be regulated by auxin transport (rather than degradation) machinery. For example, the formation of a maximum auxin concentration at the Arabidopsis root apex depends on the activity of PIN4 (Friml *et al.,* 2002a).

Effect of Auxin Signaling on Gene Expression

Auxin-dependent transcriptional activation can occur within minutes of a signal being perceived (Abel and Theologis, 1996). In the nucleus, the regulation of gene expression by auxin can be mediated by the action of two families of auxin-induced proteins: the Aux/IAA proteins and the auxin response factors (ARFs) (Hagen and Guilfoyle, 2002). ARFs bind to auxin response promoter elements upstream of genes and activate or repress their transcription. Aux/IAA proteins can dimerize with ARF proteins, thus inhibiting their activity (Tiwari *et al.,* 2003). However, they have very short half-lives, ranging from a few minutes to a few hours (Abel *et al.,* 1994; Gray *et al.,* 2001). A normal auxin response is dependent on this rapid turnover of Aux/IAA proteins, as it lowers the concentration of the inhibitory ARF-Aux/IAA dimer (Ulmasov *et al.,* 1997; Worley *et al.,* 2000).

Aux/IAA proteins are found in all higher plants and are characterized by four highly conserved domains (Abel and Theologis, 1996; Guilfoyle *et al.,* 1998). In yeast two-hybrid assays, their dimerization with ARF proteins has been shown to involve two of these domains (which are similar to those of the ARF proteins) (Ulmasov *et al.,* 1997). Another domain, domain II, is crucial for Aux/IAA function, with many lines of evidence demonstrating that it is the target for Aux/IAA protein destabilization, ensuring the rapid turnover required for a normal auxin response. The fusion of Aux/ IAA proteins to reporter proteins, such as luciferase or glucuronase (GUS), results in the destabilization of the reporter protein (Gray *et al.* 2001; Worley *et al.,* 2000).

This indicates that Aux/IAA proteins contain a transferable destabilization sequence. A nonfunctional domain II, as found in the auxin-resistant mutants *axr3-1*, *axr2- 1*, and *shy2*, dramatically increases the protein's half-life and prevents the ARF proteins from functioning (Gray *et al.,* 2001; Ouellet *et al.,* 2001; Worley *et al.,* 2000). The stabilization of an Aux/IAA-reporter fusion by inhibitors of the 26S proteasome indicates that auxin signaling requires SCFTIR1-mediated turnover of Aux/IAA proteins.

GIBBERELLIN SIGNALING

Gibberellin was first discovered in Japan by Kurusowa. He observed from his field that some of the rice seedlings had grown much taller than the others. On further observation, he found that such taller rice plants had shown unusual intermodal elongation. This internodal elongation is known as the 'bakanae' or 'foolish seedling' disease of rice. Later, it was discovered that the elongation was due to the action of a substance produced by a fungus, Gibberella fujikuroi. This substance was successfully isolated from the fungus and it was named as gibberellic acid. GAs are a large group of diterpenes comprising well over a hundred members. However, only a handful can elicit a physiological response, the others being representative of a large and complicated web of biosynthetic pathways from *ent*-kaurene, the product of the first dedicated step of GA biosynthesis. These biosynthetic pathways are now well understood. GAs regulate a wide range of physiological processes, including cell division and cell elongation, and are crucial to the control of processes as diverse as germination, stem elongation, flowering, fruit ripening, and senescence.

Like auxin, gibberellins play a major role in diverse growth processes, including seed development, organ elongation and the control of flowering time. Physiological data demonstrate that gibberellin responses are negatively regulated by DELLA proteins (The DELLA proteins get their name from the conserved N-terminal DELLA domain18.) The DELLA proteins, a class of transcription factors, has emerged as a central mediator of many GA responses, although they probably do not bind DNA directly (Dill *et al.,* 2001). Recent work exploring the integration of light and gibberellin signals during cell elongation has elucidated a model for DELLA-mediated growth regulation. DELLA proteins were shown to interact directly with the DNA-binding domain of two basic helix–loop–helix transcription factors (PIF3 and PIF4), sequestering them in inactive complexes. Gibberellin accumulation destabilizes DELLAs, freeing PIF3 and PIF4 to activate the transcription of their target genes. It is likely that DELLAs regulate the activity of several other transcription factors by this mechanism, as PIF3 and PIF4 are members of a larger subfamily of basic helix–loop–helix proteins that have similar DNA-binding domains.

It was decided that these proteins (named after a five-amino-acid N-terminal motif) are important components of the GA signal transduction pathway after the analysis of a number of dwarfed mutants from a range of species (Sun, 2000). An important example (and the mutant from which the first member of the family was cloned) is the gai1-1 mutant of Arabidopsis. Plants with a dominant mutation at the GA I allele display a semidwarf phenotype, insensitive to the exogenous application of GA

(Peng and Harberd, 1993). All results indicate that the DELLA proteins are negative regulators of GA signaling.

CYTOKININ MEDIATED SIGNALING

Cytokinin is a plant growth substance, which stimulates cell division. This was isolated by Miller and Skoog in 1954 from Herring fish. Following the discovery of kinetin many other compounds showing similar activity were discovered. These are collectively called cytokinins. The cytokinin found in the zea mays is called zeatin. Cytokinins play a major role in many different developmental and physiological processes in plants, such as cell division, regulation of root and shoot growth and branching, chloroplast development, leaf sensecence, nutrient mobilization, biomass distribution, stress response, and pathogen resistance (Willium *et al.* 2003).

In contrast to our understanding of auxin and GA signal transduction, proteins have been identified that function as cytokinin receptors. Activation tagging experiments in *Arabidopsis* identified *CKI1*, a gene encoding a receptor histidine kinase, whose overexpression was seen to induce typical cytokinin responses (Kakimoto, 1996). Although *CKI1* is able to activate the cytokinin signaling pathway, it does not bind cytokinins directly (Hwang and Sheen, 2001). Nevertheless, the discovery of *CKI1* suggested that the cytokinin transduction pathway in higher plants could be similar to the prokaryotic two-component system. This hypothesis was proved by the identification of CRE1/AHK4, another histidine kinase, as the first cytokinin receptor (Inoue *et al.,* 2001; Suzuki *et al.,* 2001; Ueguchi *et al.,* 2001). The availability of the Arabidopsis genomic sequence led to the identification of two further cytokinin receptors (AHK2 and AHK3).

After cytokinin perception, the resulting signal is transmitted by a multistep phosphorelay system through a complex form of the two-component signaling pathways that has been well characterized in prokaryotes and lower eukaryotes. Functional evidence for cytokinin sensing by the receptor CRE1/AHK4 was obtained in elegant complementation experiments in yeast and *Escherichia coli,* which rendered these heterologous hosts cytokinin sensitive. Two other histidine kinases, AHK2 and AHK3, have been shown to be active in the same complementation test system and to give protoplasts cytokinin sensitivity (Hwang and Sheen, 2001; Yamada *et al.,* 2001), indicating that these two proteins are also cytokinin receptors. Each receptor comprises an N-terminal extracellular domain, a membrane anchor, and a C-terminal transmitter domain, capable of autophosphorylation. The three cytokinin receptor genes differ in their expression pattern. *CRE1/AHK4* is mainly expressed in the roots, whereas *AHK2* and *AHK3* are present in all major organs (Inoue *et al.,* 2001; Ueguchi *et al.,* 2001). This tissue-specific expression of cytokinin receptors could be an additional layer of control to the perception of cytokinin.

It is important to understand such cross talk between cytokinins and nutrients as well as cytokinins and other phytohormones. Moore *et al.* (2003) showed that application of cytokinins as well as the use of transgenic *Arabidopsis* lines with constitutive cytokinin signaling could overcome the glucose repression response. The insensitivity of *Arabidopsis glucose insensitive2 (gin2)* to auxin and hypersensitivity to cytokinin could be the clue to understanding the antagonistic interaction between cytokinins and auxin and its dependency on the glucose status of tissues.

ABSCISIC-ACID SIGNALING PATHWAYS

Abscisic acid is a 15-C sesquiterpene compound (molecular formula $C_{15}H_{12}O_4$) composed of three isoprene residues and having a cyclohexane ring with keto and one hydroxyl group and a side chain with a terminal carboxylic group in its structure. ABA influences both physiological (gene expression in response to salt stress and drought) and developmental (e.g., germination and seedling development) processes. ABA seems to affect many different signaling pathways.The nature of the initial ABA receptor remains elusive. However, it is widely believed that the initial event in the signaling cascade is the binding of ABA to either a membrane bound or a cytosolic receptor. In many cases, this binding results in the activation of Ca^{2+}-influx channels resulting in the ABA-mediated increase in intracellular Ca^{2+} concentration (Murata *et al.,* 2001). Another important factor in this process is the altering permeability of the tonoplast to Ca^{2+}; this is influenced by intracellular lipid-derived signals and cyclic adenosine 5′-diphosphate–ribose concentration, the latter dependent on the Ca^{2+} concentration itself (Wu *et al.,* 2003). The overall increase in intracellular Ca^{2+} results first in the inhibition of K^+-influx channels and, second, in the activation of K^+-efflux channels and the inhibition of H^+-adenosine triphosphatase. Therefore it can be seen that ABA initiates a complicated network of interconnecting signals, resulting in a physiological response (Finkelstein *et al.,* 2002).

The other way by which ABA signaling takes place is a G-protein coupled receptor called GCR2. This suggestion is also controversial because the presence of a proposed transmembrane domain in GCR2 has been disputed. Furthermore, genetic analysis of gcr2 mutants failed to detect an abscisic-acid-related phenotype and some biochemical studies indicate that GCR2 does not bind abscisic acid after all. At this point, it seems unlikely that GCR2 functions as an abscisic-acid receptor.

It has become clear that the ABA signal is transduced through a number of secondary messengers, among them lipid-derived signals, H_2O_2, G proteins, and nitric oxide (Himmelbach *et al.,* 2003)The amount of free ABA able to elicit a response is thought to be dependent on many factors. These include movement of ABA through the plant, the relative rates of ABA synthesis and catabolism, and the concentration of ABA in the leaf symplast. The varying cellular concentrations of these compounds unite to influence indirectly the cytosolic concentration of Ca^{2+}, the central factor in many ABA signals (McAinsh *et al.,* 1997). It is thought that ABA has two modes of action: the first "nongenomic" effect is able to change the turgor pressure in guard cells by altering the plasma membrane's permeability to ions, and the second acts via changing the transcription levels of ABA-responsive genes. It is thought that both processes are reliant on alterations in the intracellular concentration of Ca^{2+}; however, it is not yet fully understood to what extent the pathways are separated.

ETHYLENE INDUCED SIGNALING

Neljubow in 1901 identified ethylene in laboratory air from illuminating coal gas which caused typical symptoms in etiolated pea seedlings grown in dark in the lab, *viz.,* (i) inhibition of stem elongation, (ii) stimulation of radial swelling of stems and (iii) horizontal growth of stems with respect to gravity. These symptoms were later termed as **'triple response'** and were not observed in etiolated pea seedlings grown in normal air free from coal gas. Gane (1934) clearly established that ethylene is actually a natural product of ripening fruits and is responsible for hastening ripening process.

The ethylene signaling pathway is relatively well characterized, and it has also been shown that ethylene signaling is intrinsically linked to many other phytohormonal signaling pathways. There are five ethylene receptors in Arabidopsis: ETR1, ETR2, ERS1, ERS2, and EIN4 (Stepanova and Ecker, 2000). They are all histidine kinases, the same class of two-component regulatory system as is found in the cytokinin signaling pathway. The ethylene receptors can be most easily classified in two ways. In the first, they are grouped into those with (ETR1, ETR2, and EIN4) and without (ERS1 and ERS2) a receiver domain, the domain that receives the phosphotransfer from the histidine kinase. It is thought that a signal could be transduced via a dimer of receptors (Hall *et al.,* 2000).

It has been suggested that the proteins without a receiver domain operate in a receptor complex. The second and more common classification groups the ethylene receptors according to their structure. Subfamily I (ETR1 and ERS1) has three membrane-spanning regions, whereas subfamily II (ETR2, ERS2, and EIN4) has four membrane- spanning regions and lacks conserved residues in the histidine kinase domain. Interestingly, the phenotype of receptor-deficient mutants can be rescued with an ETR1 protein with an inactivated histidine kinase domain. This work suggests that the ethylene receptor complex can transduce a signal by a mechanism other than histidine kinase–dependent phosphotransfer (Wang *et al.,* 2003). Although it can also homodimerize (Schaller and Bleecker, 1995), ETR1 has been shown to interact directly with CTR, a protein similar to the Raf family of mitogen-activated protein kinase (MAPK) kinases. The receptor complex has been shown to be located at the endoplasmic reticular membrane (Gao *et al.,* 2003), and negative regulation by a MAPK signaling cascade has been demonstrated in Arabidopsis and *Medicago* (Ouaked *et al.,* 2003). This provide another hint as to the complex relationship between what have been traditionally regarded as discrete phytohormone signaling pathways.

BRASSINOSTEROIDS

Brassinosteroids (BRs) are a class of polyhydroxysteroids that have been recognized as a sixth class of plant hormones. These were first explored nearly forty years ago when Mitchell *et al.* reported promotion in stem elongation and cell division by the treatment of organic extracts of rapeseed (*Brassica napus*) pollen Brassinolide was the first isolated brassinosteroid in 1979 when it was shown that pollen from *Brassica napus* could promote stem elongation and cell divisions, and the biologically active molecule was isolated. Brassinosteroids (BRs) have been shown to be important at a number of stages of a plant's development, including stem elongation, germination and senescence. BRs retain the basic four-ring structure of many steroid hormones; like animal steroid hormones, they are synthesized from cholesterol. Two mutants deficient in the biosynthesis of BRs, *DET2* and *CPD*, develop as if grown under light when grown in the dark (Chory *et al.,* 1991; Li and Chory, 1997). This demonstrates that, in addition to these other crucial processes, BRs play an important role in photomorphogenesis.

The identification of a BR receptor has been a recent significant advance in phytohormone biology; this section focuses on the mechanism by which BRs are initially sensed by the cell, before highlighting some interesting similarities between the perception and mode of action of BRs and the regulation of development in animals. In animals, steroids receptors are nuclear located (Marcinkowska and Wiedlocha, 2002); in plants, no nuclear steroid receptors have been found, indicating that plant cells have evolved a different method to receive the BR signal. The *bri* mutant of Arabidopsis shows a dwarfed phenotype similar to that of mutants deficient in the biosynthesis of BRs. BRI1 was thought to be involved in signaling owing to the mutant plants' unresponsiveness to applied brassinolide. Cloning of the bri1 gene revealed a leucine-rich repeat (LRR) receptor-like kinase, an immediate candidate for the BR receptor (Li and Chory, 1997). BRI1 was subsequently shown to be located at the plasma membrane (Friedrichsen *et al.,* 2000), and to have a relatively high affinity for bioactive BR (Wang *et al.,* 2001). LRR receptor kinases contain three domains: an extracellular domain comprising several leucine-rich repeating sections (in the case of BRI1, 25), a transmembrane section and a cytoplasmic kinase domain. BRI1 also contains a 70- amino-acid island in the LRR domain, necessary for the protein's function (Li and Chory, 1997).

The use of chimeric proteins has demonstrated that the BRI1extracellular domain was both necessary and sufficient. for the translation of a specific set of genes. When fused to the intracellular kinase domain of a similar receptor-like kinase, a protein involved triggering a plant's defensive response to pathogens, activation of the extracellular BRI1 domain and, hence, defense related gene expression could be induced by BR (He *et al.,* 2000). The LRR-receptor kinases are members of a very large class of proteins in Arabidopsis comprising 174 members with diverse function. Only a small number have been ascribed a function, including CLAVATA (involved in meristem development) and ERECTA (involved in organogenesis) (Dievart and Clark, 2003). The functions of LRR-receptor kinases are diverse. In Arabidopsis, three BRI-like proteins share high homology and are similar to protein sequences found in monocotyledonous species. Of these, BRL1 and BRL3 are able to rescue the *bri1* mutation, suggesting a closely related function.

CO-ORDINATION AND INTEGRATION OF HORMONE SIGNALING

Crosstalk between hormones is one of the very challenging field of research that has benefited from the recent elucidation of hormone signaling pathways. Although our knowledge of the molecular components and pathways that mediate hormone responses has improved enormously in recent years, the molecular mechanisms of hormone interaction remain poorly understood (Aaron Santner and Mark Estelle, 2009). The coming years should see a greater focus on the function of each hormone pathway in the context of larger regulatory networks. Evidence for hormone crosstalk comes largely from the analysis of mutant phenotypes. Frequently mutants that are affected in one hormone pathway also display changes in other hormone responses. For example, auxin-related mutants such as tir1, aux1 and pin2 exhibit altered response to other hormones, including ethylene and abscisic acid. From these studies and many others it is clear that hormone crosstalk is as complex as it is important. Almost certainly, all plant hormones interact with one or more additional hormones by affecting synthesis, transport or response. The type of interaction often depends on the tissue, developmental stage and environmental conditions. This presents a nearly endless number of possible opportunities for regulation. Even though the documented instances of crosstalk are numerous, our current knowledge of hormone interaction is quite sketchy.

It has become clear that diverse mechanisms have evolved to coordinate activity of hormonal pathways during development. A prominent example is the regulation of hormone metabolism by an interacting hormone. This mechanism is exemplified by ethylene and auxin. Measurement of the rate of auxin biosynthesis after ethylene treatment revealed that ethylene stimulates the auxin biosynthetic pathway. Indeed, several genes required for auxin biosynthesis are under the transcriptional control of ethylene. These genes encode both the alpha and beta subunits of anthranilate synthase (ASA1 and ASB1) and a newly identified family of tryptophan aminotransferases (TAA1). These enzymes function in tryptophan synthesis and the indole-3-pyruvic auxin biosynthetic pathway, respectively.

Auxin can influence ethylene biosynthesis as well. One of the rate-limiting enzymes in the ethylene biosynthesis pathway is 1-aminocyclopropane-1-carboxylate synthase (ACS). In Arabidopsis, there are nine ACS genes that can homodimerize or heterodimerize to form active enzymes. Several of the ACS genes have been found to be regulated by auxin treatment at the transcriptional level. Auxin has also been implicated in the production of jasmonic acid in flowers. Mutant analysis of arf6 arf8 double mutant plants demonstrated a role in the coordination transition from immature to mature flowers. During floral development, jasmonate levels peak just before anther dehiscence and then decrease. Direct measurements showed that jasmonate production was reduced in arf6 arf8 elongation. The attenuated growth response corresponded with reduced RGA (a specific DELLA protein) degradation in root cells. These observations indicate that auxin promotes the gibberellins induced destabilization of some of the DELLA proteins to affect gibberellins responses.

Similarly, ethylene may also target DELLA proteins to exert antagonistic actions with gibberellin during root growth. The gibberellin-insensitive gai/rga mutant exhibits ethylene-insensitive root growth, indicating that ethylene regulates root growth in a DELLA-dependent manner. In Arabidopsis roots, abscisic acid and gibberellin act antagonistically during root growth. Application of abscisic acid was demonstrated to stabilize the DELLA protein RGA and inhibit its gibberellin-induced degradation. Furthermore, higher-order DELLA mutants are resistant to the effects of abscisic acid on growth inhibition. DELLA proteins are not only a node for various hormone signal inputs but also mediate hormone signaling pathways in addition to gibberellin response.

DELLA proteins were recently implicated as modulators of plant immune responses. Genetic studies showed that DELLA proteins promote susceptibility to virulent biotrophs and resistance to necrotrophs. These observations were attributed, at least partially, to an alteration of the balance between salicylic acid and jasmonate signalling. These are but a few of the common strategies used to coordinate hormone signals. The regulatory network connecting individual pathways is far more complex in that hormone interactions can result in different outcomes depending on the organ, developmental stage and environmental conditions. Continued efforts to describe the plant hormone regulatory network will be essential to understand and predict plant growth and development.

CONCLUSION

Hormones involved in various physiological responses are auxins, gibberellins, cytokinins, abscisic acid and ethylene. Some new regulator of growth and organic compounds involved in signal transduction are plant steroidal hormone brassinosteroids, jasmonic acid, salicylic acid and polypeptide, systemin. The sequence of events initiated by hormones in signal transduction pathways are resolved into three sequential stages, viz., initial signal perception, amplification of the signal or signal transduction pathway and the final response. The signal perception involves the binding of the hormone with a receptor site. Cells that are destined to respond to the hormone (target cells) possess a cellular receptor, specific to the hormone molecule. Receptors are glycoproteins that bind reversibly with hormones. The hormone is considered first or primary messenger because it brings the original message to the cell surface. Diacyl glycerol (DAG), cAMP, calcium, inositol triphosphate (IP_3), cGMP, changes in intracellular pH etc. serves as secondary messengers. The function of a secondary messenger is to relay information from the plasma membrane to the biochemical machinery inside the cell. Secondary messengers also do the amplification of the original signal. Physiological data demonstrate that gibberellin responses are negatively regulated by DELLA proteins. (The DELLA proteins get their name from the conserved N-terminal DELLA domain18.) The DELLA proteins, a class of transcription factors, have emerged as a central mediator of many GA responses, although they probably do not bind DNA directly. Genetic studies showed that DELLA proteins promote susceptibility to virulent biotrophs and resistance to necrotrophs. Hence, plant hormone regulatory network will be essential to understand extensively for predicting the intricacies of plant growth and development.

REFERENCES

Aaron Santner, Mark Estelle (2009) Recent advances and emerging trends in plant hormone signaling; 459.doi: 10.1038/nature08122

Abel S, Oeller PW (1994) Theologis A. Early auxin-induced genes encode short-lived nuclear proteins. Proc Natl Acad Sci USA; 91: 326–330.doi: 10.1073/pnas.91.1.326

Abel S, Theologis (1996) A. Early genes and auxin action. Plant Physiol;111:9–17.doi: 10.1104/pp.111.1.9

Bennett MJ, Marchant A, Green HG, May ST, Ward SP, Millner PA, Walker AR, Schulz B, Feldmann KA. (1996) Arabidopsis AUX1 gene: a permease-like regulator of root gravitropism. Science; 273:948–950.doi: 10.1126/science.273.5277.948

Brenda Chow, Peter McCourt (2006) Plant hormone receptors: perception is everything. Cold Spring Harbor Laboratory Press 20: 1998-2008.doi: 10.1101/gad.1432806.

Chory J, Nagpal P, Peto CA (1991) Phenotypic and genetic analysis of DET2 a new mutant that affects light-regulated seedling development in Arabidopsis. Plant Cell;3:445–460.doi: 10.2307/3869351

Davies PJ (1995) The plant hormones: their nature, occurrence, and functions. In: Davies PJ., ed. Plant Hormones: Physiology, Biochemistry and Molecular Biology, 2nd Ed. Dordrecht, Netherlands: Kluwer Academic Publishers,:1–12.doi: 10.1006/anbo.1996.0152

Dievart A., Clark SE (2003) Using mutant alleles to determine the structure and function of leucine-rich repeat receptor-like kinases. Curr Opin Plant Biol;6:507–516.doi: 10.1016/51369-5266(03)00089-x

Dill A, Jung HS, Sun TP (2001) The DELLA motif is essential for gibberellin-induced degradation of RGA. *Proc Natl Acad Sci USA*; **98**:14,162–14,167.doi: 10.1073/pnas.251534098

Finkelstein RR, Gampala SSL, Rock CD (2002) Abscisic acid signaling in seeds and seedlings. Plant Cell;14:15–45.doi: 10.1105/tpc.12.4.599

Friml J, Palme K (2002b) Polar auxin transport—old questions and new concepts? Plant Mol Biol;49:273–284.doi: 10.1007/978-94-010-0377-3_2

Friml J, Wisniewska J, Benkova E, Mendgen K, Palme K (2002a) Lateral relocation of auxin efflux regulator PIN3 mediates tropism in Arabidopsis. Nature;415:806–809.doi: 10.1038/415806a

Gälweiler L, Guan C, Müller A, Wisman E, Mendgen K, Yephremov A, Palme K (1998) Regulation of polar auxin transport by AtPIN1 in Arabidopsis vascular tissue. Science;282:2226–2230.doi: 10.1126/science.282.5397.2226

Gao ZY, Chen YF, Randlett MD, Zhao XC, Findell JL, Kieber JJ, Schaller GE (2003) Localization of the Raf-like kinase CTR1 to the endoplasmic reticulum of Arabidopsis through participation in ethylene receptor signaling complexes. J Biol Chem; 278:34,725–34,732.doi: 10.1074/jbc.m305548200

Geldner N, Friml J, Stierhof YD, Jürgens G, Palme K (2001) Auxin transport inhibitors block PIN1 cycling and and vesicle trafficking. Nature;413:425–428.doi: 10.1038/35096571

Gray WM, Kepinski S, Rouse D, Leyser O, Estelle M (2001) Auxin regulates SCFTIR1-dependent degradation of AUX/IAA proteins. Nature 414:271–276.doi: 10.1038/35104500

Guilfoyle T, Hagen G, Ulmasov T, Murfett J (1998) How does auxin turn on genes? Plant Physiol.;118:341–347.doi: 10.1104/pp.118.2.341

Hagen G, Guilfoyle T (2002) Auxin-responsive gene expression: genes, promoters and regulatory factors. Plant Mol Biol;49: 373–385.doi: 10.1007/978-94-010-0377-3_9

Hall AE, Findell JL, Schaller GE, Sisler EC, Bleecker AB (2000) Ethylene perception by the ERS1 protein in Arabidopsis. Plant Physiol;123:1449–1457.doi; 10.1104/pp.1234.1449

He Z, Wang ZY, Li J, Zhu Q, Lamb C, Ronald P, Chory J (2000) Perception of brassinosteroids by the extracellular domain of the receptor kinase BRI1. Science;288:2360–2363.doi: 10.1126/science.288.5475.2360

Himmelbach A, Yang Y, Grill E (2003) Relay and control of abscisic acid signaling. Curr Opin Plant Biol;6:470–479.doi; 10.1016/s1369-5266(03)00090-6

Hwang I, Sheen J (2001) Two-component circuitry in Arabidopsis cytokinin signal transduction. Nature;413:383–389.doi: 10.1038/35096500

Inoue T, Higuchi M, Hashimoto Y, Seki M, Kobayashi M, Kato T, Tabata S, Shinozaki K, Kakimoto T (2001) Identification of CRE1 as a cytokinin receptor from Arabidopsis. Nature;409:1060– 1063.doi: 10.1038/35059117

Kakimoto T (1996) CKI1, a cystidine kinase homolog implicated in cytokinin signal transduction. Science;274:982–985.doi: 10.1126/science.274.5289.982

Li JM, Chory J (1997) A putative leucine-rich repeat receptor kinase involved in brassinosteroid signal transduction. Cell;90: 929–938.doi: 10.1016/s0092-8674(00)80357-8

Ljung K, Bhalerao RP, Sandberg G (2001) Sites and homeostatic control of auxin biosynthesis in Arabidopsis during vegetative growth. Plant J.;28:465–474.doi: 10.1046/j.1365-313x.2001.01173.x

Marcinkowska E, Wiedlocha A (2002) Steroid signal transduction activated at the cell membrane: from plants to animals. Acta Biochem Pol;49:735–745.

McAinsh MR, Brownlee C, Hetherington AM (1997) Calcium ions as second messengers in guard cell signal transduction. Physiol Plant;100:16–29.doi: 10.1111/j.1399-3054.1997.tb03451.x

Moore B, Zhou L, Rolland F, Hall Q, Cheng WH, Liu YX, Hwang I, Jones T, Sheen J (2003) Role of the Arabidopsis glucose sensor HXK1 in nutrient, light, and hormonal signaling. Science;300: 332–336.doi: 10.1126/science.1080585

Müller A, Guan C, Gälweiler L, Tänzler P, Huijser P, Marchant A, Parry G, Bennett M, Wisman E, Palme K (1998) AtPIN2 defines a locus of Arabidopsis for root gravitropism control. EMBO J;17: 6903–6911.doi: 10.1093/emboj/17.23.6903

Ouaked F, Rozhon W, Lecourieux D, Hirt H. A (2003) MAPK pathway mediates ethylene signaling in plants. EMBOJ; 22:1282–1288.doi: 10.1093/emboj/cdg131

Ouellet F, Overvoorde PJ, Theologis A (2001) IAA17/AXR3: biochemical insight into an auxin mutant phenotype. Plant Cell;13: 829–842.doi; 10.2307/3871343

Pao S, Paulsen I, Saier MH (1998) Major Facilitator Supperfamily. Microbiol Mol Biol Rev; 62:1–34.

Peng JR, Harberd NP (1993) Derivative alleles of the Arabidopsis Gibberellin-insensitive (Gai) mutation confer a wild-type phenotype. Plant Cell;5:351–360.doi: 10.1105/tpc.5.3.351

Schaller GE, Bleecker AB (1995) Ethylene-binding sites generated in yeast expressing the Arabidopsis ETR1 gene. *Science*; 270:1809–1811.doi: 10.1126/science.270.5243.1809

Steinmann T, Geldner N, Grebe M, Mangold S, Jackson CL, Paris S, Gälweiler L, Palme K, Jürgens G (1999) Coordinated polar localization of auxin efflux carrier PIN1 by GNOM ARF GEF. Science;286:316–318.doi: 10.1126/science.286.5438.316

Stepanova AN, Ecker JR (2000) Ethylene signaling: from mutants to molecules. Curr Opin Plant Biol;3:353–360.doi: 10.1016/s1369-5266(00)00096-0

Sun TP (2000) Gibberellin signal transduction. Curr Opin Plant Biol 3:374–380.doi: 10.1016/s1369-5266(00)00099-6

Suzuki T, Miwa K, Ishikawa K, Yamada H, Aiba H, Mizuno T (2001) The Arabidopsis sensor His-kinase, AHK4, can respond to cytokinins. Plant Cell Physiol;42 :107–113.doi: 10.1093/pcp/pce037

Tiwari SB, Hagen G, Guilfoyle T (2003) The roles of auxin response factor domains in auxin-responsive transcription. Plant Cell;15: 533–543.doi: 10.1105/tpc.008417

Ueguchi C, Sato S, Kato T, Tabata S (2001) The AHK4 gene involved in the cytokinin-signaling pathway as a direct receptor molecule in *Arabidopsis thaliana*. Plant Cell Physiol;42:751– 755.doi: 10.1093/pcp/pce094

Ulmasov T, Murfett J, Hagen G, Guilfoyle TJ (1997) AUX/IAA proteins repress expression of reporter genes containing natural and highly active synthetic auxun response elements. Plant Cell;9:1963–1971. doi: 10.2307/3870557

Wang ZY, Seto H, Fujioka S, Yoshida S, Chory J (2001) BRI1 is a critical component of a plasma-membrane receptor for plant steroids. Nature; 410:380–383.doi: 10.1038/35075629

Wang WY, Hall AE, O'Malley R, Bleecker AB (2003) Canonical histidine kinase activity of the transmitter domain of the ETR1 ethylene receptor from Arabidopsis is not required for signal transmission.ProcNatl Acad Sci USA;100:352–357.doi: 10.1073/pnas.0237085100

William Teale, Ivan Paponov, Olaf Tietz and Klaus Palme (2003) Phytohormones and Signal Transduction Pathways in Plants. Endocrinology: Basic and Clinical Principles,; 2nd Edition: 137-147.doi: 10.1007/978-1-59259-829-8-10

Worley CK, Zenser N, Ramos J, Rouse D, Leyser O, Theologis A, Callis J (2000) Degradation of Aux/IAA proteins is essential for normal auxin signalling. Plant J;21:553–562.doi: 10.1046/j.1365-313x.2000.00703.x

Wu Y, Sanchez JP, Lopez-Molina L, Himmelbach A, Grill E, Chua NH (2003) The abi1-1 mutation blocks ABA signaling downstream of cADPR action. Plant J;34:307–315.doi: 10.1046/j.1365-313x.2003.01721.x.

■■■■

14

METAL TOXICITY IN PLANTS: INDUCTION OF OXIDATIVE STRESS, ANTIOXIDATIVE DEFENSE SYSTEM, METABOLIC ALTERATIONS AND PHYTOREMEDIATION

Arti Gautam and Rama Shanker Dubey

Department of Biochemistry, Institute of Science, Banaras Hindu University, Varanasi-221005

INTRODUCTION

Exposure of plants to various kinds of harsh environmental conditions adversely affects their growth and productivity. Among various abiotic stresses to which plants are often exposed, presence of excess amounts of different essential metals and metal pollutants in agricultural soils has been the matter of greater concern during the last few decades. Metals when present in excess in the soil create considerable stress in growing plants and adversely affect crop productivity. Prolonged exposure to metals and trace elements is getting importance due to long biological half-lives and non-biodegradable nature of the metals. During the last 100 years, unrestricted mining, manufacturing and municipal waste disposal practices have resulted in addition of large amount of metals in the soil environment. Moreover due to increasing trends in industrialization and urbanization, the problem has become a global issue leading to agricultural losses and threat to human and animal health. Recent data indicate that the problem of metal exposure for growing plants is continuously increasing in many parts of the world.

Essential metals like Fe, Cu, Zn, Mn, and Ni are required by the plants in trace amounts for their normal growth and metabolism and to complete their life cycle but these metals become potentially toxic if taken up by plants in excess amounts (Kabata and Pendias, 2010). Similarly, metal pollutants such as Cd, Pb, Hg, Cr, Al, As, which are not part of the living system, when taken up by plants, cause toxicity. High level of metals inside the tissues induces the formation of reactive oxygen species (ROS) which further induce oxidative stress and cause cellular redox imbalance (Sharma *et al.,* 2012). ROS includes superoxide anion ($O_2^{\cdot -}$), hydrogen peroxide (H_2O_2) and hydroxyl radical ($OH^{\cdot}$) and are generated by one, two or three electron(s) transfer to oxygen respectively (Bergamini *et al.,* 2004). Chloroplastic, mitochondrial and plasma membrane linked electron transports invariably generate ROS as byproducts (Shah *et al.,* 2001). The oxidative damage to cellular components is mitigated by efficient quenching of ROS through a well coordinated antioxidant system (Foyer *et al.,* 2011). The uptake and accumulation of Cadmium, Lead, Chromium, Mercury, Arsenic, Aluminium, Iron, Copper, Cobalt, Manganese, Nickel, Zinc, and Boron in plants has remained focus of research by many workers. Soil properties that influence metal availability include pH, organic matter content, cation exchange capacity and biological parameters such as presence of microorganisms (Peijnenburg *et al.,* 2003). High concentration of metals within the tissues leads to many physiological constraints in plants (Shah *et al.,* 2010). Intracellular mechanisms for regulation of the concentration of metal ions in cellular compartments include a complex network of transport, chelation and sequestration process (Clemens, 2001).

Plants show different types of physiological and biochemical responses when grown at elevated levels of metals in the soil (Foyer *et al.,* 2006) which include several metabolic alterations (Verma *et al.,* 2003), accumulation or depletion of certain metabolites, synthesis of new sets of proteins and direct toxic effects evident on morphological features of plants (Sharma *et al.,* 2012). Efforts have been made during the past decades to reduce the metal pollution of soil by growing plants that accumulate metals in high concentrations. Phytoremediation is such method that involves green solution to the problem of metal pollution in the soil. Growing metal hyperaccumulator plants is a novel, cost effective, efficient and eco-friendly way of removing contaminants from the soil and water (Ismail, 2012). Hyperaccumulator plants generally accumulate the contaminants without affecting topsoil, thus conserving its utility and fertility (Ali *et al.,* 2013).

This article presents an overview of various metal pollutants and their effects on key metabolic processes in plants, focusing mainly on uptake of metals, induction of oxidative stress, role of different components of enzymic and non enzymic antioxidants under metal toxicity conditions. Phytoremediation process as a measure to decontaminate metal polluted soils has also been discussed.

SOURCES OF METALS

Metals occur naturally in the soil environment at a level around (<1000 mg kg^{-1}) which is rarely toxic (Giller *et al.,* 1989). Soil of rural and urban environment accumulates metals above a defined value and that is high enough to cause risks to human health, plants, animals, and ecosystem (Wuana *et al.,* 2011). The metals essentially become contaminants in the soil environments because (i) their rates of generation via man-made cycles are more rapid relative to natural ones; (ii) they get transferred from mines to random environmental locations where higher potentials of direct exposure occur (Kelly *et al.,* 1996).There are various sources from where metals are emanated and get access to the soil environment. These sources are mainly natural and anthropogenic.

Natural Sources

Natural sources include weathering of geologic materials and rocks during the process of soil formation (Righi *et al.,* 1995). Volcanoes are also reported to contribute for high concentration of As, Cr, Mn, Cu, Co, Ni, Zn, Cd, Pb, Sn, and Hg in the environment (Sharma *et al.,* 2005). Many igneous rocks such as olivine, augite and hornblende contribute considerable amount of Mn, Co, Ni, Cu and Zn to the soil (Mutschler *et al.,* 1993), while sedimentary rocks are responsible for contribution of Cr, Mn, Ni, Cu, Zn, Cd, Sn and Pb (Bischoff *et al.,* 1975) in the environment. Desert regions such as Sahara contribute to a high level of dust which carries Fe, Mn, Zn, Ni and Pb (Sharma *et al.,* 2005). Oceanic water and deep sea deposits are enriched with different heavy metals such as Te, Co, W, Pt, Yt, Ce, Sc, La, Fe, Mn, Cu, Ni, etc. (Marz *et al.,* 2011). Oceanic activities such as cyclone and Tsunami are responsible for deposition of metals in inland coastal areas and cause damage to the natural vegetation (Subramanian, 2011).

Anthropogenic Sources

Since the industrial revolution, human race is unable to keep pace with the increasing amount of waste materials and growing population which result in disruption of the natural biological balance. Anthropogenic inputs are mainly in the form of industrial smoke, waste disposal, waste incineration, urban effluents, traffic emissions, fertilizers, pesticides application, mining activities and production of metal containing products which include paints, dyes, preservatives, etc. (Yi *et al.,* 2011).

Industrial sources

Industrialization and urbanization have caused increased anthropogenic contribution of metals in the biosphere. Textile, tanning, petrochemicals, pharmaceuticals, smelting industries, chemical industries, refineries produce emissions and waste products which are disposed off on land and in air and serve as source of various hazardous metals and toxic compounds (Barakat *et al.,* 2011). Ore mining and smelting activities have resulted in contamination of soil that poses risk to human and ecological health. Coal mines are source of As, Cd, Fe, Zn, Pb, etc. (Bhuiyan *et al.,* 2010) whereas amount of Hg increases in soil nearby gold mines due to use of Hg in mining process (Spiegel *et al.,* 2010).

The solid particles in smoke from fires and in other emissions from factory chimneys are eventually deposited on land and cause considerable build-up of these metals in the soil (Alves *et al.,* 2011). Aerial emission of Pb from the combustion of petrol containing tetraethyl lead contributes to increased content of Pb in roadside soil of urban areas (Wong *et al.,* 2006). Domestic effluents such as soaps and detergents contain trace amounts of elements such as Fe, Mn, Cr, Zn, Sr and B. Municipal and industrial waste water gets directly mixed up with water bodies and eventually enhances metal content in the environment (Stoffers *et al.,* 1986).

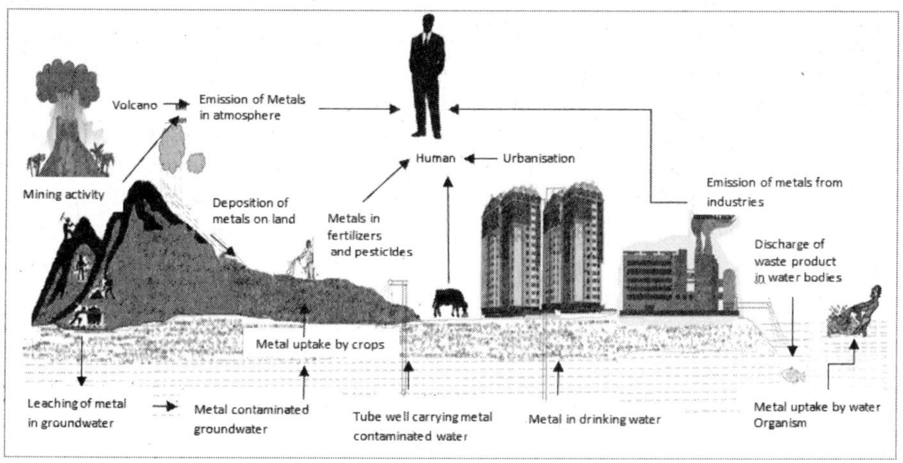

Fig 1. Different sources of metals in the environment

Agricultural sources

The pesticides, fungicides and insecticides are extensively used in agriculture, which contain Cu, Cr, As, etc. and directly get mixed into the soil. Such contamination leads to deposition of the metals/metalloids in the soil. Copper-containing fungicidal sprays such as Bordeaux mixture (copper sulphate) and copper oxychloride (Zwieten *et al.,* 2004) are widely used. Lead arsenate was used in fruit orchards for many years to control some parasitic insects (Peryea *et al.,* 1997). Arsenic-containing compounds have been also used extensively to control cattle ticks and to control pests in banana in New Zealand and Australia; timbers have been preserved with formulations of Cu, Cr, and As (CCA) (Yeates *et al.,* 1994). "Superphosphate" a fertilizer, being used in rice farming area in the north of Albufera Natural Park (Valencia Spain) contains high amounts of Cd, Co, Cu and Zn as impurities (Muramoto *et al.,* 1990).

WORLDWIDE OCCURRENCE OF SOIL METAL TOXICITY

Exposure of agricultural lands to various metals is a common problem throughout the world. In the recent years with the growth of industrialized economy, addition of metals to soil environment has increased due to human and industrial activities resulting in the deterioration of soil environment. In India the concentration of heavy metals (Cu, Zn, Cd, Pb), metalloids and various elements in agricultural soils have been elevated due to continuous use of pesticides, fertilizers, sewage sludge, municipal and industrial effluents (Sharma *et al.,* 2006). The accumulation of metals in the edible portions of vegetables has serious effects on human health and plants (Sharma *et al.,* 2006). Many industrial and agricultural processes have contributed to the contamination of fresh water systems thereby causing adverse effects on aquatic biota and human health. According to Bulletin on National Survey of Soil Contamination, jointly issued by Ministry of Environmental Protection of China and the Ministry of Land Resources of China, nearly 4 million of hectares of arable lands have been contaminated moderately or severely with metals, which accounts for about 2.9% of China's arable land (Cheng *et al.,* 2003). The monitoring of snow cover pollution by metals and elements such as zinc, copper, lead, cadmium, arsenic, nickel, chromium, strontium, manganese, fluorine, lithium performed in 20 districts of the Moscow region during 2009, 2012 and 2013 revealed substantial higher level of these elements compared to the level in the end of last century(Ermakov *et al.,* 2015). In Malaysia phosphatic fertilizers have been used for long periods and these fertilizers due to their impurities have contaminated soils with various metals (Zarcinas *et al.,* 2004). Addition of these fertilizers has resulted in concomitant increase in As, Cd, Cu and Zn in the soil. In a study by Zarcinas *et al.* (2004) Arsenic, cadmium, cobalt, chromium, copper, mercury, nickel, lead and zinc levels determined in soils of Thailand appeared to be substantially higher compared to the level three decades earlier which was primarily due to industrialization and urbanization during the past decades.

In North-Eastern United States such as in New York, Ariola, New Jersey elevated levels of Pb, Ni and Cu were reported in the soil (Pouyat *et al.,* 1991). Elevated level of Ni and Cu were also found in Sadbury, Ontario. Levels of As, Cd, Cu, Hg, Pb, Se and Zn were found above significant value in the tissues of white beaked dolphins (*Lagenorhynchus albirostris*) from Newfoundland Canada (Muir *et al.,* 1988). Pilcomayo river (South America) is strongly affected by release of mine tailings from the Potosi mines where the concentration of Pb, Cd, Cu, Zn were up to thousand times higher than the significant value (Smolders *et al.,* 2003). In a study conducted to examine heavy metal contamination in roadside soils of northern England, the elevated level of Cd, Cu, Pb and Zn were reported (Charlesworth *et al.,* 2003). Soil samplesfrom Jaworzno (South Poland) when tested for various elements, the levels of Ca, Cd, Na, Ni, Pb, Zn and Ni were found to be higher in the vicinity of Solvay soda ash plant area (Pawlowska *et al.,* 1997).The presence of Cd, Pb, Hg, methylmercury (MeHg), As, inorganic arsenic (iAs), Co, Cu, Zn, Ni, Cr and Fe was investigated in seafood collected from European marine ecosystems and higher level of these elements were detected in various organism (Lima *et al.,* 2008).

METALS INDUCED TOXICITY AND OXIDATIVE STRESS IN PLANTS

Like various abiotic environmental stresses such as salinity, drought, heat, chilling, UV-B radiation, anaerobiosis, etc., metals when present in high concentrations in plants, lead to overproduction of reactive oxygen species (ROS) such as superoxide radical ($O_2^{\cdot-}$), H_2O_2, hydroxyl radical (·OH), etc. Overproduction of ROS in the tissues induces oxidative stress marked by oxidative damage to lipids, proteins, nucleic acids. Toxicity of a metal depends upon parameters such as pH, cation exchange capacity, chemical forms of the metal and organic matter content and the biological parameters such as presence of microorganisms that determine the metal availability to the plant (Sauve *et al.,* 2000). Metals when present in high concentration in the cells interfere with activities of a number of enzymes essential for normal functioning of plants which leads to various physiological constraints (Moustakas *et al.,* 1994). Reaction of ROS with biomolecules such as lipids (peroxidation of unsaturated fatty acids in membrane), proteins (oxidation, denaturation), carbohydrates and nucleic acid causes irreversible damage to plant cells. Peroxidation of lipids decreases the fluidity of the membranes, increases leakiness and causes secondary damage to membrane proteins (Moller *et al.,* 2007) and forms 4-hydroxy-2- nonenal (HNE) and malondialdehyde (MDA), as well as formation of hydroxyl and keto fatty acids (Uchida *et al.,* 2003). These aldehydes have the potential to conjugate with DNA and proteins and thus damage and alter their normal functioning (Uchida *et al.,* 2003). Besides this, ROS cause covalent modification of protein molecules which results in protein oxidation (Cabiscol *et al.,* 2010). Protein carbonylation is another consequence of oxidative damage to proteins and it is widely used as marker of protein oxidation (Job *et al.,* 2005; Moller *et al.,* 2007). Protein carbonylation may occur due to direct oxidation of amino acid side chains (e.g. proline and arginine to γ-glutamyl semialdehyde, lysine to amino adipic semialdehyde, and threonine to aminoketobutyrate) (Shringarpure and Davies, 2002). Cysteine and Methionine are much prone towards ROS attack compared to other amino acids because ROS are likely to target proteins that contain sulfur-containing amino acids and thiol groups.

It has been reported that OH·, impairs all the components of the DNA molecule by damaging both the purine and pyrimidine bases as well as deoxyribose backbone (Halliwell and Gutteridge, 1999) whereas $^1O_2$ primarily attacks guanine (Wiseman and Halliwell, 1996). This impairment is responsible for base deletion, formation of pyrimidine dimers, cross-links, strand breaks and base modification, such as alkylation and oxidation (Tuteja *et al.,* 2001; Tuteja and Tuteja, 2001) which further affects growth and development of the entire plant or any organism (Britt, 1999).

Generation of Reactive Oxygen Species

All aerobic organisms have to deal with the highly reactive nature of oxygen derivatives, such as superoxide radicals, hydrogen peroxide and hydroxyl radicals (Halliwell *et al.,* 2005). Aerobic organisms have developed a well coordinated defense system called as antioxidative system to cope up with the reactive oxygen species which are continuously produced as byproduct of various metabolic pathways localized in different cellular compartments (Apel *et al.,* 2004). Due to exposure of plants to high levels of metals, balance between ROS production and defense system gets disturbed leading to cellular deterioration (Gill *et al.,* 2010).Various physiological and metabolic processes including photosynthesis and respiration are disturbed due to uptake of excessive amount of metals by plants (Sharma *et al.,* 2005).While converting light energy into chemical energy and chemical energy into biomass highly reactive oxygen intermediates are produced from relatively nonreactive oxygen molecules (Zhu *et al.,* 2008). Excitation of molecular oxygen produces singlet oxygen while univalent reduction of oxygen produces superoxide radicals (Apel *et al.,* 2004).

Reactive oxygen species formation in chloroplasts

Chloroplast is the primary source of ROS production in plant species as its main function is to produce reducing equivalents such as NADPH and to generate oxygen (Asada *et al.,* 2006). Photosystem I (PSI) and photosystem II (PS II) are considered as the two main sites where ROS production takes place (Prasad *et al.,* 2006). The superoxide anion is the primary reduced product of photoreduction of oxygen and its subsequent disproportionation produces H_2O_2 and O_2 in photosystem I (Asada *et al.,* 1974). However the oxygen is excited to the singlet stage in photosystem II (Triantaphylides *et al.,* 2009).

Some processes which are closely associated with photosynthesis such as (a) The oxygenase reaction of ribulose-1,5 bisphosphate carboxylase-oxygenase (Rubisco) (Law *et al.,* 2009) and (b) direct reduction of molecular oxygen by PSI electron transport are considered as oxygen consuming processes (Noctor *et al.,* 1998). In the first process the reactions linked to the photorespiratory cycle including Rubisco in the chloroplast and glycolate-oxidase and CAT-peroxidase reactions in the peroxisome are considered important for the formation of ROS (Kuchitsu *et al.,* 2013). Second process involves electron transport components associated with PS I where oxygen molecules are directly photoreduced to superoxide radical (Hemschemeier *et al.,* 2011). Excess light intensity causes over reduction of electron transport chain which leads to inactivation of PS II and suppression of photosynthesis (Choudhary *et al.,* 2001). Plants protect photosynthetic apparatus in either of two ways, first by dissipation of extra energy in the PS II antennae (non photochemical quenching) (Li *et al.,* 2003) and second by allowing electron transfer from PS II to various acceptors within chloroplast (photochemical quenching (Mohanty *et al.,* 1989). PS II continuously produces singlet oxygen (Marder *et al.,* 1987). Due to excess light energy in PS II, the QA, QB and plastoquinone are overreduced and charge separation cannot be completed and the oxidized P680 chlorophyll recombines with the reduced pheophytin (Zhu *et al.,* 2005). This leads to the formation of triplet state of p680 and generation of singlet oxygen by energy transfer. With high light intensity photo-inhibition of PSII occurs and singlet oxygen production increases (Sudhir *et al.,* 2004). Carotenoids play an important role in photoprotection as they contain a chain of isoprenic residues bearing numerous conjugated double bonds, which allow easy energy uptake from excited molecules and dissipation of excess energy as heat (Mittler, 2002).

ROS production in endoplasmic reticulum, peroxisomes and glyoxysomes

Smooth endoplasmic reticulum and microsome are the locations for various oxidative processes (Malhotra *et al.,* 2007). Mixed function oxygenases such as cytochrome p450, perform important hydroxylation reactions in the mevalonic acid pathway, adding oxygen atoms to substrate molecules. NADPH is the electron donor and superoxide may be released by such reactions (Vuori *et al.,* 1992). Peroxisomes and glyoxy-somes are single membrane organelles that compartmentalize enzymes involved in the beta oxidation of fatty acids and the C2 photorespiratory cycle, where glycolate oxidase transfers electrons from glycolate to oxygen and produces H_2O_2. Xanthine oxidase, urate oxidase and NADH oxidase generate superoxide (Gut *et al.,* 1988).

Plasma membrane and apoplastic compartment

NADPH is ubiquitously present associated with plasma membrane and may produce superoxide and H_2O_2 (Moller *et al.,* 2001). Apoplast is the first site of pathogen invasion and it is the first site of defense (Bolwell *et al.,* 2001). The most common biosynthetic pathway where phenylpropa-noid precursors of lignin are cross linked by H_2O_2 involves a reaction initiated by peroxidases (Bolwell *et al.,* 2001). Cell wall malate dehydrogenase generates NADH which is used to form H_2O_2 possibly by an NADH oxidase (Peng *et al.,* 1992).

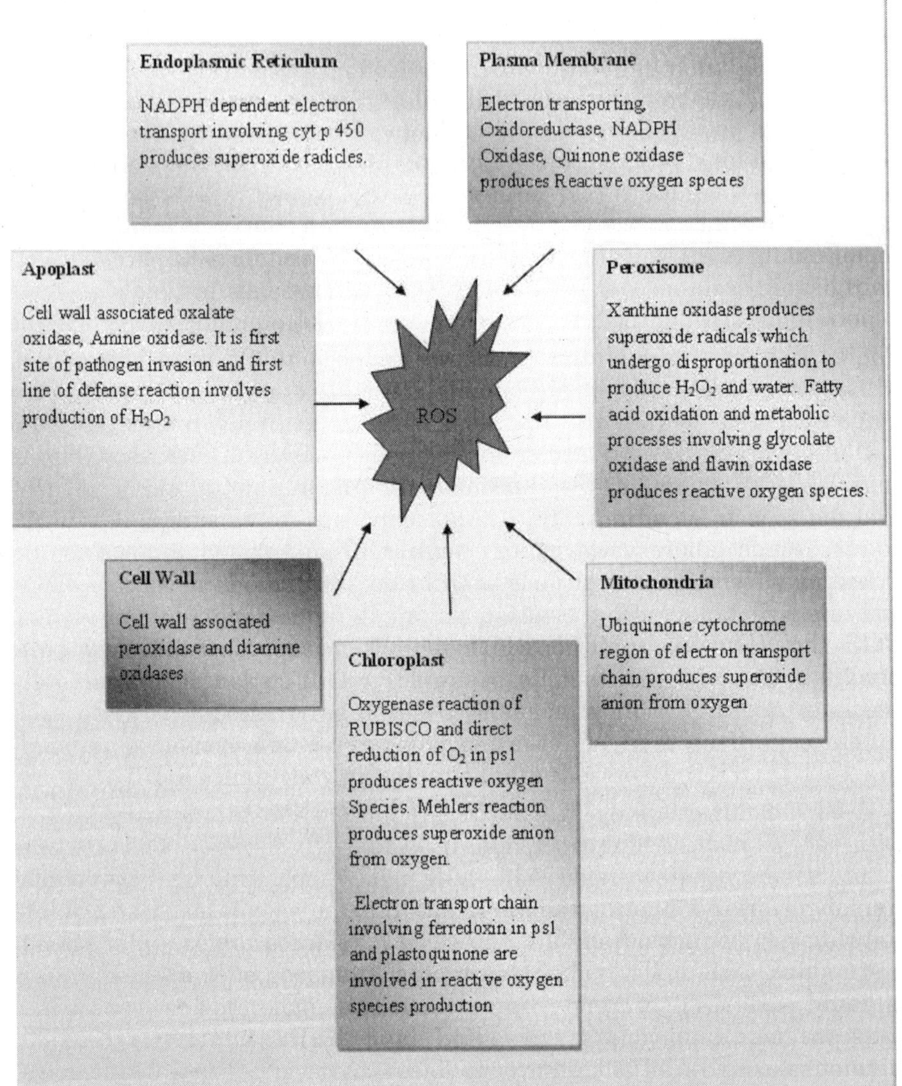

Fig 2 Production of reduced and excited species of ROS in chloroplasts, mitochondria, peroxisomes, cell wall, apoplast, endoplasmic reticulum and plasma membrane

Antioxidative Defense System in Metal Stressed Plants

When exposed to abiotic stresses, plants try to reduce the damage by efficient quenching of ROS through a well coordinated antioxidative defense system and maintain equilibrium between production and scavenging of ROS. This antioxidative system comprises of non-enzymic and enzymic components (Ali *et al.,* 2008). Non-enzymic antioxidants include (1) lipid soluble membrane associated antioxidants (e.g., α-tocopherol and β-carotene), and (2) water soluble reductants (e.g., glutathione, ascorbate and phenolics). Enzymic antioxidants comprise of the enzymes superoxide dismutase (SOD), catalase (CAT) and peroxidases (POX) and the enzymes of ascorbate-glutathione cycle (Halliwell-Asada pathway) such as ascorbate peroxidase (APX), monodehydroascorbate reductase (MDHAR), dehydroascorbate reductase (DHAR) and glutathione reductase (GR) (Karpinski and Muhlenbock, 2007; Sang *et al.,* 2007). These non-enzymic and enzymic antioxidants function in a coordinated manner to take care of the overproduced ROS.

Nonenzymic antioxidants

Non enzymic antioxidants in plant cells include ascorbic acid (AsA), glutathione (GSH), tocopherol, carotenoids and phenolic compounds among which AsA and GSH serve as cellular redox buffer (Sharma *et al.,* 2012). Ascorbic acid exists in reduced state (AsA) as well as in oxidized forms (such as monodehydroascorbate MDHA and dehydroascorbate DHA) and it plays an important role in plant cells as a powerful ROS scavenger due to its ability to donate electrons in a number of enzymatic and non enzymatic reactions (Yadav *et al.,* 2014). It provides protection to membranes by directly scavenging $O_2^{\cdot-}$and OH· and by regenerating α-tocopherol from tocopheroxy radical and removes H_2O_2 by AsA-GSH cycle. AsA plays a key role in cell growth and also influences activities of many enzymes (Karuppanapandian *et al.,* 2011).

Glutathione (GSH), a low molecular weight tripeptide thiol compound, with the sequence γ-L-glutamyl-L-cysteinyl-L-glycine is an another important antioxidant in plant cells (Kerdisk *et al.,* 2005).Glutathione acts as a disulphide reductant to protect thiol groups of enzymes, regenerates ascorbate and reacts with singlet oxygen and hydroxyl radicals (Foyer *et al.,* 2001). GSH acts as a protein disulphide reductant, it is involved in regulation of gene expression in response to environmental stresses and it serves as precursor for the biosynthesis of phytochelatins (small molecular weight compounds, which chelate heavy metals and sequester them in vacuoles) (Foyer *et al.,* 2001; Navrot and Rouhier, 2007). GSH also participates in the regeneration of AsA from DHA via the enzyme dehydroascorbate reductase (Li *et al.,* 2010).

Proline, the amino acid is considered as an antioxidant and potential inhibitor of programmed cell death (Chen *et al.,* 2005). It accumulates in stressed plants and serves as an osmoprotectant, a protein stabilizer, a metal chelator, an inhibitor of LPO and OH· and O_2 scavenger (Hayat *et al.,* 2012). Tocopherols are other important lipid soluble antioxidants which scavenge ROS and lipid radicals and function as molecular chaperons by promoting protein stability and acting as membrane stabilizer (Hasanzaman *et al.,* 2013). Tocopherols protect pigments and proteins of photosynthetic apparatus (Hasanzaman *et al.,* 2013).

Enzymic antioxidants

The antioxidative enzyme defense system in plant cells comprises of the enzymes superoxide dismutase (SOD), catalase (CAT), peroxidases and the enzymes of Ascorbate-Glutathione Cycle (Halliwell-Asada Pathway). Superoxide dismutase scavenges superoxide ion by dismutating $O_2^{\cdot-}$ into H_2O_2 and O_2. $H_2O_2$is degraded by catalase (Imlay *et al.,* 2011). Isoforms of SOD include copper/zinc (Cu/Zn-SOD), manganese (Mn-SOD) and iron (Fe-SOD), which are localized in different cellular compartments (Mittler, 2002). Catalase is a tetrameric, heme containing enzyme with the potential ability to directly dismutate H_2O_2 into H_2O and O_2 (Garg and Manchanda, 2009). It scavenges H_2O_2 generated by mitochondrial electron transport, ß-oxidation of fatty acids and photorespiratory oxidation (Karuppanaapandian *et al.,* 2011). Ascorbate peroxidase (APX) and Guaiocol peroxidase (GPX) are two another antioxidants in which APX detoxifies H_2O_2 using ascorbate for reduction while GPXs have been proposed to participate in the lignification of cell wall, degradation of IAA, biosynthesis of ethylene, wound healing, and defense against pathogens (Sharma *et al.,* 2012).

Enzymes of Halliwell-Asada Pathway

Halliwell- Asada pathway is also called glutathione ascorbate cycle and it occurs at different subcellular locations like in cytosol, chloroplast, mitochondria and peroxisomes and detoxifies ROS produced as a byproduct in various reactions (Karuppanapandian *et al.,* 2011). To scavenge ROS this cycle involves successive oxidation and reduction of AsA, GSH and NADPH catalyzed by the enzymes APX, monodehydroascorbate reductase (MDHAR), dehydroascorbate reductase (DHAR) and glutathione reductase (GR) (Sharma *et al.,* 2012). Initially H_2O_2 gets reduced to water by APX and ascorbate acts as electron donor. The oxidized ascorbate is regenerated by monodehydroascorbate reductase (MDAR). However, monodehydroascorbate (MDHA) is a radical and if not rapidly reduced it disproportionates into ascorbate and dehydroascorbate (Asada *et al.,* 1997). Dehydroascorbate is reduced to ascorbate by DHAR at the expense of GSH, yielding oxidized glutathione (GSSG) (Dalton *et al.,* 1986). Finally GSSG is reduced by the enzyme GR using NADPH as electron donor (Morales *et al.,* 2015). Thus ascorbate and glutathione are not consumed; the net electron flow is from NADPH to H_2O_2. The reduction of dehydroascorbate may be non-enzymatic or it can be catalyzed by the enzymes DHAR, glutathione S-transferase, omega-1 or glutaredoxins, etc. (Kim *et al.,* 2012).

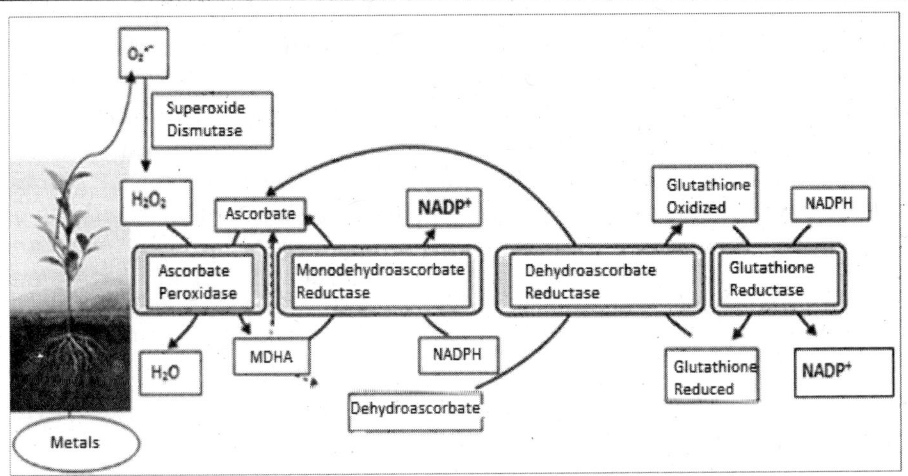

Fig 3. Non enzymic and enzymic antioxidants involved in detoxification of ROS in plants.

PHYTOTOXIC METALS: UPTAKE, TRANSPORT AND METABOLIC ALTERATIONS IN PLANTS

Cadmium

Cadmium(Cd) is one of the most potent heavy metal pollutants and its concentration in the soil is increasing with time, because its addition becomes greater than its removal through leakage and plant harvesting (Naidu *et al.,* 1994). Major anthropogenic sources which have increased Cd emissions in the environment are Cd containing phosphate fertilizers, sewage, sludge and industrial emissions (Adriano, 1986). *Cadmium* is long range transported air pollutant which has been deposited for decades in the close vicinity of industrial areas (Hagenmeyer *et al.,* 1986). *Cadmium* is presently being used by various industries in large number of products which have resulted in release and deposition of this element to aquatic and soil environment leading to serious environmental cadmium contamination (Vangronsveld *et al.,* 1995).

Cd after being absorbed by roots gets rapidly transported to shoots through xylem (Page *et al.,* 2005). The ability of xylem-mediated Cd translocation into shoots is regarded as a major determinant for shoot Cd accumulation in plants (Mori *et al.,* 2009; Uraguchi *et al.,* 2009). There are many transporters which transport divalent transition metals in different species of plant and animals. One of them is AtIRT1, a ZIP family transporter for Fe^{2+}, Zn^{2+} and Mn^{2+} that mediates Cd uptake in roots of *Arabidopsis thaliana* (Vert *et al.,* 2002).

After being taken up by the plants Cd adversely effects germination of seeds, seedling vigour, plant growth and metabolism (Shah and Dubey,

1995) by decreasing water transport to leaves, impairing transpirational rate causing ultrastructural changes in cell organelles and altering the behaviour of key metabolic pathways.

Symptoms caused by elevated levels of Cd include growth retardation and root damage, chlorosis of leaves and red brown coloration of leaf margins or veins (Kabata Pendias and Pendias, 1992). In higher plants, Cd significantly reduces the normal H^+/K^+ exchange and the activity of plasma membrane ATPase (di Toppi and Gabbreilli, 1999). *Cadmium* decreases water transport to the leaves (Chardonnens *et al.,* 1998) and alters the nutrient status in both roots and leaves of the plants (Scandalio *et al.,* 2001).

Cadmium (Cd) inhibits photosynthetic rate and chlorophyll content (Scandalio *et al.,* 2001). It alters the activities of proteolytic, nucleolytic and phosphorolytic enzymes in growing rice plants (Shah and Dubey, 1997) and inhibits the activities of several enzymes such as acid phosphatase, ribonuclease (Shah and Dubey, 1995), glucose 6 phosphate dehydrogenase, Malic enzyme, isocitrate dehydrogenase (Van Assche and Clijsters, 1990), Rubisco and carbonic anhydrase (Siedlecka *et al.,* 1997) in growing plants. *Cadmium* toxicity causes oxidative stress in the tissues by causing increased generation of ROS (Hendry *et al.,* 1992).

Lead

Lead has gained importance as one of the most important metal pollutants of the environment due to industrial emissions and automobile exhausts. Lead contamination of the soil has resulted from mining and smelting activities, burning of coals, Pb containing paints, gasoline and explosives, as well as from the disposal of municipal sewage sludges enriched in Pb (Gracia *et al.,* 2007). Short and long range atmospheric deposition of lead is a significant contributor to soil contamination. The global average Pb concentration of soil ranges from 2 to 100 mg kg-1 (Siegel *et al.,* 1998).

Plant usually accumulates Pb but it is also transported to different plant organs (Donelly *et al.,* 1990), however, the mobility of Pb from root to shoot is lower than Cd. Lead has been shown to enter and move within the cytoplasm and its uptake has been thought to be passive (Tung and Temple, 1996), although certain proteins have been identified that mediate cross membrane movement of Pb (Arazi *et al.,* 1999).The absorption of lead by roots occurs via the apoplastic pathway or via Ca^{2+}permeable channels (Wang *et al.,* 2006). The factors that regulate lead uptake from soil include soil pH, soil particle size, cation exchange capacity, root surface area, root exudation and degree of mycorrhizal transpiration (Miller *et al.,* 1976). After uptake lead gets concentrated primarily in root cells due to blockage by casperian strips within the endodermis (Lee *et al.,* 2013). Pb decreases the germination percentage, root and shoot length and damages root tips (Sharma *et al.,* 2005). Pb affects water relations in plants (Leakey *et al.,* 2009).

An increase in leakage of electrolytes was observed in two rice cvs. Ratna and IR-36 when subjected to Pb treatment (Bajji *et al.,* 2002). Pb alters granal structure of chloroplasts and as a result the photosynthetic capacity is decreased. Pb causes several physiological changes like decrease in transpirational intensity and increase in stomatal resistance (Azmat *et al.,* 2009). Decline in the activities of nitrate reductase, mitochondrial and extramitochondrial isocitrate dehydrogenase and malate dehydrogenase, α-ketoglutarate dehydrogenase and succinate dehydrogenase has been observed in Pb treated plants (Van der merwe, 2009). Lead progressively stimulates the activities of catalase, ascorbic acid oxidase, peroxidase as well as lipoxygenase (Rashid and Mukherjee, 1991). Accumulation of proline is observed in Pb stressed plants (Sharma *et al.,* 2005). Further, increased lipid peroxidation as well as enhanced activities of antioxidative enzymes has been observed in rice seedlings grown in presence of Pb (Verma and Dubey, 2003).

Chromium

Wide industrial application of Chromium has made it one of the most common metal contaminants in ground water, soil and sediments (Gao *et al.,* 2011). Chromium exists in several oxidation states among which, Cr (III) and Cr (VI) are the most stable forms (Rock *et al.,* 2001). Cr (VI) has the highest potential of persistency in sediments (Alfonsi *et al.,* 2008) and therefore it is able to impose more toxicity to biota in comparison to other forms (Singh *et al.,* 2013).

Cr is taken up by plants via sulphate transporters along with essential nutrient sulphate (Vidmar *et al.,* 2000). After uptake Cr is largely retained in the roots as it is poorly translocated to other organs of the plant (Zayed *et al.,* 1998). However, it is taken up by other non-specific carriers along with the uptake of essential elements and water (Shanker *et al.,* 2005). The factors which determine Cr uptake are pH, oxidation states, concentration, salinity and the presence of dissolved salts (Babula *et al.,* 2008). Maximum amount of Cr is accumulated by sulphur loving plants of family Brassicacea, which suggests involvement of sulphur uptake pathway, i.e., sulphate carriers in Cr translocation (Ramiraz Diaz *et al.,* 2008).

Cr induces toxicity in plants at morphological, physiological, biochemical, and molecular levels. Chromium interferes with various metabolic processes that are vital for plant growth and development. Chromium also alters the expression of various stress and defense-related genes. Toxicity symptoms of Cr include reduction in growth and decline in biomass accumulation, structural alterations, alteration in activities of enzymes related to starch and nitrogen metabolism, alteration in water and mineral uptake mechanism (Singh *et al.,* 2013). Various enzymatic activities related to photosynthesis and respiration are decreased under Cr toxicity either due to direct interference of Cr with the enzymes or through the production of ROS (Shanker *et al.,* 2005). Cr toxicity induces oxidative damage by causing destruction of membrane lipids and DNA damage (Stohs *et al.,* 2001).Various metabolic alterations due to Cr (III, VI) include direct effect on enzymes and/or metabolites or increased ROS generation resulting in lipid peroxidation and oxidative damage (Montes-Holguin *et al.,* 2006).

Copper

Copper (Cu) is an essential micronutrient for plants (Cooper *et al.,* 2005). It is a component of the protein structure of a range of enzymes involved in electron transport and redox reactions in mitochondria, chloroplast, cell wall and the cytoplasm of the cell (Bartoli *et al.,* 2000). Copper is an essential element in plant nutrition being required as a cofactor for a number of enzymes and electron transport proteins, thus playing a key role in metabolic and biosynthetic processes in higher plants (Yruela *et al.,* 2005). Cu is also an essential component of various proteins like plastocyanin of photosynthetic system and cytochrome oxidase of respiratory electron transport chain (Yadav *et al.,* 2010).

Cu has a tendency to bind with organic matters and form organic complexes which play an important role in the regulation of Cu mobility in the soil (Cabanis *et al.,* 1988). As Cu serves as a micronutrient its concentration in plants is usually low which is commonly in between 2-20 ppm (Panda *et al.,* 2007). It has been shown that phytometallophores play an important role in Cu absorption from the soil (Adhikari *et al.,* 2015). Reichman (2002) showed that doubly radiolabelled 64Cu-(14C) tetraethylenepentamine was directly absorbed by the plant which proves importance of phytometallophores in Cu absorption. Cu if present in greater than the adequate amount in soil causes chlorosis and inhibits growth in many plant species (Gaetke, 2003). Cu binds to the sulphydryl groups of membrane proteins and causes damage to the cell membrane and induces lipid peroxidation which leads to the membrane disruption (Stauber *et al.,* 1987).

Cu toxicity symptoms in plants include lowering of leaf chlorophyll content, reduction of thylakoid membrane network, increase in the leakage of K^+ ions, decrease in the total lipid content of plasma membrane, a higher phospholipid content and reduction of steryl lipids (Zhang *et al.,* 2008). Cu toxicity reduces photosynthesis and electron transport activities modifies pigment and protein composition of photosynthetic membrane (Baszynski *et al.,* 1995). Runner bean plants treated with Cu showed significant changes in acyl lipid content and loss of core antenna PS2 polypeptide. Exposure of plants to excess Cu leads to overproduction of ROS and induction of oxidative stress in the tissues (Maksymiec *et al.,* 1994).

Iron

Iron is an essential micro-nutrient for plants and is required in small amounts for normal growth of plants (Thomine *et al.,* 2011). Iron has important role in many key plant processes such as photosynthesis, chloroplast development and chlorophyll biosynthesis (Myouga *et al.,* 2008). Iron is major component of the cell redox system because it is integral part of heme proteins such as cytochromes, catalase, peroxidase, leghemoglobin, iron sulphur proteins, ferredoxin, aconitase and superoxide dismutase (SOD) (Marschner, 1995).

Despite the fact that mineral soils contain sufficient iron, its toxicity is expressed mainly in waterlogged soils due to microbial reduction of insoluble Fe^{3+} into soluble Fe^{2+} (Becker and Asch, 2005). The factors responsible for iron toxicity include Fe^{2+} uptake by roots and its transportation to leaves via transpiration stream. Two different strategies exist in plants for iron uptake (Kochian, 1993). In dicots and nongraminaceous monocot plants which do not absorb Fe chelates or Fe complexes, a series of physiological changes such as increased root ferric chelate reductase activity, acidification of rhizosphere, acquisition of Fe^{+2} across the plasma membrane of root epidermal cells mediate the uptake of iron from roots (Olsen and Brown, 1980; Welch *et al.,* 1993). In graminaceous monocot plants which excrete chelates such as mugeneic and avenic acids (Kochian, 1993; Fan *et al.,* 1997) which are known as phytosiderophores or phytometallophores rapidly chelate iron and other metals, and ascertain ready supply of metals for transport across the plasma membrane (Grusak *et al.,*1999).

Excess Fe when present in the soil, becomes phytotoxic and induces formation of ROS. Typical iron toxicity symptoms in plants include reduction in leaf size; deepening of green leaf color (particularly in the younger leaves); reddening or purpling of stems and older leaves; wilting of shoots; dieback of older leaves (especially from the tips or margins); larger necrotic patches on leaves; blackening of leaf tips, stem bases and roots; reduction in root branching and formation of precipitates on roots (Snowden and Wheeler, 1993). Many cellular structures and membranes get irreversibly damaged under Fe toxicity. In tobacco, canola, soybean and *Hydrilla verticillata* plants reduced photosynthesis and increase in oxidative stress have been observed under Fe toxicity (Sinha *et al.,* 1997).

Manganese

Manganese is an abundant element in the earth crust and is found in soil in various ionic forms (Yabuchi *et al.,* 2012). It is an essential micronutrient for plants. The availability of Mn in the soil depends on soil pH, temperature and moisture (Godo *et al.,* 1980). Mn has low solubility at alkaline and neutral pH, but it is readily soluble and available under acidic pH (pH 5). Therefore Mn toxicity is a common problem in acidic or poorly drained soils where pH is less than 5 (Bradl *et al.,* 2004). Acidic soils serve as potential constraint to crop production due to high solubility of Mn and Al. On global scale nearly 70% of cultivable land is acidic (Vera *et al.,* 2007). Mn exists in a range of oxidation states (Mn^{1+}, Mn^{2+}, Mn^{3+}, Mn^{4+}, Mn^{6+}, and Mn^{7+}) and the most commonly found forms of Mn in biological systems are Mn^{2+}, Mn^{3+} and Mn^{4+} with Mn^{4+} being the least stable (Marschner, 2012).

Divalent Mn (Mn^{2+}) is the most soluble species as well as most efficiently accumulated form of Mn in the soil. Soil pH primarily determines the oxidation state of Mn in soil (Morgan *et al.,* 2005). At neutral or higher pH, Mn^{3+} and Mn^{4+}species predominate and insoluble Mn oxides are formed (Rengel, 2000; Marschner, 2012). Only Mn (II) can be taken up via an active transport system by root epidermal cells (Sasaki *et al.,* 2012). Many genes are annotated to be Mn transporters as they have amino acid sequences homologous to those of Nramp proteins in animals (Kehres *et al.,* 2000). These proteins are localized in root tissues which reveal their participation in uptake of soil Mn (Cailate *et al.,* 2010). Mn transportation occurs via high affinity Mn^{2+} transporter, SMF1 of the NRAMP family (Supek *et al.,* 1996) and a phosphate transporter, PHO84. A PHO84-type phosphate transporter basically forms neutral metal-phosphate complexe and forms the substrate i.e $MnHPO_4$. Chlamydomonas, a photosynthetic algae which requires high affinity Mn^{2+} efflux for its normal physiological processes also encodes NRAMP protein (Allen *et al.,* 2007). Cyanobacteria require Mn for PSII assembly and a gene product with sequence similarity to an ABC (ATP binding cassette)-type transporter is implicated in Mn uptake in this organism (Bartsevich and Pakrasi, 1995). Mn participates in various key processes such as photosynthesis, nitrogen metabolism (for sequential reduction of nitrate), in maintaining chloroplast structure, and in the biosynthesis of aromatic amino acids, hormones (Auxins), phenols, lignins, etc. (Millaleo *et al.,* 2010). Mn^{2+} is a component of oxygen evolving complex of photosystem II (PSII) which catalyzes the water splitting reaction to produce oxygen and electrons required for photosynthetic electron transport chain (Yi *et al.,* 2005). Mn is readily absorbed by plant roots and is transported and accumulated in shoots.

Excess Mn in plant cells is stored in vacuoles, cell walls or golgi vesicles. Therefore toxicity of Mn is more evident in shoots than in roots. Shoots show characteristic Mn toxicity symptoms. The key symptoms of Mn toxicity in plants are chlorosis, dark browning and necrotic spots on leaves, streaking, mainly curling and rolling of growing leaves (Hue *et al.,* 2001). Mn-excess alters the absorption, translocation and utilization of various mineral elements essential for plant growth such as Ca, Mg, Fe and Zn. Chlorosis is a common symptom of Mn toxicity, which is due to reduced absorption of Fe under excess Mn (Srivastava *et al.,* 2011).

Nickel

Nickel (Ni) is primarily prevalent in serpentine soils but it is also found in natural soils in trace amounts. Concentration of Ni is increasing in certain areas by human activities such as mining works, emissions, burning of coal and oil, sewage, use of phosphate fertilizers and pesticides (Gimeno-Garcia *et al.,* 1996). Polluted soils may contain upto 200–26,000 mg kg^{-1} Ni which is 20 to 30 fold higher than the normal range (10–1,000 mg kg^{-1}) found in natural soils (Izosimova 2005).

Factors which influence Ni uptake by plants include Ni concentration in the soil, soil pH, presence of other metals and organic matter composition (Chen *et al.,* 2009). At alkaline pH range, with increase in pH Ni uptakc by plants decreases due to the formation of less soluble complexes (Temp1991). Many workers have shown that Ni is readily absorbed by the plant roots when supplied in the ionic form (Aschmann and Zasoski, 1987). Turina *et al.,* (1968) suggested important role of root caps in absorption of Ni in some monocot plants such as rye (*Secale cereale*), wheat (*Triticum vulgare*) and maize (*Zea mays*). Ni is taken up from the soil by plant roots by an active process and the uptake is affected by temperature, anaerobic conditions and respiratory inhibitors such as dinitrophenol (Aschmann and Zasoski, 1987). Toxicity symptoms of Ni in plants include chlorosis and necrosis of leaves, reduced root development, inhibited germination and reduced plant growth (Rajpoot *et al.,* 2015). Excessive dose of Ni^{2+} in lichen causes considerable damages to the cell membrane by affecting reduction in K^+ concentration. Increasing concentrations of Ni also affects the restoration of reabsorption of K^+ and Mg^{2+} and the release of Ca^{2+} (Espen *et al.,* 1997). It is suggested that Ni affects the mechanism of transport by interfering with the proton-extruding mechanism at the plasma membrane which is known to play a central role in ion transport and in the control of homeostatic cell pH.

Excessive level of Ni in the tissues causes decrease in leaf chlorophyll, reduced CO_2 assimilation and inhibition in the synthesis of DNA, RNA and proteins. High Ni levels cause alteration in several metabolic and physiological processes such as absorption, translocation of mineral nutrients, photosynthesis, respiration, and carbohydrate and protein metabolism. Excess Ni in plant tissues triggers oxidative damage which relates to the observed diverse toxic effects of the metal. Therefore, large quantities of ROS are produced that cause damage to cellular organelles and oxidize proteins and lipids and degrade chlorophyll pigments (Sharma *et al.,* 2012). However, plants are well equipped with an organized defense system to counter the toxic effects of Ni that includes exclusion/restriction of entry of the metal into the cell through plasma membrane and chelation of the metal followed by its sequestration into the vacuole.

Zinc

Zinc is an essential nutrient for plants and it is required in adequate concentration for normal growth and functioning of plants (Saper *et al.*, 2009). However, it is also considered as a major pollutant of terrestrial and aquatic environment (Fosmire *et al.*, 1990). Its increased use in various industries has led to increased environmental zinc concentration (Krishna *et al.*, 2004). Atmospheric zinc is mainly bound to aerosol particles that may travel long distances (Pacyna *et al.*, 2001). A typical soil contains Zinc concentration in the range 10 to 300 ppm but at contaminated places near metal smelters high concentrations of Zn have been reported (Pacyna *et al.*, 1972). Being an essential micronutrient Zn is readily up taken by plants and has high mobility in soil which is one of the main reasons for its excess uptake and toxicity (White *et al.*, 2011). Zinc is a structural component of ribosome and contributes to its integrity (Ban *et al.*, 2000). Zinc is involved in DNA replication, protein synthesis and in control of gene expression (Lin *et al.*, 1998). It also serves as cofactor for many enzymes (Harris *et al.*, 1992).

At high concentration Zn becomes strongly toxic and impairs plant growth and metabolism (Fosmire *et al.*, 1990). High Zn treatment causes growth reduction in pants and alters the activity of enzymes of chlorophyll biosynthesis (Prasad *et al.*, 1999), inhibits Hill reaction activity and leads to decline in photochemical activities associated with PS2 (Rout *et al.*, 2009). Zn toxicity inhibits activities of ribonuclease and nitrate reductase and induces synthesis of heat shock proteins and lowers phosphorus content (Burd *et al.*, 2000). Zinc causes root inhibition in many plant species (Lin *et al.*, 2007). Zn inhibits CO_2 assimilation at relatively low concentration whereas at higher concentrations it initiates the loss of photosynthetic pigments and a decline in the chlorophyll a/b ratio and also results in inhibition of photosynthetic electron transport and photophosphorylation (Rout *et al.*, 2009). Zn excess significantly inhibits RuBP carboxylase activity by substitution of Zn^{2+} for Mg^{2+} (Lin *et al.*, 2007).

Arsenic

Arsenic is known to mankind since the beginning of human civilization due to its highly toxic properties. In recent years arsenic has become a serious environmental pollutant in many parts of the world. Although it occurs naturally in a wide range of minerals but release and mobilization of arsenic from its sources constitute the availability of this element in the soil, water and air (Arehart *et al.*, 2006). Arsenic is present in high amounts in deep water in many parts of the world. Arsenate and arsenite are the two major forms of arsenic and both of these forms are readily taken up by the plants (Ali *et al.*, 2009).

Arsenate enters the root as a phosphate analogue and then it gets reduced to arsenite (Meharg *et al.*, 2004). This uptake is inhibited by phosphate, suggesting that phosphate and arsenate are taken up by the same uptake system. A number of aquaporins, nodulin 26 like intrinsic proteins are able to transport arsenite, the predominant form of arsenic in reducing environment (Kamiya *et al.,* 2009). In rice (*Oryza sativa*), arsenite uptake shares the highly efficient silicon (Si) pathway of entry to root cells (Zhao *et al.,* 2009). In root cells arsenate is rapidly reduced to arsenite, which is effluxed to the external medium, complexed by thiol peptides or gets translocated to shoots (Zhao *et al.,* 2009). Arsenite inhibits activities of many enzymes which contain protein sulphydryl groups (Miller *et al.,* 2002). Arsenate after its uptake interferes with aerobic phosphorylation following reduction to arsenite (Bellosta *et al.,* 2008).

Increased As level inside the tissues decreases plant growth and reduces fruit yield, causes root discoloration, root plasmolysis, wilting of leaf tips and leaf margins and causes decrease in photosynthetic capacity (Malik *et al.,* 2011). Exposure to inorganic arsenic leads to increased generation of ROS in the tissues leading to induction of oxidative stress and increased activity of many antioxidative enzymes (Mishra *et al.,* 2011). Arsenic depolarizes membrane potential and decreases DNA and protein synthetic activities (Mishra *et al.,* 2006).

Aluminium

Aluminium (Al) is a major constituent of earth crust and becomes constraint to crop productivity in acid soils because solubility of Al increases at acidic pH. Aluminium (Al) becomes growth limiting factor for plants when soil pH becomes less than 5.0 (Mahapatra *et al.,* 2015). Al occurs in various minerals such as beryl, cryolite, garnet, spinel, and turquoise zumdahl (2012).

Aluminium is rhizotoxic. Characteristic feature of Al toxicity in plants is inhibited root growth. Root elongation inhibition is the immediate response of Al toxicity in plants (Henning 1998). Much of the Al taken up by plant roots penetrates the boundary between root apex and root cap and accumulates in the nuclei and cytoplasm of cells adjacent to this zone (Henning 1998). Aluminium (Al) is disruptive to the uptake of Ca^{2+} therefore prolonged Al stress leads to Ca^{2+} deficiency (Walton *et al.,* 2012). Disturbance of cytoplasmic Ca^{2+} homeostasis leads to the break-down in Ca^{2+} mediated signal transduction which is a primary mechanism of Al rhizotoxicity.

Aluminium toxicity leads to overproduction of ROS in the tissues and thereby causes oxidative stress and alters expression of antioxidative enzymes (Pandey *et al.,* 2014). It has been observed that enhanced lipid peroxidation mediated by Fe is the consequence of primary effects of Al on membrane structure and integrity in tobacco cells (Yamamoto *et al.,* 2001). Al disturbs cell division particularly in root tips and lateral roots, increases cell wall rigidity by increasing cross linking pectins (Rout *et al.,* 2001). DNA replication is reduced due to increase in the rigidity of DNA double helix (Mohanty *et al.,* 2004). Aluminium toxicity not only fixes phosphorus in less available forms on root surfaces (Rout *et al.,* 2001) but also decreases root respiration (Hinsinger *et al.,* 2003). The enzyme activity controlling sugar phosphorylation (Graham *et al.,* 2002), deposition of cell wall polysaccharides, uptake, transport and processing of several essential nutrients (Ca,Mg, K, P and Fe) are adversely affected under Al toxicity (Silva *et al.,* 2010).

The accumulation of Al in tobacco cells causes instantaneous repression of mitochondrial activities. Al excess in the tissues causes respiration inhibition, ATP depletion and inhibition of cell growth (Yamamoto *et al.,* 2001). A marked reduction of protein content in cytoplasm and an increase in peroxidase activity have been observed due to Al toxicity in rice plants (Panda *et al.,* 2009).

Boron

Boron (B) is an essential micronutrient for plants and is released naturally into the atmosphere and aquatic environment from geothermal steams, and weathering of clay-rich sedimentary rocks, coal and shale (Simsek *et al.,* 2003). Boron has intermediate properties between metals and non-metals (Alloway *et al.,* 2013) and it is widely distributed in rocks (5–10 mg kg^{-1}), rivers (3–30 µg kg^{-1}) and oceans (~4.5 mg L^{-1}) (Camacho Cristobel *et al.,* 2008). Boron, primarily present in soil as boric acid $B(OH)_3$, can be easily leached under high rainfall conditions and thereby it causes retarded growth and alteration in metabolism of plants that grow there (e.g. many regions in Japan, China, USA, and Brazil) (Camacho Cristobel *et al.,* 2008). Quite opposite, in arid and semi arid regions under low rainfall conditions, B cannot be sufficiently leached and therefore may accumulate to toxic levels (Hasnain *et al.,* 2011).

Soluble undissociated boric acid $B(OH)_3$ is the most widely available form of B in soils with pH values 5.5 to 7.5 and plants readily uptake B in this form (Miwa *et al.,* 2010). There are three different molecular mechanisms by which boric acid is readily taken up by plants (i) passive diffusion across lipid bilayer (Tanaka *et al.,* 2008); (ii) facilitated transport by major intrinsic protein (NIP) channel (Takano *et al.,* 2006); and (iii) an energy-dependent high-affinity transport system induced in response to low B supply, which is mediated via BOR transporters (Tanaka *et al.,* 2008). To fulfill the requirement of B by plants, boric acid absorption by roots primarily involves B diffusion across the lipid bilayer (Tanoi *et al.,* 2011).

The fact that excessive B accumulates at the end of the transpiration stream is theoretically proved by the evidence that typical chlorotic and necrotic patches exist at the margins and the tips of the older leaves in plants suffering from B toxicity (Sutton *et al.,* 2007). These patches have elevated levels of B concentrations (Oertli and Roth, 1969) and their genotypes also show characteristically different pattern as well. B toxicity reduces plant growth, causes alteration in metabolism, reduces root cell division, lowers leaf chlorophyll contents and photosynthetic rates and decreases lignin and suberin levels (Wang *et al.,* 2010). Despite the fact that the physiological basis for B toxicity is not clear, three main causes have been proposed taking into account our knowledge of B chemistry, i.e. the ability of B to bind compounds with two hydroxyl groups in the *cis*-configuration (Lacina *et al.,* 2011). These are (i) alteration of cell wall structure (Hu *et al.,* 1994); (ii) metabolic disruption by binding to the ribose moieties of molecules such as adenosine triphosphate (ATP) (Ried *et al.,* 2004), nicotinamide adenine dinucleotide reduced form (NADH) or nicotinamide adenine dinucleotide phosphate reduced form (NADPH); and (iii) disruption of cell division and development by binding to ribose, either as the free sugar or within RNA (Ried *et al.,* 2004).

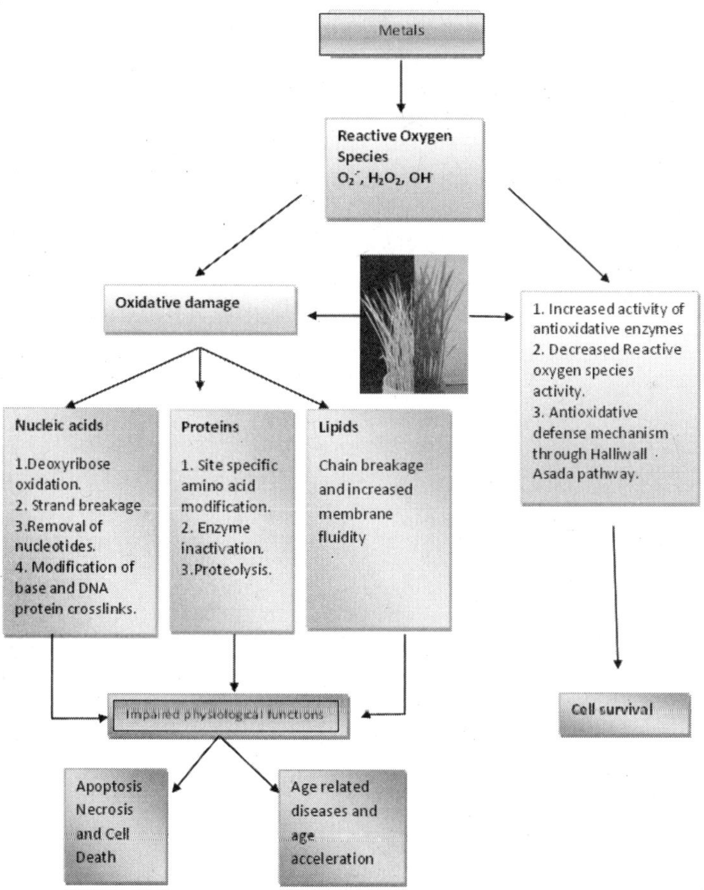

Fig 4. A flow chart showing response of plants to excess levels of metals. ROS produced in the cells get scavenged through antioxidative defense system or it causes oxidative damage to cellular biomolecules.

PHYTOREMEDIATION

Phytoremediation is a cost effective and environment friendly process to use plants and plant products to decrease metal load, and to restore or stabilize metal contaminated sites. Phytoremediation takes advantage of the natural ability of plants to take up, accumulate, and store contaminant substances which include metals, metalloids, radionuclides, chlorinated solvents, petroleum hydrocarbons, organophosphate insecticides, explosives and surfactants (Khan *et al.,* 2004). More than 45 families of plants have been identified which hyperaccumulate metals in which some of the families are Brassicaceae, Fabaceae, Euphorbiaceae, Asteracea, Lamiaceae and Scrophulariaceae (Prasad *et al.,* 2003).

Table 1. Plants commonly used for phytoremediation of various metals

S.N	Plants	Heavy metals	Mode of action	References
1.	*Brassica juncea*	Co, Cr, Ni, Pb, As, Cd.	Complexation, sorption and precipitation	Singh *et al.,* 2012
2.	*Thalspi caerulescens*	Zn, Au, Hg, Mo	Hyper-accumulation	Robinson *et al.,* 1998
3.	*Helianthus annus*	Ag, Cd	Hyper-accumulation	Adesodun *et al.,* 2010
4.	*Nicotiana tobaccum, Secale cereale, Spinacia oleracea*	Cd, Cu, Ni,Zn, Cr	Rhizosphere accumulation through sorption, concentration and precipitation	Liu *et al.,* 2000
5.	*Arabidopsis thaliana*	Organics and inorganics, chlorinated solvents	Volatilization by leaves through transpiration	Khoudi *et al.,* 2013
6.	Hybrid poplars, stone wart, *Salix nigra*, algae	Organic compounds, chlorinated solvents, phenols, herbicides.	Pollutant eradication	Jatav *et al.,* 2015
7.	*Tithonia diversifolia*	Uranium	Concentration in root	Adewole *et al.,* 2010
8.	Gossypium spp.	Pb , Zn	Accumulation in root, leaves and stems	Wilberforce *et al.,* 2015
9.	*Lemna minor*	Trinitrotoluene	Degradation	Saha *et al.,* 2015
10.	*Medicago sativa*	Trichloroethlene, Zn , Hg	Volatilization by leaves through transpiration	Gordan *et al.,* 1998

S.N	Plants	Heavy metals	Mode of action	References
11.	*Jatropha curcus*	Pb, Cd, As, Cr	Concentration in shoot	Chang *et al.*, 2014
12.	*Populus tremula*	Cd, Zn	Accumulation in leaves	Lee *et al.*, 2012
13.	*Hordeum vulgare*	Cd, Cu , Zn	Phytostabilization and phytoextraction	Guerra *et al.*, 2011
14.	*Juncus maritimus*	Cu	phytostabilization	Montenegro *et al.*, 2016
15.	*Zea mays*	Cd, Cu	Chelation	Hechmi *et al.*, 2013

Brassica juncea, Indian mustard, is a promising plant which is being used widely for the purpose of phytoremediation due to its good ability to transport Pb, Ni,Cu and Cd from root to shoot (Bayon *et al.*, 2002). Aquatic plants such as the floating *Eichornia crassipes* (water hyacinth) (Singhal *et al.*, 2003), *Lemna minor* (duckweed) (Appenroth *et al.*, 2010), and pistia (Odjegba *et al.*, 2004) are used for rhizofilteration. *Pteris vittata* and *Pityrogamma calomelanos* plants possess the potential to accumulate arsenic (Salido *et al.*, 2003). Jayweera *et al.* (2007) reported phytoremediation efficiencies of water hyacinth (*Eichhornia crassipes*) for removal of Al. Common weed species such as *Sorghum halepense, Cyperus rotundus, Cynodon dactylon, Amaranthus retroflexus, Echinochloa crussgali and Chenopodium album* are efficient accumulators of Boron (Aydin and Cakir, 2009). Spinach (*Spinacia oleraceae*), Fenugreek (*Trigonella foenum-graecum*) have natural potential to remove Cr from Cr contaminated soil (Dheri *et al.*, 2007). Robinson *et al.* (1999) demonstrated Ni and Co uptake properties of *Berkheya coddi*. Mojiri *et al.* (2011) showed the potential of corn (*Zea mays*) for phytoremediation of soil contaminated with Cd and Pb. Shengguo *et al.* (2003) reported the remarkable ability of *Phytolacca acinosa* plants to concentrate Mn from the soil. *T. caerulescens* plants have been shown to possess high affinity for the uptake of Zn and low affinity for uptake of Cd (Alkorta *et al.*, 2004).

Phytoremediation does not have any negative impact on soil fertility and can be applied to remediate shallow soil, ground water and surface water bodies. Most of the metal contaminated sites contain hardy, tolerant, weed species and phytoremediation through these and other non edible species can restrict the contaminant from being introduced into the food chain.

CONCLUSIONS AND FUTURE PROSPECTS

A wide range of studies suggest that metals occur naturally as well as they are contributed anthropogenically to the environment. Some of the metals are essential elements as they are required by the plants for their normal growth and metabolism, whereas many metals which are non-essential for plants and are added to the soil environment by various activities are regarded as pollutants. However, both essential and non-essential metals when present in high amounts in the soil, cause adverse effects on plant growth and metabolism. Increasing uses of metals in industrial, agricultural, domestic and technological applications have contributed to a greater amount of the metals in the soil environment which has now become a matter of global concern. Plants readily absorb various elements from the soil that have either known biological functions or which are non-essential. Both essential and non-essential metals when taken up by plants in large amounts show toxicity symptoms. A number of individual studies of metal toxicity have shown different aspects of physiological constraints and metabolic alterations. Some debate still exists as to exactly how the uptake of various metals is regulated in plants. Though the literature supports uptake in response to concentration gradient induced by selective uptake of ions by roots or by diffusion of elements in the soil, our knowledge has advanced much during the recent years related to involvement and role played by various transporters at molecular level.

Different metals after being taken up by plant roots, cause increased production of reactive oxygen species (ROS). Due to excessive production of ROS, plants undergo a condition called as oxidative stress marked by oxidation of cellular biomolecules and disturbance in redox balance. In plant tissues high levels of metals affect cellular organelles such as cell membrane, mitochondria, lysosomes, endoplasmic reticulum, nuclei, and many enzymes involved in metabolism, detoxification, and damage repair. Besides reducing crop productivity, the elements which are often associated with toxicity in plants, such as Cd, Pb, Mn, Zn, Cr, Cu, Fe, As, Al and B, accumulate within the plant tissues and then move to humans via the food web. Although the mode of transfer of toxic elements from plants to animals, particularly humans and their toxicity effects in humans are poorly studied, evidences indicate that consumers are at risk when agriculture produce contains excess amounts of toxic elements. With the advent of phytoremediation a lot of fieldwork has been done in the recent years to remove metal load from the soil in developed and developing countries which has proved to be as a viable alternative to conventional remediation methods.

Simultaneous exposure of plants to multiple metals may produce toxic effects which are additive, antagonistic or synergistic. Therefore, it is essential to intensify well designed and quantitative research programmes for better understanding of the effects of metal toxicity on plants, and to look for different strategies including phytoremediation to decrease metal content of soils.

REFERENCES

Adhikari T, Sarkar D, Mashayekhi H (2013) Growth and enzymatic activity of maize (*Zea mays* L.) plant: solution culture test for copper dioxide nano particles. J. Plant. Nutr. 99-115.

Al-asheh S, Banat F (2001) Adsorption of zinc and copper ions by the solid waste of the olive oil industry. Adsorpt. Sci. Technol. 19:117-129.

Alfonsi K, Colberg J, Dunn PJ, Fevig T, Jennings S, Johnson T (2008) Green chemistry tools to influence a medicinal chemistry and research chemistry based organization. Green Chem. 10:31-36.

Ali W, Isayenkov SV, Zhao FJ, Maathuis FJM (2009) Arsenite transport in plants. Cell Mol Life Sci. 66: 2329-2339.

Alkorta I, Allica J, Becerril JM, Amezaga I, Albizu I (2004) Chelate enhanced phytoremediation of soils polluted with heavy metals. Environ. Sci. Biotechnol. 3:55-70.

Alkorta I, Harnandez-Allica J, Becerril J (2004) Recent finding on the phytoremediation of soil contaminated with environmentally toxic heavy metal and metalloids such as zinc, cadmium, lead and arsenic. Rev. Environ. Sci. Biotechnol. 3: 71-90.

Alves C, Vicente A, Nunes T, Goncalves C, Fernandes AP, Mirante F, Tarelho L, Querol X, Careiro A, Monteiro C (2009) Summer 2009 wildfires in Portugal: Emission of trace gases and aerosol composition. Atmos. Environ. 45:641-649.

Amundsen PA, Kashulin NA, Teretjeev P, Gjelland KO, Koroleva IM, Dauvlter VA (2011) Heavy metal contents in whitefish (*Coregonus lavaretus*) along a pollution gradient in a subarctic watercourse. Environ. Monit. Assess 182:301-316.

Amundsen PA, Staldvik FJ, Lukin AA, Kashulin NA, Popova OA, Reshetnikov YS (1997) Heavy metal contamination in freshwater fish from the border region between Norway and Russia. Sci Total Environ 201:211-224.

Apel K, Hirt H (2004) Reactive oxygen species: Metabolism, Oxidative stress, and signal transduction. Annu. Rev. Plant Biol. 55:373-399.

Arazi T, Sunkar R, Kaplan B, Fromm H (1999) A tobacco plasma membrane calmodulin- binding transporter confers Ni^{2+} tolerance and Pb^{2+} hypersensitivity in transgenic plants. Plant J 20:171-182.

Arehart GB, Chryssoulis L, Kesler SE (2006) Gold and arsenic in iron sulphides from sediments-hosted disseminated gold deposits; implications for depositional processes. Econ Geol 88:171-185.

Asada H, Kawamaru Y, Maruyama K, Kume H, Ding RG, kanbary, Kuzume H, Sanbo M, Yagi T Obata K (1997) Cleft palate and decreased brain gamma-aminobutyric acid in mice lacking the 67-kDa isoform of glutamic acid decarboxylase. Proc Natl Acad Sci USA. 94: 6496-6499.

Aydin M, Cakir F (2009) research on weed species for phytoremediation of boron polluted soil. Afr J Biotechnol 8:4514-4618.

Babula P, Adam V, Opatrilova R, Zehnalek J, Havel L, Kizek R (2008) Uncommon heavy metals, metalloids and their plant toxicity: a review environmental chemistry letter Environ. Chem. Lett. 6:189-213.

Bajji M, Kinet JM, Lutts S (2002) Osmotic and ionic effects of NaCl on germination, early seedling growth, and ion content of *Atripplex halimus* (Chenopodiaceae). Can J Bot 80:297-304.

Ban N, Nissen P, Hansen J, Moore PB, Steitz TA (2000) The complete atomic structure of the large ribosomal subunit at 2.4 A° resolution. Science 289:905-920.

Bartsevich VV, Pakrasi HB (1995) Molecular identification of an ABC transporter complex for manganese: analysis of a cyanobacterial mutant strain impaired in the photosynthetic oxygen evolution process. EMBO J 14:1845-1853.

Bayon MM, Meija J, Leduc DL, Terry N, Caruso JA Medal A (2003) HPLC-ICP-MS and ESI-Q-TOF analysis of biomolecules induced in *Brassica juncea* during arsenic accumulation. J. Anal. At. Spectrum 19:153-158.

Bellosta RV, Sorribas V (2008) Role of rat sodium/phosphate cotransporters in the cell membrane transport of arsenate. Toxicol Appl Pharmacol 232:125-134.

Bergamini, Carlo M,́ Stefania, Dondi, Alessia, Cervellati, Carlo (2004) Oxygen, reactive oxygen species and tissue damage. Curr Pharm Des 10: 1611- 1626.

Bhuiyan MAH, Parvez L, Islam MA, Dampare SB, Suzuki S (2010) Heavy metal pollution of coal mine affected agricultural soils in the northern part of Bangladesh. J **Hazard Mater** 173:384-392.

Bischoff JL, Dickson FW (1975) Sea water – basalt interaction at 200°c and 500 bars : implications for origin of sea floor heavy metal deposits and regulation of seawater chemistry. Earth Planet Sci Lett 25:385-397.

Cabiscol E, Tamarit J, Ros J (2000) Oxidative stress in bacteria and protein damage by reactive oxygen species. Int Microbiol 3: 234-245.

Cacmak I, Yilmaz A, Akiz H, Torun B (1996) Zinc deficiency as a critical problem in wheat production in central Anatolia. Plant Soil 180:165-172.

Camacho Cristobel J J, Rexach J, Fontes G, Fontes A (2008) Boron in plants: Deficiency and toxicity. J Integr Plant Biol 50:1247-1255.

Chaney RL, Brown JC, Tiffin LO (1972) Obligatory reduction of ferric chelates in iron uptake by soybeans. Plant Physiol 50:208-213.

Chardonnens AN, Bookum WMT, Kuijper LDJ (2002) Distribution of cadmium in leaves of cadmium tolerant and sensitive ecotypes of *Silene vulgaris*. Physiol Plant 104: 75-80.

Charlesworth S, Everett M, Mc Carthy R, Ordonez A, Migual Ed (2003) A Comparative study of heavy metal concentration and distribution in deposited street dusts in a large and a small urban area: Birmingham and Coventry, West Midland. UK Environ Int 29:563-573.

Chen C, Huang D, Liu J (2009) Functions and toxicity of nickel in plants: Recent advances and future prospects. Clean Soil Air Water 37: 304-313.

Cheng S (2003) Heavy metal pollution in china: Origin, pattern and control. Environ Sci Pollut Res 10:192-198.

Clarkson DT (1966) Effect of Aluminium on the uptake and metabolism of phosphorus in barley seedlings. Plant Physiol 41:165-172.

Clemens S (2001) Molecular mechanism of plant metal tolerance and homeostasis. Planta 212:475-486.

Cooper JSG, Chan YK, Dissanayake AM, Leahy FE (2005) Demonstration of a hyperglycemia driven pathogenic abnormality of copper homeostasis in diabetes and its reversibility by selective chelation : quantitative comparisons between the biology of copper and eight other nutritionally essential elements in normal and diabetic individuals. Diabetes 54: 1468-1476.

Crooke WM, Hunter JG, Vergnano O (2008) The relationship between nickel toxicity and iron supply. Ann Appl Biol 41:311-324.

Dalton DA, Russell SA, Hanus FJ, Pascoe GA, Evans HJ (1986) Enzymatic reactions of ascorbate and glutathione that prevent damage in soybean root nodules. Proc Natl Acad Sci USA 83:3811-3815.

Davies FT, Puryear JD, Newton RJ, Egilla JN, Grossi AS (2002) Mycorrhizal fungi increase chromium uptake by sunflower plants: influence on tissue mineral concentratin growth and gas exchange. J Plant Nutr 25:2389-2407.

Dekock PC (1956) Heavy metal toxicity and iron chlorosis. Ann Bot 20:133-141.

Dheri GS, Brar MS, Malhi SS (2006) Comparative phytoremediation of chromium contaminated soils by fenugreek, spinach, and raya. Commun Soil Sci Plant Anal 38:11-12.

Dominiak BC, Ekman JH (2013) The rise and demise of control option for fruit fly in Australia. **Crop Prot** 51:57-67.

Dringen R, Scheiber IF, Mercer JFB (2013) Copper metabolism of astrocytes. Front Aging Neurosci. 5:9.

Duruibe JO, Ogwuegbu MOC (2007) Heavy metal pollution and human biotoxic effects. Int J Phys Sci 2 : 112-118 .

Ermakov AA, Karpova EA, Malysheva AG, Mikhaylova RI, Ryzhova IN (2015) Monitoring of the content of heavy metals and elements in the snow cover in agricultural soils at the territory of the Moscow region. Gig Sanit 94:31-36.

Espen L, Pirovano L, Cocucci SM (1997) Effects of Ni^{2+} during the early phases of radish (*Raphanus sativus*) seed germination. Environ Exp Bot 38:187-197.

Foyer CH (1998) Ascorbate and glutathione: keeping active oxygen under control. Annu Rev Plant Biol.49: 249-279.

Foyer CH, Lelandais M, Kunart KJ (1994) Photooxidative stress in plants. **Physiol Plant** 92: 696-717.

Foyer CH, Shigeoka S (2011) Understanding oxidative stress and antioxidant functions to enhance photosynthesis. Plant Physiol 155:93-100.

Foyer, Christine H (2001) Prospects for enhancement of the soluble antioxidants, ascorbate and glutathione. Biofactors 15:75-78.

Gao Y, Xia J (2011) Chromium contamination accident in china: viewing environment policy of china. Environ Sci Technol 45: 8605-8606.

Garg N, Manchanda G (2009) ROS generation in plants: Boon or bane? Plant Biosyst 143:81-96.

Godo GH, Reisenaur HM (1980) Plant effects on soil manganese availability. J Am Soc Agron 44: 993-995.

Gracia RC, Snodgrass WR (2007) Lead toxicity and chelation therapy Am J Health Syst Pharm 64: 45-53.

Gunawardana B, Singhal N, Johnson A (2011) Effects of amendments on copper, cadmium and lead phytoextraction by *Lolium perenne* from multiple-metal contaminated solution. Int J Phytoremediation 13:215-232.

Hafedh YSA, Ali SA (2004) Effect of feeding on survival, cannibalism, growth and feed conversion of African catfish, *Clarias gariepinus* (Burchell) in concrete tanks. J Appl Ichthyol 20: 225-227.

Hagenmeyer J, Kahle H, Breckle SW, Waisel Y (1986) Cadmium in *Fages sylvatica* L. trees and seedlings: leaching, uptake and interconnection with transpiration. Water Air Soil Pollut 29:347-359.

Harris ED (1991) Copper as a cofactor and regulator of copper zinc superoxide dismutase. J Nutr 122(S):636-640.

Hasnain A, Mahmood S, Akhtar S, Malik SA, Bashir N (2011) Tolerance and toxicity levels of Boron in Mung bean (*Vigna Radiata*) cultivars at early growth stages. Pak J Bot 43:1119-1125.

Hasnain A, Mahmood S, Akhtar S, Malik SA, Bashir N (2011) Tolerance and toxicity levels of Boron in Mung Bean (*Vigna radiata* (L) Wilczek) cultivars at early growth stage. Pak J Bot 43: 1119-1125.

Hinsinger P, Plassard C, Tang C, Jaillard B (2003) Origins of root mediated pH changes in the rhizosphere and their response to environmental constraints: A review. Plant Soil 248:43-59.

Hue NV, Vega S, Silva JA (2001) Manganese toxicity in a hawaiian oxisol affected by soil pH and organic amendments. Soil Sci Soc Am J 65:153–160.

Jayweera MW, Kasturiarachi KKA, Wijeyekoon SLJ (2007) Removal of aluminium by constructed wetlands with water hyacinth (*Eichornia crassipes*) grown under different nutritional conditions J Environ Sci Health 42:185-193.

Kachenko AG, Singh B (2006) Heavy metal contamination in vegetable grown in urban and metal smelters contaminated sites in Australia. Water Air Soil Pollut 169:101-123.

Kamiya T, Tanaka M, Mitani N, Ma JF, Maeshima M, Fujiwara T (2008) Nip1;1, an aquaporin homolog, determines the arsenite sensitivity of *Arabidopsis thaliana.* J Biol Chem. 284:2114-2120.

Karuppanapandian T, Moon JC, Kim C, Manoharan K, Kim W (2011) Reactive oxygen species in plant: their generation, signal transduction, and scavenging mechanism. Aust. J Crop Sci. 5: 709-725.

Kasai K, Takano J, Miwa K, Toyada A, Fujiwara T (2010) High boron induced ubiquitination regulates vacuolar sorting of the BOR1 borate transporters in *Arabidopsis thaliana* . J Biol Chem 286: 6175-6183.

Kelly CK, Woodward FI, Crawley MJ (1996) Ecological correlates of plant range size: taxonomies and phylogenies in the study of plant commonness and rarity in great Britain Phil Trans R Soc, Series B 23:234-245.

Kelly J, Thornton L, Simpsom PR (1996) Urban geochemistry: A study of the nfluence of anthropogenic activity on the heavy metal content of soils in traditionally industrial and non industrial areas of Britain. Environ Geochem 11:363-370.

Khan AG (2006) Mycorrhizoremediation- An enhanced form of phytoremediation. J Zhejiang Univ Sci 7: 503-514.

Krishna S S, Grishin NV (2004) The finger domain of the human Deubiquitinating enzyme HAUSP is a zinc ribbon. Cell Cycle 3:1044-1047.

Lima D, Santos MM, Ferreira AM, Micaelo C (2008) The use of the shanny *Lipophrys Pholis*for pollution monitoring: A new sentinel species for the northwestern European marine ecosystems. Environ Int 34:94-101.

Lin W, Evans OR, Xiong RG, Wang Z (1998) J Am Chem Soc 120:13272-13273.

Marschner H, Romheld V (1995) Strategies of plants for acquisition of iron. Plant Soil 59:375-388.

Marz C, Stratmann A, Matthiessen J, Meinhardt AK, Eckert S, Schnetger B,Vogt C, Stein R, Brumsack HJ (2011) Manganese-rich brown layers in Arctic Ocean sediments:composition, formation mechanisms, and diagenetic overprint. Geochim Cosmochim Acta 75:7668-7687.

Meharg AA (2004) Arsenic in rice-understanding a new disaster for south east asia. Trends in Plant Sci 9:415-417.

Millaleo R, Diaz R, Ivanov AG, Mora ML, Alberdi M (2010) Manganese as essential and toxic element for plants: transport, accumulation and resistance mechanisms. J Soil Sci Plant Nutr 10:470-481.

Miller JD, Wier CC, Pastore RE, Kelly WJ, Dooling RJ (1976) Discrimination and labelling of noise buzz sequence with varying noise lead times: An example of categorical perception. J Acoust Soc Am 60:410.

Miller WH, Schipper HM, Lee JS, Waxman S (2002) Mechanism of action of arsenic trioxide. Am Assoc **Cancer Res** 62:234-245.

Mishra S, Dubey RS (2006) Heavy metal uptake and detoxification mechanism in plants. **Int J Agric Res** 2:122-141.

Mishra S, Jha AB, Dubey RS (2010) Arsenite treatment induces oxidative stress, upregulates antioxidant system, and causes phytochelatin synthesis in rice seedlings. Protoplasma 248:565-577.

Miwa K, Fujiwara T (2010) Boron transport in plants : coordinated regulation of transporters. Ann Bot 105:1103-1108.

Moller IM, Jensen PE, Hansson A (2007) Oxidative medications to cellular components in plants. Annu Rev Plant Biol 58: 459-481.

Muir DCG, Wagemann GNP, Norstrom RJ, Simon M, Lien J (1988) Organochlorine chemical and heavy metal contaminant in white-beaked dolphins (*Lagenorynchuc albirostris*) and pilot whales (*Globicephala melaena*) from the coast of Newfoundland, Canada Arch. Environ Contam 17:613-629.

Muramoto S, Aoyama I (1990) Effects of fertilizers on the vicisstude of cadmium in rice plant. **Environ Sci Health A** 25:629-635.

Mutschler FE, Mooney TC (1993) Precious metal deposits related to alkali igneous rocks: provisional classification, grade-tonnage data and exploration frontiers. In: Kirkham RV, Sinelair WD, Thorpe RI, Duke JM (eds) Mineral Deposit Modeling. Geol Assoc Can Pap 40:479–520.

Naidu R,Bolan NS, Kookana RS, Tiller KG (2006) Ionic strength and pH effects on the sorption of cadmium and the surface charge of soils. Eur J Soil Sci 45: 419-429.

Nasreddine L, Massin P (2002) Food contamination by metals and pesticides in the European union, Should we worry ? Toxicol Lett 127:29-41.

Navrot N, Rouhier N, Gelhaye E, Jacquot JP (2006) Reactive oxygen species generation and antioxidant system in plant mitochondria. Physiol Plant 129:185-195.

Panda SK, Baluska F, Matsumoto H (2009) Aluminium stress signalling in plants. Plant Signal Behav 4:592-597.

Pandey P, Srivastava RK, Dubey RS (2014) Water deficit and aluminium tolerance are associated with a high antioxidative enzyme capacity in Indica rice seedlings. Protoplasma 251:147-160.

Pawlowska TE, Blaszkowski J, Ruhling A (1997) The mycorrhizal status of plants colonizing a calamine spoil mound in southern Poland. Mycorrhiza 6: 499-505.

Peijneneburg WJGM, Jager T (2003) Monitoring approaches to assess bioaccessibility and bioavailability of metals: matrix issue. Ecotoxicol Environ Saf 56:63-77.

Peryea FJ, Kammereck R (1997) Phosphate enhanced movement of arsenic out of lead arsenate contaminated topsoil and through uncontaminated subsoil. Water Air Soil Pollut 93:243-254.

Peryea FJ, Kammereck R (1997) Phosphate enhanced movement of arsenic out of lead arsenate contaminated topsoil and through uncontaminated subsoil. Water Air Soil Pollut 93:243-254.

Pouyat RV, McDonnell MJ (1991) Heavy metal accumulations in forest soils along an urban-rural gradient in South eastern New York, USA. Water Air Soil Pollut 57:797-807.

Prasad MNV (2003) Metal hyperaccumulation in plants- Biodiversity prospecting for phytoremediation technology. Electron J Biotechnol 6:234-245.

Rajpoot R, Rani A, Srivastava RK, Pandey P (2015) *Terminalia arjuna* bark extract alleviates nickel toxicity by suppressing its uptake and modulating antioxidative defense in rice seedlings. Protoplasma pp 1-14. DOI:10.1007/s00709-015-0899-x.

Ramirez-Diaz MI, Perez CD, Vargas E, Rosas HR, Cervantes C (2008) Mechanism of bacterial resistance to chromium compounds. Biometals 21:321-332.

Richman E (2002) Does copper have a role in ALS? Lancet Neurol.1: 1-6.

Ried RJ, Hayes JE, Post A, Stangoulis JCR, Graham RD (2004) A critical analysis of the cause of boron toxicity in plants. Plant Cell Environ 27: 1405-1414.

Righi D, Meunier A (1995) Origin of clays by rock weathering and soil formation. In : Origin and Mineralogy of Clays, Ed. Velde B. Springer, pp. 43-161.

Rock ML, James BR, Helz G (2001) Hydrogen peroxide effects on chromium oxidation state and solubility in four diverse chromium enrinched soils. Environ Sci Technol 35: 4054-4059.

Rout G, Samantaray S, Das P (2001) Aluminium toxicity in plants: a review. Agronomie EDP Sciences 21:3-21.

Salido AL, Hasty KL, Lim JM, Butcher DJ (2003) Phytoremediation of arsenic and lead in contaminated soil using Chinese brake ferns (*Pteris vittata*) and Indian mustard (*Brassica juncea*). Int J Phytoremediation 5:89-103.

Sand W, Gerke T, Hallmann R, Schippers A (1995) Sulphur chemistry, biofilm , and the (in) direct attack mechanism- a critical evaluation of bacterial leaching. Appl Microbiol 43: 961-966.

Saper RB, Rash R (2009) Zinc: an essential micronutrient. Am Fam Physian 79:768-772.

Sauve S, Hendershot W, Aleen HE (2000) Solid solution partitioning of metal in contaminated soils : Dependence on pH, total metal burden and organic matter. Environ Sci Technol 34:1125-1131.

Shah K, Kumar RG, Verma S, Dubey RS (2001) Effect of cadmium on lipid peroxidation, superoxide anion generation and activities of antioxidant enzymes in growing rice seedlings. Plant Sci 161:1135-1144.

Sharma P, Jha AB, Dubey RS, Pessarakali M (2012) Reactive oxygen species, oxidative damage, and antioxidative defense mechanism in plants under stressful conditions. J Bot 1-26, article id 217037.

Sharma RK, Agrawal M (2005) Biological effects of heavy metals: An overview. J Environ Biol 26:301-313.

Sharma RK, Agrawal M, Marshall F (2006) Heavy metal contamination in vegetables grown in wastewater irrigated areas of Varanasi India. Bull Environ Contam Toxicol 77:312-318.

Sharma SS, Dietz KJ (2005) The significance of amino acids and amino acid derived molecules in plant responses and adaptation to heavy metal stress. J Exp Bot 57:711-726.

Sharma VK, Kazama F, Jiangyong H, Ray AK (2005) Ferrates iron (VI) and iron (V)): Environmentally friendly oxidants and disinfectants. J Water Health 3: 45-58.

Siedlecka A, Krupa Z (1999) Cd/Fe interaction in higher plants- its consequences for the photostnthetic apparatus. Photosynthetica 36:321-331.

Singh HP, Mahajan P, Kaur S, Batish DR, Kohli RK (2013) Chromium toxicity and tolerance in plants. Environ Chem Lett 11:229-254.

Sirnoff N (1996) Botanical briefing: the function and metabolism of ascorbic acid in plants. Ann Bot 78:661-669.

Smolders APJ, Lock RAC, Hoyos RIM, Roelofs JGM (2003) Effect of mining activities on heavy metal concentrations in water, sediment, and macroinvertebrate in different reaches of the Pilcomayo river, south amerce. **Arch Environ Contam Toxicol** 44:314-323.

Snowden RED, Wheeler BD (1993) Iron toxicity to fen plant species. J Ecol 81:35-46.

Spiegel SJ, Veiga MM (2010) International guidelines on mercury management in small scale gold mining. **J Clean Prod** 18:375-385.

Spiegel VD, Noordam MY, Klerx HJF (2013) Safety of novel protein sources (insects, microalgae, seaweed, duckweed and rapeseed) and legislative aspects for their application in food and feed production. Compr Rev Food Sci Food Saf 12:662-678.

Srivastava S, Dubey RS (2011) Manganese excess induces oxidative stress, lowers the pool of antioxidants and elevates activities of key antioxidative enzymes in rice seedlings. Plant Growth Regul 64: 1-16.

Stauber JL, Florence TM (1987) Mechanism of toxicity of ionic copper and copper complexes to algae. Mar Biol 94: 511-519.

Stoffers P, Glasby GP, Wilson CJ, Davis KR, Walter P (1986) Heavy metal pollution in Wellington Harbour. New Zealand **J Marine Freshwater Res** 20:495-512.

Stohs SJ, Bagchi D, Hassoun E, Bagchi M (2001) Oxidative mechanism in the toxicity and chromium and cadmium ions. **J Environ Pathol Toxicol Oncol** 20: 234-245.

Subramanian V (2011) Coastal Environments - Focus on Asian Coastal Regions 1-18.

Sudhir P, Murthy SDS (2004) Effects of salt stress on basic processes of photosynthesis. Photosynthetica 42:481-486.

Sutton T, Baumann U, Hayes J, Collins NC, Shi BJ, Schnurbusch T, Hay A, Mayo G, Pallota M, Tester M, Langridge P (2007) Boron toxicity tolerance in barley arising from efflux transporter amplification. Science 318:1446-1449.

Tanaka M, Fujiwara T (2007) Physiological roles and transport mechanism of boron: perspectives from plants. Eur J Appl Physiol 456:671-677.

Tanoi K, Saito T, Iwata N, Kobayashi N 1, Nakanishi TM (2011) The analysis of magnesium transport system from external solution to xylem in rice root. J Soil Sci Plant Nutr 57: 265-271.

Tuteja N, Singh MB, Mishra MK, Bhalla PL, Tuteja R (2001) Molecular mechanism of DNA damage and repair: Progress in plants. Crit Rev Biochem Mol Biol 36: 337-397.

Uraguchi S, Mori S, Kuramata M, Kawasaki A, Arao T, Ishikawa S (2008) Root-to-shoot Cd translocation via the xylem is the major process determining shoot and grain cadmium accumulation in rice. J Exp Bot 60:2677-2688.

Verma S, Dubey RS (2001) Effect of cadmium on soluble sugars and enzymes of their metabolism in rice. Biol Plant 44:117-123.

Verma S, Dubey RS (2003) Lead toxicity induces lipid peroxidation and alters the activities of antioxidant enzymes in growing rice plants. Plant Sci 164:645-655.

Wang Y, Zhang J, Wang X, Antoetti M, Li H (2010) Boron and fluorine containing mesoporous carbon nitride polymers: metal free catalysts for cyclohexane oxidation. Angewandte Chemie 49: 3356-3359.

White PJ, Broadley MR (2011) Physiological limits to zinc biofortification of edible crops. Front Plant Sci 44:231-245.

Wong JWC, Mak NK (1997) Heavy metal pollution in children playgrounds in Hong Kong an its health implications. Environ Technol 18:109-115.

Yabuchi N, Kajiyama M, Iwatate J, Nishikawa H, Hitomi S, Okuyama R, Usui R, Yamada Y, Komaba S (2012) P2 type $Na(Fe_{1/2}\ Mn_{1/2})O_2$ made from earth abundant elements for rechargeable Na batteries. Nat Mater 11:512-517.

Yadav M, Jhunjhunwala S, Phung QT, Lupardus P, Tanguay J, Bumbaca S, Franci C, Fritsche J, Weinschenk T, Modrusan Z, Mellman I (2014) Predicting immunogenic tumour mutations by combining mass spectrometry and exome sequencing. Nature 515:572-576.

Yamamato Y, Kobayashi Y, Matsumoto H (2001) Lipid peroxidarion is an early symptoms triggered by Aluminium, but not the primary cause of elongation inhibition in pea roots. Plant Physiol 125:199-208.

Yeats GW, Orchard VA, Speir TW, Hunt JL, Hermans MCC (1994) Impact of pasture contamination by copper, chromium, arsenic timber preservative on soil biological activity. Biol Fert Soil 18:200-208.

Yi Y, Yang Z, Zhang S (2011) Ecological risk assessment of heavy metals in sediment and human health risk assessment of heavy metal in fishes in the middle and lower reaches of the Yangtze river basin. Environmen Poll 159:2575-2585.

Yruela I (2005) Copper in plants. Braz. J. Plant Physiol. 17: 231-245.

Zarcinass BA, Ishak CF, McLaughlin MJ, Cozens G (2004) Heavy metals in soils and crops in southeast Asia Environ. Geochem Health 26:343-357.

Zayed A, Lytle M, Terry N (1998) Chromium accumulation, translocation and chemical speciation in vegetable crops. Planta 206:293-299.

Zhao FJ, Meharg AA, Mcgrath SP (2009) Arsenic uptake and metabolism in plants. New Phytol 181:777-794.

Zhou Y, Wang S, Zhang K (2008) Visual detection of copper by azide and alkyne functionalized gold nanoparticles using click chemistry. Angewandte Chemie 120: 7564-7566.

Section II:

Plant Trace Elements in Plant Physiology

15

NICKEL AND ITS ROLE IN PLANT PHYSIOLOGY

Enrique J. Baran

Centro de Química Inorgánica (CEQUINOR/CONICET,UNLP)
Facultad de Ciencias Exactas, Universidad Nacional de La Plata,
C. Correo 962, 1900-La Plata, Argentina
E-mail: baran@quimica.unlp.edu.ar

INTRODUCTION

During a long time, nickel was the only element of the second half of the 3d transition metal series, for which a biological role could not be clearly established (Kaim and Schwederski, 1991). Notwithstanding, the importance of nickel depending enzymes in Archea, bacteria, plants and primitive eukariotes is now very well documented (Lancaster, 1988; H Sigel and A Sigel, 1988; A Sigel, H Sigel and RKO Sigel, 2007). Despite the fact that no enzymes utilizing nickel have been found in mammalian species, the impact of nickel biochemistry on human health is also significant (Zambelli and Ciurli, 2014).

Nickel abundance in the Earth's crust is ca. 0.02%, about 500 times less than iron (Ciurli and Mangani, 2001). In Nature it exists in the form of sulfides, oxides or silicates and also as insoluble particles, which are components of dusts and fumes. Soluble and insoluble nickel compounds are also found in soils and in waters. In aqueous solutions it is usually found in the form of the green hexahydrate, $[Ni(H_2O)_6]^{2+}$ (Zambelli and Ciurli, 2014).

ESSENTIALITY OF NICKEL IN PLANTS

The simplest definition of an essential element is, obviously, that it is an element required for the maintenance of life. It absence results in death or severe malfunction of the organism. Experimentally, this rigorous criterion cannot be always satisfied and has led to broader definitions of essentiality. For example, as follows: An element is considered essential when a deficient intake produces an impairment of function and when restoration to physiological levels of that element prevents or relieves the deficiency. The organism can neither grow nor complete its life cycle without the element in question. The element should have a direct influence on the organism and be involved in its metabolism. The effect of the essential element cannot be totally replaced by any other element (Frieden, 1984; Baran, 2004).

Evidences for the essentiality of nickel go back to the 70's of the 20th. Century, when it was found that the growth of different plant species was promoted by the addition of nickel to growth media and/or soil cultures. Furthermore, its possible role in nitrogen metabolism, due to its presence in urease was suggested (Farago and Cole, 1988; López and Magnitskiy, 2011). Some time later, this role was demonstrated for leguminous plants (Eskew *et al.,* 1983) and rapidly extended the idea that nickel is also essential for higher plants (Eskew *et al.,*1983,1984). Some years later, Brown *et al.* (1987) established the essential role of nickel in non-leguminous plants, specifically in barley.

These studies and other related to this matter, finally allowed to include nickel within the group of essential mineral nutrients for plants and, therefore, it is understood that plants may not complete their life cycle without this nutrient (Farago and Cole, 1988; Seregin and Kozhevnikova, 2011; López and Magnitskiy, 2011).

Finally, some recent studies clearly suggest that nickel, together with boron, chloride, copper, iron, zinc, manganese and molybdenum can be considered as essential micronutrients for all higher plants (Hänsch and Mendel, 2009).

UPTAKE AND DISTRIBUTION OF NICKEL IN PLANTS

The major way of nickel uptake from soil is its absorption by plant roots. Some of the soil-accumulated metal becomes bound to its organic components and remains inaccessible to plants, whereas Ni(II) ions can enter the roots. The mechanism and kinetics for Ni(II) absorption by plants are not clearly understood but, in the same way as for other metals it may proceed by passive diffusion and active transport (Seregin and Kozhevnikova, 2011).

The major environmental factors that affect metal uptake by plants are soil acidity, its cation exchange capacity, the content of organic material and lime, moisture potential, granulometric composition and the concentrations of micro- and macronutrients. The effects of all these factors on metal uptake are largely nonspecific.

The uptake of nickel is enhanced by low pH values, and the available nickel has shown to increase at pH values less than 6.5 by the breakdown of nickel complexes with iron and manganese hydrous oxides. Besides, in soil Ni(II) can also be strongly absorbed by magnesium and calcium oxides and at higher pH values the surface of roots and soil particles would be coated with films of insoluble Fe(III) compounds, which can bind Ni(II) and make it inaccessible to plants (Farago and Cole, 1988, Seregin and Kozhevnikova, 2011). Competition kinetic studies showed Cu(II) and Zn(II) to inhibit Ni(II) absorption competitively, suggesting that the three elements are absorbed using the same carrier site (Cataldo *et al.,* 1978).

What's more, it has been suggested that phytosiderophores and related molecules, which play a central role in iron uptake by plants (Baran, 2013), may also be involved, to a certain degree, in nickel uptake (Haydon and Cobbett, 2007).

Once nickel is absorbed by the root, its movement to the upper parts of the plant is closely related to the possible formation of complexes with different organic molecules. In general, potential ligands for metals in plants ca be classified into three classes: 1) oxygen donor ligands (carboxylates: malate, citrate, succinate, oxalate); 2) sulfur donor ligands (metallothioneins and phytochelatins) and 3) nitrogen donors (mainly amino acids) (López and Magnitskiy, 2011).

Plants, like all other organisms, possess homeostatic mechanisms to maintain the correct concentration of essential metal ions in different cellular compartments and to minimize the damage from exposure to non-essential metal ions. A regulated network of metal transport, complexation, trafficking and sequestration activities functions to provide the uptake, distribution and detoxification of metal ions (Clemens, 2001).

Systems available for the acquisition of metal ions by the roots, transport and distribution around the plants, and regulation of their cytosolic concentrations are clearly integral to normal plant growth and development and transport proteins must play a central role in metal homeostasis.

The complexity of metalloenzyme active sites ranges from one unique metal ion bonded with several protein ligands to polynuclear clusters including protein side-chains, and exogenous ligands such water or sulfide ions. Assembly of these metal centers is a very complicated process, involving the participation of many accessory or helper proteins. One recently identified class of such accessory proteins are the so called metallochaperones that binds metal cations and delivers them directly to target enzymes, *via* protein-protein interactions (Rosenzweig, 2002)

Metallochaperones are soluble metal receptor proteins which garner metals from their very small homeostatic concentrations inter- or intracellularly and deliver those metals to the metalloenzyme that requires them for activity. The chaperone must be a thermodynamically stable complex as it approaches its target apoprotein; however, the metal within the complex must be kinetically labile to allow fast metal ion exchange for its proper transport and distribution (Roat-Malone, 2002). The way metallochaperones bind metal ions, recognize specific partner proteins and facilitate metal transfer are still open questions in the field of bioinorganic chemistry. For nickel, the best so far recognized metallo-enzyme present in plants is urease (cf. next Section) and, interestingly, a Ni(II)-binding protein could be identified in soybean that acts as Ni-metallochaperone essential for urease activity (Freyermuth *et al.,* 2000). Various other Ni-chaperones have recently been found and characterized in different bacterial species (Park *et al.,* 1994; Ciurli and Mangani, 2001).

Ni-METALLOENZYMES OF PLANTS

General Aspects of Ni Metalloenzymes

Nickel containing enzymes are especially prominent in the metabolism of anaerobic bacteria. Several of their substrates as well as some of the reaction products of reactions catalyzed by Ni-enzymes are gaseous molecules (H_2, CO, CH_4, CO_2, NH_3). Nickel containing catalysts appear as remnants from primitive bioinorganic systems (reducing atmosphere, greater abundance of sulfur, presence of H_2, SH_2 and CH_4).

Another remarkable aspect of Ni-containing enzymes is that different systems containing this metallic center present no common structural peculiarities (Hausinger, 1997).

The so far known Ni-dependent enzymes include urease, hydro-genase, carbon monoxide dehydrogenase (and CO dehydrogenase/ acetyl-coenzyme A synthase), methyl-S-coenzyme M reductase, glyoxalase I, one class of superoxido dismutase (Ermler *et al.,* 1996; Hausinger, 1997; Maroney, 1999, Ciurli and Mangani, 2001) and acireductone dioxygenase (Maroney and Ciurli, 2014).

The first Ni-enzyme to be identified was urease, which is of widespread occurrence and plays a central role in the metabolism of plants, fungi and bacteria.

Hydrogen gas is formed or consumed in microorganisms as part of their energy metabolism by using hydrogenase enzymes ($H_2 \leftrightarrow 2\ H^+ + 2e^-$), some of which possess nickel.

In the methanogenic bacteria, nickel is also involved in a process, in which a complex series of reactions leads to the final release of methane gas. The enzyme involved, methyl-S-coenzyme M reductase, contains a cofactor F_{430}, which is a nickel-porphinoid complex. On the other hand, carbon monoxide dehydrogenase catalyzes the reversible conversion of carbon monoxide and water to carbon dioxide and 2 H^+ ($CO + H_2O \leftrightarrow CO_2 + 2H^+ + 2e^-$).

The Ni-depending superoxide dismutase shows no homologies with the other three known superoxide dismutases (Cu/Zn, Fe, and Mn depending systems), although the reaction mechanism is probably similar to those of these other enzymes, involving the Ni(III)/Ni(II) redox couple.

From all the mentioned Ni-depending enzymes, only urease and glyoxalase I have been found and characterized in plants.

Urease: A Key Enzyme for Plants

Historical perspective

In 1926, James B. Sumner (1887-1955), professor of Biochemistry at Cornell University, isolated and crystallized the enzyme urease from the ground meal of jack beans (*Canavalia ensiformis*). This was the first enzyme which could be crystallized and Sumner also demonstrated that it was a protein, a finding which was in disagreement with the ideas usually accepted at this time.

For this work, Sumner received the 1946 Nobel Prize in Chemistry, shared with John H. Northrop (1891-1987) and Wendell M. Stanley (1904-1971) for their contributions to the isolation, crystallization, and study of enzymes, proteins and viruses.

During its studies with urease, Sumner considered the possibility that the protein might contain metal ions, but his analyses for iron and manganese were negative. It was not until 1975 that careful metal analysis and evaluation of the equivalent weight of the enzyme clearly demonstrated the presence of stoichiometric nickel ion in jack bean urease (2.00 ± 0.12 g-atoms of nickel per 96.6 kDa subunit) (Andrews *et al.,* 1988b) In this way, urease became the first example of a nickel containing enzyme, providing also the first specific biological role for this element (Dixon *et al.,* 1975a,b; Andrews *et al.,* 1988a,b; Lippard, 1995).

Finally, twenty years later, the crystal structure of the enzyme (in this case isolated from *Klebsiella aerogenes*) was solved at a resolution of 2.0 Å revealing the intimate details of the molecular geometry at the active site (Jabri *et al.,* 1995; Lippard, 1995).

Biochemical and physiological characteristics of urease

At present, three basic mechanisms of urea biosynthesis are known to function in higher plants (Seregin and Kozhevnikova, 2006). First, it is the terminal step of the ornithine cycle catalyzed by the mitochondrial arginase; a step which involves arginine hydrolysis to ornithine and urea. Second, it is the purine catabolism leading to the formation of ureides, such as allantoin and allantoinic acid; the latter is hydrolyzed to glyoxylate and urea in a reaction catalyzed by allantoicase. Finally, an additional pathway of urea biosynthesis from the nonprotein amino acid canavanine (structurally related to the proteinogenic amino acid L-arginine) was found in *Canavalia* and soybean.

Urea derived from these reactions is hydrolyzed by urease (urea amidohydrolase, EC 3.5.1.5), the most efficient hydrolase known (Maroney and Ciurli, 2014), which primary role is to allow organisms to use external or internally generated urea as a nitrogen source (Mobley *et al.,* 1995). The ammonium liberated in this process participates in various anabolic processes, particularly in glutamine synthesis from glutamate run by glutamate synthetase, a crucial reaction in the glutamine synthetase/ glutamate synthase pathway of ammonium assimilation (Seregin and Kozhevnikova, 2006).

Urease is an important enzyme in the global nitrogen cycle and is synthesized by many organisms, including plants, some bacteria, fungi, algae and invertebrates. An important number of plant ureases are so far very well characterized. Bacterial ureases play an important role in the pathogenesis of certain species including *Proteus mirabilis*, *Staphilococcus saprophiticus*, *Ureaplasma urealiticum* and others. In other cases (for example, *Klebsiella aerogenes*), urease activity transforms urea in the sole nitrogen source. One of the most frequently examples found in the recent literature is the urease from *Helicobacter pylori* because of its essential role in the pathogenesis of this microorganism and the high prevalence of this human pathogen. In the invertebrate *Aplysia californica*, urease, together with carbonic anhydrase, is employed for the generation and homeostasis of statoconia, calcium carbonate inclusions in the lumen of the gravity-sensing organ, the statocyst (Sirko and Brodzik, 2000).

The plant and fungal ureases are homo-oligomeric proteins (identical subunits), whereas those of bacterial origin are multimers formed from a complex of two or three subunits. Significant amino-acid similarities were found among all known ureases. Amino-terminal residues of the monomers of plant and fungal enzymes are similar to the small subunits of bacterial enzymes. The large subunits of bacterial ureases resemble the carboxy-terminal portions of plant and fungal subunits (Sirko and Brodzik, 2000). The high sequence similarity indicates that all ureases are variants of the same enzyme and are likely to possess similar tertiary structures and catalytic mechanisms (Mobley *et al.,* 1995).

Enzymology

The enzymatic hydrolysis of urea occurs in two steps: in the first one ammonia and carbamate are generated (eq. (1)), which then spontaneously reacts in water to yield a second molecule of ammonia and bicarbonate (eq. (2)) (Ciurli and Mangani, 2001; Maroney and Ciurli, 2014):

$$O{=}C(NH_2)_2 + H_2O \rightarrow (NH_2)\text{-}COO^- + NH_4^+ \quad (1)$$

$$(NH_2)\text{-}COO^- + H_2O \rightarrow CO_3H^- + NH_3 \quad (2)$$

The first process occurs 10^{14} times faster than the noncatalized reaction at pH 7-8, with a half time of the order of microseconds (Ciurli and Mangani, 2001).

Urease exhibit typical Michaelis-Menten behavior. The observed values of K_m fall, in the 0.2 - 32 mM range, with typical values around 1-4 mM, and are largely independent of pH. Weak substrate and product inhibition are generally observed. Furthermore, urea is not the only substrate for the enzyme, and hydrolytic activity is also observed for urea analogues such as formamide, acetamide, N-methylurea, semicarbazide and thiourea, as well as different kinds of phosphoric acid esters and amides. The values of k_{cat} for these substrates are ca. 2-3 orders of magnitude lower than that observed for urea (Maroney and Ciurli, 2014).

Spectroscopic and magnetic studies

The presence of six-coordinate octahedral Ni(II) in jack bean urease was initially revealed by optical absorption spectroscopic studies. These spectra show close similarities to those of Ni(II)-phosphoglucomutase and Ni(II)- carboxypeptidase A and were reversibly modified by β-mercaptoethanol (Dixon *et al.,* 1975). Further studies of this interaction, clearly confirmed the presence of Ni(II) at the active site of the enzyme (Dixon *et al.,* 1980a,b), as well as the octahedral coordination of the metal center and its direct interaction with β-mercaptoethanol (Blakeley *et al.,* 1983).

On the other hand, EXAFS and X-ray absorption near edge structure studies suggested a pseudooctahedral environment of Ni(II) conformed by Ni-N (probably from histidine residues) and Ni-O bonds (Hasnain and Piggott, 1983; Alagna *et al.,* 1984).

Magnetic susceptibility experiments on jack bean urease indicated the existence of weakly antiferromagnetically coupled ($J = - 6.3\ cm^{-1}$) high-spin ($S = 1$) closely spaced octahedral Ni(II) ions, demonstrating the presence of a dinuclear Ni(II) active site (Clark and Wilcox, 1989).

Other interesting spectroscopic results were obtained from variable-temperature MCD studies of jack bean urease and of the enzyme inhibited with β-mercaptoethanol. The results provide another unambiguous evidence for the presence of a dinuclear Ni(II) center with strong antiferromagnetic exchange coupling ($J = -40\ cm^{-1}$) in the inhibited enzyme (Finnegan *et al.,* 1991).

The variable-temperature MCD spectra of the native enzyme exhibit two negative bands centered at 420 and 745 nm that increase in intensity with decreasing temperature. These bands were assigned to the two highest energy d-d transitions of Ni(II) in an octahedral ligand field, i.e. $^3A_{2g} \rightarrow 3T_{1g}(P)$ (420 nm) and $^3A_{2g} \rightarrow 3T_{1g}(F)$ (745 nm). The MCD spectrum of the inhibited enzyme presents negative temperature-dependent MCD bands centered at 324, 380 (shoulder), 422 and 750 nm. By analogy to the native enzyme, the bands at 422 and 470 nm were assigned again to the two highest energy Ni(II) d-d transitions, indicating that the magnitude of the ligand field is not significantly perturbed by β-mercaptoethanol binding. Besides, the new MCD bands at 324 nm and 380 (shoulder) nm can be attributed to thiolate S → Ni(II) charge-transfer transitions (Finegan *et al.,* 1991).

The structures of the dinuclear sites of urease from jack bean and *Klebsiella aerogenes* was later investigated by EXAFS and no significant differences were found between the plant and bacterial enzyme. After addition of β-mercaptoethanol to both ureases, a new peak was found in the Fourier transform that could be fit using a model that involved the presence of two Ni(II) ions separated by ~ 3.26 Å and bridged by the thiolate sulfur of β-mercaptoethanol (Wang *et al.,* 1994).

Structure of the active site

As mentioned in Sect. 2.1., the first report of the crystal structure of an urease dates back to 1995, in the case of *Klebsiella aerogenes* urease (Jabri *et al.,* 1995) and now the total number of full structures of ureases in the Protein Data Bank amount to 43. Of these, 38 are bacterial ureases, four are from jack bean urease (*C. ensiformis*) and one is from pigeon pea (*Cajanus cajan*) (Maroney and Ciurli, 2014).

Most bacterial ureases present a quaternary structure composed of a trimer of trimers of the type $(\alpha\beta\gamma)_3$, with α, β and γ being three different subunits. The active site is found in the α subunit, giving rise to three active sites per biological subunit. In some other bacterial ureases, the quaternary structure is made of only two subunits to form $(\alpha\beta)_3$ trimers. In the case of *H. pylori* urease the $(\alpha\beta)_3$ trimer further tetramerizes, producing the spherically shaped tetramer of trimers $[(\alpha\beta)_3]_4$, containing 12 active sites. Finally, plant ureases are generally made up of a dimmer of homotrimers $(\alpha_3)_2$, wehre the α subunit is derived from the fusion of the corresponding αβγ subunits found in bacteria. In all cases, the secondary and tertiary structures of all ureases are very similar (Maroney and Ciurli, 2014).

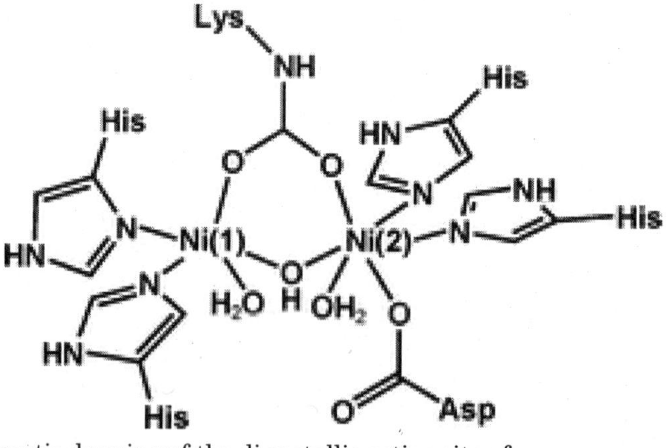

Fig. 1. Schematic drawing of the di-metallic active site of urease.

A consensus has been reached on the structure of the coordination sphere of the Ni(II) ions in the active site of ureases. All the so far determined structures indicate the presence of a dinuclear metallic site, with the two Ni(II) ions separated by 3.5 - 3.7 Å, bridged by the oxygen atoms of a carbamylated lysine residue, and bound to two N-atoms from histidine residues. As shown in Fig. 1, one of the Ni(II) ions is additionally bound to an aspartate carboxylate oxygen and the coordination of each Ni(II) is completed by a terminally bound H_2O molecule and by a nickel-bridging hydroxide anion (Maroney and Ciurli, 2014).

The presence of two types of bridging ligands in the dinuclear metal center of the native enzymes, the carboxylate group of carbamylated lysine and the OH^- anion, explain the above commented finding of a weak antiferromagnetic coupling (Clark and Wilson, 1989).

As it is also evident from Fig. 1, the described ligand arrangement yields one pentacoordinated Ni(II) ion (Ni(1)) with a distorted square-pyramidal geometry and one hexacoordinated Ni(II) ion (Ni(2)) with a distorted octahedral geometry.

Interestingly, an extended network of hydrogen bonds stabilizes the nickel bound solvent molecules in the active site including an additional free water molecule as a part of the hydrogen-bonding network completing a tetrahedral cluster of four water/hydroxide molecules in close proximity of the Ni(II) ions, hinting at the existence of an active site cavity designed to stabilize a tetrahedral transition state and/or intermediate (Maroney and Ciurli, 2014).

Another interesting structural determination was that of the β-mercaptoethanol-inhibited urease from *Bacillus pasteurii*, solved at a resolution of 1.65 Å using synchrotron X-ray cryogenic diffraction data (Benini *et al.,* 1998). The results clearly confirm the above discussed spectroscopic and magnetic data. In this case, the two H_2O molecules, bonded to each of the Ni(II) ions as well as the bridging OH^- anion (cf. again Fig. 1) found in the active site of native ureases are absent. The Ni(II) ions are now bridged by the sulphur atom of the inhibitor which also chelates one of the Ni(II) by the OH functionality. The coordination of Ni(1) is distorted square-pyramidal, with the equatorial plane constituted by the S and O atoms of β-mercaptoethanol, by the O-atom of the bridging carbamylated lysine and by the N-atom of one of the histidine residues, the apical ligand is the N-atom of the other histidine. On the other hand, the coordination geometry of Ni(2) is best described as distorted trigonal-bipyramidal, with the equatorial plane constituted by the bridging S-atom of β-mercaptoethanol and by the two N-atoms from the histidine residues, the opposite apical ligands being the O-atom of the bridging carbamylated lysine and the aspartato O-atom. The equatorial planes of the square and trigonal pyramids are joined through the bridging S-atom of β-mercaptoethanol, with a dihedral angle between the two average equatorial planes of 63°. Interestingly, another molecule of inhibitor forms a mixed disulfide with a cysteine residue, thus sealing the entrance to the active site cavity by steric hindrance.

It must also be remarked that the complete three-dimensional structure of jack bean urease, at 2.05 Å, has also been recently reported, confirming again most of all the above made general comments on this enzyme (Balasubramanian and Ponnuraj, 2010).

Mechanism of action of the enzyme

The reaction mechanism of enzyme-catalyzed urea decomposition has been the subject of discussions along the last century and still continues to be of interest. Three main reaction pathways have been proposed:

1. The carbamic acid (hydrolytic) mechanism which involves a carbamic acid (or carbamate) intermediate.
2. A carbon dioxide mechanism, in which urea is directly hydrolyzed to CO_2 and NH_3.
3. A cyanic acid (elimination) mechanism, associated with the formation of cyanate intermediates.

These three possible mechanisms are schematized in Fig. 2 (Estiú and Merz, 2007). The so far performed biochemical studies show that there is no agreement on whether the reaction involves a carbamic acid or a cyanate intermediate or if it proceeds via direct decomposition.

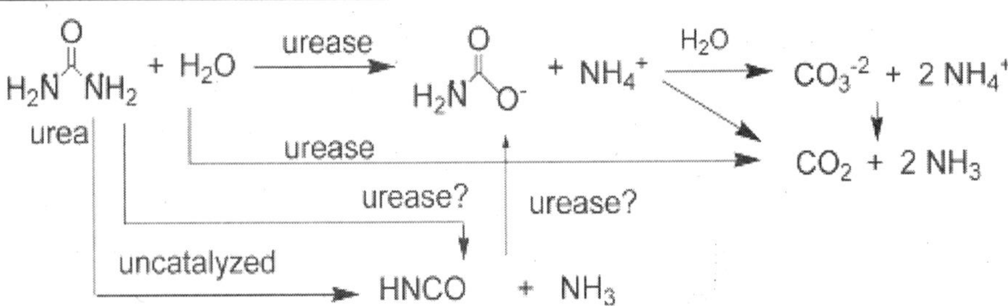

Fig. 2. Simplified overview of the different pathways proposed for catalyzed and uncatalized urea decomposition.

It is usually assumed that in the active site of urease, urea binds to the more electrophilic Ni(1) ion with the O-atom of its carbonyl group, owing to which the carbonyl carbon becomes more electrophilic, hence more susceptible to nucleophilic attack. Upon replacing the two coordinated H_2O molecules, urea is further bound to Ni(2) through the nitrogen of one of its amino groups, making its binding overall bidentate (Benini *et al.,* 1999; Muralidhar, 2014). This binding is believed to facilitate the nucleophilic attack of water on the carbonyl carbon, resulting in the formation of a tetrahedral intermediate from which NH_3 and carbamate are released. This nucleophilic attack may be performed by the bridging hydroxide (Benini *et al.,* 1999). Another possibility is that this attack involves directly the His320 residue located in the mobile flap of the active site (Karplus *et al.,* 1997; Muralidhar, 2014).

Related to enzymatically catalyzed reaction, several questions which remain open are mainly related to the coordination mode of urea to the Ni centers and to the potential protonation states of His219 and His320. With the help of molecular dynamics simulation, the critical role played by residues of the mobile flap, namely His320, Cys319, Ala363, His219, Asp221 and Asp 360, as well as the relevance of the OH bridging group, become apparent (Estiu and Merz, 2006).

Urease inhibitors

The increase of pH arising from urea hydrolysis causes a broad range of deleterious effects. Bacterial ureases are involved in many diseases and presence of urease activity in soils is also a major concern for agriculture. Excessive levels of urease in soil degrade urea used as a fertilizer too rapidly and result in phytopathic effects and loss of volatilized ammonia. Hence, because of their negative side effects in agriculture and health, the introduction of urease inhibitors has been proposed as an alternative to control unwanted ureolytic activity.

In agriculture, use of urease inhibitors in conjunction with urea fertilizers can lead to a reduction in environmental pollution and to an enhancement in the efficiency of nitrogen uptake by plants.

On the other hand, urease inhibitors are also of great pharmacological interest (Amtul *et al.,* 2002); Muralidhar *et al.,* 2014).

An analysis of the so far prepared and investigated inhibitors show that they can be divided into two major classes, based on their binding modes (Amtul, *et al.,* 2002; Upadhyay, 2012):

(a) Inhibitors which bind in a substrate or active-site directed mode.

(b) Inhibitors which bind in a non-substrate like or mechanism-based directed mode.

The active site directed inhibitors are the current chemotypes and their analogs. They share a common feature: each inhibitor bridges the two paramagnetic Ni(II) ions in the active site of the enzyme. Thiourea and hydroxyurea (Fig. 3) are the main examples of these types of inhibitors. Hydroxamic acid derivatives and phosphazenes are also substrate-like inhibitors.

Non-substrate-like inhibitors do not have any structural similarity with the enzyme substrate (urea), rather they are designed to interfere with the enzymatic catalysis mechanism leading to enzyme inactivation. Phosphorodiamidate and imidazoles are typical examples of these types of inhibitors.

Some heavy metals are also potent urease inhibitors. In particular, metals forming insoluble sulfides are particularly potent inhibitors. Ag(I) and Hg(II) inhibited urease completely in the range of 10^{-6}. Also fluoride, boric acid, boronic acids and phosphate are strong inhibitors (Upadhyay, 2012). Most recently it has been found that amide and ester derivatives of phosphoric and thiophosphoric acid are also very effective inhibitors (Maroney and Ciurli, 2014).

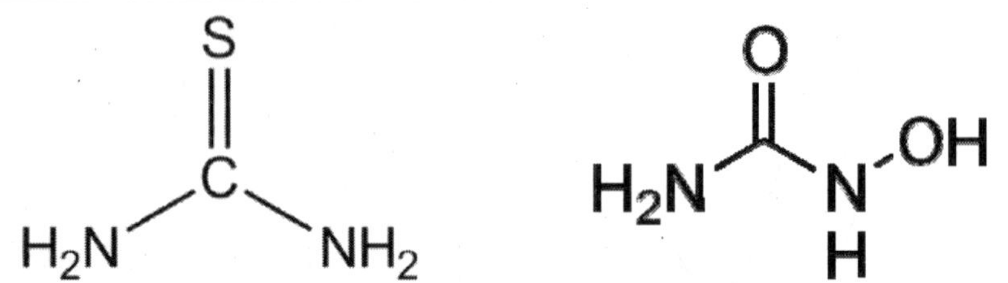

Fig. 3. Schematic structures of thiourea (left) and hydroxyurea (right).

In this context it is also interesting to mention that certain Bi(III) compounds are also very efficient urease inhibitors. Bismuth compounds are widely used for the treatment if peptic ulcers and *Helicobacter pylori* infections (Baran and Tobón Zapata, 1997) and it has been shown that the complexes Bi(EDTA) and $Bi(Cys)_3$ and the-ulcer drug ranitidine-bismuth-citrate are good inhibitors, opening new views in relation to the action of Bi(III) pharmaceuticals for the treatment of gastric disorders (Zhang *et al.,* 2006).

GLYOXALASE I

Glyoxilase I (Glo I, EC 4.4.1.5) is a widely distributed enzyme found in bacteria, fungi, protoctista, plants and animals (Maroney and Ciurli, 2014) that catalyzed the conversion of a hemithioacetal to a 2-hydroxythioester (cf. Fig. 4).

The hemithioacetal is derived from a spontaneous reaction of of an α-ketoaldehyde and a thiol. The most important α-ketoaldehyde is methyl-glyoxal, which is a naturally occurring toxic metabolite that accumulates when Glo I is supressed or when GSH (γ-glutamyl-L-cysteinylglycine) is depleted, leading to glycation products. Methylglyoxal is capable of forming adducts with nucleic acids and amino acids leading to protein, DNA and RNA modifications and a number of cytotoxic effects. In these ways, Glo I has been linked to oxidative stress and to aging processes.

Glo I is the first of two enzymes in the glyoxalase system. The second enzyme in this pathway, Glo II (EC 3.1.2.6), hydrolizes the thioester produced by Glo I to generate the corresponding α-hydroxy acid (cf. again Fig. 4). Thus, the glyoxalate system is accepted as an important pathway for detoxifying and controlling the level of methylglyoxal in cells by converting it to D-lactate, as opposed to L-lactate arising from glycolysis (Maroney and Ciurli, 2014).

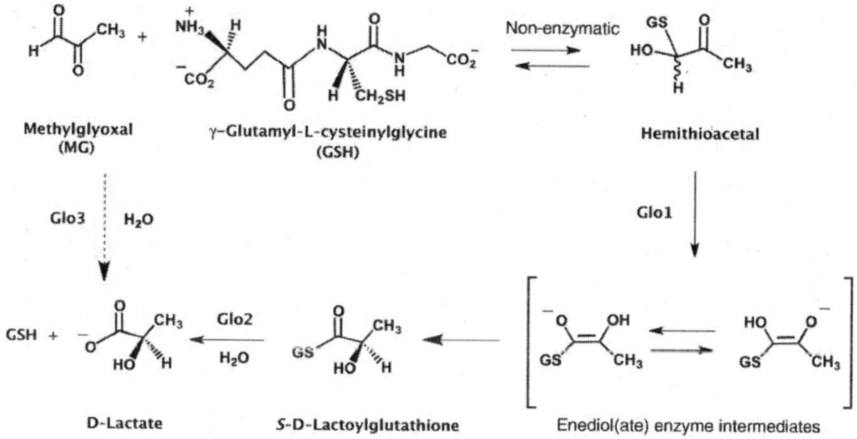

Fig. 4. Transformation of methylglyoxal to D- lactate catalyzed by the glyoxalase system (Glo 1 and Glo 2).

Early work on the bioinorganic chemistry of of Glo I enzymes using metal activation assays of apoenzyme showed that the enzymes from mammals, yeast, and *Pseudomonas putida* were activated by Zn(II) as well as by other divalent metals. Based on these precedents, all Glo I enzymes were assumed to be Zn(II) metalloenzymes until the discovery that *E. coli* Glo I was not active with Zn(II) bound, but was maximally activated by Ni(II) ions, and to a lesser extent by Co(II), Mn(II) and Cd(II). This report was followed by others resulting from studies in Gram-negative bacteria, Gram-positive bacteria and the trypanosome protozoans *Leishmania major* and *Trypanosoma cruzi* that found metal activation characteristics similar to that of *E. coli* enzyme. Consequently, two classes of Glo I have emerged based on metal activation: class I, which can be activated by Zn(II) and class II, which cannot be activated by Zn(II) and is maximaly activated by Ni(II) ions (Maroney and Ciurli, 2014).

The origin of the active site metal preferences in the two classes of enzyme derives from subtle structural differences between the primary structures and the differences between the dimmers formed in both structures.

In the Ni(II) complex of *E. coli* Glo I, the nickel is six-coordinate with octahedral geometry, bonded to the N-atoms of two histidine residues, two O-atoms from carboxylate groups of glutamate moieties and two water molecules. The same overall structure was also observed for the metal sites in the less active Co(II) complex and the weakly active Cd(II) complex (He *et al.,* 2000; Maroney and Ciurli, 2014). However, the Zn(II), which is inactive, forms a five coordinate complex with trigonal-bipyramidal geometry, missing one aqua ligand (He *et al.,* 2000; Maroney and Ciurli, 2014).

Glyoxalase I in plants

In general, plant glyoxalases remain poorly characterized from the biochemical point of view and also regarding their possible molecular functions (Deswal *et al.*, 1993).

The first report of isolation and purification of a plant glyoxalase I was that of *Brassica juncea*. Although no presence of any transition metal was reported, it was found that Mg(II) strongly stimulate the enzyme activity (Deswal and Sopory, 1991).

In the case of rice (*Oryza sativa*) a Ni-dependent glyoxalase I (named OsGLYI-11.2) was found and characterized as a monomeric enzyme, possessing a single Ni(II) coordination site, despite containing two Glo I domains. It was found that the expression of this enzyme is highly substrate inducible, suggesting an important mode of regulation for its cellular levels (Mustafiz *et al.*, 2014).

It is also interesting to mention that transgenic tobacco plants (*Nicotiana tabacum*) overexpressing glyoxalase pathway enzymes resist an increase in the levels of methylglyoxal that increases under salinity stress. These plants suffered minimal salinity stress induced oxidative damage and maintained a high reduced glutathione (GSH) content (Yadav *et al.*, 2005a). These results clearly demonstrate the importance of the glyoxalase system under stress conditions (cf. also Yadav *et al.*, 2005b) .

Also in the case of tomato (*Lycopersicon esculentum*), glyoxalase I was shown to be upregulated under salt and water stresses (Espartero *et al.*, 1995).

Most recently three active glyoxalase I enzymes were found in *Arabidopsis thaliana* (AtGLY12, AtGLY13 and AtGLY16) belonging to different metal activation classes coexisting in this plant. These three new enzymes have been characterized in terms of their activity, metal dependency, kinetic parameters and their role in conferring tolerance to multiple abiotic stresses in *E. coli* and yeast. AtGLY12 was found to be Zn(II) dependent, whereas the other two were Ni(II) dependent. This is the first report of a higher eukaryotic species having multiple active glyoxalase I enzymes belonging to different metal activation classes (Jain *et al.*, 2016).

NICKEL HYPERACCUMULATING PLANTS

Some plants have evolved the ability to not only to survive in metal-rich soils but to sequester and store exceptionally high levels of metals and metalloids in their tissues. These plants are called hyperacumulators (Brooks, 1994; Callahan *et al.*, 2006). Hyperaccumulation has been recognized as an extreme physiological response in heavy metal tolerance (Baker, 1981). This trait allows these plants to respond positively to the presence of high concentrations of metal ions in soils. When grown in soils with normal metal concentrations, hyperaccumulator species sequester much higher metal concentrations in their shoots than do normal plants. However, when transplanted to soils containing low levels of these ions, hyperaccumulators display normal growth patterns, indicating that there is no essential physiological requirement for high metal concentrations (Reeves and Baker, 1984).

The most extreme example of hyperaccumulation reported to date is the Ni-hyperaccumulating tree from New Caledonia, *Sebertia acuminata* Pierre ex Baillon. The nickel concentration in the latex of this plant was determined to be ~25% of the dry mass and this is the highest concentration of nickel so far found in any biological fluid (Jaffré *et al.*, 1976; Callahan *et al.*, 2006). The mass of nickel contained in one single tree was estimated to be 37 kg (Sagner *et al.*, 1998).

Nickel constitutes by far the most often and commonly hyperaccumulated metallic element. These hyperaccumulating plants are distributed mainly in eight major geographic regions: New Caledonia, Australia, Southern Europe and Asia Minor, Malaysia, Cuba, Western United States, Central Brazil and Southern Africa (Brooks, 1994). The greatest concentration and diversity of these plants are found in New Caledonia. About 50 hyperaccumulators from 14 genera and eight different families have been reported from this isolated Pacific island. The plants belong to the following genera and families: *Agatea* (Violaceae), *Argophyllum* (Escalloniaceae), *Casearia* (Flacourtiaceae), *Cleidion* (Euphorbiaceae), *Geissois* (Cunoniaceae), *Homalium* (Flacourtiaceae), *Hybanthus* (Violaceae), *Lasiochlamys* (Flacourtiaceae), *Oncontheca* (Oconthecaceae), *Pancheria* (Cunoniaceae), *Phyllantus* (Euphorbiaceae), *Psychotria* (Rubiaceae), *Sebertia* (Sapotaceae) and *Xylosma* (Flacourtiaceae) (Brooks, 1994).

On the other hand, the genus *Alyssum* contains so far the greatest number of individual Ni-hyperaccumulating plants. Some 48 members of section Odontarrhena have this ability and Anatolia (Asia Minor) is the centre of maximum multiplicity and diversity (Brooks, 1994).

One very interesting problem in relation to metal hyperaccumulation is the form in which the metallic species are complexed within the plant. It is clear that only a small fraction of the metal content in plants is present in the form of free aqua ions, i.e. for example $[M(H_2O)_6]^{n+}$. It is assumed that most metallic cations are bound to low molecular mass ligands or to proteins (Callahan *et al.*, 2006). Plants produce a number of possible ligands, including organic acids, amino acids, peptides and proteins.

Nitrogen donor ligands are assumed to play an important role in hyperaccumulation. In particular, the amino acid histidine is considered of central importance in relation to nickel hyperaccumulation (Krämer *et al.*, 1996). Histidine (Fig. 5) can act as a tridentate ligand via its carboxylato, amino and imidazole functions but it is a very versatile ligand with a number of known coordination modes (Szabó-Plánka *et al.*, 2000). On the other hand, stability constants of Ni(II)/histidine complexes are higher than those of similar complexes of Fe(II), Co(II) and Zn(II); only Cu(II) binds stronger to histidine than Ni(II) (Martell and Smith, 1974).

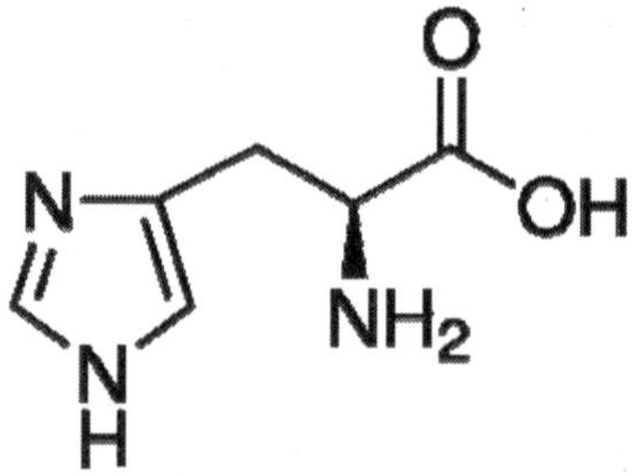

Fig. 5. Schematic structure of histidine.

In the synthetic complex [Ni(hys)$_2$]·H_2O the Ni(II) ions lie on a two-fold axis and are octahedrally coordinated by the amino nitrogen, the imidazole nitrogen and a carboxyl O-atom of each histidine group at 2.11, 2.09 and 2.11 Å, respectively (Fraser and Harding, 1967).

Different studies with an important number of plant species (hyperaccumulators and non-hyperaccumulators) clearly suggested that histidine is involved in Ni-hyperaccumulation. However, its detailed role remains to be defined (Callahan *et al.,* 2006).

It is also well-known that some simple carboxylic acids are present at relatively high concentrations in the cell vacuoles of photosynthetic tissues, including citric and isocitric acid as well as tartaric, malic, oxalic, malonic and aconitic acids. Many studies have shown that these acids may be play a role in hyperaccumulation, as often an increase in the levels of some of these acids were observed in response to increasing metal concentrations in the soil (Callahan *et al.,* 2006).

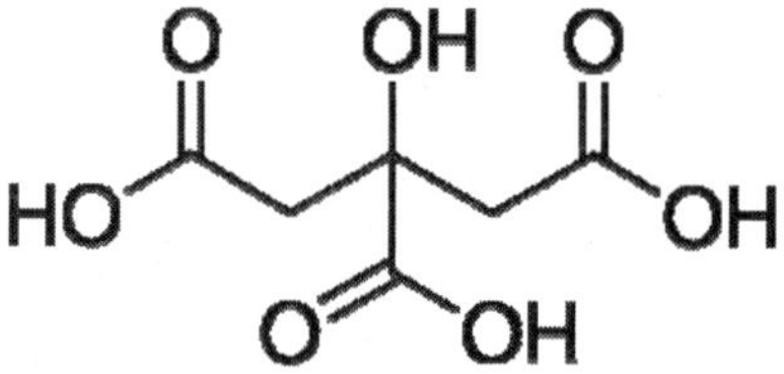

Fig. 6. Schematic structure of citric acid.

In the particular case of nickel, it has been found that a clear correlation between Ni(II) and citric acid (Fig. 6) exists in the case of the strongest nickel hyperaccumulating plants of New Caledonia. Citric acid levels of 1000-4000 μg/g are characteristic of those New Caledonian species which usually have less than 100 μg/g nickel. Where nickel is present at 100-1000 μg/g, elevated citric acid concentrations (4000-8000 μg/g) are found, and the Ni-hyperaccumulators show citric acid levels of 8000-20000 μg/g (Lee *et al.*, 1978). In the extreme case of the *Sebertia acuminata* latex, one of the most remarkable of all known biological fluids, removal of water gives a solid typically containing 16.7% nickel and 25% citrate. In the latex, nickel is present both as the $[Ni(H_2O)_6]^{2+}$ cation and as an anionic citrato-nickelate complex (Lee *et al.*, 1977, 1978).

A synthetic dimeric Ni(II)/citrate complex of composition $K_2[Ni(C_6H_5O_7)(H_2O)_2]_2 \cdot 4H_2O$ has been prepared. It exists as centrosymmetric dimmers, with each triionised citrate ion bound as a tridentate ligand trough the hydroxyl oxygen, one O-atom of the central carboxylate group and another O-atom from an end carboxylate. The other end carboxylate bridges to the second Ni(II) ion. Octahedral coordination of each Ni(II) cation is completed by two water molecules (Baker *et al.*, 1983). This complex was also investigated by vibrational (IR and Raman) spectroscopy and by magnetic susceptibility measurements (Bovi Mitre *et al.*, 2000).

Interestingly, in the case of the hyperaccumulator *Thlaspi goesingense*, citrate, together with histidine, were found as ligands for nickel in leaves (Krämer *et al.*, 2000), showing again the importance of these two organic ligands in the accumulation of nickel.

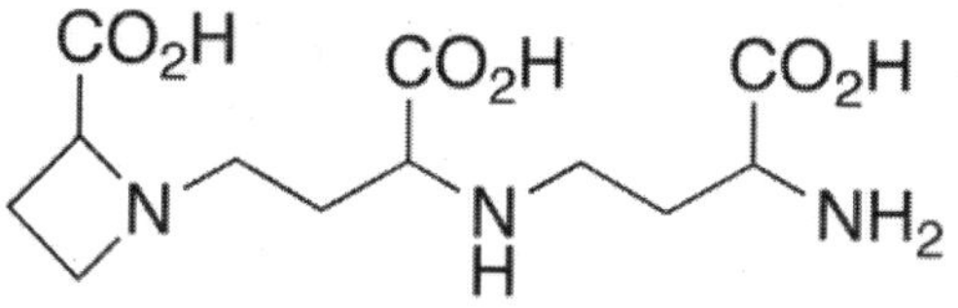

Fig. 7. Schematic structure of nicotianamine.

Another interesting organic molecule present in plants is nicotianamine (Fig. 7), the precursor of phytosiderophores and strongly related to iron metabolism in plants (Baran, 2013). Notwithstanding, the stability constants of M(II)/nicotianamine complexes follows the well-known Irving-Williams series, i.e. Mn(II) < Fe(II) < Co(II) < Ni(II) < Cu(II) > Zn (II), whereas the Fe(III) complex shows a much higher value (Baran, 2013).

In the case of nickel, it has been shown that an increase in the biosynthesis of this molecule enhance tolerance of this metal in plants (Kim *et al.*, 2005). On the other hand, nickel complexes with nicotianamine were identified by mass spectrometry in extracts of the Ni-hyperaccumulating plant *Thlaspi caerulescens* (Vacchina *et al.*, 2003). In another study, it was also found that in the case of *Sebertia acuminata*, apart from the well-established citrate-nickelato complex (see above) a small fraction of nickel (~0.3%) is also complexed by nicotianamine (Schaumlöffel *et al.*, 2003).

Finally, it must be mentioned that on the basis of all the so far accumulated evidences, phytochelatins (Baran, 2012) appear not involved in metal hyperaccumulation, whereas a possible role of metallothioneins is also not clear (Callahan *et al.,* 2006).

SOME COMMENTS ON NICKEL TOXICITY IN PLANTS

Although nickel is an essential micronutrient for plants, the required amount for this element is relatively low. Hence, with the continuous increase of the levels of environmental Ni pollution (Harasim and Filipek, 2015) it becomes essential to understand its toxic effects on plants and other biological systems.

Adverse effects derived from nickel toxicity are observed in the form of growth and development inhibition, induction of leaf chlorosis, necrosis and wilting and damage in the photosynthetic apparatus (Chen *et al.,* 2009; Bhalerao *et al.,* 2015). As shown by different recent studies the toxicity is likely to be caused by indirect mechanisms, basically: (a) interference with other essential metal ions and (b) induction of oxidative stress (Chen *et al.,* 2009).

As Ni has some similar characteristics to other biometals such as Mg, Ca, Mn, Fe, Cu and Zn, it can compete with these metals in absorption processes. As a result of this competition, Ni at high concentrations may inhibit the absorption of these metals, decrease their concentration and even lead to their deficiency in plants. Subsequently, this may affect important physiological processes, and ultimately result in toxic effects.

On the other hand, many important plant enzymes are metalloenzymes depending on Fe, Mn, Cu or Zn (Hay, 1994). Since excess Ni decrease the content of these metals in plant tissues, one can speculate that different physiological processes depending on these metals may also be affected. Besides, it has been shown that excessive Ni leads to significant increase in the concentration of OH radicals, superoxide anions and H_2O_2. Since Ni is not a redox-active metal, it cannot directly generate these reactive oxygen species (ROS). However, as it interferes indirectly with a number of antioxidant enzymes (superoxide dismutases, catalases, glutathione peroxidase, etc.) the plant's capability to scavenge ROS is dramatically reduced, leading to ROS accumulation and finally oxidative stress in plants. Additionally, Ni induced depletion of low molecular weight proteins, such as GSH, may also contribute to the induction of oxidative stress in plants (Chen *et al.,* 2009).

Acknowledgement

The author is an Emeritus Professor from the Universidad Nacional de La Plata and the continuous support from this University is gratefully acknowledged.

REFERENCES

Alagna, L, SS Hasnain, B Piggott and DJ Williams. 1984. The nickel ion environment in jack bean urease. *Biochem. J.* **220:** 591-5.

Amtul, Z, A-ur-Rahman, RA Siddiquit and MI Choudhari. 2002. Chemistry and mechanism of urease inhibition. *Curr. Med. Chem.* **9**, 1323-48.

Andrews, RK, RL Blakeley and B Zerner. 1988a. Nickel in proteins and enzymes. In: *Metal Ions in Biological Systems* (eds Sigel, H and A Sigel), vol. 23, pp. 165-284, Marcel Dekker, New York.

Andrews, RK, RL Blakeley and B. Zerner. 1988b. Urease - A Ni(II) metalloenzyme. In: *The Bioinorganic Chemistry of Nickel* (ed Lancaster, Jr., JR), pp. 141-65, VCH Publishers, New York.

Baker, AJM. 1981. Accumulators and excluders-strategies in the response of plants to heavy metals. *J. Plant Nutrit.* 3: 643-54.

Baker, EM, HM Baker, BF Anderson and RD Reeves. 1983. Chelation of nickel(II) by citrate. The crystal structure of a nickel-citrate complex, $K_2[Ni(C_6H_5O_7)(H_2O)_2]_2 \cdot 4H_2O$. *Inorg. Chim. Acta* **78:** 281-5.

Baran, EJ. 2004. Trace elements supplementation: Recent advances and perspectives. *Mini Rev. Med. Chem.* **4:** 1-9.

Baran, EJ. 2012. Phytochelatins: Natural chelating agents involved in plant protection. In: *Advances in Plant Physiology* (Ed. Hemantaranjan, A), Vol.13, pp. 3889-414, Scientific Publishers, Jodhpur.

Baran, EJ. 2013. Phytosiderophores and related systems: Metal uptake by plants. In: *Advances in Plant Physiology* (Ed. Hemantaranjan, A), Vol.14, pp. 1-27, Scientific Publishers, Jodhpur.

Baran, EJ and GE Tobón Zapata. 1997. La Nueva Farmacoterapia Inorgánica. XVI. Compuestos de bismuto. *Acta Farm. Bonaerense* **14:** 133-8.

Balasubramanian, A and K Ponnuraj. 2010. Crystal structure of the first plant urease from jack bean: 83 years of journey from its first crystal to molecular structure. *J. Mol. Biol.* **400:** 274-83.

Benini, S, WR Rypniewski, KS Wilson, S. Ciurli and S Mangani. 1998. The complex of *Bacillus pasteurii* urease with β-mercaptoethanol from X-ray data at 1.65 Å resolution. *J. Biol. Inorg. Chem.* **3:** 268-73.

Benini, S, WR Rypniewsky, KS Wilson, S Miletti, S. Ciurli and SA Mangani. 1999. A new proposal for urease mechanism based on the crystal structures of native and inhibited enzyme from *Bacillus pasteurii*: why urea hydrolysis costs two nickels. *Structure* **7:** 205-16.

Bhalerao, SA, AS Sharma and AC Poojari. 2015. Toxicity of nickel in plants. *Int. J. Pure Appl. Biosci.* **3:** 345-55.

Blakeley, RT, Dixon, NE and Zerner, B. 1983. Jack bean urease. VII. Light scattering and nickel(II) spectrum. Thiolate → nickel(II) charge-transfer peaks in the spectrum of the β-mercaptoethanol-urease complex. *Biochim. Biophys. Acta* **744:** 219-29.

Brooks, RR. 1994. Plant that hypperaccumulate heavy metals. In: *Plants and the Chemical Elements* (ed Farago, ME), pp.87-105, Verlag Chemie, Weinheim.

Bovi Mitre MG, NR Wierna, CC Wagner and EJ Baran. 2000. Spectroscopic and magnetic properties of a Ni(II) complex with citric acid. *Biol. Trace Elem. Res.* **76:** 183-90.

Brown, PH, RM Welch and EE Cary. 1987. Ni: A micronutrient essential for higher plants. *Plant Physiol.* **85:** 801-3.

Callahan, DL, AJM Baker, SD Kolev and AG Wedd. 2006. Metal ion ligands in hyperaccumulating plants. *J. Biol. Inorg. Chem.* **11:** 2-12.

Cataldo, D, TR Garland and RE Wildung. 1978. Nickel in plants. I. Uptake kinetics using intact soybean seedlings. *Plant Physiol.* **62:** 563-5.

Chen, C, D Huang and J Liu. 2009. Functions and toxicity of nickel in plants: Recent advances and future prospects. *Clean* **37:** 304-13.

Ciurli, S and S. Mangani. 2001. Nickel-containing enzymes. In: *Handbook of Metalloproteins* (eds Bertini, I, A Sigel and H Sigel), pp. 669-707, Marcel Dekker, New York.

Clark, PA and DE Wilcox. 1989. Magnetic properties of the nickel enzyme urease, nickel substituted carboxypeptidase A, and nickel-substituted carbonic anhydrase. *Inorg. Chem.* **28:** 1326-33.

Deswal, R and SK Sopory. 1991. Purification and partial characterization of glyoxalase I from a higher plant *Brassica juncea*. *FEBS Lett.* **282:** 277-80.

Deswal, R., TN Chakaravarty and SK Sopory. 1993. The glyoxalase system in higher plants: Regulation in growth and differentiation. *Biochem. Soc. Trans.* **21:** 527-30.

Dixon, NE, C Gazzola, JJ Waters, RL Blakeley and B Zerner. 1975a. Inhibition of jack bean urease (EC 3.5.1.5) by acetohydroxamic acid and by phosphoramidate. Equivalent weight for urease. *J. Am. Chem. Soc.* **97:** 4130-1.

Dixon, NE, C Gazzola, RL Blakeley and B. Zerner. 1975b. Jack bean urease (EC 3.5.1.5). A metalloenzyme. A simple biological role for nickel? *J. Am. Chem. Soc.* **97:** 4131-2.

Dixon, NE, RL Blakeley and B. Zerner. 1980. Jack bean urease (EC 3.5.1.5). III. The involvement of active-site nickel ion in inhibition by β- mercaptoethanol, phosphoramidate and fluoride. *Canad. J. Biochem.* **58:** 481-8.

Ermler, U, W Grabarse, S Shima, M Goubeaud and RK Thauer. 1998. Active sites of transition-metal enzymes with a focus on nickel. *Curr. Op. Struct. Biol.* **8:** 749-58.

Eskew, DL, RM Welch and EE Cary. 1983. Nickel: An essential micronutrient for legumes and possibly all higher plants. *Science* **222:** 621-3.

Eskew, DL, RM Welch and WA Norvell. 1984. Ni in higher plants: further evidence for an essential role. *Plant Physiol.* **76:** 691-3.

Espartero, J, I Sánchez-Aguayo and JM Pardo. 1995. Molecular characterization of glyoxalase I from a higher plant; upregulation by stress. *Plant Mol. Biol.* **29:** 1223-33.

Estiu, G and KM Merz, Jr. 2006. Catalyzed decomposition of urea. Molecular dynamics simulation of the binding of urea to urease. *Biochemistry* **45:** 4429-43.

Estiu, G and KM Merz. Jr. 2007. Competitive hydrolytic and elimination mechanisms in the urease catalyzed decomposition of urea. *J. Phys. Chem.* **B 111**, 10263-74.

Farago, ME and MM Cole. 1988. Nickel in plants. In: *Metal Ions in Biological Systems* (eds Sigel, H and A Sigel), vol. 23, pp. 47-90, Marcel Dekker, New York.

Fraser, KA and MM Harding. 1967. The crystal and molecular structure of bis(histidino)nickel(II) monohydrate. *J. Chem. Soc.* **A1967:** 415-20.

Freyermuth SK, M Becanamwo and JC Polacco. 2000. The soybean Eu3 gene encodes an Ni-binding protein necessary for urease activity. *Plant J.* **21:** 53-60.

Finegan, MG, AT Kowal, MT Werth, PA Clark, DE Wilcox and MK Johnson. 1991. Variable-temperature magnetic circular dichroism spectroscopy as a probe of the electronic and magnertic properties of nickel jack bean urease. *J. Am. Chem. Soc.* **113:** 4030-2.

Frieden, E. 1984. A survey of the essential biochemical elements. In: *Biochemistry of the Essential Ultratrace Element* (ed Frieden, E) pp. 1-15, Plenum Press, New York.

Hänsch, R and RR Mendel. 2009. Physiological functions for mineral micronutrients (Cu, Zn, Mn, Fe, Ni, Mo, B, Cl). *Curr. Op. Plant Biol.* **12:** 259-66.

Harasim, P and T Filipek. 2015. Nickel in the environment. *J. Elem.* **20:** 525-34.

Hasnain, SS and B Piggott. 1983. An EXAFS study of jack bean urease, a nickel metalloenzyme. *Biochem. Biophys. Res. Commun.* **112:** 279-83.

Hausinger, RP. 1997. Metallocenter assembly in nickel-containing enzymes. *J. Biol. Inorg. Chem.* **2:** 279-86.

Hay, RW.1994. Plant metalloenzymes. In: *Plants and the Chemical Elements* (ed Farago, ME), pp.107-48, Verlag Chemie, Weinheim.

Haydon, MJ and ChS Cobbett. 2007. Transporters of ligands for essential metal ions in plants. *New Phytologist* **174:** 499-506.

He, MM, SL Clugston, JF Honek and BW Matthews. 2000. Determination of the structure of *Escherichia coli* Glyoxalase I suggests a structural basis for differential metal activation. *Biochemistry* **39:** 8719-27.

Jabri, E, MB Carr, RP Hausinger and PA Karplus. 1995. The crystal structure of urease from *Klebsiella aerogenes*. *Science* **268:** 998-1004.

Jaffré, T, RR Brooks, J Lee and RD Reeves. 1976. *Sebertia acuminata*: A hyperaccumultor of nickel from New Caledonia. *Science* **193:** 579-80.

Jain, M, R Batth, S Kumari and A Mustafiz. 2016. *Arabidopsis thaliana* contains both Ni^{2+} and Zn^{2+} dependent glyoxalase I enzymes and ectopic expression of the latter contributes more towards abiotic stress in *E. coli*. *PLOS ONE*, in the press.

Kaim, W and B Schwederski. 1991. *Bioanorganische Chemie*, B.G. Teubner, Stuttgart.

Karplus, PA, MA Pearson and RP Hausinger. 1997. 70 Years of crystalline urease: What have we learned? *Acc. Chem. Res.* **30:** 330-337.

Krämer, U, JD Cotter-Howells, JM Charnock, AJM Baker and JAC Smith. 1996. Free histidine as a metal chelator in plants that accumulate nickel. *Nature* **379:** 635-8.

Krämer, U, IJ Pickering, RC Prince, I Raskin and DE Salt. 2000. Subcellular localization and speciation of nickel in hyperaccumulator and non-accumulator *Thlapsi* species. *Plant Physiol.* **122:** 1343-53.

Lancaster, Jr., JR (Ed.). 1988. *The Bioinorganic Chemistry of Nickel*, VCH Publishers, New York.

Lee, J, RD Reeves, RR Brooks and T Jaffré. 1977. Isolation and identification of a citrate complex of nickel from nickel-accumulating plants. *Phytochemistry* **16:** 1503-5.

Lee, J, RD Reeves, RR Brooks and T Jaffré. 1978. The relation between nickel and citric acid in some nickel-accumulating plants. *Phytochemistry* **17:** 1033-5.

Lippard, SJ. 1995. At last - the crystal structure of urease. *Science* **268:** 996- 7.

López, MA and S. Magnitskiy. 2011. Nickel: The last of the essential micro-nutrients. *Agron. Colomb.* **29:** 49-56.

Maroney, MJ. 1989. Structure/function relationships in nickel metallobioche-mistry. *Curr. Op. Chem. Biol.* **3:** 188-99.

Maroney, MJ and S Ciurli. 2014. Nonredox nickel enzymes. *Chem. Rev.* **114:** 4206-28.

Martell AE and RM Smith. 1974. *Critical Stability Constants*, Plenum Press, New York.

Mobley, HL, MD Island and RP Hausinger. 1995. Molecular biology of microbial ureases. *Microbiol. Rev.* **59:** 451-80.

Muralidhar, K, J Khurana and V Sharma. 2014. Urease (EC 3.5.1.5): A perspective. *World J. Pharm. Pharmaceut. Sci.* **3:** 923-48.

Mustafiz, A, A Ghosh, AK Tripathi, C. Kaur, AK Ganguly, NS Bhavesh, JK Tripathi, A Pareek, AK Sopory and SL Singla-Pareek. 2014. A unique Ni^{2+} - dependent and methylglyoxal-inducible rice glyoxalase I possesses a single active site and functions in abiotic stress response. *Plant J.* **78:** 951-63.

Park, IS, MB Carr and RP Hausinger. 1994. *In vitro* activation of urease apoprotein and role of UreD as a chaperone required for nickel metallocenter assembly. *Proc. Natl. Acad. Sci. USA* **91:** 3233-7.

Reeves, RD and AJM Baker. 1984. Studies on metal uptake by plants from serpentine and non-serpentine populations of *Thlaspi Goesingense* Hálácsy (Cruciferae). *New Phytol.* **98:** 191-204.

Roat-Malone, RM. 2002. *Bioinorganic Chemistry. A Short Course*, J. Wiley, Hoboken, NJ.

Rosenzweig, A. 2002. Metallochaperones: Bind and deliver, *Chem. Biol.* **9:** 673-677.

Sagner, S, R Kneer, G Wanner, JP Cosson, B Deus-Neumann and MH Zenk.1998. Hyperaccumulation, complexation and distribution of nickel in *Sebertia acuminata. Phytochemistry* **47:** 339-47.

Schaumlöffel, D, L Ouerdane, B Bouyssiere and R Lobinski. 2003. Speciation analysis of nickel in the latex of a hyperaccumulating tree *Sebertia acuminata* by HPLC and CZE with ICP MS and electrospray MS-MS detection. *J. Anal. Atom. Spectrom.* **18:** 120-7.

Seregin, IV and AD Kozhevnikova. 2006. Physiological role of nickel and its toxic effects on higher plants. *Russ. J. Plant Physiol.* **53:** 257-77.

Sigel, A, H Sigel and RKO Sigel (Eds). 2007. *Metal Ions in Life Sciences*, Vol. 2: Nickel and its surprising impact on Nature. J. Wiley, Chichester.

Sigel, H and A. Sigel (Eds). 1988. *Metal Ions in Biological Systems*, Vol. 23: Nickel and its role in biology, Marcel Dekker, New York.

Sirko, A and R Brodzik. 2000. Plant ureases: Roles and regulation. *Acta Biochim. Polon.* **47:** 1189-95.

Szabó-Plánka T, A Rockenbauer, L Korecz and D Nagy. 2000. An electron spin resonance study of coordination modes in the copper(II)-histamine and copper(II)-L-histidine systems in fluid aqueous solution. *Polyhedron* **19:** 1123-31.

Upadhyay, LSB. 2012. Urease inhibitors: A review. *Indian J. Biotechnol.* **11**, 381-8.

Vacchina, V, S Mari, P Czernic, L Marqués, K Pianelli, D Schaumlöffel, M Lebrun and R Lobinski. 2003. Speciation of nickel in a hyperaccumulating plant by high-performance liquid chromatography-inductively coupled plasma mass spectrometry and electrospray MS/MS assisted by cloning using yeast complementation. *Anal. Chem.* **75:** 2740-5.

Wang, S, MH Lee, RP Hausinger, PA Clark, DE Wilcox and RA Scott. 1994. Structure of the dinuclear active site of urease. X-ray absorption spectroscopic study of native and 2-mercaptoethanol-inhibited bacterial and plant enzymes. *Inorg. Chem.* **33:** 1589-93.

Yadav, SK, SL Singla-Pareek, MK Reddy and SK Sopory. 2005a. Transgenic tobacco plants overexpressing glyoxalase enzymes resist an increase in methylglyoxal and maintain higher reduced glutathione levels under salinity stress. *FEBS Lett.* **579:** 6265-71.

Yadav, SK, SL Singla-Pareek, M Ray, MK Reddy and SK Sopory. 2005b. Methylglyoxal levels in plants under salinity stress are dependent on glyoxalase I and glutathione. *Biochem. Biophys. Res. Commun.* **337:** 61- 7.

Zambelli, B and S Ciurli. 2014. Nickel and human health. In: *Metal Ions in Life Sciences* (eds Sigel, A, H Sigel H and RKO Sigel), Vol. 13: Interrelations between essential metal ions and human diseases, pp 321- 57, Springer, Dordrecht.

Zhang, L, SB Mulrooney, AFK Leung, Y Zeng, BBC Ko, RP Hausinger and H Sun. 2006. Inhibition of urease by bismuth(III): Implications for the mechanism of action of bismuth drugs. *Biometals* **19:** 503-11.

■■■■

16

IODINE: A UNIQUE ELEMENT WITH SPECIAL REFERENCE TO SOIL-PLANT-AIR SYSTEM

Prasann Kumar and A. Hemantaranjan

Department of Plant Physiology, Institute of Agricultural Sciences, Banaras Hindu University, Varanasi, 221005, U.P., India
Email: prasann0659@gmail.com

INTRODUCTION

Iodine is unique amongst the elements as while most elements in soils are derived from weathering of parent material, iodine mainly derived from the oceans through the atmosphere. Soils are generally rich in iodine as compared to soil parent materials. Organic matter and clay help in its retention in the soil. The distribution of iodine deficiency disorders (IDD) generally reflects the soil geochemistry of iodine, with areas remote from marine influence, such as central continental and rain shadow areas of high mountain ranges, being seriously affected. Animals and plants in the food chain ultimately depend upon soil, water and rock in the environment for minerals (Ammerman et al. 2000). Therefore, supplementation of iodine in food chains through plant uptake is believed to be a cost-effective way to improve human nutrition and help to reduce the worldwide incidence of IDD (Jopke *et al.,* 1996). The linkage between iodine deficiency and the onset of health problems such as goiter, growth impairment, mental retardation and even cretinism has been known since 1895. These health problems are generally regarded as IDD, and as per the estimates of WHO about 1 billion people around the globe today are at risk from IDD (Hetzel, 1983). Soluble forms of I seem to be easily available to plants, therefore, terrestrial plants contain much less I than do marine plants which are known to concentrate I from 53 to 8800 mg kg^{-1} DW

(Shacklette and Cuthbert, 1967). Plants absorb iodine in iodide (I^-) and iodate (IO_3^-) both forms, but iodate can be considered as potential iodine fertilizer to increase iodine content in vegetables (Dai *et al.*, 2006). Increasing iodine applications to the soil could enhance iodine accumulation in the edible parts of vegetables (Dai *et al.*, 2004). The inorganic iodine fertilizer provides a quicker supply for plant uptake, but the higher level of iodine could be toxic to plant growth. The organic iodine fertilizer (seaweed composite) provided a more sustainable iodine supply for plants (Weng *et al.*, 2008). However, excessive iodine in the soil can be toxic to plants, such as 'akagare' disease of rice, a physiological disorder caused by flooding paddy fields on iodine-rich soils. Iodine toxicity to the plant may be caused by the intracellular oxidation of I^- to I_2, resulting in thc inhibition of photosynthetic processes (Mynet and Wain, 1973). High levels of Cl^- can reduce the toxic effect of I^- (Lewis and Powers, 1941). Iodine was discovered by Bernard Courtois in 1811 while making gun powder. He was burning seaweed to collect KNO_3 when noticed that a beautiful violet vapor was produced. When the vapor cooled, it changed into dark, shiny, metallic- like crystal. Humphry Davy later suggested the name iodine for the element from the Greek word 'iodes' for violet.

Table 1: Basic facts about the Iodine

S.N.	Facts	Remarks
1.	Symbol	I
2.	Name origin	Greek-lodes (violet coloured)
3.	Classification	Halogen
4.	Group	17
5.	Period	5
6.	Atomic number	53
7.	Atomic weight	126.90447
8.	Oxidation level	-1,0,+1,+3,+5,+7
9.	Crystal structure	Orthorhombic
10.	Colour	Bluish-black, lustrous solid

(Source: http://en.wikipedia.org/wiki/)

Baumann in 1895 discovered it in the human thyroid gland and its necessity in the human diet was discovered shortly thereafter. Crops grown in iodine deficient soil are also iodine deficient. Because animal and human are dependent on crops for iodine, they too can become iodine deficient if they eat vegetables etc. grown in iodine deficient soil. Iodine is present in seafoods and in foods grown in iodine rich soil. Dairy products are a significant source of iodine in American diet due to the iodine containing chemicals used to clean dairy equipment and to the addition of iodine to the feed of dairy cows. Realizing the dangers of iodine deficiency most countries throughout the world now fortify their salt supply with iodine to prevent deficiency. Iodine was first added to the salt supply with iodine to prevent deficiency. Iodine was first added to the salt supply in Switzerland and in Michigan in the 1920's. This was done to combat iodine deficiency. Figure A describes the potential transfer of iodine in soil-plant–air system where it is clear that iodine transfer from liquid phase to plant root easily rather than solid and gas. Table 2 indicates the different sources of iodine in the soil.

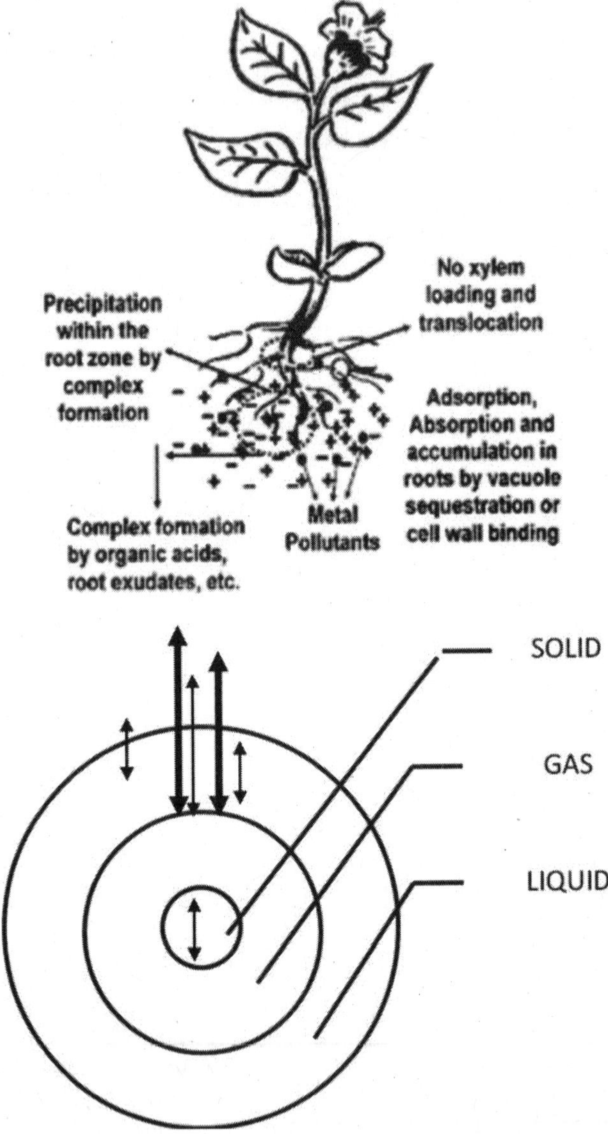

Figure A: Schematic diagram of potential transfer of iodine in soil-plant-air system (Source: Diagram drawn by authors, unpublished, 2016)

Table 2: Source of Iodine in Soil

Sl. No.	Name of Minerals	Iodine percentage
1.	Marshite	66.63
2.	Lautrite	65.10
3.	Bruggenite	62.22
4.	Miersite	56.73
5.	Coccinite	55.86
6.	Lodargyrite	54.05

(Source: http://en.wikipedia.org/wiki/)

For I, the stimulating effect has been observed at iodine levels: 0.1 μg kg^{-1} in soil and nutrient solution toxic effects occur in excess of about 0.5-1.0 μg kg^{-1} in soil (Martin, 1966). The toxic concentration is considerably higher than normal soluble soil I levels so that I toxicity has not been reported in agricultural soils. Inorganic ions of I are: I^-, IO_3^-, I_3^-, IO^-, IO_6^{3-} $H_4IO_6^-$; Organic I compounds: CH_3I, CH_2I_2 and CH_2ClI

SIGNIFICANT FACTORS DETERMINE IODINE CONTENT IN SOIL

The weathering of the rocks having iodine content will acts as the key driver for the supply of iodine into the soil. The physical, chemical and biological processes of the soil control the iodine enrichments and its fate of behavior with plants and parent materials. It can be distributed into the different components of the soil at the same time i.e., partitioning is critical for the iodine for its availability to the plants. Within the soil there are large numbers of factors which are responsible for the mobility of iodine such as;

- **Texture:** Heavy textured soils like loam, clay loam, sandy loam and silt loam are richer in iodine than light textured soils.
- **Anion exchange capacity (AEC):** Due to AEC, I^- exchange on the colloidal surface with other anions like Cl^- takes place thus I retention in the soil takes place.
- **Organic matter and clays:** Organic matter and clay form a complex and iodine is the part of that complex thus iodine retention in the soil takes place.
- **Sesquioxides (Al_2O_3, Fe_2O_3):** Sesquioxides form hydrous oxides at low pH and I binds to these Al and Fe hydroxides thus I retention in soil takes place.
- **pH:** With increase acidity of clay suspension, there is a progressive increase in I sorption.

Most of the iodine in nature resides in the ocean. It was present during primordial development of the earth. Iodine was leached from the surface soil by glaciers, snow and rain and was carried by wind, river and floods into the sea. Iodine ions are oxidized by sunlight into elemental I_2 in the atmosphere and returned to the soil by rain. In this way the cycle is completed and this cycle goes on in the nature. The scientist Whitehead in the year 1984, reported the iodine concentration in the various components of our surrounding environments.

Table 3: Iodine concentration in various components of the environment

S.N.	Environmental components	Iodine concentration
1.	Coal*	1-15 mg kg^{-1}
2.	Soils*	0.5-20 mg kg^{-1}
3.	Sea water	45-60μg l^{-1}
4.	Marine fish*	0.5-6.0 mg kg^{-1}
5.	Igneous rocks*	0.08-0.50 mg kg^{-1}
6.	Sedimentary rocks*	2-6.0 mg kg^{-1}
7.	Atmosphere	0.01-0.02 μg l^{-1}
8.	Higher plants*	0.05-5.0 mg kg^{-1}
9.	Marine algae*	90-2500 mg kg^{-1}
10.	Mammalian tissue*	0.05-0.5 mg kg^{-1}
11.	Fresh water fish*	0.06-0.2 mg kg^{-1}
12.	Sedimentary rocks*	2-6.0 mg kg^{-1}
13.	Marine sediments*	3-400 mg kg^{-1}

(Source: Whitehead, 1984)
(*Where, for solid material the concentration was based on dry weight basis)

Table 3 indicates that marine algae have the highest I concentration with respect to sedimentary rocks than igneous rocks.

PLANTS AND IODIEN UPTAKE WITH SPECIAL REFERECE TO PLANT NUTRITION

It is very interesting to say that, the roots of plant take the iodine from the liquid phage of the soil. This is considered as an important process because since it is a mean by which iodine may enter into the food chains. Iodine fails to prove the criteria of essentiality so it is not been considered as the essential elements of the plant. Scientist reported that, high concentration of iodine within the plants shows the toxic effect (Figure B).

1. Normal level of I in healthy plants: $\leq$ 0.5 μg kg^{-1}
2. Iodide toxicity has been found where the level in plant is in excess of 8 μg kg^{-1}
3. Toxic effects of high I levels: Chlorosis and Abscission (in older leaves) while the younger leaves remain a very dark green color. Growth is severely restricted and the leaves curl back and

necrosis occurs at the tip and edges. In severe cases the plant dies.

4. Terrestrial plants contain much less I than do marine plants which are known to concentrate I from 53-8800 μg kg^{-1} dw (Shacklette and Cuthbert, 1967).

High levels of Cl^- reduce the toxic effect of I^- which suggests that there is a competitive effect between both elements (Lewis and Powers, 1941).

Figure B: Agricultural crops having capacity to take the iodine. These plants also suffered from toxicity with iodine at higher concentration (Source: http://en.wikipedia.org/wiki/Image.jpg)

IODINE TRANSFER FACTOR IN SOIL –PLANT SYSTEM

Plant uptake to iodine can be measured with the help of iodine transfer factor or TF. Since a long time it inherently takes into the account for the relation between soil and elemental concentration within the plants. It can be determine with the formula like

$$\text{TF} = \frac{\text{Dry plant iodine concentration (mg per kg)}}{\text{Dry soil iodine concentration (mg per kg)}}$$

So on the basis of above equation it can be conclude that, high TFs leads to the high level of transfer from the soil tom the plant and vice – versa. We want to indicate one thing that, TF is the unit less. So don't confuse with the units of related concern for the sake of calculation. Dai *et al.,* in the year 2004 reported that in the case of most of vegetables the TF will increases with the increase in the concentration of iodine in the soil.

Table 4: Transfer factor for the different crops

Sl. No.	Plants species	Transfer Factor	References
1.	Vegetables	0.1 to 10	Dai *et al.*, 2004
2.	Corn	0.024 to 0.19	Sheppard *et al.*, 1993
3.	Beets	0.024 to 0.19	Sheppard *et al.*, 1993
4.	Cabbage	0.024 to 0.19	Sheppard *et al.*, 1993
5.	Radish root	0.01-0.03	Kashparow *et al.*, 2005
6.	Lettuce leaves	0.01-0.03	Kashparow *et al.*, 2005
7.	Bean pods	0.003-0.004	Kashparow *et al.*, 2005
8.	Wheat grains	0.001	Kashparow *et al.*, 2005

From the Table 4 it is clear that, TFs of iodine are low, indicating that the element is not readily transferred from the soil to plants. The significant cause behind the fact is that; strong soil adsorption in the solid phage of the soil. So on the behalf of this statement we can say that the contribution of human diet of iodine derived from crop plants is very limited. In general it is reported that those plants which can tolerate the anaerobic condition in the root zone can have the more transfer ratio of iodine like rice. But this thought was in general and not applicable for whole plants. Soil organic matter such as humic acid has a role in reducing IO_3^- to I_2. The highly volatile I cannot remain in soil when it is present in the elemental form. Since I is highly reactive with organic compound, the formed I likely associates with SOM as organo- iodine in upland fields. In upland fields, SOM is responsible for the reductive reaction. In submerged paddy field, development of reduction condition attributable to microbial activity. Thus organo-iodine is a potential source of I absorbed by plants. **Weng and coworkers in 2008** conducted an experiment to study Mechanism of Iodine Uptake by Cabbage: Effects of Iodine Species and where it is store

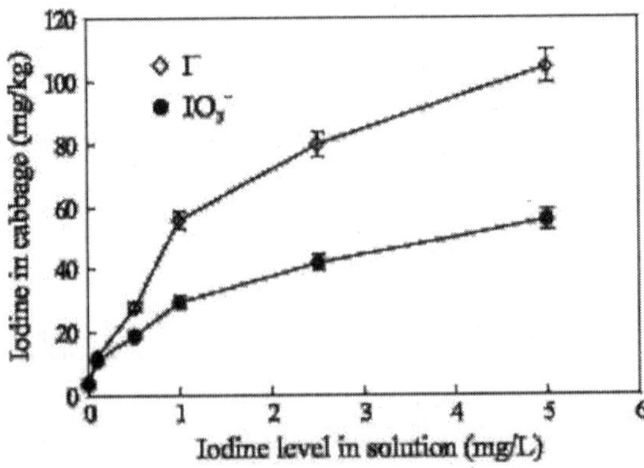

Fig. C: Iodine uptake responding to amount of the applied iodine fertilizer (I^- to IO_3^-) under solution culture (Source: Weng *et al.*, 2008)

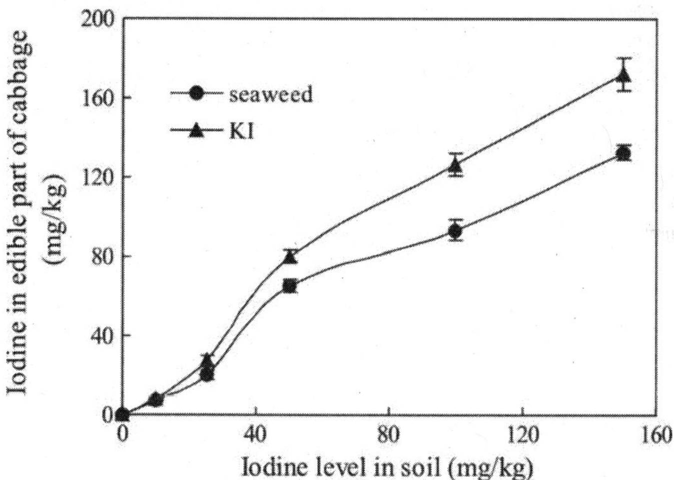

Figure D: Iodine uptake responding to amount of the applied iodine fertilizer (KI or sea weed composite) in pot planting (Source: Weng *et al.*, 2008)

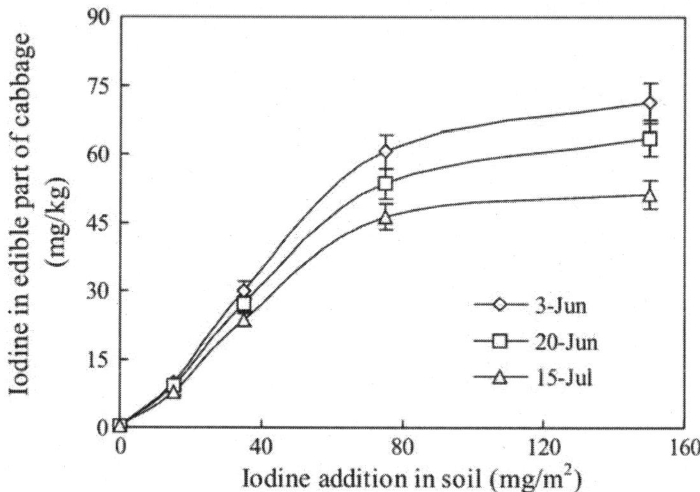

Figure E. Iodine uptake responding to amount of the applied iodine fertilizer (Sea weed composite) in field condition (Source: Weng *et al.*, 2008)

They found that iodine uptake by Chinese cabbage increases with the amount of the applied iodine fertilizer (Fig. C). However, there is a turning point at about 1.0 mg L^{-1} concentration in the solution; below this point the rate of iodine increase is much greater than above this concentration for both iodine species. The results from the pot planting (Fig. D) and the field experiments (Fig. E) exhibited similar patterns. The turning point was at 50 mg kg^{-1} in the pot planting (Fig. D) and 75 mg m^{-2} (Fig. E) in field experiments. This change in the iodine uptake of the plant with response to the increase of iodine concentration in the culture solutions and soils is partly caused by the toxic effect of the high iodine concentration on the growth of Chinese cabbage. They also found that the biomass of the cabbage increases with the iodine concentration in the culture solution when the concentration remains low, but decreases with the increase of the concentration. The turning point for the effects on growth is lower than that for iodine uptake, although the particular values of the turning point and the magnitude of the effects are different for different iodine species. This phenomenon was confirmed in pot planting experiments too. In the pot planting experiments, at the exogenous iodine level of 10 mg kg^{-1}, the biomass of Chinese cabbage increased somewhat compared to the control group. However, when exogenous iodine was higher than 25 mg kg^{-1}, there were significant negative correlations ($r=-0.96$ for KI and -0.93 for seaweed composite, $P<0.01$) between the biomass and the iodine level in soil. In addition, growth problems in Chinese cabbage occurred when the iodine level in soil was higher than 50 mg kg^{-1} such as chlorosis in young leaves, brown, stunted, and coralloid roots. Iodate shows much less toxicity to the cabbage than iodide, partly because of its lower uptake in higher concentration (Fig. F).

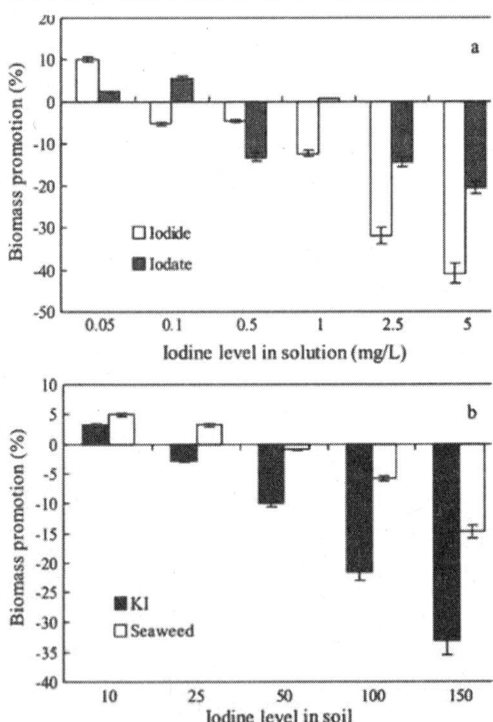

Figure F: Comparison of cabbage biomass growth as response to iodine concentration in solution and pot culture (Source: Weng *et al.,* 2008)

VOLATILIZATION OF IODINE FROM PLANTS ANS ITS PARTS

One thing very interesting to say that, plant can convert elemental iodine into volatile form and release through its pore on the surface of leaves and others. Several scientists reported that, volatilization of iodine from the shoots is significant than the soil in the case of rice with flooded field. In the year 1995, Muramatsu and Yoshida reported that several things related to volatilization of the iodine from the plant parts as well as soil surface. These points are as follow [1] iodine emission were highly stimulated by the presence of oats on no flooded soil and, particularly, by the presence of rice on the flooded soil; [2] in the year 1999, he reported that, volatilization of the microbial generated methyl iodide gas from rice plants as a process by which iodine concentrations of soils may be reduced. The process of giving rise to enhanced methylation, and hence emission, from rice-planted and flooded soil was postulated by Muramatsu and Yoshida. We think he has credited first to postulate such an idea.

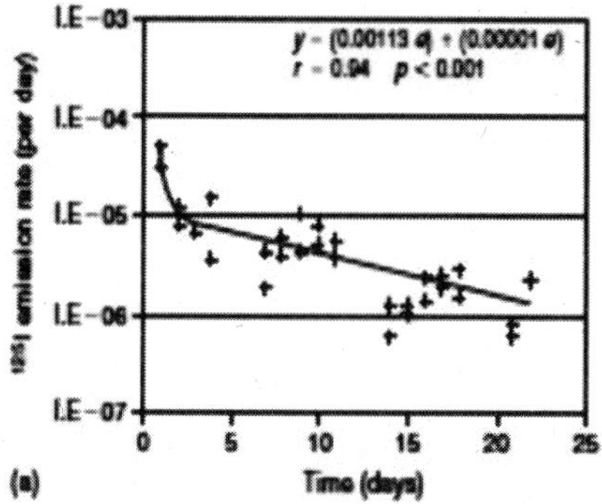

Fig. G: Emission of volatile ^{125}I from soils of Coniferous forest with 22 days duration (Source: Asworth DJ, 2006)

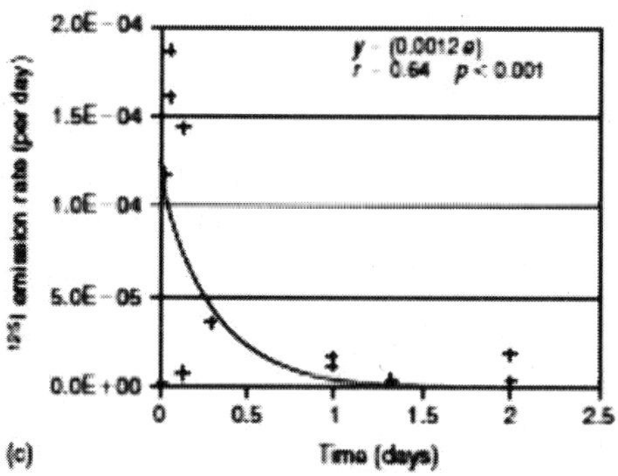

Fig. H: Emission of volatile ^{125}I from soils of Coniferous forest after 48h duration (Source: Asworth DJ, 2006)

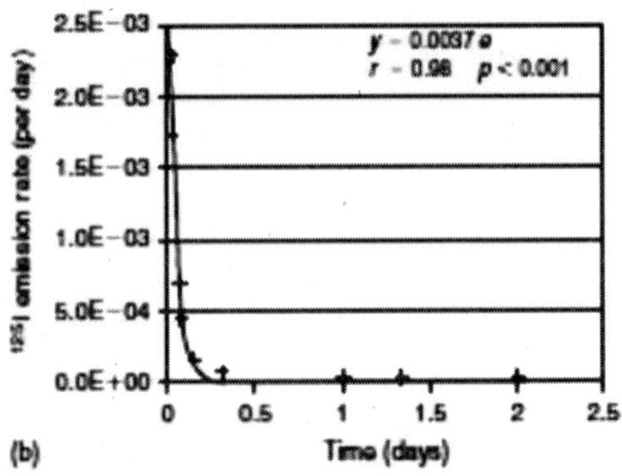

Fig. I: Emission of volatile ^{125}I from soils of Grassland after 48 h duration (Source: Asworth DJ, 2006)

Figure G, H and I represent the emission of volatile ^{125}I from the soils of different time periods at different location. Figure G represent the emission of volatile iodine through soils of coniferous forest over a 22- days period. This was measured with the per day basis. Figure H, represent the emission of the volatile iodine form the coniferous forest soils over a period of 48h with per day basis and in last the figure I represent the emission of the volatile Iodine form the soils of the grassland forest over a 48 h period. This was measured with per day basis.

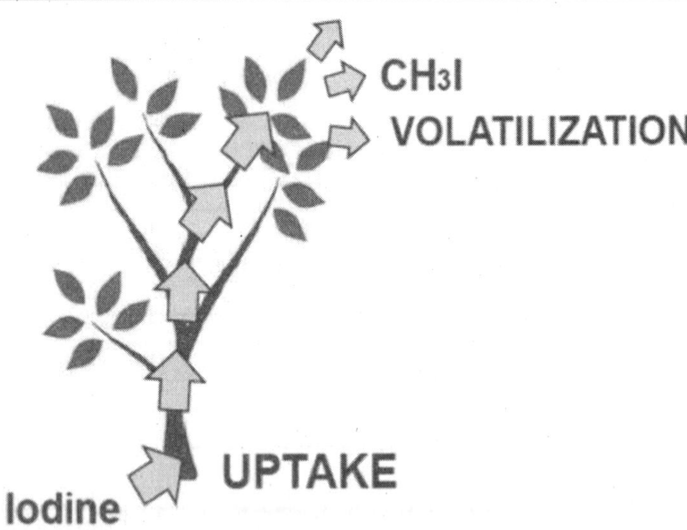

Fig. J: Schematic diagram of Iodine volatilization (Source: http://en.wikipedia.org/wiki/Image.jpg)

Figure J shows the iodine uptake and its volatilization from the aerial; parts of the plant. It is understood that, a mechanism exists by which soil iodine can become volatile and transfer to the aboveground atmosphere. So in general, volatilization of iodine, primarily as methyl iodide, occurs form both soil and plants. Again, the presence of anoxic soil conditions is likely to enhance this process. This occurs due to fact that, plants with root able to tolerate anoxic condition tend to take up greater quantities of iodine from the soil. Iodine mobility under anoxic soil condition is significant greater than under oxic condition.

IODINE AND PHYTO-THERAPY

It is reported that, iodine has also been utilized to contain the following plant disease. Iodine has been an established biocide in the medicinal and food industry for the many decades. It is considered as the number one disinfectant in use because of its antipathogenic properties and having high redox potential. In the Table 5 we are giving the few names of diseases and pathogens, which can be controlled by the application of iodine significantly.

Table 5: Iodine phyto-therapy

Sl. No.	Plant species	Pathogen/Disease control
1.	Cut flower	Powdery mildew and rust, *Puccinia* sp.
2.	Cucumber	Anthracnose or *Colletotrichum orbicular* and angular spot or *Pseudomonas syringe*
3.	Tomatoes	Late blight or *phytophotra infestans* II, early blight or *Alternaria solani*
4.	Wine grapes	Black Rot, Botrytis, Downy Mildew
5.	Blueberries	Anthracnose
6.	Ginger root	Nematodes
7.	Potato seeds	*Rhizoctonia solani*

(Source: Review literature)

IODINE TRADITIONAL APPLICATION IN THE FIELD OF AGRICULTURE

It is reported that, Iodine application were posed risks of phyto-toxicity on crops due to the insolubility of the crude materials as well as the hazards associated with the solvents used to make iodine solution (Table 6). So on the basis of the test conducted at the independent laboratories, experts recommended the some important guideline related to the traditional use of iodine for the purpose of remediation from disease or other thing else.

Table 6: Iodine application in the field of agriculture

S.N.	Purpose	Remarks related to the recommendation
1.	Soil Nematodes	The soil of agricultural field must be up to the field capacity and having good drainage system to avoid the water logging condition prior to treatment
2.	Seed Treatments	We use the iodine for seed treatment in the case of tubers and bulbs to prevent the contamination from eelworm. For this purpose we just soak the bulbs in an iodine solution at a minimum of 500ppm for at least one hour. The treated seed sows the enhanced percentage of germination with respect to untreated seeds.
3.	Orchids & Vineyards	150-300 ppm solution of Iodine using 100 gallons of water per acre with a fine mister/ blower can be used
4.	Field vegetables & Fruits	Apply 150-300 ppm solution of Iodine using a minimum of 40 gallons/acre of water
5.	Hard fruits and Tubers	Deep the concerned fruits and tubers into the solution having 12 ppm iodine concentration
6.	Soft fruits	For this purpose we spray thoroughly with a fine mist of 12 ppm of iodine solution and then allow to air dry before packing

(Source: Review Literature)

CONCLUSION

Iodine is considered as one of the non essential element for plants but is an essential element for animals with special reference to human. One very interesting characteristic with iodine is that, it can move in all three phase of the soil, i.e., solid, liquid and gas. That's why the exposure to human depends on the mobility in to the different phases of soil existing. Iodine has been used as a potential biocide in different sector since a long time. It is due to its disinfectant in nature. High REDOX potential of iodine leads to its significant disinfectant property. In the soil system it is considered as immobile because its attraction with the solid phage is very high. One thing is very interesting to say that under the anaerobic condition its presence appears into the liquid and gas phage also. Which indicate that, combined effects of physical, chemical and biological behavior of soil are the key driver for iodine behaviour?

Iodine partitioning between all three soil phases, i.e., solid, liquid and gas leads to potential transfer away from the soil body, e.g., by leaching, plant uptake, and volatilization. Although iodized salt is widely used to prevent and control IDD, and this is the simplest, most economical and practical method, iodized salt has its own disadvantages. The inorganic iodine has the double threshold values, that is, if it is either too high or too low, iodine may cause negative biological effects to the organism. In other words, intake of too much iodine may also cause the same diseases as the deficiency. Because inorganic iodine is volatile, it is difficult to control the loss of iodine during storage, transportation and cooking with high temperature oil. Therefore, producing iodine- enhanced vegetables is an attractive way to control IDD and it is investigated recently, that iodine in foodstuffs can be easily assimilated by human and its bioavailability is upto 99%. The organic iodine fertilizer provided a more sustainable iodine supply for plants than inorganic iodine fertilizer.

ACKNOWLEDGEMENT

Authors are very thankful to UGC, New Delhi for providing Senior Research Fellowship to the first author.

REFERENCES

Ammerman, C. B., Baker, D. H. and Lewis, A. J. 2000. Bioavailability of Nutrie-nts for Animals. New York: Academic Press; National Research Council.

Ashworth, D.J. and Shaw, G. 2006. Sci. total environ., 359: 244-254.

Dai, J. L., Zhu, Y. G., Zhang, M. and Huang, Y. Z. 2004. Selecting iodine-enriched vegetables and residual effectiveness of iodate application to soil. *Biol Trace Elem Res.*101:265–276.

Dai, J. L., Zhu, Y. G., Zhang, M. and Huang, Y.Z. and Song, J. L. 2006. Availability of iodide and iodate to spinach (*Spinacia oleracea* L.) in relation to total iodine in soil solution. *Plant and Soil.* 289:301–308.

Hetzel, B. S. 1983. Iodine deficiency disorders (IDD) and their eradication. The Lancet, London 2: 1126-1129.

http://en.wikipedia.org/wiki/Image.jpg

Jiang T. H., Zheng S. J. and Shi J. Q. 1995. Several considerations in kinetic research on nutrients uptake by plants. *Plant Nutr. Fertil. Sci* ., 1:11–16.

Jopke, P., Bahadir, M., Fleckenstein, J. and Schnug, E. 1996. Iodine determination in plant materials. *Commun. Soil Sci. Plant Anal.*, 27:741–751.

Kashparov, V., Colle, C., Zvarich, S., Yoschenko, V., Levchuk, S. and Lundin, S. 2005. J. Environ. Radiact., 79: 187-204.

Lewis, J. C. and Powers, W. L. 1941. Antagonistic action of chlorides on the toxicity of iodides to corn. *Plant Physiol.*, 16: 393-398.

Martin, J. P. 1966. Bromine. In: H. D. Chapman, Diagnostic Criteria for Plants and Soils. Univ. of California, *Div. of Agric. Scien.*, p. 62-64.

Mc Dowell, L. R., 1997. Minerals for Grazing Ruminants in tropical Regions. 3rd ed. University of Florida Cooperative Extension Service.

Muramatsu, Y. and Yoshida, S. 1995. J. Radionanal. Nucl. Chem Atom, Environ., 29: 21-25.

Mynet, A. and Wain, R. L .1973. Herbicidal action of iodide: effect on chlorophyll content and photosynthesis in dwarf bean *Phaseolus vulgaris. Weed Res.,* 13:101–109.

Pendias, A. K. and Pendias, H. 1989. Trace elements in soils and plants Elements of Group VII. p. 220.

Shacklette, H. T. and Cuthbert, M. E. 1967. Iodine content of plant groups as influenced by variation in rock and soil type. *Geol. Soc. Am. Spec. Pap.*, 90: 31.

Sheppard, S.C., Evenden, W.G. and Amiro, B.D. 1993. J. Environ. Radioact., 21:9-32.

Weng, H.X., Hong, C.L., Yan, A.L., Qin, Y.C., Bao, L.T., Xie, L.L. and Pan, L.H. 2008. Mechanism of Iodine Uptake by Chinese Cabbage: Effects of Iodine Species and Where It is Stored. *Biol Trace Elem Res.,* 125:59–71.

Whitehead, D.C. 1984. The distribution and transformation of iodine in the environment. *Journal of Environmental Science.* 10 : 321-339.

WHO, 1984. Techn. Ret. Ser., 713.

WHO, 1985. IDD in S E Asia, SEARO Reg. Health Paper No. 10 New Delhi.

World Health Organization. 2001. Trace elements in Human nutrition and health. p. 49-71 Geneva: World Health Organization Publication, Geneva.

■■■■

Section III:

Plant Functions Research in Agricultural Progression

17

PLANT PURPLE ACID PHOSPHATASES: STRUCTURE AND FUNCTIONS

Enrique J. Baran

Centro de Química Inorgánica (CEQUINOR/CONICET,UNLP) Facultad de Ciencias Exactas, Universidad Nacional de La Plata, C. Correo 962, 1900-La Plata, ARGENTINA

INTRODUCTION

Phosphatases are enzymes that remove a phosphate group from an organic compound by hydrolysis, generating a phosphate anion, PO_4^{3-}, and a molecule containing a hydroxyl group. This reaction is opposite to the action of phosphorylases and kinases, which attaches phosphate groups to their substrates, usually with the participation of energy-rich molecules such as ATP (Williams and Fraústo da Silva, 1996).

One of the best known and widely investigated phosphatase is the so called alkaline phosphatase (AP; EC 3.1.3.1.) which is found in many organisms from bacteria to man (Coleman, 1992; Millán, 2006). It is a nonspecific phosphomonoesterase that catalyzes the hydrolysis reaction via a phosphoseryl intermediate to produce inorganic phosphate and the corresponding alcohol, with maximum activity at pH 8 or above. The enzyme can also transfer the phosphate group to a hydroxyl group of an acceptor (transferase activity) (Hughes, 1981).

On the other hand, the so called purple acid phosphatases (PAPs) represent a group of enzymes that also hydrolyze phosphoric acid esters but show optimal catalytic activity at low pH-values (Klabunde and Krebs, 1997; Schenk *et al.*, 2013).

In the first part of this overview we present briefly the most important aspects related to the structure and mechanism of action of alkaline phosphatases; whereas the main part of them is devoted to the discussion of purple acid phosphatases found and isolated in the plant kingdom.

ALKALINE PHOSPHATASES

Structural and Mechanistic Aspects

The dimeric AP from *Escherichia coli* is one of the most extensively investigated enzymes of this type (for a review, cf. Coleman, 1992). The enzyme hydrolyzes not only orthophosphate monoesters, but also a variety of O- and S-phosphorothioates, phosphoroamidates, thiophosphate and phosphate; hydrolysis of this latter group is reflected by the catalysis of the exchange of ^{18}O from $H_2{}^{18}O$ into inorganic phosphate or the release of H_2S from thiophosphate. The enzyme has an alkaline pH maximum and the rate follows an approximately sigmoid pH-rate profile with an apparent pK_a of ~ 7.5 (Coleman, 1992).

The crystal structure of native *E. coli* AP complexed with inorganic phosphate has been solved and refined to 2.0 Å resolution; also structural data of the Cd-substituted enzyme, complexed with inorganic phosphate (2.5 Å resolution), and the phosphate-free native enzyme (2.8 Å resolution) were reported (Kim and Wykoff, 1991).

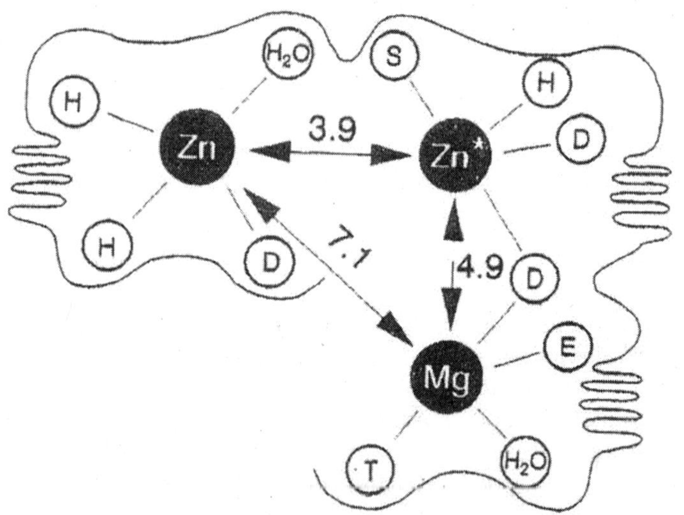

Fig. 1. Schematic drawing of the catalytic site of *E. coli* alkaline phosphatase. (H = Histidine, D = Aspartic acid; S = Serine; E = Glutamic acid; T = Threonine)

The active site of each monomer contains two Zn(II) ions and one Mg(II) ion in a pocket created by the termination of a number of helices and sheets that is open to the surface. The active sites of the two subunits are separated by about 30 Å (Kim and Wykoff, 1991). As shown in Fig. 1, the three metal cations form the apices of a non-equilateral triangle; the two Zn(II) ions are separated by 3.94 Å and the Mg(II) ion is separated by 7.09 Å from Zn1 and by 4.88 Å from Zn2. Zn1 is coordinated to an imidazole nitrogen of both His-331 and His-412 and to both carboxylate O-atoms of Asp-327. Minimally, one H_2O molecule completes the coordination of this metal center. The Zn2 cation (marked as Zn* in Fig. 1) is tetracoordinated to one carboxylate oxygen each of Asp-51 and Asp-369, the imidazole nitrogen of His-370 and the OH^- group of Ser-102. Finally, Mg(II) is linked to Zn2 through the Asp-51 bridge and coordinated to one of the O-atoms of Glu-322 and the hydroxyl of Thr-155, plus three H_2O molecules (Kim and Wykoff, 1991; Vallee and Auld, 1993).

The importance of the three metallic centers is well-documented. Mg(II) does not directly participate in the reaction but is relevant for the structural stabilization of the enzyme. On the other hand, the two Zn(II) cations are directly involved in the catalytic process as they are adequately positioned to activate the serine and water for nucleophilic attacks and they are involved in holding of the phosphate portion of the substrate. Zn1 is directly required for catalysis and binds both, the substrate and the phosphate; whereas Zn2 interacts with the hydroxyl group of the active serine to stabilize its deprotonated form, necessary for the nucleophilic attack on the phosphate (Bortolato *et al.,* 1999).

In the usually accepted kinetic scheme for the enzymatic hydrolysis/ transphosphorilation of phosphate monoesters by alkaline phosphatase (Fig. 2), the nucleophile, Ser-102, is transiently phosphorilated giving a covalent phosphoseryl intermediate (**E-P**). This intermediate is subsequently hydrolyzed to give a non-covalent enzyme phosphate complex (**E·P**$_i$). In the presence of a phosphate acceptor (for example an alcohol), the enzyme exhibits transphosphorilation activity transferring phosphate to form another phosphate monoester (Coleman, 1992; Holtz and Kantrowitz, 1999).

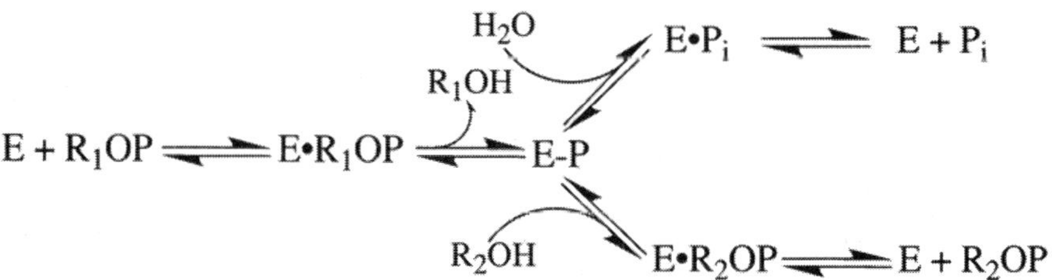

Fig. 2. Mechanistic pathways for alkaline phosphatase.

The **E-P** / **E·P**$_i$ equilibrium in Fig. 2 has been subjected to several NMR investigations. ^{31}P NMR studies demonstrated that the hydrolysis of the **E-P** intermediate is rate-limiting under acidic conditions, while the hydrolysis of **E·P**$_i$ is rate limiting under alkaline conditions. The rate-limiting dissociation is essentially pH-independent between pH 6 and 9. Below pH 6, dephosphorilation becomes rate-limiting (Holtz and Kantrowitz, 1999). Additional support to the mechanistic scheme were derived from crystallographic studies with the cadmium substituted enzyme (Kim and Wyckoff, 1991; Coleman, 1992; Holtz and Kantrowitz, 1999), by structural studies of the enzyme in the presence of VO_4^{3-} ions (Holtz, *et al.,* 1999) and by site-specific mutations of the enzyme (Coleman, 1992; Ma *et al.,* 1995; Holtz and Kantrowitz, 1999).

Interestingly, a later crystallographic study, of the wild-type enzyme, in the presence and absence of inorganic phosphate and refined to 1.75 Å, suggested that also the Mg(II) cation may be involved in the catalytic cycle (Stec *et al.,* 2000).

PURPLE ACID PHOSPHATASES

General Aspects

Acid phosphatases (EC 3.1.3.2) represent a group of enzymes that hydrolyze phosphoric acid esters and show optimal catalytic activity at low pH-values. A purple-colored, iron containing subclass of acid phosphatases was first isolated from bovine spleen (Campbell and Zerner, 1973) and porcine uterine fluid (Chen *et al.,* 1973). Because of their intense color these acid phosphatases have come to be known as purple acid phosphatases (PAPs) (Klabunde and Krebs, 1997; Schenk *et al.,* 2013).

In recent years PAPs have been isolated from a variety of animal, plant and bacterial sources. Mammalian PAPs have been characterized from different human, cow, pig, mouse and rat tissues (Schenk *et al.,* 2013). Despite the characteristic features that support their designation as purple acid phosphatases, these enzymes from mammalian, plant and bacterial sources differ considerably in their molecular mass, subunit arrangement and amino acid sequence. Whereas those of mammalian origin are monomeric glycoproteins with a molecular mass of approximately 35 kDa, the molecular mass of the dimeric plant enzymes ranges from 100 to 120 kDa. Besides, bacterial PAPs are highly glycosylated monomeric proteins with molecular masses around 80 kDa (Klabunde and Krebs, 1997).

Mammalian and plant PAPs have similar active site structures despite the low sequence identity (<20%). In the case of bacterial enzymes, the analysis of their deduced amino acid sequences and predicted secondary structures also suggest relations to the mammalian and plant enzymes and homology models indicate close similarities in the immediate vicinity of the active site (Schenk *et al.,* 2000).

Another characteristic feature of all these enzymes is the presence of two closely spaced metal ions forming a dinuclear reaction center. In the mammalian enzymes this center is conformed by two iron ions (a Fe(III)/Fe(II) pair), whereas the majority of plant PAPs have Fe(III)/Zn(II) or Fe(III)/Mn(II) centers (Schenk *et al.,* 2013).

PAPs have been implicated in a number of different biological functions. The physiological role of mammalian PAPs have been associated with iron transport, generation of reactive oxygen species (ROS) as an immune response and increased bone resorption. Mammalian PAPS are bifunctional, catalyzing both hydrolytic reactions and peroxidations (Schenk *et al.,* 2013). The latter reactions are facilitated by the redox-active iron in the dinuclear site, and the mechanism is believed to be based on Haber-Weiss-Fenton-type reactions:

$$Fe^{III}—Fe^{II} + H_2O_2 \rightarrow Fe^{III}—Fe^{III} + OH\bullet + OH^- \quad (1)$$

$$Fe^{III}—Fe^{III} + O_2^{\bullet -} \rightarrow Fe^{III}—Fe^{II} + O_2 \quad (2)$$

Due to its bifunctionality (hydrolysis and peroxidation), a variety of biological roles have also been proposed for plant PAPs but the assignment of specific functions have been rendered difficult due to the occurrence of multiple isoforms (see below).

Apart from the requirement for two closely spaced metal ions in their active sites (Sträter *et al.,* 1996; Klabunde and Krebs, 1997; Schenk *et al.,* 2013), PAPs present several other characteristic features: a) their typical purple color (Antanaitis *et al.,* 1983; Klabunde and Krebs, 1997; Schenk *et al.,* 2013); b) between 5% and 10% of their molecular mass is due to their carbohydrate content (Baumbach *et al.,* 1991; Wang *et al.,* 2005; Schenk *et al.,* 2013); c) their resistance towards inhibition by L-tartrate (for this reason PAPs are frequently referred as to tartrate-resistant acid phosphatases (TRAPs) (Klabunde and Krebs, 1997; Schenk *et al.,* 2013)).

Origin of the Characteristic Color

The characteristic purple color, related to an absorption band at 510-560 nm in PAPs is due to a charge transfer transition from the O-atom of a conserved tyrosine residue to a Fe(III) ion present in the active site.

On the basis of EPR, Mössbauer and magnetic susceptibility measurements the mammalian PAPs isolated from porcine uterus and bovine spleen have shown to contain an antiferromagnetically coupled binuclear iron site. These PAPs can be obtained in two oxidation states. The purple oxidized Fe(III)/Fe(III) form is catalytically inactive and EPR silent with a maximum absorption at 550 nm ($\varepsilon \approx 4000\ M^{-1}.cm^{-1}$). The exchange coupling constant has been estimated to be between 80 and 300 cm^{-1}, suggesting strong antiferromagnetic coupling between the two high-spin ferric ions. The pink colored, reduced form (Fe(III)/Fe(II)), presents catalytic activity and has an absorption maximum shifted to 510 nm ($\varepsilon \approx 4000\ M^{-1}.cm^{-1}$). The EPR *g*-values are at 1.94, 1.76 and 1.56; the latter results from antiferromagnetic coupling between the high-spin Fe(III) and the high-spin Fe(II), giving an S_{tot} =1/2 ground state (Yang *et al.,* 1997). The fact that the purple color derives from a charge transfer transition from a tyrosine ligand to the ferric center has been demonstrated by resonance Raman spectroscopy (Averill *et al.,* 1987; Yang *et al.,* 1997). Features assigned to a Fe(III)-bound tyrosinate are also observed in the ^{1}H-NMR spectrum (Yang *et al.,* 1997). No dramatic changes are observed in absorbance and resonance Raman spectroscopy upon one-electron reduction of the Fe(III)/Fe(III) enzyme, which implies that the tyrosine ligand binds only to the ferric center (Averill *et al.,* 1987; Dietrich *et al.,* 1991, Yang *et al.,* 1997).

The electronic structure of the characteristic charge transfer transition was thoroughly investigated using a combination of absorbance, circular dichroism and magnetic circular dichroism spectroscopy (Yang *et al.,* 1997). The combined data of these studies indicated that the broad charge transfer energy region may be resolved into several separate transitions. The theoretical analysis predicts six possible transitions. The bond formed between the tyrosinate O-ligand and Fe(III) ion strongly depends on the two donor 2p orbitals located on the O-atom. With the Fe^{III}-O_{tyr} axis defining the z-axis, the relevant acceptor orbitals in the charge transfer transitions are the $d_{z}2$, d_{xz} and d_{yz} metal orbitals. From these, essentially the energy transitions involving the last two *d*-orbitals are responsible for the observed purple coloration.

Usually, *d-d* transitions of high-spin Fe(III) sites have very low extinction coefficients because they are spin-forbidden and, therefore, are not detectable by absorption spectroscopy in proteins. However, these transitions can gain intensity from mixing with the intense low-energy charge transfer transitions and in this case, a contribution of these transitions could also be detected in the spectra.

PLANT PURPLE ACID PHOSPHATASES

Although it is well known that PAPs are abundantly distributed in the plant kingdom, detailed studies have so far only been performed in a limited number of cases, for example red kidney bean (*Phaseolus vulgaris*), sweet potato (*Ipomea batatas*), lupin (*Lupinus albus* and *Lupinus luteus* L.), soybean (*Glycine max*), mung beans (*Vigna radiata*), tomato (*Lycopersicon esculentum*), duckweed (*Spirodela oligorrhiza*) and thale cress (*Arabidopsis thaliana*) among a few others. Notwithstanding, the nature and structure of the active enzymatic site has only been determined for kidney bean, sweet potato and yellow lupin.

Purple Acid Phosphatase from Red Kidney Bean

The red kidney bean (*Phaseolus vulgaris*) purple acid phosphatase was the first structure of a PAP to be determined (Nordlund, 2001). Thus, this study allowed, for the first time, a deeper and very detailed insight into the structural characteristic of the enzyme and provided the necessary background for more elaborated mechanistic studies and for the analysis of other related structures.

The enzyme was initially described by Nochumson (1975) and first insights into its overall characteristics were reported by Beck *et al.* (1986). They developed a purification procedure which allows the preparation of relatively large amounts of essentially homogeneous phosphatase samples. It was found that it is a dimmer with an approximate molecular weight of 130 kD, containing one atom of iron and one atom of zinc per subunit. The visible absorption spectrum shows an intense absorption band at 560 nm ($\varepsilon = 3360\ M^{-1}.cm^{-1}$). The presence of the Fe/Zn pair was additionally confirmed by spectroscopic studies, comparing the behavior of the EPR spectra of the native enzyme with that of the corresponding Fe(II)/Fe(III) derivative (Beck *et al.*, 1998a).

When the apoenzyme, prepared from the native enzyme, is treated separately with Fe(III) and Zn(II), very little activity is observed; whereas treatment with both Fe(III) and Zn(II) resulted in complete restoration of activity, indicating that both metals are essential for the catalytic process. Essentially full reactivation was also achieved by addition of Fe(III) together with Fe(II) or Co(II) to the apoenzyme; Fe(III)/Cd(II) gave 27% restoration of activity, whereas Fe(II) in combination with Cu(II), Ni(II), Mn(II) or Hg(II) gave little or no increase in activity (Beck *et al.*, 1988b).

First information into the coordination sphere of both metallic centers and on the metal-metal distances was obtained through a XANES study of the native enzyme and the corresponding Fe(III)/Fe(III) metal exchanged form (Ahlers *et al.,* 1995). Besides, the N-linked glycan structures were characterized by selective enzymic degradation in combination with matrix-assisted laser desorption/ionization mass spectrometry (MALDI-MS). The purified protein was cleaved by cyanogen bromide. The peptides were separated by HPLC and the glycosilated species were identified both by their heterogeneous mass spectra and by an immunoassay. None of the glycopeptides proved to have more than one glycosylation site. The composition of the carbohydrate moieties were calculated by comparing the mass spectra of the glycopeptides before and after enzymic deglycosylation. These results were complemented by data from a carbohydrate composition analysis. There are five heterogeneous glycosilation sites per monomer of the enzyme (Stahl *et al.,* 1994).

Determination of the three-dimensional crystal structure of the kidney bean PAP revealed the molecular architecture of this 111-kD plant enzyme (Sträter *et al.,* 1995; Klabunde and Krebs, 1997). The structure was solved by multiple isomorphous replacement with wolfamate, a strong inhibitor, bound to the dinuclear metal center as one heavy-atom derivative. It was refined at 2.9 Å resolution in space group $C222_1$. The homodimeric enzyme, with a cystine disulfide bridge connecting the two subunits, was found to have the shape of a twisted hearth with overall dimensions 40 x 60 x 75 Å. The two dinuclear metal centers are 35 Å apart and lie at the bottom of a broad pocket formed by both monomers and are exposed to the solvent. The enzyme contains 432 amino acid residues and, as mentioned before, five heterogeneous glycosilation sites per monomer. Each subunit contains two domains: an N-terminal domain (residues 1-120) of unknown function and a C-terminal domain (residues 121-432) which contains the catalytic reaction center.

The active dimetalic center is shown in Fig. 3. The Fe(III) is coordinated by Tyr(167), which is obviously responsible for the characteristic charge transfer band at 560 nm and the purple color, by the N_{ε} of His(325), and monodentatly by one O-atom of the carboxylate group of Asp(135). The Zn(II) is ligated the Nε of His(286), the N_{δ} of His(323), and the amide oxygen of Asn(201). The two metal ions are bridged by one O-atom of the carboxylate group of Asp(164). The Fe-Zn distance was refined to 3.1 Å.

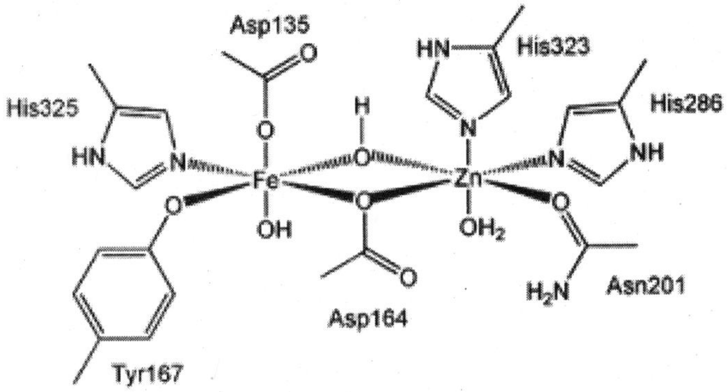

Fig. 3. Schematic drawing of the dinuclear metal center of read kidney bean purple acid phosphatase.

As it was expected at the resolution of 2.9 Å, the electron density map revealed no unambiguously the presence of water, hydroxide or oxide ligands. However, on the basis of the observed coordination geometry of the metal ions and on spectroscopic studies three exogenous ligands were modeled into the coordination spheres, as shown in Fig. 3: a terminal hydroxo ligand coordinated to iron, a terminal aqua ligand coordinated to zinc and a μ-hydroxo bridge, generating a practically perfect octahedron around Fe(III) and a distorted one around Zn(II).

The presence of a terminal hydroxo ligand on Fe(III) was disputed in later studies which suggested its absence and a trigonal pyramidal coordination around this metal center (Schenk *et al.*, 2008).

On the other hand, the presence of the postulated μ-hydroxo bridge is strongly supported by magnetic susceptibility measurements (Gehring *et al.*, 1996).

Such types of bridged dimetallic systems are not uncommon as they are also found in many other bioinorganic systems (Vincent *et al.*, 1990; Nordlund, 2001).

Purple Acid Phosphatase from Sweet Potato

The purification and physical properties of this phosphatase were reported more than forty years ago (Uehara *et al.*, 1974ab). However, experiments determining the metal content and composition of the active site have given contradictory results. Based on atomic absorption measurements and EPR-spectroscopic results the metal center has been identified as containing a Mn(III) dimmer (Fujimoto *et al.*, 1980; Sugiura *et al.*, 1980, 1981)) and one Fe(III) per subunit (Hefler and Averill, 1987), respectively.

More recent studies clearly pointed out the existence of three different isoforms of sweet potato PAP and two of these forms have been initially purified (Schenk *et al.*, 1999a, 2001). One form was shown to contain iron and zinc, present in the form of a dinuclear Fe(III)/Zn(II) site (Durmus *et al.*, 1999) in the other one the dinuclear site is conformed by a Fe(III)/Mn(II) pair (Schenk *et al.*, 1999a).

A detailed investigation of the isoform containing Fe(III)/Zn(II) in the active site using matrix-assisted laser desorption/ionization mass spectrometry analysis revealed a molecular mass of *ca.* 112 kD (Durmus *et al.*, 1999). The characteristics of the dimetal center were initially derived from an EXAFS study (Sift *et al.*, 1999) and the metal content was also reconfirmed by X-ray fluorescence, using synchrotron radiation (Durmus *et al.*, 1999). An analysis of the amino acid sequences suggests that the residues involved in metal-binding are identical to those found in read bean PAP and tyrosine functions as one of the ligands for the chromophoric Fe(III). The observed pH dependence of both, the charge transfer transition and the characteristics of the EPR signal, strongly suggests the presence of an ionizable group with a pK_a of 4.7, arising from an aquo ligand coordinated to Fe(III), which presence has strong mechanistic implications (Durmus *et al.*, 1999).

The investigation of the other isoform, containing manganese, shows that the enzyme is also conformed by two subunits with an approximate molecular weight of 55 kD each and confirms clearly the presence of one iron and one manganese ion in each subunit. The characteristic charge transfer transition is observed at 550-560 nm ($\varepsilon \approx 3000\ M^{-1}.cm^{-1}$). The metal environments are identical to those found in kidney bean PAP (Schenk *et al.*, 1999a). On the other hand, low temperature EPR measurements indicated strong antiferromagnetic coupling between the two metal centers, suggesting that these centers are linked via an μ-oxo bridge, in contrast to the μ-hydroxo bridge found in kidney bean PAP (Schenk *et al.*, 2001).

A number of possible explanations can be given for the discrepancies between the results commented above for the two well-characterized isoforms of sweet potato PAP and the previous reported data: a) partial loss of divalent metal during purification of the enzyme; b) exchange of one divalent metal cation for another; c) different metal contents depending on cultivar, soil composition, culture method and isolation procedure.

Furthermoe, for sweet potato PAP, an important advance for a deeper characterization of the enzyme was achieved by means of crystallographic studies. A preliminary X-ray diffraction study allows a first insight into the unit cell dimensions, refined under space group $P6_522$, and overall geometric characteristics (Schenk *et al.,* 1999b). A complete crystallographic study was performed some years later (Schenk *et al.,* 2005). The enzyme consists of two identical polypeptide chains of 435 amino acid residues. The crystal structure with bound phosphate was solved by molecular replacement to 2.5 Å resolution. The metal ions (Fe(III) and Mn(II)) are well resolved in both subunits and two phosphate molecules are observed in the structure as "islands" of electron density coordinated to the dimetallic center.

Although the overall structures of red kidney bean and sweet potato PAP are highly conserved, there are several regions where the structures are substantially different. An important difference is that the size of the active site pocket is considerably smaller in the sweet potato enzyme than that in kidney bean (Schenk *et al.,* 2005).

In the active site, the Fe(III) and Mn(II) ions are separated by a distance of 3.26 Å and both metals present coordination number six. The coordination of iron involves the N-atom of His324, a deprotonated tyrosine oxygen (Tyr166) and two aspartic acid O-atoms (Asp163 and Asp136), where the former bridges the two metal ions. The manganese is coordinated to two histidine N-atoms (His285 and His322), an asparagines O-atom (Asn200) and the mentioned bridging O-atom of Asp163. Three of the oxygen atoms of the bound phosphate anion are also involved in bonding. One of these atoms (O3) acts as a second bridge between the two metal ions, whereas O1 and O2 complete the coordination sphere of the two metal ions, generating a relative unusual μ-η^2-η^2-phosphate coordination (Schenk *et al.,* 2005).

The above mentioned third isoform was finally described in a most recent study as containing the Fe(III)/Fe(II) pair as the active center. This enzyme is a dimeric, disulfide linked PAP with a total molecular weight of *ca.* 110 kD, and is similar to that known for kidney bean. Enzymatic studies, spectroscopic properties (maximum absorption at 550-565 nm) and quantitative analytical analyses clearly confirmed a diiron enzyme. The coordination sphere of the two metal centers resulted identical to that reported for kidney bean and for the Fe/Mn pair of sweet potato, analyzed above, involving the same amino acid residues around the two irons. Metal replacement in the second metal binding site (Fe(II)) upon preparation of the semiapo-enzyme with Fe(II), Zn(II) or Mn(II) showed highest activities with Fe(II) (Waratrujiwong *et al.,* 2006).

Purple Acid Phosphatase from Yellow Lupin

A first study on the isolation, purification and general properties of PAP of white lupin (*Lupinus albus*) was obtained investigating the roots-secretion of PAP and its enzymatic activity, under phosphorus-deficiency conditions. It was found that the enzyme is a homodimer with an approximate molecular weight of 72 kD for each subunit and that it contains manganese (Ozawa *et al.,* 1995). A somewhat lower molecular weight of 49 kD, was found some years latter in another, similar, study (Miller *et al.,* 2001).

More detailed and definitive structural studies were performed latter with yellow lupin (*Lupinus luteus*). Initial studies, allowed the characterization of two unspecific PAPs in yellow lupin seeds (Olczak *et al.,*1997; Olczak and Watorek, 2003) as well as an enzyme that specifically hydrolyzed diphosphonucleotides and phosphodiesters (called PPD1) (Olczak *et al.,* 2000). Later studies of this enzyme have shown that it contains iron and manganese and presents a strong absorption maximum at 530 nm ($\varepsilon \approx 3600$ $M^{-1}.cm^{-1}$). Despite low amino acid sequence homology between PPD1 and the subunits of well-characterized plants purple acid phosphatases, their overall fold, topology of active center and metal content are highly similar (Olczak *et al.,* 2009).

The complete structure of this enzyme was determined recently by means of single-crystal X-ray diffractometry, using synchrotron radiation (Antonyuk *et al.,* 2014). The overall structure is defined by two domains: a sweet-potato like domain containing the Fe/Mn active site, and previously uncharacterized N-terminal domain with unknown function. The crystallographic asymmetric unit contains three identical subunits, which together form a homotrimer. Five cystine disulfide bridges are present in each subunit, three of which are involved in positioning the structural elements that form the trimer interfaces. The N terminal region (residues 1-150) is longer than in other PAPs. A further analysis of the structural data suggests that the most probable quaternary structure is a hexamer, with an estimated dissociation free energy of 42 kcal/mol, compared with 24 kcal/mol for the trimer. The trimers are rotated by approximately 60 ° with respect to each other, giving a staggered appearance to the N-terminal domains when viewed along the threefold axis. Experimental confirmation of the hexameric form of PPD1 in the solution state was obtained by small-angle X-ray scattering measurements. Such a homohexameric quaternary structure is unique for a PAP, since as previously shown, only homodimers have been reported in all other plant PAPs.

Purple acid phosphatases from different sources, including now PPD1, show a high degree of conservation among the residues present in the catalytic centre. The reported crystal structure shows the active site of PPD1, with a phosphate ion bound to both the metal atoms, similar to that previously described to be unique for sweet potato PAP (see above). The coordination environment of the metals is identical in each of the three subunits of the trimer. The Fe(II) is coordinated by protein residues His478, Tyr315, Asp271 and Asp312, which bridges the two metal ions. Mn(II) is coordinated by His 434, His476, Asn345 and the bridging Asp312. One oxygen-atom of the bound phosphate anion acts as an additional bridge between the two metal ions, and two more O-atoms from the phosphate group complete the coordination spheres of both metals (Antonyuk *et al.,* 2014).

Purple Acid Phosphatase from Soybean

A violet colored acid phosphatase from soybean (*Glycine max*) was initially reported by Fujimoto *et al.* (1977a). They could isolate and purify the enzyme, which shows the characteristic charge transfer band of PAPs at 540 nm. The only metal detected as associated with the protein was manganese, accompanied by trace amounts of copper and zinc.

Independently, some years later a PAP secreted by soybean suspension-culture cells was investigated and is, apparently, a homodimer of 58 kD monomers, containing one manganese ion in each subunit, presenting also the characteristic absorption band at 556 nm. The presence of iron in this enzyme was conclusively excluded (LeBansky *et al.,* 1992).

Later it was established that this PAP, isolated from soybean seeds, has a very close relation to the red kidney bean enzyme. It is conformed by two subunits, each with a molecular weight of *ca.* 55 kD and the active center is conformed again by a Fe(III)/Zn(II) pair. It also presents the characteristic purple color, with absorption maximum at 550-560 nm ($\varepsilon \approx 3000\ M^{-1}.cm^{-1}$) (Schenk *et al.,* 1999a).

It is possible again, that the PAP reported by Fujimoto *et al.* and LeBansky *et al.,* containing manganese, is one isoform whereas the Fe/Zn enzyme represents a second isoform of this PAP.

Purple Acid Phosphatases from Other Plants

Putative PAP genes have been described in an important number of different plants but only in a few cases the corresponding enzyme has been isolated and partially, or totally, characterized.

In the case of rice (*Oriza sativa*) six violet-colored acid phosphatases associated with cell wall preparations of cultured cells could be isolated and characterized. Their molecular weight was determined in each case to be approximately 65 kD. The metal analysis of all the purified isozymes showed the presence of significant amounts of manganese. Furthermore, when each enzyme was denaturated by guanidine-HCl, the bound manganese was released from the protein (Igaue *et al.,* 1976).

Three PAPs from potato (*Solanum tuberosum* L) have been characterized. Protein sequence analysis revealed that they contain structural features typical for PAPs. The analysis of the N-terminal sequences of the three potato amino acid sequences indicated that all three proteins contain a putative secretory signal sequence, a structure that is common to many described PAPs. Five conserved motifs containing the seven described residues involved in metal binding are found throughout all compared sequences. One of these phosphatases has a low molecular weight (ca. 35 kDa), and is similar to other PAPs found in mammals, plants and cyanobacteria. It is homologous to the PAPs decribed in sweet potato, rice, kidney bean and soybean. The other two phosphatases present higher molecular weights (ca. 55 kDa) and are typical for plants. No information of the metal or metals present in the active site of potato PAPs has so far been reported (Zimmermann *et al.,* 2004).

Different studies have been performed in relation to purple acid phosphatases from tomato (*Lycopersicon esculentum*). Three phosphate starvation-responsive PAPs have been identified from tomato suspension cell cultures; two secreted acid phosphatases (named SAP1 and SAP2), are monomeric glycoproteins with molecular weights of 84 and 57 kDa, that displayed the characteristic PAP absorption maxima at 518 and 538 nm, respectively. The third is an intracellular acid phosphatase (named IAP), and is a heterodimeric protein composed of 63 and 57 kDa subunits (Bozzo *et al.,* 2002, 2004). Later, detailed analysis found different spatial and temporal expression patterns of SAP1, SAP2 and IAP in phosphate-deficient conditions (Bozzo *et al.,* 2006).

The mentioned studies were performed on suspension cell cultures, rather than on whole tomato plants. A recent investigation of the response of the plant to phosphate-starvation, especially during seed germination and early seed growth, which are plant's first interactions with the environment, allowed the characterization of another phosphatase (named SIPAP1). The amino acid sequence of the protein showed 56-89% similarity with other plants PAPs and contain the conserved amino acid motifs required for metal binding in the active site. Based on its putative amino acid sequence, SIPAP1 is predicted to be a secretory glycoprotein with a molecular weight of 50.7 kDa. As mentioned above, tomato suspension cell culture secreted SAP2 (molecular weight 57 Da) during phosphate-starvation. Therefore, it may be possible that SIPAP1 and SAP2 are identical PAPs (Suen *et al.,* 2015). Also in this case, no information on the metal composition of these metalloproteins is so far available.

A purple acid phosphatase isolated from leaves of spinach (*Spinacia oleracea*) was described by Fujimoto *et al.* (1977b). The purified preparation shows the characteristic PAP charge-transfer absorption at 530 nm. The molecular weight of the enzyme was found to be approximately 92 kDa and it is composed by two identical subunits. The metal analysis shows the presence of an important amount of manganese.

A partially purified PAP from mung bean (*Vigna radiata*) seeds was reported some years ago (Surchandra *et al.,* 2012). Its molecular weight was estimated as 34.5 kDa. The optimal pH-value for enzyme activity was found to be 5.2 and maximum activity was observed at 70 °C. Some time later, a similar enzyme, with a molecular weight of 29 kDa, was isolated from leaves of germinating mung bean seeds, and some of its general chemical and biochemical properties were discussed (Saeed *et al.,* 2014).

Although glycosylphosphatidylinositol (GPI)-anchored proteins occur frequently on the surface of animal cells, the existence of a GPI-anchored protein in plant cells was established for the first time by the presence of a supposed alkaline- phosphatase anchored in this way in the aquatic plant *Spirodela oligorrhiza* (Morita *et al.,* 1996). Notwithstanding, further studies on this plant enzyme conclusively showed that it is a purple acid phosphatase (Nakazato *et al.,* 1998). The GPI-anchored phosphatase, retaining part of its GPI anchor is a 100 kDa homodimer consisting of 57 kDa subunits (Nakazato *et al.,* 1997).

Metal analysis shows the presence of iron and manganese. Solutions of the purified enzyme were purple-colored and the visible absorption spectrum presents a maximum at approximately 556 nm. Additional data confirm that this PAP closely resembles the PAP of kidney bean, *Arabidospis* and soybean (Nakazato *et al.,* 1998).

The *Arabidopsis thaliana* purple acid phosphatase family (AtPAP) is encoded by 29 genes (Li *et al.,* 2002; Tran *et al.,* 2010). To date, only AtPAP15, AtPAP23 and AtPAP26 have been functionally characterized in transgenic *Arabidopsis* (Tran *et al.,* 2010).

A series of recent studies provide convincing evidences that AtPAP26 is the predominant intracellular, as well as a major secreted acid phosphatase isozyme upregulated by phosphate-starved *Arabidopsis.* It exhibits the typical purple color (absorption maximum at 520 nm) and exists as a 100 kDa homodimer composed of 55 kDa glycosilated subunits (Veljanovski *et al.,* 2006).

AtPAP15 is the only member of this family that has thus far been shown to possess significant phytase activity and it appears also involved in ascorbate synthesis (Tran *et al.,* 2010)

In another *Arabidopsis thaliana* PAP (AtPAP23), the presence of important amounts of iron and manganese was established (Zhu *et al.,* 2005), suggesting that this may be the metal couple present in the purple acid phosphatase of this plant.

AtPAP17 is one of the few phosphate-starvation inducible purple acid phosphatases that has been purified and characterized from phosphate-starved *Arabidopsis* seedlings. It presents a molecular weight of *ca.* 34 kDa, shows significant sequence similarities to other plant PAPs and it shares all the residues implicating in metal binding, including the tyrosinate residue specific of PAPs (del Pozo *et al.,* 1999; Tran *et al.,* 2010). It was also demonstrated that AtPAP17 is a binfunctional enzyme that has both acid phosphatase and alkaline peroxidase activity, suggesting that it may be involved in phosphate mobilization and in the metabolism of reactive oxygen species (del Pozo *et al.,* 1999).

SOME COMPARISONS OF PLANT PAPs WITH MAMALIAN PAPs

The primary structure of uteroferrin (Uf), a 35 kDa monomeric mammalian PAP containing a Fe(III)/Fe(II) dimetal center, has been compared with the sequence of the homodimeric Fe/III(/Zn(II) kidney bean PAP. The alignment suggests that the amino acid residues coordinating the dimetal center are identical in both enzymes, although the geometry of the coordination sphere might be slightly different. Secondary structure predictions suggest that also the folding topologies of both PAPs are similar (Klabunde *et al.,* 1995).

What's more, in the plant enzyme Zn(II) can be exchanged for Fe(II) as well as in the mammalian enzyme Fe(II) can be exchanged for Zn(II) without affecting the activity. The similarity between Fe(III)/Zn(II) kidney bean enzyme and the native Fe(III)/Fe(II) mammalian PAP regarding the visible absorption spectra, EPR spectra and Mössbauer data and regarding the interaction with substrates and oxoanions inhibiting the phosphatase reaction indicates a high degree of conformity in the active site. Since Uf and other mammalian PAPs show over 90% homology, all the mentioned conclusions hold for all mammalian PAPs so far characterized (Klabunde *et al.,* 1995).

On the other hand, the characteristic N-domain found in plant PAPs is absent in Uf PAP. Since this domain does not contain amino acid involved in metal binding and is probably not involved in the catalytic process, it seems reasonable that this domain is absent in the smaller mammalian enzymes (Klabunde *et al.,* 1995).

The 1.55 Å resolution structure of pig Uf has been solved, using the oxidized Fe(III)/Fe(III) form, in the presence of phosphate, and by multiple isomorphous replacement (Guddat *et al.,* 1999). The determined Fe(III)-Fe(III) distance is 3.31 Å, comparable to that found for the Fe(III)-Zn(II) distance in kidney bean. The coordination spheres of both iron ions are close to octahedral and phosphate bridges the two metals symmetrically. The same group of ligands as in the other analyzed PAPs are fond coordinated to the two iron centers and one μ-oxo bridge, together with the usual aspartate residue, bridges the two centers. Coordination spheres are completed by O-atoms of the phosphate group (Guddat *et al.,* 1999).

Interestingly, there are significant differences between the pig Uf and the red kidney bean enzymes beyond the immediate vicinity of the metal ions. The most obvious difference is the presence of a loop, which is an insert in the Uf sequence compared with that of kidney bean (Guddat *et al.,* 1995). This loop has, eventually, an effect on the catalytic efficiency and on the substrate-specificity of the enzyme (Guddat *et al.,* 1995; Sträter *et al.,* 2005).

A 2.7 Å resolution crystal structure of a recombinant tartrate resistant acid phosphatase from rat, in its oxidized form and in the presence of phosphate, and using multiple isomorphous replacements, was also described. The side-chains coordinating the iron ions are equivalent to those found in the red kidney bean enzyme. The distance between the two iron ions was refined to 3.1 Å and the phosphate ion was modeled to coordinate these ions in a bidentate fashion. Also in this case the presence of a surface loop near the active site was described (Uppenberg *et al.*, 1999).

The crystal structure of a PAP from rat bone has been determined and refined to 2.2. Å resolution. In this case, the two Fe(III) ions are separated by 3.5 Å and their environment is similar to that found in the other described PAPs with an approximately octahedral coordination of the two metallic centers. It is not totally clear if the μ-oxo bridge is an oxygen atom or an OH-group; and one of the metal ions bound one solvent molecule whereas the other incorporate an O-atom of a sulfate anion (a consequence of the crystallization process) (Lindqvist *et al.*, 1999).

One interesting and important conclusion derived from these structural experimental studies of mammalian PAPs, in comparison with those of kidney bean PAP and other plant PAPs is that mammalian enzymes show a high degree of similarity in their amino acid sequences (90% identities); whereas the sequence identity between mammalian and plant enzymes is only around 20%. Notwithstanding, comparison of the structures of both enzymes revealed that, in spite of this low overall similarity in sequence, the overall 3D structures are very similar and, in particular, all residues involved in coordination of the metal ions are invariant.

AN INSIGHT INTO THE MECHANISM OF PURPLE ACID PHOSPHATASES

Numerous studies have been performed to understand the role of the two metals at the active site in the reaction path. The pH-dependence as well the EPR and visible spectra suggest that an Fe(III) bounded OH-group functions as the nucleophile attacking the phosphate group (Dietrich *et al.*, 1991; Klabunde *et al.*, 1997). Moreover, the independence of stopped-flow rate constants from substrate concentration suggests that a rapid binding of the substrate to the M(II) center precedes the rate controlling attack of the nucleophilic hydroxide ion.

Interactions of different tetrahedral oxoanions with PAPs have been investigated in several studies in order to attain a wider insight into the detailed mechanism of phosphate ester hydrolysis. PAPs exhibit product inhibition by phosphate, as well as by arsenate, vanadate, molybdate and wolframate (Klabunde *et al.*, 1996; Klabunde and Krebs, 1997).

Interesting advances were obtained by means of a structural study of kidney bean PAP complexed with phosphate as an enzyme product model and with the tight binding inhibitor wolframate bound to the active site. Also a newly refined structure of the free enzyme was determined, based on a 2.65 Å resolution data set collected with synchrotron radiation (Klabunde *et al.*, 1996). It is found that phosphate interacts in a bidentate fashion with both metal centers, displacing the presumed solvent molecules. The bridging mode of phosphate is nearly symmetrical with Fe-O-P and Zn-O-P angles of 120° and 127°, respectively. The inhibitory wolframate anion was found to bind in a nearly identical form as phosphate, with Fe-O-W and Zn-O-W angles of 112° and 117°, respectively (Klabunde *et al.,* 1996).

On the basis of these and other previous results, it is possible to propose that the phosphate group of the substrate is bound to the Zn(II) ion via one of the non-esterified O-atoms, under exchange of the water ligand. Then, the substrate is oriented for an in-line attack of the Fe(III) bound OH^- anion from a position opposite the esterified oxygen. The configuration at the phosphorous is inverted via a pentacoordinate transition state in which the leaving group and the attacking nucleophile are in apical positions. In order to lower the energy of such a transition state, in addition to the Lewis acidity of the zinc ion, electrostatic interactions with two protonated side chains occur and withdraw negative charge from the phosphorus atom. Subsequently, the alcohol group of the substrate is released generating a phosphate ion bound to the two metal centers via two O-atoms (Klabunde *et al.,* 1996; Klabunde and Krebs, 1997).

As the phosphate bound kidney bean phosphatase illustrates the product-bound state rather than providing a mimic for the pre-catalytic substrate bound complex, the interaction of the enzyme with fluoride and sulfate was investigated to attain a deeper insight into this situation. While no experimental data for the structure of the pre-catalytic complex are currently available, the crystal structure of the kidney bean PAP-sulfate complex provides some information and identifies interactions that are likely to be of importance in stabilizing this pre-catalytic state (Schenk *et al.,* 2013). In contrast to the PO_4-bound kidney bean-PAP structure, the sulfate ion in the active site does not coordinate directly to the metal ions. Instead it is positioned in the second coordination sphere, stabilized via an extensive hydrogen bond network, which involves the μ-hydroxo group and two of the sulfato oxygens (Schenk *et al.,* 2008).

Besides, a recent structural study of kidney bean-PAP in the presence of fluoride, known as an enzymatic inhibitor (Elliot *et al.,* 2006), has shown that fluoride is able to displace the OH-group bridging the two metal centers, thus forming a μ-fluoro binuclear site. The alternative proposed mechanism, derived from these studies, suggest that the bridging OH-group acts as the nucleophile. Therefore, following the binding of the substrate in the pre-catalytic complex, structural rearrangements lead to the bidentate interaction of the phosphate with the two metallic centers and to a "quasi monodentate" interaction with the μ-hydroxide. Nucleophilic attack by this group is followed by the release of the leaving group and the active site is returned to its resting state by the exchange of the bound phosphate by two water molecules (Schenk *et al.,* 2008).

Depending on the type of substrate used in the reaction, and perhaps also on the metal composition of the active site, it may be possible that PAPs employ different nucleophiles to initiate hydrolysis. The nucleophilic hydroxyde has been proposed to be: i) terminally bound to Fe(III); ii) bridging the two metal centers; iii) residing in the second coordination sphere (Mitić *et al.,* 2010; Schenk *et al.,* 2013).

POSSIBLE FUNCTIONS OF PLANT PURPLE ACID PHOSPHATASES

Despite extensive knowledge on substrate specificity, action of inhibitors, subcellular localization, active site structure and geometry, relationship to other enzymes, protein sequence, and the chemical mechanism, the physiological function of plant PAPs continues relatively unclear (Klabunde and Krebs, 1997; Olczak *et al.,* 2003; Schenk *et al.* 2013). Notwithstanding, there is increasing agreement that one general role of these PAPs may be the mobilization and/or scavenging of inorganic phosphate from organophosphates in the soil.

Orthophosphate (Pi) plays a central role in virtually all major metabolic processes in plants, including photosynthesis and respiration. It is also a key structural component of important biomolecules such as nucleic acids, phospholipids, ATP, NADPH and sugar-phosphates (Bezkorovainy and Rafelson, 1996). However, the soluble Pi concentration of many soils is appreciably lower than the intracellular Pi concentration required for optimal plant growth and development. Maximizing the yield of most modern crop varieties therefore, relies on the extensive application of phosphate fertilizers (Tran *et al.,* 2010).

Phosphorus is assimilated by plants in its fully oxidized state, $H_2PO_4^-$ or HPO_4^{2-}; whereas PO_4^{3-} is the least accessible macronutrient in many natural soils because it readily forms insoluble compounds. Extensive soil Pi reserves exist in the form of phosphate-esters, derived from decomposed organic material. Plant has two possible mechanisms to acquire Pi from the soil: i) they can solubilize free Pi by secretion of large amounts of organic acids into the rhizosphere or ii) they can mobilize Pi from soil organic Pi-pools secreting hydrolytic enzymes that free esterified phosphate. Notwithstanding, where soils organic Pi-pools are accessible to roots remains unclear, and most plants cannot use phytate (*myo*-inositol hexaphosphate), an abundant organic-Pi component of certain soils, as a source of nutritional phosphate (Tran *et al.,* 2010).

Within species-dependent limits, plants have evolved the ability to tolerate extended periods of Pi deprivation by eliciting a complex array of morphological, physiological and biochemical adaptations, collectively known as the Pi-starvation response. The Pi-starvation response arises in part from the coordinated induction of hundreds of Pi-starvation inducible genes that reprioritize internal Pi use and maximize external Pi acquisition. The evolution of elaborate strategies to enhance the acquisition and use of Pi from the environment allows many plants to effectively acclimate to periods of nutritional Pi deficiency (Tran *et al.*, 2010).

The involvement of PAPs in these processes has been clearly established as it has been shown that a variety of intracellular and secreted acid phosphatase isozymes that are upregulated following plant Pi deprivation are purple acid phosphatases. As mentioned above, such PAPs have been biochemically characterized from different species, including bean, tomato, lupin, sweet potato, and *Arabidopsis*, between others. Most of the known PAPs are secreted, soluble proteins, except for the mentioned glycosylphosphatidylinositol-anchored phosphatase isolated from *Spirodela oligorrhiza* (Olczak *et al.*, 2003).

Another interesting aspect is the fact that most PAPs are unspecific, hydrolyzing a broad spectrum of phosphate esters, including ATP, sugar phosphates, mononucleotides and inorganic diphosphates (Olczak *et al.*, 2003). Differential glycosilation might represent a subtle form of cellular control, either in targeting phosphatases to different compartments, or increasing their affinity for specific proteins or enzymes (Duff *et al.*, 1994).

Apart from its role in phosphate acquisition, other possible functions remain obscure (Olczak *et al.*, 2003; Tran *et al.*, 2010). An auxiliary role in the utilization of phytate has been suggested (Duff *et al.*, 1994; Olczak *et al.*, 2003). During germination, the PAP isolated from soybean cotyledons efficiently hydrolyses phytate (Hegeman and Grabau, 2001). Homologous phytase sequences also have been found in *Arabidopsis* genome (Li *et al.*, 2002). On the other hand, a mutant from another *Arabidopsis* PAP, with phytase activity, has shown to increase foliar ascorbate, one of the most important antioxidants in plant cells, and also cofactor of a relevant number of key enzymes (Zhang *et al.*, 2008). It has become increasingly evident that PAPs with phytase activity are very common in different plant species (Kuang *et al.*, 2009; Schenk *et al.*, 2013).

The peroxidase activity detected in some plant PAPs may suggest that these enzymes are also involved in the removal of reactive oxygen species in stressed or senescent plant organs (del Pozo *et al.*, 1999).

Additionally, it has been suggested that these enzymes may also be involved in the transport and storage of iron and zinc (Olczak *et al.*, 2003).

Evidently, more studies and investigations are necessary to attain a complete understanding on the possible functions of these enzymes and the relations and possible interdependence among all of them.

MODEL STUDIES RELATED TO PURPLE ACID PHOSPHATASES

The synthesis and physicochemical characterization of model compounds analogous to important natural systems is one of the main challenges for bioinorganic chemists. In this context the bioinorganic chemistry of iron, zinc and manganese has often been explored in great detail using a variety of model systems (Kaim and Schwederski, 1991; Lippard and Berg, 1994; Baran, 1995; Casas *et al.*, 2002; Roat-Malone, 2002).

Since the detection of PAPs and of the nature of their active metallic sites, a considerable number of synthetic and physicochemical studies have been reported on the structural and functional properties of biomimetic model compounds. Besides various models with Fe/Fe sites a special preparative and relative complex challenge was the synthesis of heterodinuclear models for the Fe/Zn site in PAPs from plants (Krebs, 2006). On the other hand, an important number of model studies have also been devoted to the hydrolysis of phosphates, esters and related molecules (Bashkin, 1999).

For the synthesis of diiron complexes, many research groups employed tripodal ligands. In most cases such ligands lead to μ-oxo bridged complexes. Because of their chelating effect and the predictable formation of dinuclear centers the use of dinucleating ligands is a promising and relatively powerful strategy for the synthesis of dinuclear iron complexes.

In the case of PAP models dinucleating ligands having an oxygen/nitrogen ligand sphere are most suitable for the preparation of biomimetic model complexes. For example, the ligand types depicted in Fig. 4 generate μ-alkoxo and μ-phenoxo bridged complexes, respectively. Variation of the residues R-R''' generate different Lewis acidities at the metal centers. Using benzimidazole, phenol, pyridine and alkyl groups as residues, an important number of different complexes were obtained. All these complexes were characterized by single crystal X-ray structure analysis and Mössbauer spectroscopy. Also their magnetic and electrochemical behavior was investigated (Than *et al.*, 1999).

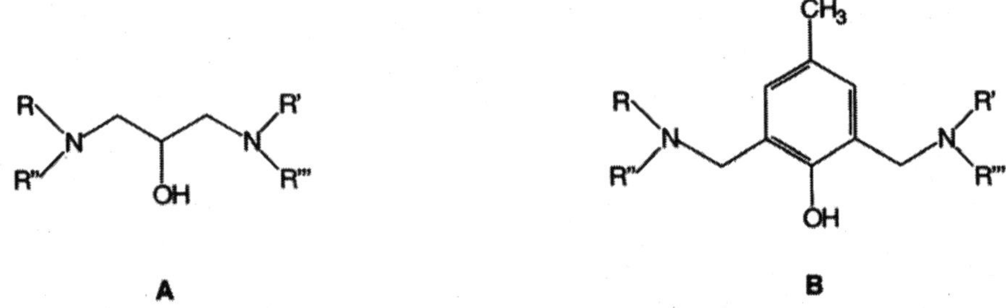

Fig. 4. Structure of dinucleating alkoxo- (**A**) and phenoxo-bridging (**B**) ligands.

The first two model complexes closely related to the active site of PAPs were obtained with one of the derivatives of ligand **B** of Fig. 4, namely, 2,6-bis[bis(2-pyridylmethyl) aminomethyl]-4-methylphenol (Hbpmp). Reaction of equimolecular amounts of $Fe(ClO_4)_3{\cdot}9H_2O$ and $ZnBr_2$ in methanol with Hbpmp leads, after addition of diphenyl phosphate, bluish black crystals of [Fe(III)Zn(II) (bpmp) $(O_2P(OPh_2)]$

$(ClO_4)_2 \cdot 1.5CH_3OH \cdot H_2O$. The analogous reaction with two equivalents of $Fe(ClO_4)_3 \cdot 9H_2O$ yields dark green crystals of [Fe(III)Fe(II)(bpmp) $(O_2P(OPh_2)]$ $(ClO_4)_2 \cdot CH_3OH \cdot H_2O$. Both complexes contain a μ-phenoxo bridged M(III)M(II) unit in which the metal centers are additionally linked by two μ-diphenylphosphato bridges. The nitrogen donors of the organic ligand bpmp⁻ complete an octahedral environment of the two metal centers. Mössbauer spectra and electrochemical behavior confirms the oxidation state of the iron cations present in these structures (Schepers *et al.,* 1990; Than *et al.,* 1999).

Analogous Fe(III)/Zn(II) and Fe(III)/Fe(II) complexes were also prepared with the ligand 2,6-bis({bis(2-pyridylmethyl)amino}methyl]-4-methoxyphenol. The resulting metal coordination is similar to those described for Hbpmp, as the metal centers exhibit octahedral coordination and the metal atoms are linked by a μ-phenoxo group of the ligand and by two μ-diphenylphosphato bridges; the coordination sphere is complemented by N-donor atoms of the ligands (Albedyhl *et al.,* 2001).

The μ-alkoxobis(μ-hydrogenphosphato)diiron(III) core in $[Fe_2$(tbpo) $(O_2PO(OH)_2]$ $\cdot ClO_4 \cdot CH_3OH \cdot 8.5H_2O$ (Htbpo, N,N,N',N'-tetrakis(2-benzimidazolylmethyl)-1,3-diamino-2-propanol) was the first example for a type of 'inorganic' phosphate bridging a biomimetic Fe(III)–Fe(III) complex confirmed by single crystal X-ray analysis (Krebs *et al.,*1997). Consequently, it may be considered as a model compound for the product complex of the hydrolysis (Than *et al.,* 1999).

Mixed-valent (μ-alkoxo)diiron(II,III) complexes with the ligand *N,N,N',N'*-tetrakis(2-pyridylmethyl)-1,4-diamino-2-butanol were characterized by X-ray crystallography as well as by Mössbauer and electronic spectroscopy and electrochemical measurements. The asymmetric bridging unit leads to five- and six-membered chelate rings (Suzuki *et al.,* 1998).

In another study, heterodinuclear Zn(II)/Fe(III) complexes with different coordination spheres, μ-alkoxo-bridging symmetric and asymmetric dinucleating ligands were prepared and structurally characterized. The Zn(II) center exhibits a trigonal-bipyramidal coordination sphere whereas that of Fe(III) is octahedral. Also two Zn(II)/Zn(II) complexes with similar dinucleating ligands were prepared, showing both metallic centers in a trigonal-bipyramidal environment (Albedyhl *et al.,* 2002).

As most of the prepared heterodinuclear Zn(II)/Fe(III) model complexes have been prepared using symmetrical ligands, an interesting advance in the design of such models was the regio-directed synthesis of such a complex, derived from an unsymmetrical ligand. The use of an unsymmetrical ligand, possessing both a hard and a soft site may be considered as an interesting synthetic strategy to prepare heterobimetallic complexes in a directed, non statistical way. This was achieved using the binucleating ligand shown in Fig. 5, which involves two different coordination sites connected by a phenol spacer: a hard site, a [bis2-

hydroxybenzyl)amino]methyl moiety, suited for Fe(III) complexation and a softer site, a [bis(2-pyridinylmethyl)-amino]methyl moiety, suited for Zn(II) complexation. In the obtained complex, both metallic centers show octahedral coordination. The NO_5 coordination around the Fe(III) models the oxygen rich coordination sphere of this ion in plants purple acid phosphatases (Belle *et al.,* 2002).

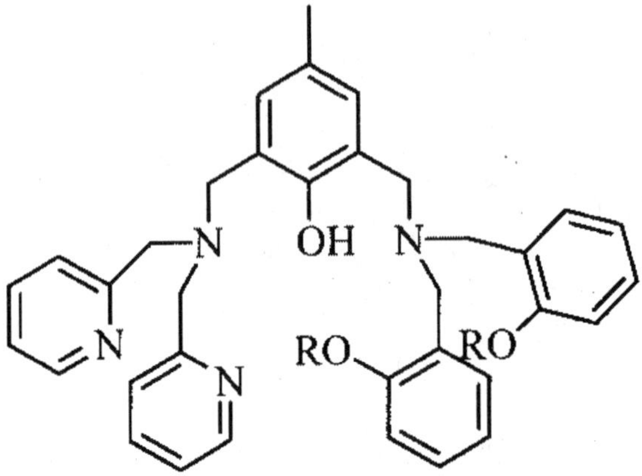

Fig. 5. Structure of the dinucleating unsymmetrical ligand 2-{[bis(2-hydroxybenzyl)]amino}methyl-6-{[2-pyridinylmethyl)amino]methyl}-4-methylphenol.

What's more, the crystallographic study of a recently reported tetranuclear Fe(III) complex with another asymmetric multidentate ligand, has shown that two of the iron centers present a similar NO_5 coordination as those mentioned above; whereas the other two have a N_2O_4 environment. Detailed information of this complex was obtained by UV-vis and MCD spectroscopy, as well as by magnetic susceptibility measurements and variable-temperature Mössbauer spectroscopy, giving a very interesting insight into the overall characteristics of these metal-to-ligand environments (Boudalis *et al.,* 2007).

Finally, it is also interesting to comment that different Fe(III)/Fe(III) and Fe(III)/Fe(II) model complexes have also been prepared and characterized in the last years as useful models for the active site of mammalian purple acid phosphatases (Bernhardt *et al.* 2015; Comba *et al.,* 2012a,b). Also these model systems present valuable and useful information in the context of plant PAPs.

Acknowledgement

The author is an Emeritus Professor from the Universidad Nacional de La Plata and the continuous support from this University is gratefully acknowledged.

REFERENCES

Ahlers, F; Zippel, F; Klabunde, T; Krebs, B; Löcke, R; Witzel, H and Nolting HF. 1995. X-ray absorption studies of the purple acid phosphatase from red kidney beans (native enzyme, metal exchanged form). *Physica B* **208/209:** 731-2.

Albedyhl, S; Averbuch-Pouchot, MT; Belle, C; Krebs, B; Pierre, JL; Saint-Aman, E and Torelli, S. 2001. Dinuclear zinc(II)-iron(III) and iron(II)-iron(III) complexes as models for purple acid phosphatases. *Eur. J. Inorg. Chem.* **2001:** 1457-64.

Albedyhl, S; Schnieders, D; Jancsó, A; Gajda, T and Krebs, B. 2002. Heterodinuclear Zinc(II)-iron(III) complexes and dinuclear zinc complexes as models for zinc-containing phosphatases. *Eur. J. Inorg. Chem.* **2002:** 1400-9.

Antanaitis BC; Aisen, P and Lilienthal, HR. 1983. Physical characterization of two- iron uteroferrin. *J. Biol. Chem.* **258:** 3166-72.

Antonyuk, SV; Olczak, M; Olczak; T; Ciuraszkiewicz, J and Strange, RW. 2014. *IUCrJ* **1:** 101-9.

Averill, BA; Davis, JC; Burman, S; Zirino, T; Sanders-Loehr J; Loeher, TM; Sage, JT and Debrunner, PT. 1987. Spectroscopic and magnetic studies of the purple acid phosphatase from bovine spleen. *J. Am. Chem. Soc.* **109:** 3760-7.

Baran, EJ. 1995. *Química Bioinorgánica*, Mc.Graw-Hill Interamericana de España, S.A., Madrid.

Bashkin, JK. 1999. Hydrolysis of phosphates, esters and related substrates by models of biological catalysts. *Curr. Opin. Chem. Biol.* **3:** 752-8.

Baumbach, GA; Saunder, PKT; Ketcham, CM; Bazer, FW and Roberts, RM. 1991. Uteroferrin contains complex and high mannose-type oligosaccharides when synthesized *in vitro*. *Mol. Cell. Biochem.* **105:** 107-17.

Beck, JL; McCornachie, LA; Summors, AC; Arnold, WN; de Jersey, J and Zerner, B. 1986. Properties of a purple phosphatase from red kidney bean: a zinc-iron metalloenzyme. *Biochim. Biophys. Acta* **869:** 61-8.

Beck; JL; de Jersey, J; Zerner, B; Hendrich, MP and Debrunner, PG. 1988a. Properties of the Fe(II)-Fe(III) derivative of red kidney bean purple phosphatase. Evidence for a Zn-Fe center in the native enzyme. *J. Am. Chem. Soc.* **110:** 3317-8.

Beck, JL; McArthur, MJ; de Jersey, J and Zerner, B. 1988b. Derivatives of purple phosphatase from red kidney bean: replacement of zinc with other divalent metal cations. *Inorg. Chim. Acta* **153:** 39-44.

Belle,C.; Gautier-Luneau, I; Karmazin, L; Pierre, JL; Albedyhl, S; Krebs, B and Bonin, M. 2002. Regio-directed synthesis of a $Zn^{II}Fe^{III}$ complex from an unsymmetrical ligand and its relevance to purple acid phosphatases. *Eur. J. Inorg. Chem.* **2002:** 3087-90.

Bernhardt, PV; Bosch, S; Comba, P; Gahan, LR; Hanson, GR; Mereacre, V; Noble, CJ; Powell, AK; Schenk, G and Wadepohl, H. 2015. An approach to more accurate model systems for purple acid phosphatases (PAPs). *Inorg. Chem.* **54:** 7249-63.

Bezkorovainy, A and Rafelson, ME. 1996. *Concise Biochemistry*, Marcel Dekker, New York.

Bortolato, M; Besson, F and Roux, B. 1999. Role of metal ions on the secondary and quaternary structure of alkaline phosphatase from bovine intestinal mucosa. *Proteins: Struct., Funct. Genet.* **37:** 310-18.

Boudalis, AK; Aston, RE; Smith, SJ; Mirams, RE; Riley, MJ; Schenk, G; Blackman, AG; Hanton, LR and Gahan, LR. 2007. Synthesis and characterization of a tetranuclear iron(III) complex of a new asymmetric multidentate ligand. A structural model for purple acid phosphatases. *J. Chem. Soc. Dalton Transact.* **2007:** 5132-9.

Bozzo, GG; Raghothama, KG and Plaxton, WC. 2002. Purification and characterization of two secreted purple acid phosphatase isozymes from phosphate-starved tomato (*Lycopersicon esculentum*) cell cultures. *Eur. J. Biochem.* **269:** 6278-86.

Bozzo, GG; Raghothama, KG and Plaxton, WC. 2004. Structural and kinetic properties of a novel purple acid phosphatase from phosphate-starved tomato (*Lycopersicon esculentum*) cell cultures. *Biochem. J.* **377:** 419-28.

Bozzo, GG; Dunn, EL and Plaxton, WC. 2006. Differential synthesis of phosphate-starvation inducible purple acid phosphatase isozymes in tomato (*Lycopersicon esculentum*) suspension cells and seedlings. 2006. *Plant, Cell and Environm.* **29:** 303-13.

Campbell, HD and Zerner, B. 1973. A low-molecular weight acid phosphatase which contains iron. *Biochem. Biophys. Res. Commun.* **54:** 1498-503.

Casas, JS; Moreno, V; Sánchez, A; Sánchez, JL and Sordo, J. 2002. *Química Bioinorgánica*, Editorial Sintesis, Madrid.

Chen, TT; Baser, FW; Cetorelli, JJ; Pollard, WE and Roberts, RM. 1973. Purification and properties of a progesterone-induced basic glycoprotein from the uterine fluids of pigs. *J. Biol. Chem.* **248:** 8560-6.

Coleman, JE. 1992. Structure and mechanism of alkaline phosphatase. *Ann. Rev. Biophys. Biomol. Struct.* **21:** 441-83.

Comba, P; Gahan, LR; Mereacre, V; Hanson, GR; Powell, AK; Schenk, G and Zajaczkowski-Fischer, M. 2012a. Spectroscopic characterization of the active $Fe^{III}Fe^{III}$ and $Fe^{III}Fe^{II}$ forms of a purple acid phosphatase model system. *Inorg. Chem.* **51:** 12195-209.

Comba, P; Gahan, LR; Hanson, GR; Mereacre, V; Noble, CJ; Powell, AK; Prisecaru, I; Schenk, G and Zajaczkowski-Fischer, M. 2012b. Monoesterase activity of a purple acid phosphatase mimic with a cyclam platform. *Chem.-Eur. J.* **18:** 1700-10.

del Pozo, JC; Allona, I; Rubio, V; Leyva, A; de la Peña, A; Aragoncillo, C and Paz-Ares, J. 1999. A type 5 acid phosphatase gene from Arabidopsis thaliana is induced by phosphate starvation and by some other types of phosphate mobilizing/oxidative stress conditions. *The Plant J.* **19:** 579-89.

Dietrich, M; Münstermann, D; Suerbaum, H and Witzel, H. 1991. Purple acid phosphatase from bovine spleen. Interactions at the active site in relation to the reaction mechanism. *Eur. J. Biochem.* **199:** 105-13.

Duff, SMG; Sarath, G and Plaxton, WC. 1994. The role of acid phosphatases in plant phosphorus metabolism. *Physiol. Plant.* **90:** 791-800.

Durmus, A; Eicken, C; Sift, BH; Kratel, A; Kappi, R; Hüttermann, J and Krebs, B. 1999. The active site of purple acid phosphatase from sweet potatoes (*Ipomea batatas*) *Eur. J.Biochem.* **260:** 709-16.

Elliot, TW; Mitić, N; Gahan, LR; Guddat, LW and Schenk, G. 2006. Inhibition studies of purple acid phosphatases: Implications for the catalytic mechanism. *J. Braz. Chem. Soc.* **17:** 1558-65.

Fujimoto, S; Nakagawa, T and Ohara, A. 1977a. Isolation of violet-colored acid phosphatase from soybean. *Agric. Biol. Chem.* **41:** 599-600.

Fujimoto, S; Nakagawa, T; Ishimitsu, S and Ohara, A. 1977b. Purification and some properties of violet-colored acid phosphatase from spinach leaves. *Chem. Pharm. Bull.* **25:** 1459-62.

Fujimoto, S; Ohara, A and Uehara, K. 1980. Carbohydrate and metal analyses of violet-colored acid phosphatase of sweet potato. *Agric. Biol. Chem.* **44:** 1659-60.

Gehring, S; Fleischhauer, P; Behlendorf, M; Hüber, M; Lorösch, J; Haase, W; Dietrich, M; Witzel, H; Löcke, R and Krebs, B. 1996. Magnetic susceptibility studies on the diiron forms of the metalloprotein purple acid phosphatase from bovine spleen and kidney bean. *Inorg. Chim. Acta* **252:** 13-7.

Guddat, LW; McAlpine, AS; Hume, D; Hamilton, S; de Jersey, J and Martin, JL. 1999. Crystal structure of mammalian purple acid phosphatase. *Structure* **7:** 757-67.

Hefler, S and Averill, B. 1987. The manganese(III)-containing purple acid phosphatase from sweet potatoes is an iron enzyme. *Biochem. Biophys. Res. Commun.* **146:** 1173-7.

Hegeman, CE and Grabau, EA. 2001. A novel phytase with sequence similarity to purple acid phosphatases is expressed in cotyledons of germinating soybean seedlings. *Plant Physiol.* **126:** 1598-608.

Holtz, KM and Kantrowitz, ER. 1999. The mechanism of the alkaline phosphatase reactions: insights from NMR, crystallography and site-specific mutagenesis. *FEBS-Lett.* **462:** 7-11.

Holtz, KM; Stec, B. and Kantrowitz, ER. 1999. A model of the transition state in the alkaline phosphates reaction. *J. Biol. Chem.* **274:** 8351-4.

Hughes, MN. 1981. *The Inorganic Chemistry of Biological Processes*, 2nd. Edit., Wiley, Chichester.

Igaue, I; Watabe, H; Takahashi, K; Takekoshi, M and Morota, A. 1976. Violet-colored acid phosphatase isozymes associated with cell wall preparations from rice plant cultured cells. *Agr. Biol. Chem.* **40:** 823-5.

Kaim, W and B. Schwederski 1991. *Bioanorganische Chemie*, B.G. Teubner, Stuttgart.

Kim, EE and Wyckoff, HW. 1991. Reaction mechanism of alkaline phosphatase based on crystal structure: Two-metal ion catalysis, *J. Mol. Biol.* **218:** 449-64.

Klabunde, T; Sträter, N; Krebs, B and Witzel, H. 1995. Structural relationship between the mammalian Fe(III)-Fe(II) and the Fe(III)-Zn(II) plant purple acid phosphatases. *FEBS-Lett.* **367:** 56-60.

Klabunde, T; Sträter, N; Fröhlich, R; Witzel, H and Krebs, B. 1996. Mechanism of Fe(III)-Zn(II) purple acid phosphatase based on crystal structures. *J. Mol. Biol.* **259:** 737-48.

Klabunde, T and Krebs, B. 1997. The dimetal center in purple acid phosphatases. *Struct. Bonding* **89:** 177-198.

Krebs, B; Klabunde, T; Sträter, N; Witzel, H; Than, R and Sift, B. 1997. Mechanism of plant and mammalian purple acid phosphatases based on crystal structures. *J. Inorg. Biochem.* **67:** 322.

Krebs, B. 2006. Metals in biology: Metalloenzymes and their biomimetic model compounds. Purple acid phosphatase and catechol oxidase. *Anales Acad. Nac. Cs. Ex. Fís. Nat.* **58:** 39-55.

Kuang, R; Chan, KH; Yeung, E and Lim, BL. 2009. Molecular and biochemical characterization of AtPAP15, a purple acid phosphatase with phytase activity, in Arabidopsis, *Plant Physiol.* **151:** 199-209.

LeBansky, BR; McKnight, TD and Griffing, LR. 1992. Purification and characterization of a secreted purple phosphatase from soybean suspension cultures. *Plant Physiol.* **99:** 391-5.

Li, D; Zhu, H; Liu, K; Leggewie, G; Udvardi, M and Wang, D. 2002. Purple acid phosphatases of *Arabidopsis thaliana.* Comparative analysis and differential regulation by phosphate deprivation. *J. Biol. Chem.* **277:** 27772-81.

Lindqvist, Y; Johansson, E; Kaija, H; Vihko, P and Schneider, G. 1999. Three-dimensional structure of a mammalian purple acid phosphatase at 2.2 Å resolution with a μ-(hydr)oxo bridged di-iron center. *J. Mol. Biol.* **291:** 135-47.

Lippard, SJ and Berg, JM. 1994. *Principles of Bioinorganic Chemistry*, University Science Books, Mill Valley, CA.

Ma, L; Tibbitts, TT and Kantrowitz, ER. 1995. *Escherichia coli* alkaline phosphatase: X-ray structural studies of a mutant enzyme (His-412 → Asn) at one of the catalytically important zinc binding sites. *Protein Sci.* **4:** 1498-506.

Millán, JL. 2006. Alkaline Phosphatases: Structure, substrate specificity and functional relatedness to other members of a large superfamily of enzymes. *Purinergic Signalling* **2:** 335-41.

Miller, SS; Liu, J; Allan, DL; Menzhuber, CJ; Fedorova, M and Vance, CP. 2001. Molecular control of acid phosphatase secretion into the rhizosphere of proteoid roots from phosphorus-stressed white lupin. *Plant Physiol.* **127:** 594-606.

Mitić, N; Hadler, KS; Gahan, LR; Hengge, AC and Schenk, G. 2010. The divalent metal cation in the active site of uteroferrin modulates substrate binding and catalysis. *J. Am. Chem. Soc.* **132:** 7049-54.

Morita, N; Nakazato, H; Okuyama, H; Kim, Y and Thompson, GA. 1996. Evidence for a glycosylinositolphospholipid-anchored alkaline phosphatase in the aquatic plant *Spirodela oligorrhiza. Biochim. Biophys. Acta* **1290:** 53-62.

Nakazato, J; Okamoto, T; Ishikawa, K and Okuyama, H. 1997. Purification and characterization of phosphatase inducibly synthesized in *Spirodela oligorrhiza* under phosphate deficient conditions. *Plant Physiol. Biochem.* **35:** 437-46.

Nakazato, H; Okamoto, HT; Nishikoori, M; Washio, K; Morita, N; Haraguchi, K; Thompson, GA and Okuyama, H. 1998. The glycosylphosphatidylinositol-anchored phosphatase from *Spirodela oligorrhiza* is a purple acid phosphatase. *Plant Physiol.* **118:** 1015-20.

Nochumson, S. Ph.D. Thesis. 1975. The Hahnemann Medical College; Philadelphia; U.S.A.

Nordlund, P. 2001. Structure-function of nonheme iron proteins with oxygen- and nitrogen- dominated coordination. In *Handbook of Metalloproteins* (Eds. I. Bertini, A. Sigel and H. Sigel) pp.461-570, Marcel Dekker, New York.

Olczak, M; Watorek W and Morawiecka, B. 1997. Purification and characterization of acid phosphatase from yellow lupin (*Lupinus luteus*) seeds. *Biochim. Biophys. Acta* **1341:** 14-25.

Olczak, M; Kobialka, M and Watorek, W. 2000. Characterization of diphosphonucleotide phosphatase/phosphodiesterase from yellow lupin (*Lupinus luteus*) seeds. *Biochim. Biophys. Acta* **1478:** 239-47.

Olczak, M and Watorek, W. 2003. Two subfamilies of plant purple acid phosphatases. *Physiol. Plant.* **118:** 491-8.

Olczak, M; Morawiecka, B and Watorek, W. 2003. Plant purple acid phosphatases - genes, structures and biological function. *Acta Biochim. Polon.* **50:** 1245-56.

Olczak, M; Ciuraszkiewicz, J; Wójtowicz, H; Maszczak, D and Olczak, T. 2009. Diphosphonucleotide phosphatase/phosphodiesterase (PPD1) from yellow lupin (*Lupinus luteus* L) contains an iron-manganese center. *FEBS Lett.* **583:** 3280-4.

Ozawa, K; Osaki, M; Matsui, H; Honma, M and Tadano, T. 1995. Purification and properties of acid phosphatase secreted from lupin roots under phosphorus-deficiency conditions. *Soil Sci. Plant Nutr.* **41:** 461-9.

Roat-Malone, RM. 2002. *Bionorganic Chemistry, A Short Course*, Wiley, Hoboken, NJ.

Saeed, A; Salim, M; Naz, Z; Zaman, U; Baloch, AL; Nadir, S and Saeed, A. 2014. Partial purification, characterization and some kinetic properties of low molecular weight acid phosphatase from leaves of germinating *Vigna radiata* seeds. *J. Anim. Plant Sci.* **24:** 1466-77.

Schenk, G; Ge, Y; Carrington, LE; Wynne, CJ; Searle, IR; Carroll, BJ; Hamilton, S and de Jersey, J. 1999a. Binuclear metal centers in plant purple acid phosphatases: Fe-Mn in sweet potato and Fe-Zn in soybean. *Arch. Biochem. Biophys.* **370:** 183-9.

Schenk, G; Carrington, LE; Hamilton, SE; de Jersey, J and Guddat, LW. 1999b. Crystallization and preliminary X-ray diffraction data for a purple acid phosphatase from sweet potato. *Acta Crystallogr. D* **55:** 2051-2.

Schenk, G.; Korsinczky, MLJ; Hume, DA; Hamilton, SA and de Jersey, J. 2000. Purple acid phosphatases from bacteria: similarities to mammalian and plant enzymes. *Gene* **255:** 419-24.

Schenk, G; Boutchard, CL; Carrington, LE; Noble, CJ; Moubaraki, B; Murray, KS; de Jersey, J; Hanson, GR and Hamilton, S. 2001. A purple acid phosphatase from sweet potato contains an antiferromagnetically coupled bimunclear Fe-Mn center. *J. Biol. Chem.* **276:** 19084-8.

Schenk, G; Gahan, LR; Carrington, LE; Mitić, N; Valizadeh, M; Hamilton, SE; de Jersey, J and Guddat, LW. 2005. Phosphate forms an unusual tripodal complex with the Fe-Mn center of sweet potato purple acid phosphatase. *Proc. Nat. Acad. Sci. USA* **22:** 273-8.

Schenk, G; Elliot, TW; Leung, E; Carrington, LE; Mitić, N; Gahan, LR and Guddat, LW. 2008. Crystal structures of a purple acid phosphatase, representing different steps of this enzyme's catalytic cycle. *BMC Struct. Biol.* **8:** 6.

Schenk, G; Mitić, N; Hanson, GR and Comba, P. 2013. Purple acid phosphatase: A journey into the function and mechanism of a colorful enzyme. *Coord. Chem. Rev.* **257:** 473-82.

Schepers, K; Bremer, B; Krebs, B; Henkel, G; Althaus, E; Mosel, B and Müller-Warmuth, W. 1990. $Zn^{II}Fe^{III}$ and $Fe^{II}Fe^{III}$ complexes containing a novel (μ-phenoxo)bis(μ-diphenylphosphato)-dimetal(II,III) core as model complexes for active sites of purple phosphatases. *Angew. Chem. Int. Ed. Engl.* **29:** 531-3.

Sift, BH; Durmus, A; Meyer-Klaucke, W and Krebs, B. 1999. EXAFS studies on the active site of purple acid phosphatase from sweet potato, *Ipomea batatas*. *J. Synchrotron Rad.* **5:** 421-2.

Stahl, B; Klabunde, T; Witzel, H; Krebs, B; Steup, M; Karas, M and Hillenkamp, F. 1994. The oligosaccharides of the Fe(III)-Zn(II) purple acid phosphatase of the red kidney bean. *Eur. J. Biochem.* **220:** 321-30.

Stec, B; Holtz, KM and Kantrowitz, ER. 2000. A revised mechanism for the alkaline phosphatase reaction involving three metal ions. *J. Mol. Biol.* **299:** 1303-11.

Sträter, N; Klabunde, T; Tucker, P; Witzel, H and Krebs B. 1995. Crystal structure of a purple acid phosphatase containing a dinuclear Fe(III)-Zn(II) active site. *Science* **268:** 1489-92.

Sträter, N.; Lipscomb, WN; Klabunde, T and Krebs, K. 1996. Two-metal ion catalysis in enzymatic acyl- and phosphoryl- transfer reactions. *Angew. Chem. Int. Ed.* **35:** 1024-55.

Sträter, N; Jasper, B; Scholte, M; Krebs, B; Duff, AP; Langley, DB; Han, R; Averill, BA; Freeman, HC and Guss, JM. 2005. Crystal structures of recombinant human purple acid phosphatase with and without an inhibitory conformation of the repression loop. *J. Mol. Biol.* **351:** 233-46.

Sue, PK; Zhang, S and Sun, SSM. 2015. Molecular characterization of a tomato purple acid phosphatase during seed germination and seedling growth under phosphate stress. *Plant Cell Rep.* **34:** 981-92.

Sugiura, Y; Kawabe, H and Tanaka, H. 1980. New manganese (III)-containing acid phosphatase. Evidence for charge transfer band and tyrosine phenolate coordination. *J. Am. Chem. Soc.* **102:** 6581-2.

Sugiura, Y; Kawabe, H; Tanaka, K; Fujimoto, S and Akira, O. 1981. Purification, enzymatic properties, and active site environment of a novel manganese(III) containing acid phosphatase. *J. Biol. Chem.* **256:** 10664-70.

Surchandra, T; Roy, SS; Rakesh Singh, N; Sahoo, MR and Prakash, N. 2012. Partial purification and biochemical characterization of acid phosphatase from germinating mung bean (*Vigna radiata*) seeds. *African J. Biotechnol.* **11:** 16777-82.

Suzuki, M; Fujinami, S; Hibino, T; Hod, H; Maeda, Y; Uehara, A and Suzuki, M. 1998. Synthesis and characterization of mixed valence μ-alkoxo-diiron(II,III) complexes with an unsymmetric dinucleating ligand. *Inorg. Chim. Acta* **283:** 124-35.

Than, R; Feldmann, AA and Krebs, B. 1999. Structural and functional studies on model compounds of purple acid phosphatases and catechol oxidases. *Coord. Chem. Rev.* **182:** 211-41.

Tran, HT; Hurley, BA and Plaxton, WC. 2010. Feeding hungry plants: The role of purple acid phosphatases in phosphate nutrition. *Plant Sci.* **179:** 14-27.

Uehara, K; Fujimoto, S and Taniguchi, T. 1974a. Studies on violet colored acid phosphatase of sweet potato. I. Purification and some physical properties. *J. Biochem.* **75:** 627-38.

Uehara K; Fujimoto, S; Taniguchi, T and Nakai, K. 1974b. Studies on violet-colored acid phosphatase of sweet potato.II. Enzymatic properties and amino acid composition. *J. Biochem.* **75:** 639-49.

Uppenberg, J; Lindqvist, F; Svensson, C; Ek-Rylander, B and Andersson, G. 1999. Crystal structure of a mammalian purple acid phosphatase. *J. Mol. Biol.* **290:** 201-11.

Vallee, BL and Auld, DS. 1993. New perspective on zinc biochemistry: cocatalytic sites in multi-zinc enzymes. *Biochemistry* **32:** 6493-500.

Veljanovski, V; Vanderbeld, B; Knowles, VL; Snedden, WA and Plaxton, WC. 2006. Biochemical and molecular characterization of AtPAP26, a vacuolar purple acid phosphatase up-regulated in phosphate-deprived Arabidopsis suspension cells and seedlings. *Plant Physiol.* **142:** 1282-93.

Vincent, JB; Olivier-Lilley, GL and Averill BA.1990. Proteins containing oxo-bridged dinuclear iron centers: A bioinorganic perspective. *Chem. Rev.* **90:** 1447-67.

Wang, Y; Norgård, M and Andersson, G. 2005. N-glycosilation influences the latency and catalytic properties of mammalian purple acid phosphatase. *Arch. Biochem. Biophys.* **435:** 147-56.

Waratrujiwong, T; Krebs, B; Spener, F and Visoottiviseth, P. 2006. Recombinant purple acid phosphatase isoform 3 from sweet potato is an enzyme with a diiron metal center. *FEBS J.* **273:** 1649-59.

Williams, R.J.P. and J.J.R. Fraústo da Silva. 1996. *The Natural Selection of the Chemical Elements*, Clarendon Press, Oxford.

Yang, YS; McCormick, JM and Solomon, E.I. 1997. Circular dicroism and magnetic circular dicroism studies of mixed-valence binuclear non-heme iron active site in uteroferrin and its anion complexes. *J. Am. Chem. Soc.* **119:** 11832-42.

Zhang, W; Gruszewski, HA; Cevone, BI and Nessler, CL. 2008. An Arabidopsis purple acid phosphates with phytase activity increases foliar ascorbate. *Plant Physiol.* **146:** 431-40.

Zhu, H; Qian, W; Lu, X; Li, D; Liu, X; Liu, K and Wang, D. 2005. Expression patterns of purple acid phosphatase genes in Arabidopsis organs and functional analysis of AtPAP23 predominantly transcribed in flower. *Plant Mol. Biol.* **59:** 581-94.

Zimmermann, P; Regierer, B; Kossmann, J; Frossard, E; Amreihn, N and Bucher, M. 2004. Differential expression of three purple acid phosphatases from potato. *Plant Biol.* **6:** 519-28.

■■■■

18

PLANT LECTINS, AGRICULTURAL ADVANCEMENTS AND MAMMALIAN TOXICITY

Prasann Kumar and Padmanabh Dwivedi

Department of Plant Physiology, Institute of Agricultural Sciences, Banaras Hindu University, Varanasi, 221005, U.P., India
Email address: prasann0659@gmail.com

INTRODUCTION

Plant lectins, the versatile proteins have not only demonstrated their role in possessing cytotoxic, fungitoxic, anti-nematode and other properties but are also considered to be the promising candidates against many insect pests. This provides an opportunity, creates a hope towards increasing the crop yield and provides a scientific basis for the development of new pest management strategies, but it is necessary to check the possible harmful effects in a plant for dietary uptake by animals/ humans and developing genetically modified crop plants. Keeping these things in mind, this paper describes plant lectins, properties, their capacity towards agricultural development and toxic/non-toxic nature towards mammals/higher animals including humans.

PLANT LECTINS PROPERTIES, THEIR CAPACITY TOWARDS AGRICULTURAL DEVELOPMENT AND TOXIC/NON-TOXIC NATURE TOWARDS MAMMALS/HIGHER ANIMALS INCLUDING HUMANS

Until the mid 1990s, lectins, belonging to only a few lectin families were known to be the proteins present in seeds or vegetative storage tissues of plants. Studies conducted later also found the potential of these plant lectins in bearing insecticidal activities against certain plant insect pests.

This approach as a result strongly encouraged the search of plant lectins possessing insecticidal properties. In the present situation plant lectins are considered to be the promising agents against some insect pests for which even *Bacillus thuringiensis* (Bt) gene has been found ineffective. Studies have indicated that the production of transgenic plants reduces the load of using chemical pesticides, insecticides, herbicides (Velkov *et al.,* 2005) in plants, thereby lend a hand in improving corresponding environmental conditions and human health. Many of the plant lectin genes till date have been transformed successfully through genetic engineering into a variety of crops including wheat, rice, tobacco and potatoes in order to use them as a part of integrated pest management (IPM) strategies.

It is also believed that lectins are although considered to be toxic, but it is worth mentioning that the Mannose binding lectins (MMBLs) even after carrying insecticidal properties are considered non toxic for mammals and birds (Ripoll *et al.,* 2003), thus these types of more genes or related genes can be searched for using them in future research with an aim of controlling some important insect pest, fungal diseases and that too without harming human health. A protein named "Ricin" found and isolated from the castor bean (*Ricinus communis*) supports the discovery of plant lectins which dates back to the end of the nineteenth century. It also gave birth to the word "lectin" which was originally derived from the word" Legere" meaning "to select" in Latin. Later on, plant lectins were being represented as the plant proteins possessing one (at least) non-catalytic domain which reversibly binds specifically to a monosaccharide/ an oligosaccharide (Peumans and Van Damme, 1995). Lectins are although abundantly present in plant seeds but these have also been found to be localized in different vegetative tissues like flowers, rhizomes, roots, leaves, bulb and bark (Ratanapo *et al.,* 1998; Van Damme *et al.,* 2000). Classifying plant lectins is a tedious task. The classification of plant lectins into 4 major groups on the basis of their overall domain architecture (Van Damme *et al.,* 1998) is as follows (1) Merolectins, are basically described as the proteins containing only one carbohydrate binding domain. It is due to their monovalent nature that they are not able to agglutinate cells (2) Hololectins, possess two or more identical/homologous carbohydrate binding domains and thus have the capability to agglutinate the cells and precipitate glycoconjugates (3) Superlectins, are composed of at least two carbohydrate binding domains and have the capability of recognizing structurally unrelated carbohydrate structures (4) Lastly, Chimerolectins are classified as lectins present abundantly in plants. These represent almost all the plant lectins which comprise of one or more than one carbohydrate binding domain fused to a domain that renders them an independent biological activity. In a different classification of plant lectins, they have been subdivided into seven structurally and evolutionary related proteins *viz.* Chitin binding, Legume binding lectins, Monocot mannose binding, Jacalin related,

Amaranthin family lectins, Cucurbit phloem type, type-2 RIP (Ribosome inactivating protein) and other lectins (Murdock and Shade, 2002). The classification was based on the basis of their biochemical characteristics, physicochemical sequence data, structural information and molecular aspects. In an another classification related to the subdivision of plant lectins, lectins (on the basis of monosaccharide for which they exert highest affinity) are divided into five major groups like D-mannose/D-glucose, D-galactose/N-acetyl-D-galactosamine, N-acetyl-D-Glucosamine, L-fucose and N-acetylneuraminic acid (Goldstein *et al.,* 1997). Plant lectin's unique carbohydrate binding capacity is due to the presence of water bridges, oligomerization process and post transitional modifications. Even a small alteration in the tertiary structure may lead to a change in sequence, which may further affect the formation of quaternary structure in a major way (Vijayan and Chandra, 1999). It has been reported earlier that the extent of the insecticidal activity of lectins was found to be directly proportional to the number of subunits possessed by lectins, for example: Tetrameric lectins (containing four subunits) generally had the highest insecticidal activities than Dimeric lectins (Barre *et al.,* 2001) (containing only two subunits). There is a presence of an additional mannose binding region located in each subdomain of the two interacting lectins along with the already present six binding sites (which were identified earlier). The presence of this additional seventh high affinity carbohydrate recognition domain not only helps lectins in recognizing the high chain oligosaccharides but also acts as a plus point by increasing the efficiency of garlic lectins in controlling insect pests, even at a lower dose (Bouckaert *et al.,* 1999; Chandra *et al.,* 1999). The target sites for plant lectins present inside the insect body are numerous. The Concavalin A, (ConA lectin) was observed to be present in fat tissues while both *Galanthus nivalis* (GNA lectin) and Concavalin A lectins were found to accumulate in the Hemolymph and the Malphigian tubules of *Lacanobia oleracea.* After feeding *Nilaparvata Lugens* (Rice Brown Plant Hopper) with *Galanthus nivalis* lectin (Powell *et al.,* 1998), it was pointed out that *Galanthus nivalis* lectin (GNA) was found to be present in fat tissues and ovarioles. Most of the analysis illustrates that apart from the above mentioned, the additional target sites of plant lectins inside the body of an insect are most likely the Malphigian tubules, fat tissues and ovaries. Plant lectins, in addition to the disruptive effects on the cell membrane, have also been shown to inhibit the activity of various intestinal and brush border enzymes, for example, *Allium sativum* lectin (ASA lectin) present in garlic showed a relatively increased activity of disaccharides and acid phosphatases (ACP) as compared to the normal in rat jejunum while the activity of enzymes like alkaline phosphatases (ALP), lactate dehydrogenase (LDH) and adenosine triphosphatase (ATPase) was found to be relatively less (Gupta and Sandhu, 1998). The inclusion of *Pinellia ternata* agglutinin (PTA) in basal diet offered to rats causes a significant decrease in the activities of enzymes like Maltase (MA), Alkaline phosphatase (ALP), Glutamyl transferase (GT) and Lectin aminopeptidase (AP) (Higuchi *et al.,* 1984). The use of genetically engineered crop plants with an enhanced insect resistance activity has brought revolution in agriculture by reducing the usage of pesticides and lowering the production costs of many crops worldwide. Today, the use of these transgenic plants expressing the insecticidal toxins mostly depend on the typical presence of toxins basically derived from *Bacillus thuringiensis* (Bt) bacterium. Bt proteins have although shown the potential to protect several crops from many chewing type of insect pests belonging to lepidopteran and coleopteran classes (Wunn *et al.,* 1996; Nayak *et al.,* 1997), but according to the past findings, the phloem sap sucking type of insect pests (those belonging to Homeopteran class), which are known to nourish on free amino acids and amides of plant sap are, however, not affected by *B. thuringiensis* (Bt) proteins (Sogawa, 1970). It was confirmed (Malone *et al.,* 2008) that there are yet no Bt toxin which could be used to control the plant sap sucking insect pests such as thrips, bugs, mirids, hoppers or aphids, which not only cause severe physiological damage, but also act as vectors for transmission of viruses like grassy stunt virus, rice tungro virus and ragged stunt virus to important crop plants including rice. Therefore, the search of curative agents against these pests and an alternative against Bt toxin has received a lot of attention. Many scientists in this regard have reported plant lectins to be one of the promising candidates of proteins exhibiting insecticidal properties against these insect pests. Lectins from *Pinellia pedatisecta* and *Pinellia ternata,* when incorporated into the artificial feed diets at the concentrations of 1.2 and 1.5 grams/Litre, showed to possess insecticidal activities against cotton aphids and peach potato aphids (*Aphis gossypii* and *Myzus persicae* respectively) (Huang *et al.,* 1997; Pan *et al.,*1998). Results of some experiments based on the artificial diets showed that some of the plant lectins were found to be toxic to two most serious sap sucking insect pests like Brown plant hopper (*Nilaparvata lugens*) and Green leaf hopper (*Nephotettix virescens*). Snowdrop lectin - *Galanthus nivalis* (GNA) was found to be one of the most toxic lectinx (Powell *et al.,* 1995) amongst all lectins. A reduced nymph survival (Powell *et al.,* 1995a; 1995b) in green plant hoppers and brown plant hoppers confirmed the enhanced resistance due to these lectins against such insect pests when they were fed on an artificial diet containing lectins (Rao *et al.,* 1998; Stoger *et al.,* 1999). The carbohydrate binding proteins (Lectins) e.g. *Galanthus nivalis* agglutinin (GNA) and Wheat germ agglutinin (WGA) are known to be effective in controlling a wide range of insect pests (Hilder *et al.,* 1995; Powell *et al.,* 1995a; 1995b; Down *et al.,* 1996; Rao *et al.,* 1998) in plants. The results from earlier experiments using bioassays based on an artificial diet feeding system also mentioned the toxic nature of lectins against these insect pests. They further reported *Galanthus nivalis* (GNA) to be the most toxic lectin which was found to be responsible for decreasing the Brown Plant Hopper (BPH) survival rate of 50% at concentrations as low as 6 millimolar. *Allium sativum* lectin ASAL (a Mannose binding leaf garlic lectin) is a mannose binding protein which binds to the carbohydrate residue of 55 and 45 kilodaltons brush border membrane vesicle (BBMV) receptor proteins (Bandyopadhyay *et al.,* 2001). The binding of *Allium sativum* lectin (ASAL) to the gut receptors present in the luminal epithelium of the two important homopteran insect pests viz. Mustard aphids (*Lypaphis erysim*), Red cotton bug (*Dysdercus cingulatus*), decrease the permeability of the membrane and can thus be co-related to the insecticidal activities of lectins inside the insect's gut environment (Peumans and Van Damme, 1995a; Zhu-Salzman *et al.,* 1998). The lethal doses of lectins prescribed in insect bioassays are 2 nanomol in case of *Dysdercus cingulatus* and 4 nanomol for aphids. These doses are even relatively lower in comparison to *Galanhus nivalis* GNA lectin dose against brown plant hoppers (Powell *et al.,* 1995b). In an experiment, the transgenic rice crop plants transformed with *Allium sativum* lectin ASAL, led to a significant increase in the nymphal mortality rate of Hemipteran insect pests such as Brown leaf hopper (*Nilaparvata lugens*) and Green leaf hopper (*Nephotettix virescesns*). Exceptions to this are the insects of Homoptera (which suck away the phloem sap full of free amino acids and sugars) lacking peritrophic membrane, normally do not care about the enzymes and thus rules out the possibility of binding of lectins to any of the glycosylated digestive enzymes present in insect gut. Earlier it was believed plant lectins bind to the amylase enzyme present in the digestive tract (Macedo *et al.,* 2007; Sadeghi *et al.,* 2008b) but in a recent report, after analysis of target proteins by *Galanthus nivalus* (GNA lectin) in case of *Drosophila*, it was reported that lectins probably act on multiple target glycoproteins of insects but the irony is that the process was not only shown to happen in target insect pests such as beetles, aphids and caterpillars but also in non targeted insects like honey bees (Vandenborre *et al.,* 2010; 2011). The mode of action of lectins varies according to the type of lectin used, for example (1) An anti-insect activity was found in *Arisaema helleborifolium* lectin towards the second instar larvae of *Bactrocera cucurbitae* (Kaur *et al.,* 2006b). The insecticidal activity of lectins in this case may be due to the decreased activity of enzymes like acid phosphatase (ACP) and alkaline phosphatase (ALP) which may be further due to the orchestration of enzyme activity of larvae after treatment with different types of lectins, (2) *Dioscorea bataas* lectin was found avidly binding to the brush border and peritrophic membrane of *Helocoverpa armigera* larvae thereby inhibiting its development from larval to adult form (Ohizumi *et al.,* 2009), (3) A treatment of *Plutella xylostella* larvae with Galectin-1 lectin brought disruption of the microvilli with increased abnormalities in the epithelial cells (Chen *et al.,* 2009b), (4) *Arum maculatum* tuber lectin binds to the brush border membrane vesicle proteins present in the insect's gut and cause *Aphis craccivora* and *Lipophis erysimi* to capitulate (Majumder *et al.,* 2005) (5) Binding of *Olyneya resota* lectin to the midgut (glycoconjugated) and microvillae of

Zabrotes subfaciatus larvae, led to reduce opposition which ultimately resulted in a failure of emergence of adult beetles (Lagarda-Diaz *et al.*, 2009), (6) The toxicity of *Annona coriacea* lectin for *Anagasta kuehniella* apparently resulted in interruption of digestive enzymes recycling mechanism by binding to the mid gut protein and led to the changes in the gut membrane environment (Coelho *et al.*, 2007), (7) It was reported by Hopkins and Harper (2001) that European corn borer (*Ostrinia nubilalis*) showed a breakdown of microvilli and hyper secretion of disorganized plasma membrane layers into the midgut lumen, when it was fed on a Wheat germ agglutinin (WGA) lectin containing diet and led to accelerated cell loss, shedding of brush border membranes and the formation of shortened, sparse, irregular enterocyte microvilli (Bardocz *et al.*, 1995; Herzig *et al.*, 1997). Many of the lectin genes transformed in different plants are presented in Table 1.

Table 1. Plant species transformed with Insecticidal lectin genes

Transformed plant	**Lectin**	**Target pest**	**Reference**
Maize	WGA	*Ostrinia nubilaris*	Maddock *et al.*, 1991.
		Diabrotica undecimpunctata	
Mustard	WGA	*Lipaphis erysimi*	Kanrar *et al.*, 2002.
Arabidopsis	PHA	*Lacanobia oleracea*	Fitches *et al.*, 2001b.
Potato	GNA	*Aulacorthum solani*	Down *et al.*, 1996.
		Myzus persicae	Gatehouse *et al.*, 1996.
		Lacanobia oleracea	Gatehouse *et al.*, 1997.
		Lacanobia oleracea	Down *et al.*, 2001.
		Aphidius ervi	Couty *et al.*, 2001b.
	ConA	*Lacanobia oleracea; Myzus persicae*	Gatehouse *et al.*, 1999.
Rice	GNA	*Nilaparvata lugens*	Maqbool *et al.*, 2001.
		Cnaphalocrocis medinalis (rice leaf folder)	Maqbool *et al.*, 2001. Foissac *et al.*, 2000.
		Nephotettix virescens (green leafhopper)	Sun *et al.*, 2002; Wu *et al.*, 2002.
		Scirpophaga incertulas (yellow stem borer)	
		Laodelphax striatellus (rice small brown planthopper)	
Sugarcane	GNA	*Eoreuma loftini* (Mexican rice borer)	Setamou *et al.*, 2002. Setamou *et al.*, 2002.
		Diatraea saccharalis (sugarcane borer)	Tomov and Bernal, 2003.
		Parallorhogas pyralophagus	
Tobacco	PSA	*Heliothis virescens* (tobacco budworm)	Boulter *et al.*, 1990.

Transformed plant	Lectin	Target pest	Reference
	GNA	*Helicoverpa zea* (cotton bollworm)	Wang and Guo, 1999.
Wheat	GNA	*Sitobion avenae* (grain aphid)	Stoger *et al.*, 1999.

[Source: Vasconcelos *et al.*, 2004]

WGA- Wheat germ agglutinin, PHA- Phytohemagglutinin, ConA-Concavalin A, GNA-*Galanthus nivalis* Agglutinin, PSA-*Pisum sativum* agglutinin.

Studies found that the fruiting bodies from some of the species of fungus, for example, Basidiomycetes or other species belonging to the Lepista or Cantharellus genera are never inhabited by insects. This is because the insecticidal properties of these fungi are attributed to lectins. The first protein, isolated from a mushroom Red cracking bolete (*Xerocomus chrysenteron*), bearing insecticidal activity was a lectin named *Xerocomus chrysenteron* lectin (XCL) (Trigueros *et al.,* 2003). A Lectin named *Clitocybe nebularis* lectin (CNL), isolated from *Clitocybe nebulis* was found to recognize N-acetylgalactosamine (Horejsi and Kocourek, 1978). The insecticidal activity of this lectin (*i.e.* CNL), along with its property to impart resistance to proteolytic degradation (Pohleven *et al.,* 2009) has thus opened a new hope in the production of natural insecticides. Lectins belonging to the superfamily of lectins, MMBLs (Mannose binding lectins) e.g. *Gastrodia elata* (GAFP-1) which is an antifungal protein believed to confer resistance to pathogenic fungi like *Botrytis cinerea, Gibeberella zaea, Ganoderma lucidum, Rhizoctonia solani* and *Valsa ambient* (Xu *et al.,* 1998; Wang *et al.,* 2001). The lectins from the seeds of red cluster pepper (*Capsicum frutesceae* L var. *Fasciculatum*) have been found to inhibit conidial germination and hyphal growth in *Aspergillus flavus* and *Fusarium moniliforme* during *in-vitro* conditions (Ngai and Ng, 2007). The exogenous saccharides including glycoproteins present on the surface of several plant viruses, fungi or in the midgut of epithelial membrane of insects/plant predators are the most possible receptors of lectins. Apart from the antifungal activity of lectins, research findings have also demonstrated that some lectins are responsible for inducing bacterial cell proliferation. It was found that the secondary toxic effect of undigested lectin Phytohemagglutinin (PHA) may reduce the efficiency of absorption and digestion of food in small intestine, resulting in an abnormal overgrowth of coliform bacteria (Bardocz *et al.,*1996; Wilson *et al.,*1980), which is followed by the appearance of symptoms like mucus secretion, serum protein leakage, reduce digestion of bacterial proteins and loss of epithelial cells which acts as a good source of nutrients (Grant, 1999) for the bacterial (*Escherichia coli*) growth and thereby aid bacterial cell proliferation. Rice, also known as the staple food in major part of the world, is taken by more than 2.5 billion people at present. With an ever increasing population, the yield of rice also needs to be increased by an additional at least 30% in the next 30 years to meet the growing demands (Rosegrant *et al.,* 1995). In order to achieve this dream, the use of nitrogen fertilizer used by farmers will automatically have to be increased by two folds, which is not a suitable option both economically and environmentally (Ladha and Reddy, 2003). Many Scientists believe that plant lectins may have the capability of inducing the nitrogen fixation process in crops and that they can be used as another alternative option in this respect. These sugar binding proteins (Sharon, 2004) have the potential to interact with the sugar moieties present on the surface of the nitrogen fixing bacterium (*Rhizobium*). The interaction is hypothesized (Hirsch, 1999; Kijne, 1992) in order to enable the two symbionts recognize each other and to promote the close association between them. Nitrogen fixation in legumes is a normal phenomenon. Kalsi and Etzler (2000) described the formation of bacterial aggregation and their attachment to the root hair zones of legumes in an experiment, when pea lectin (*Pisum Sativum* Lectin PSL; Diaz *et al.,* 1995) and soybean lectin [Le1, DB46-LNP; Van Rhijn *et al.,* 1998] were targeted to the plant root hair tips. Apart from these, lectin nucleotide phosphohydrolases [LNPs] of DB46 and GS52 *Dolichos biflorus* and *Glycine soja,* respectively, have also been implicated to play a crucial role in host and rhizobia recognition (Etzler *et al.,* 1999; Day *et al.,* 2000). But the question which arises here is that can the process take place in monocots too? Thus, increasing interest has been focused in order to determine whether the nitrogen fixing capability like that found in legumes can be transferred in monocot crop such as rice, to render the crop independent of nitrogen (Ladha and Reddy, 2003) which is provided to the crops from outside in the form of fertilizers at present. Van Rhijn et al. (1998) through an experiment demonstrated that the transformation of Soybean lectin (*Le1*) gene in *Lotus corniculatus* induced infection along with the development of nodule structure like formations upon inoculation of *Lotus corniculatus* with *Bradyrhizobium japonicum* (a soybean specific microsymbiont) which normally does not modulate *L. corniculatus.* Researchers have always remained doubtful about the nature of plant lectins to be toxic/anti-nutritional in case of both man and animals. A TV program on BBC in the year 1981, reported 330 outbreaks involving 880 people who suffered from kidney bean food poisoning (Bender and Reaidi, 1982). Apart from this, seven incidents involving 43 persons once were found related to the toxins present in uncooked/partially cooked kidney beans (*Phaseolus vulgaris*). A hospital in 1988, once organized "Healthy Eating Day" for the staff where red kidney beans and 31 portions were served during the lunch, as a result of which, 11 customers were reported to suffer from vomiting and diarrhoea after some time. Although no pathogen was isolated from the food they had eaten, but the reasons found later was due to the presence of high concentrations of Phytohemagglutinin lectin (PHA), also known as *Phaseolus vulgaris* agglutinin (Gilbert, 1988). These examples are although all related to kidney beans, but it is possible that there may be many other problems associated with lectins which may not have been reported yet. This may be because the symptoms related to the lectin toxicity are so common that it becomes difficult to make out the real cause of the problem. Lectins such as 'Ricin' and 'Abrin' present in castor beans (*Ricinus communis*) and the seeds of jequirity bean (*Abrus precatorius*) are also well known examples of plant lectins responsible for causing severe toxicity towards mammals. In an experiment, the toxic effect of lectins on experimental animals fed on diets containing plant lectins, showed the symptoms of decreased body weight, loss of appetite and eventual death (Lajolo and Genovese, 2002). The interaction of lectins with the glycoproteins present in the digestive tract results in both local and systemic reactions. Symptoms like vomiting, nausea and diarrhoea characterize the oral acute toxicity of lectins for mammals (including humans) whenever ingested or are exposed to lectins. It was also found that lectins in some cases (e.g. rat) mediated changes in the glycosylation pattern of gut resulting in an increase in the number of binding sites in which these lectins may attach (Pusztai *et al.,* 1995). Now, as it is believed that the glycosylation state of gut shows many similarities among higher animals, thus it increases the probability to suppose that the adverse effects of lectins demonstrated on the animals (experimental) can be comparable in human beings too. Thus, it was concluded that the plant lectins (either *in-vitro/in-vivo*) which are known to possess cytotoxic, fungitoxic, anti-insecticidal, anti-nematodes properties and have also been found to be toxic for higher animals (Oka *et al.,* 1997; Ripoll *et al.,* 2003) including humans but the question is does that mean all lectins are toxic? The transgenic plants expressing a high level of plant lectins, although exhibit some degree of resistance against the targeted insect pests, but the researchers have great interest in lectins belonging particularly to the mannose binding superfamilies (MMBLs) possessing insecticidal properties. This resides in a fact that lectins belonging to these superfamilies show toxicity against various orders of insect species, but they are found to be nontoxic to mammals and birds (Powell *et al.,* 1995; Down *et al.,* 1996; Bandyopadhyay *et al.,* 2001). MMBLs, particularly derived from Garlic lectins (*Alliaceae*) and *Galanthus nivalis* lectin (*Amaryllidaceae*), have been found effective in controlling the targeted insect pests but are non-toxic to mammals and birds. They are believed to be the safest lectin for diet (Peumans and Van Damme, 1996). For the past years *Galanthus nivalis* lectin protein has already been transferred and expressed in several crop plants and has found to be suitable for the development of transgenic plants because there is no evidence of toxic nature of GNA towards higher animals (Pusztai *et al.,* 1990). It is although considered to be toxic to insect pests, but has been found to be nontoxic to mammals (Pusztai, 1991).

So far many lectins have been tested in rats and mice in order to assess their toxic effects by both intraperitoneal and intradermal administration in insects but there are results which have interpreted that the parentally toxic lectins were found to be orally toxic while a few even exhibited no oral toxicity, for example: (1) *Cratylia argentea* (CAA) was reported to possess a strong agglutination activity for a wide range of erythrocytes but by being essentially non toxic (Rios *et al.,* 1996). (2) In an experiment, clear abnormalities in rats after kidney bean lectin protein/Phytohaemagglutinin (PHA-E) ingestion was observed in an experiment which was designed to assess the safety of a genetically modified rice crop plant expressing (PHA-E) (Poulsen *et al.,* 2007a, 2007b), (3) While in a similar feeding case study using the same crop (Transgenic rice) expressing *Galanthus nivalis* lectin (GNA), revealed no adverse effects on rats even after continuous dietary uptake (Poulsen *et al.,* 2007b). This clearly illustrates that the case sensitivity of plant lectins not only gives an idea about the concerning safety assessments in the case of genetically engineered crops, but also creates a hope of agricultural advancements in the future and that too without harming human health.

CONCLUSION AND FUTURE SCOPE

Plant Lectins are no doubt versatile proteins with numerous properties. The immense amount of research has been done in the past to elucidate the role of lectins in plant defence, but more studies related to the insecticidal activity of lectins belonging to the superfamily of mannose binding lectins and their toxic effect on mammals (especially humans) needs to be attempted. It would be an additional advantage. The presence of lectins naturally in edible plants may increase the chances of consumer's acceptance and create a hope for future genetically engineered pesticides free crop plants. Overall, it can be concluded that the plant lectins may help in understanding the roles and distribution of lectins in the environment and can be used for future research in controlling pests and fungal diseases without risking the human health.

References

Bandyopadhyay, S., Roy, A., and Das, S. (2001). Binding of garlic (*Allium sativum*) leaf lectin to the gut receptors of homopteran pests is correlated to its insecticidal activity. *Plant Science* **161:** 1025-1033.

Bardocz, S., Ewen, S.W.B., Grant, G., and Pusztai, A. (1995). Lectins as growth factors of the small intestine and the gut. *Lectins: Biomedical Perspectives* 4:103-116.

Bardocz, S., Grant, G., Pusztai, A., Franklin, M.F., and Carvalho, A.F.F.U. (1996). The effect of phytohaemagglutinin on the growth, body composition and plasma insulin of the rat at different dietary concentrations. *British Journal of Nutrition* 76: 613-626.

Barre, A., Bourne, Y., Van Damme, E.J.M., Peumans, W.J., and Rouge, P. (2001). Mannose-binding plant lectin: different structural scaffolds for a common sugar-recognition process. *Biochimie* 83: 645-51.

Bender, A.E., and Reaidi, G.B. (1982). Toxicity of kidney beans (*Phaseolus vulgaris*) with particular reference to lectins. *Journal Plant Foods* 4: 15-22.

Bouckaert, J., Hamelryck, T., Wyns, L., and Loris, R. (1999). Novel structures of plant lectins and their complexes with carbohydrates. *Current Opinion of Structural Biology* **9:** 572-577.

Boulter, D., Edwards, G.A., Gatehouse, A.M.R., Gatehouse, J.A., and Hilder, V.A. (1990). Additive protective effects of incorporating two different higher plants derived insect resistance genes in transgenic tobacco plants. *Crop Protection* 9: 351-354.

Chandra, N.R., Ramachandraiah, G., Bachhawat, K., Dam, T.K., Surolia, A., and Vijayan, M. (1999). Crystal structure of a dimeric mannose specific agglutinin from garlic: quaternary association and carbohydrate specificity. *Journal of Molecular Biology* 285: 1157-1168.

Chen, S.J., Chen, N.T., Wang, S.H., Hsu, J.C., Ding, W.H., Kuo-Huang, L.L., and Huang, R.N. (2009b). Insecticidal action of mammalian galectin-1 against diamond back moth (*Plutella xylostella*). *Pest Management Science* 65: 923-930.

Coelho, M.B., Marangoni, S., and Macedo, M.L. (2007). Insecticidal action of *Annona coriacea* lectin against the flour moth *Anagasta kuehniella* and the rice moth *Corcyra cephalonica. Comparative Biochemistry and Physiology Part-C. Toxicology Pharmacology* 146: 406-414.

Couty, A., Down, R.E., Gatehouse, A.M.R., Kaiser, L., Pham-Delegue, M.H., and Poppy, G.M. (2001b). Effects of artificial diet containing GNA and GNA-expressing potatoes on the development of the aphid parasitoid *Aphidius ervi haliday. Journal of Insect Physiology* 47: 1357-1366.

Day, B.R., McAlvin, C.B., Loh, J.T., Denny, R.L., Wood, T.C., Yosung, N.D., and Stacey, G. (2000). Differential expression of two soybean apyrases, one of which is an early nodulin. *Molecular Plant-Microbe Interaction* 13:1053-1070.

Diaz, C.L., Logman, T.J.J., Stam, H.C., and Kijne, J.W. (1995). Sugar-binding activity of pea lectin expressed in white clover hairy roots. *Plant Physiology* 109: 1177-1197.

Down, R.E., Ford, L., Bedford, S.J., Gatehouse, L.N., Newell, C., Gatehouse, J.A., and Gatehouse, A.M.R. (2001). Influence of plant development and environment on transgene expression in potato and consequences for insect resistance. *Transgenic Research* 10: 223-236.

Down, R.E., Gatehouse, A.M.R., Hamilton, W.D.O., and Gatehouse, J.A. (1996). Snowdrop lectin inhibits development and decreases fecundity of glasshouse potato aphid (*Aulacorthum solani*) when administered *in-vitro* and *via.* transgenic plants in laboratory and glasshouse trials. *Journal of Insect Physiology* 42: 1035-1045.

Etzler, M.E., Kalsi, G., Ewing, N.N., Roberts, N.J., Day, R.B., and Murphy, J. B. (1999). A nod factor binding lectin with apyrase activity from legume roots. *Proceedings of the National Academy of Sciences* 96: 5856-5861.

Fitches, E., Ilett, C., Gatehouse, A.M.R., Gatehouse, L.N., Greene, R., Edwards, J.P., and Gatehouse, J.A. (2001b). The effects of *Phaseolus vulgaris* erythro and leuco-agglutinating isolectins (PHA-E and PHA-L) delivered *via.* artificial diet and transgenic plants on the growth and development of tomato moth (*Lacanobia oleracea*) larvae; lectin binding to gut glycoproteins *in-vitro* and *in-vivo. Journal of Insect Physiology* 47: 1389-1398.

Foissac, X., Loc, N.T., Christou, P., Gatehouse, A.M.R., and Gatehouse, J.A. (2000). Resistance to green leafhopper (*Nephotettix virescens*) and brown planthopper

(*Nilaparvata lugens*) in transgenic rice expressing snowdrop lectin (*Galanthus nivalis*). *Journal of Insect Physiology* 46: 573-583.

Gatehouse, A.M., Davison, G.M., Newell, C.A., Hamilton, W.D.O., Burgess, E.P.J., Gilbert, R.J.C., and Gatehouse, J.A. (1997). Transgenic potato plants with enhanced resistance to the tomato moth *Lacanobia oleracea*: growth room trials. *Molecular Breeding* 3: 49-63.

Gatehouse, A.M., Davison, G.M., Stewart, J.N., Gatehouse, L.N., Kumar, A., Geoghegan, I.E., Birch, A.N.E., and Gatehouse, J.A. (1999). Concanavalin A inhibits development of tomato moth (*Lacanobia oleracea*) and peach-potato aphid (*Myzus persicae*) when expressed in transgenic potato plants. *Molecular Breeding* 5:153-165.

Gatehouse, A.M.R., Down, R.E., Powell, K.S., Sauvion, N., Rahbe, Y., Newell, C.A., Merryweather, A., Hamilton, W.D.O., and Gatehouse, J.A. (1996). Transgenic potato plants with enhanced resistance to the peach potato aphid *Mirzus persicae*. *Entomologia Experimentalis et Applicata* 79: 295-307.

Gilbert, R.J. (1988). Healthy eating day. *Communicable Disease Report* 33: 3-4.

Goldstein, I.J., Winter, H.C., and Poretz, R.D. (1997). Plant lectins: tools for the studying of complex carbohydrates. *Glycoproteins, Elsevier* 2: 403-474.

Grant, G. (1999). Plant lectins. In: Caygill, J.C., Mueller-Harvey, I. (Eds.), Secondary Plant Products: Antinutritional and Beneficial Actions in Animal Feeding, Nottingham University Press, Nottingham, UK: 87-110.

Gupta, A., and Sandhu, R.S. (1998). Effect of garlic agglutinin and garlic extracts on the rat jejunum. *Nutrition Research* 18: 841-850.

Herzig, K.H., Bardocz, S., Grant, G., Nustede, R., Folsch, U.R., and Pusztai, A. (1997). Kidney bean lectin is a potent cholecystokinin releasing stimulus in the rat inducing pancreatic growth. *Gut* 41: 333-338.

Higuchi, M., Tsuchiya, I., and Iwai, K. (1984). Growth inhibition and small intestinal lesions in rats alter after feeding with isolated winged bean lectin. *Agricultural and Biological Chemistry* 48: 695-701.

Hilder, V.A., Powell, K.S., Gatehouse, A.M.R., Gatehouse, J.A., Gatehouse, L.N., Shi, Y., Hamilton, W.D.O., Merryweather, A., Newell, C.A., and Timans, J.C. (1995). Expression of snowdrop lectin in transgenic tobacco plants resulted in added protection against aphids. *Transgenic Research* 4: 18-25.

Hirsch, A.M. (1999). Role of lectins and rhizobial exo-polysaccharides in legume nodulation. *Current Opinion in Plant Biology* **2:** 320-326.

Hopkins, T.L., and Harper, M.S. (2001). Lepidopteran peritrophic membranes and the effect of dietary germ agglutinin on their formation and structure. *Archives of Insect Biochemistry and Physiology* 47: 100-109.

Horejsí, V., and Kocourek, J. (1978). Studies on lectins, properties of some lectins prepared by affinity chromatography on oglycosyl polyacrylamide gels. *Biochimica et Biophysica Acta* 538 (2): 299-315.

Huang, D.F., Pan, Y.H., Zhang, S.X., Cao, J.P., Yang, X.M., Zhang, J., and Yin, W.Z. (1997). The discovery of insecticidal protein against aphids from *Pinellia pedatisecta* and *P. ternata*. *Science Agriculture Sinica* 30: 94.

Kalsi, G., and Etzler, M.E. (2000). Localization of a Nod factor-binding protein in legume roots and factors influencing its distribution and expression. *Plant Physiology* 124: 1039–1048.

Kanrar, S., Venkateswari, J., Kirti, P.B., and Chopra, V.L. (2002). Transgenic Indian mustard (*Brassica juncea*) with resistance to the mustard aphid (*Lipaphis erysimi*). *Plant Cell Reports* 20: 976-981.

Kaur, M., Singh, K., Rup, P.J., Saxena, A.K., Khan, R.H., Ashraf, M.T., Kamboj, S.S., and Singh, J. (2006b). A tuber lectin from *Arisaema helleborifolium Schott* with anti-insect activity against melon fruit fly, *Bactrocera cucurbitae* (Coquillett) and anti-cancer effect on human cancer cell lines. *Archives of Biochemisty and Biophysics* 445: 156-165.

Kijne, J.W. (1992). The Rhizobium infection process. In: G. Stacey, R. Burris, H. Evans (Eds.), Biological Nitrogen Fixation, Chapman & Hall, New York, pp. 349-398.

Ladha, J.K., and Reddy, P.M. (2003). Nitrogen fixation in rice systems: state of knowledge and future prospects. *Plant Soil* 252: 151-167.

Lagarda-Diaz, I., Guzman-Partida, A.M., Urbano-Hernandez, G., Ortega-Nieblas, M.M., Robles-Burgueño, M.R., Winzerling, J., and Vazquez- Moreno, L. (2009). Insecticidal action of PF2 lectin from Olneya tesota (*Palo Fierro*) against *Zabrotes subfasciatus* larvae and midgut glycoconjugate binding. *Journal of Agricultural and Food Chemistry* 57: 689-694.

Lajolo, F.M., and Genovese, M.I. (2002). Nutritional significance of lectins and enzyme inhibitors from legumes. *Journal of Agricultural and Food Chemistry* 50: 6592-6598.

Macedo, M.L., Das Gracas Machado Freire, M., Da Silva, M.B., and Coelho, L.C. (2007). Insecticidal action of *Bauhinia monandra* leaf lectin (BmoLL) against *Anagasta kuehniella* (Lepidoptera: Pyralidae), *Zabrotes subfasciatus* and *Callosobruchus maculates* (Coleoptera: Bruchidae). *Comparative Biochemistry and Physiology Part-A Molecular and Integrated Physiology* 146: 486-498.

Maddock, S.E., Hufman, G., Isenhour, D.J., Roth, B.A., Raikhel, N.V., Howard, J.A., and Czapla, T.H. (1991). Expression in maize plants of wheat germ agglutinin, a novel source of insect resistance. In: Third International Congress in Plant Molecular Biology, Tucson, Arizona, USA.

Majumder, P., Mondal, H.A., and Das, S. (2005). Insecticidal activity of *Arum maculatum* tuber lectin and its binding to the glycosylated insect gut receptors. *Journal of Agricultural and Food Chemistry* 53: 6725-6729.

Malone, L.A., Gatehouse, A.M.R., and Barratt, B.I.P. (2008). Beyond Bt: alternative strategies for insect resistant genetically modified crops. In: Romeis, J., Shelton, A.M., Kennedy, G.G. (Eds.), Integration of Insect-Resistant Genetically Modified Crops within IPM Programs. *Springer* 357–417.

Maqbool, S.B., Riazuddin, S., Loc, N.T., Gatehouse, A.M.R., Gatehouse, J.A., and Christou, P. (2001). Expression of multiple insecticidal genes confers broad resistance against a range of different rice pests. *Molecular Breeding* **7:** 85-93.

Murdock, L.L., and Shade, R.E. (2002). Lectins and protease inhibitors as plant defense against insects. *Journal of Agricultural and Food Chemistry* 50 (22**):** 6605-6611.

Nayak, P., Basu, D., Das, S., Basu, A., Ghosh, D., Ramakrishnan, N.A., Ghosh, M., and Sen, S.K. (1997). Transgenic elite indica rice plants expressing CryIAc endotoxin of *Bacillus thuringiensis* are resistant against yellow stem borer

(*Scirpophaga incertulus*). *Proceedings of National Academy of Sciences* 94: 2111-2116.

Ngai, P.H., and Ng, T.B. (2007). A lectin with antifungal and mitogenic activities from red cluster pepper (*Capsicum frutescens*) seeds. *Applied Microbiology and Biotechnology* 74: 366-371.

Ohizumi, Y., Gaidamashvili, M., Ohwada, S., Matsuda, K., Kominami, J., Nakamura-Tsuruta, S., Hirabayashi, J., Naganuma, T., Ogawa, T., and Muramoto, K. (2009). Mannose-binding lectin from yam (*Dioscorea batatas*) tubers with insecticidal properties against *Helicoverpa armigera*. *Journal of Agricultural and Food Chemistry* 57: 2896-2902.

Oka, Y., Chet, I., and Spiegel, Y. (1997). Accumulation of lectins in cereal roots invaded by the cereal cyst nematode *Heterodera avenae*. *Physiology Molecular Plant Pathology* 51: 333-345.

Pan, Y.H., Zhang, S.X., Cao, J.P., and Huang, D.F. (1998). The isolation, purification of *Pinellia pedatisecta* lectin and its activity on aphid resistance. *Progress in Natural Science* 8: 502-5.

Peumans, W.J., and Van Damme, E.J.M. (1995a). Lectins as plant defence proteins. *Plant Physiology* 109: 347-352.

Peumans, W.J., and Van Damme, E.J.M. (1996). Prevalence, biological activity and genetic manipulation of lectins in foods. *Trends in Food Science and Technology* **7:** 132-38.

Peumans, W.J., and Van Damme, J.M. (1995). Lectins as plant defence proteins. *Plant Physiology* 109: 347-352.

Pohleven, J., Obermajer, N., Sabotič, J., Anžlovar, S., Sepčić, K., Kos, J., Kralj, B., Štrukelj, B., and Brzin, J. (2009). Purification, characterization and cloning of a ricin-B like lectin from mushroom *Clitocybe nebularis* with antiproliferative activity against human leukemic T cells. *Biochimica et Biophysica Acta* 1790 (3): 173-181.

Poulsen, M., Kroghsbo, S., Schrøder, M., Wilcks, A., Jacobsen, H., Miller, A., Frenzel, T., Danier, J., Rychlik, M., Shu, Q., Emami, K., Sudhakar, D., Gatehouse, A., Engel, K.H., and Knudsen, I. (2007a). A 90-day safety study in Wistar rats fed genetically modified rice expressing snowdrop lectin *Galanthus nivalis* (GNA). *Food and Chemical Toxicology* 45: 350-363.

Poulsen, M., Schrøder, M., Wilcks, A., Kroghsbo, S., Lindecrona, R.H., Miller, A., Frenzel, T., Danier, J., Rychlik, M., Shu, Q., Emami, K., Taylor, M., Gatehouse, A., Engel, K.H., and Knudsen, I. (2007b). Safety testing of GM rice expressing PHA-E lectin using a new animal test design. *Food and Chemical Toxicology* 45: 364-377.

Powell, K.S., Gatehouse, A.M.R., Hilder, V.A., and Gatehouse, J.A. (1995a). Antifeedant effects of plant lectins and an enzyme on the adult stage of rice brown plant hopper (*Nilaparvata lugens*). *Entomologia Experimentalis et Applicata* 75: 51-59.

Powell, K.S., Gatehouse, A.M.R., Hilder, V.A., Van Damme, E.J.M., Peumans, W.J., Boonjawat, J., Horsham, K., and Gatehouse, J.A. (1995b). Different antimetabolic effects of related lectins towards nymphal stages of *Nilaparvata lugens*. *Entomologia Experimentalis et Applicata* 75: 61–65.

Powell, K.S., Gatehouse, A.M.R., Hilder, V.A., Van Damme, E.J.M., Peumans, W.J., Boonjawat, J., Horsham, K., and Gatehouse, J.A. (1995). Different antimetabolic effects of related lectins towards nymphal stages of *Nilaparvata lugens*. *Entomologia Experimentalis et Applicata* 75: 61–65.

Powell, K.S., Spence, J., Bharathi, M., Gatehouse, J.A., and Gatehouse, A.M.R. (1998). Immunohistochemical and developmental studies to elucidate the mechanism of action of the snowdrop lectin on the rice brown planthopper (*Nilaparvata lugens*). *Journal of Insect Physiology* 44: 529-539.

Pusztai, A. (1991). General effects on animal cells. In: Plant Lectins (ed. Psusztai A.), Cambridge University Press, Cambridge. pp, 167-180.

Pusztai, A., Ewen, S.W.B., Carvalho, A.F.U., Grant, G., Brantner, K., and Bardocz, S. (1995). Dietary lectins affect hormone balance of body and modulate its general metabolism. In: Singer, M.V., Ziegller, R., Rohr, G. (Eds.), Gastrointestinal Tract and Endocrine System, Kluwer Academic Publishers, Lancaster, pp. 457-463.

Pusztai, A., Ewen, S.W.B., Grant, G., Peumans, W.J., Van Damme, E.J.M., Rubio, L., and Bardocz, S. (1990). The relationship between survival and binding of plant lectins during small intestinal passage and their effectiveness as growth factors. *Digestion* 46: 308-316.

Rao, K.V., Rathore, K.S., Hodges, T.K., Fu, X., Stoger, E., Sudhakar, D., Williams, S., Christou, P., Bharathi, M., Bown, D.P., Powell, K.S., Spence, J., Gatehouse, A.M.R., and Gatehouse, J.A. (1998). Expression of Snowdrop lectin (GNA) in transgenic rice plants confers resistance to rice brown plant hopper. *The Plant Journal* 15: 469-477.

Ratanapo, S., Ngamjunyaporn, W., and Chulavatnatol, M. (1998). Sialic acid binding lectins from leaf of mulberry (*Morus alba*). *Plant Science* 139: 141-148.

Rios, F.J.B., Cavada, B.S., Medeiros, D.A., Moreira, R.A., Vasconcelos, I.M., and Oliveira, J.T.A. (1996). Digestibility of plant lectins from *Canavalia,Cratylia, Dioclea* and *Artocarpus* genera. In: Lectins, Biology, Biochemistry and Clinical Biochemistry (E. Van Driessche, J. Fisher, S. Beeckmans, TC Bog-Hansen, eds.), Textop, Denmark, pp. 277-284.

Ripoll, C., Favery, B., Lecomte, P., Van Damme, E., Peumans, W., Abad, P., and Jouanin, L. (2003). Evaluation of the ability of lectin from snowdrop (*Galanthus nivalis*) to protect plants against root-knot nematodes. *Plant Science* 164: 517-523.

Rosegrant, M.V., Sombilla, M.A., and Perez, N. (1995). Global food projections to 2020: implications for investment, food, agriculture and the environment. *International Food Policy Research Institute* 5: 1-54.

Sadeghi, A., Smagghe, G., Proost, P., and Van Damme, E.J.M. (2008b). Ferritin acts as a target site for the snowdrop lectin (GNA) in the midgut of the cotton leaf worm *Spodoptera littoralis*. *Insect Science* 15: 513-519.

Setamou, M., Bernal, J.S., Legaspi, J.C., Mirkov, T.E., and Legaspi, B.C. Jr. (2002). Evaluation of lectin expressing transgenic sugarcane against stalkborers: effects on life history parameters. *Journal of Economic Entomology* 95: 469-477.

Sharon, N., and Lis, H. (2004). Lectins. *Springer-Verlag* **6:** 1-470.

Sogawa, K. (1970). Effects of nitrogen deficiency of host plant insect feeding. *Japanese Journal of Applied Entomology and Zoology* 14: 101-106.

Stoger, E., William, S., Christou, P., Down, R.E., and Gatehouse, J.A. (1999). Expression of the insecticidal lectin from snowdrop (*Galanthus nivalis*) in transgenic wheat plants: effects on predation by the grain aphid *Sitobion avenae*. *Molecular Breeding* 5: 65-73.

Sun, X., Wu, A., and Tang, K. (2002). Transgenic rice lines with enhanced resistance to the small brown planthopper. *Crop Protection* 21: 511-514.

Tomov, B.W., and Bernal, J.S. (2003). Effects of GNA transgenic sugarcane on life history parameters of *Parallorhogas pyralophagus* (Marsh), a parasitoid of Mexican rice borer. *Journal of Economic Entomology* 96: 570-576.

Trigueros, V., Lougarre, A., Ali-Ahmed, D., Rahbé, Y., Guillot, J., Chavant, L., Fournier, D., and Paquereau, L. (2003). *Xerocomus chrysenteron* lectin: Identification of a new pesticidal protein. *Biochimica at Biophysica Acta* 1621(3): 292-298.

Van Damme, E.J.M., Hao, Q., Charels, D., Barre, A., Rouge, P., Van Leuven, F., and Peumans, W.J. (2000). Characterization and molecular cloning of two different type 2 ribosome inactivating proteins from the monocotyledonous plant *Polygonatum multiflorum*. *European Journal of Biochemistry* 267: 2746-2759.

Van Damme, E.J.M., Peumans, W.J., Barre, A., and Rougé, P. (1998). Plant lectins: a composite of several distinct families of structurally and evolutionary related proteins with diverse biological roles. *Critical Reviews in Plant Sciences* 17: 575-692.

Van Rhijn, P., Goldberg, R.B., and Hirsch, A.M. (1998). *Lotus corniculatus* nodulation specificity is changed by the presence of soybean lectin gene. *Plant Cell* 10: 1233-1249.

Vandenborre, G., Smagghe, G., Gesquière, B., Menschaert, G., Rao, R.N., Gevaert, K., and Van Damme, E.J.M. (2011). Diversity in protein glycosylation among insect species. *PLoS ONE* 6(2): e16682. doi: 10.1371/journal.pone.0016682.

Vandenborre, G., Van Damme, E.J.M., Ghesquière, B., Menschaert, G., Hamshou, M., Rao, R.N., Gevaert, K., and Smagghe, G. (2010). Glycosylation signatures in Drosophila: fishing with lectins. *Journal of Proteome Research* **9:** 3235-3242.

Vasconcelos, I.M., and Oliveira, J.T.A. (2004). Anti-nutritional properties of plant lectins, *Toxicon* 44: 385.

Velkov, V.V., Medvinsky, A.B., Sokolov, M.S., and Marchenko, A.I. (2005). Will transgenic plants adversely affect the environment? *Journal of Biosciences* 30: 515-548.

Vijayan, M., and Chandra, N. (1999). Lectins. *Current Opinion in Structural Biology* 9: 707-714.

Wang, X.C., Bauw, G., Van Damme, E.J.M., Peumans, W.J., Chen, Z.L., Van Montagu, M., Angenon, G., and Dillen, W. (2001). Gastrodianin like mannose-binding proteins: a novel class of plant proteins with antifungal properties. *Plant Journal* **25:** 651-61.

Wang, Z.B., and Guo, S.D. (1999). Expression of two insect resistant genes cryIA and GNA in transgenic tobacco plants results in added protection against both cotton bollworm and aphids. *Chinese Science Bulletin* 44: 2051-2058.

Wilson, A.B., King, T.P., Clarke, E.M.W., and Pusztai, A. (1980). Kidney bean (*Phaseolus vulgaris*) lectin induced lesions in the small intestine. *Journal of Comparative Pathology* 90: 597-602.

Wu, A., Sun, X., Pang, Y., and Tang, K. (2002). Homozygous transgenic rice lines expressing GNA with enhanced resistance to the rice sap-sucking pest *Laodelphax striatellus*. *Plant Breeding* 121: 93-95.

Wunn, J., Kloti, A., Burkhardt, P.K., Biswas, G.C., Ghosh Launis, K., Iglesias, V.A., and Potrykus, I. (1996). Transgenic indica rice breeding line IR58 expressing a synthetic *CryIA* (*b*) gene from *Bacillus thuringiensis* provides effective insect pest control. *Biotechnology* 14: 171-176.

Xu, Q., Liu, Y., Wang, X.C., Gu, H., and Chen, Z. (1998). Purification and characterization of a novel antifungal protein from *Gastrodia elata*. *Plant Physiology* and *Biochemistry* 36: 899-905.

Zhu-Salzman, K., Shade, R.E., Koiwa, H., Salzman, R.A., Narasimhan, M., Bressan, R.A., Hasegawa, P.A., and Murdoch, L.A. (1998). Carbohydrate binding and resistance to proteolysis control the insecticidal activity of *Griffonia simplicifolia* lectin II. *Proceedings of the National Academy of Sciences* 95: 15123-15128.

■■■■

Section IV:

Physiological Basis of Yield

19

PHYSIOLOGICAL BASIS OF YIELD VARIATION

A.K. Trivedi, G. Pandey and V.K. Singh

ICAR-Central Institute for Subtropical Horticulture, Rehmankhera, P.O. – Kakori, Lucknow – 226 101

INTRODUCTION

Radical variation exists between actual and potential yield of crops. Therefore, to ensure food security, understanding of factors affecting yield are pivotal for agricultural research. Precise information about potential yield of a crop and the practical limitations to its realization in the field is essential for economically feasible and meaningful research. In addition, identification of morphological as well as physiological traits closely associated with high crop yields is crucial for crop improvement. It is important to understand how they are affected by environmental conditions as well as cultural practices. There are several methods to find out impact of environmental factors as well as cultural practices on plants. However, in conventional method, a researcher has to grow crop under a wide range of contrasting environmental conditions, then compare its performance under each growth condition by direct observation. Contrary to this, comparatively easy techniques are available to infer the expected out come such as by correlation analysis and correlating environmental conditions, morphological or physiological characters with observed difference in crop yield. For all the cellular activities, energy is basic input and solar radiation is the ultimate source of energy for crop production. Net daily rate of crop dry matter production depends on the amount of light energy intercepted by the crop during the day, the efficiency with which the crop uses that light energy in the production of new dry matter as well as daily rates of loss of crop dry matter. As well, the yield of the harvestable component is equivalent to the proportion of the increment in new dry matter partitioned to the harvestable component in the unit time minus rates of dry matter loss. Considerable dry matter loss occurs through physiological, pathological and mechanical means. Dense crop is considered as closed crop system, where the spatial heterogeneity in the down ward light flux density in any horizontal plane within the volume of the crop canopy can be ignored. In sparse vegetation each plant receives different light flux. Hence, light interception and light use efficiency of closed crop and sparse plants will be different. Difference in light utilization of a crop will affect potential yield of the crop.

Energy of solar radiation is vital for plant processes viz., growth (Penning de Vries *et al.,* 1983; Thornley and Johnson, 1990); nitrate reduction, symbiotic dinitrogen fixation (Simpson, 1987); root N-uptake (Bloom *et al.,* 1992); other ion uptake (Thornley and Johnson, 1990); phloem loading (Geiger, 1975; Bouma, 1995), protein turnover (Vierstra, 1993); maintenance of cell ion concentrations and gradients (Bouma, 1995); and apparently wasteful, heat producing respiration following the alternative (cyanide resistant) pathway or futile cycles (Hue, 1982; Lambers, 1997). Moreover, energy is needed for secondary metabolism, detoxification of pollutants as well as all forms of damage repair. In the context of crop production, light means that portion of the incident radiant energy that lies within the 400 - 700 nm wave band. Variation in the yield of a crop may be due to variation in canopy structure needed for harvesting the light, light interception, availability of direct or diffused light, light utilization efficiency, dry matter production, crop duration, respiration, dry matter partitioning towards harvestable part and harvest index etc.

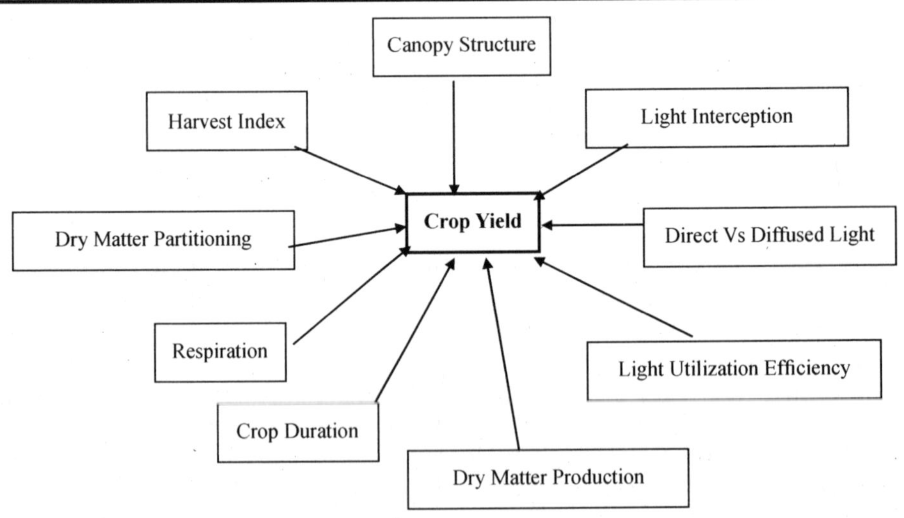

Fig. 1. Components of crop yield.

CANOPY STRUCTURE

Plant canopy structure is the spatial arrangement of above ground organs of the plants in a plant community (Campbell and Norman, 1989). Leaves and other photosynthetic organs on a plant help to harvest the solar energy and exchange of gases. Stems and branches support to orient leaves in such a way that radiative and convective exchange can occur in an efficient manner. Moreover, canopy structure influences temperature, vapour concentration and radiation regime in the plant environment. Interception and transmission of precipitation as well as soil temperature and soil heat flow are also affected by canopy structure. Directly or indirectly, canopy structure affects most of the vital plant processes viz., photosynthesis, transpiration, cell enlargement, disease and insect pest infestation, photomorphogenesis as well as competition between species in a plant community. In a plant community canopies modify their own microclimate. The heat and vapour released into the atmosphere at plant surfaces changes the temperature and humidity of the air in contact with those surfaces. The importance of this atmospheric feedback depends amongst other things on the area of the plant canopy. Small area of vegetation modifies shallow layers of the atmosphere and local changes in microclimate are small. However, large fields have considerable effect upwards as well as horizontally. Differences in canopy form may affects how much photosynthetically active radiation is intercepted by plants as well as regulates the spectral composition of radiation that filters to lower levels in the canopy. Light plays key role in the net primary productivity (Dewar, 1996), light availability varies with spatial arrangements of plant population, especially with canopy structures. The variation in canopy light availability is a result of foliage structural and canopy architectural characteristics (Maddonni *et al.,* 2001; Acreche *et al.,* 2009). The extent of shading will also be affected by canopy shape. In fact, canopy arrangement influences even the extent to which spores of a disease or the droplets of a chemical sprayed to control them can enter infected zone.

Canopy structure of a crop stand can be determined either by direct methods or through indirect methods. The direct methods involve the measurement of leaf area, angle and position of leaves in the canopy. It may involve destructive sampling. However, there are several methods to measure leaf area with optical area meter devices, scanners, hand-held plani meters as well as weighing of paper replicates (Nobel *et al.,* 1993, Asner *et al.,* 2003). These methods allow the computation, separately, of the form, size and number of leaves (Breda, 2003). Leaf area can be measured by correlating either green or dry biomass to leaf area, to find a conversion factor i.e. the specific leaf area (m^2 kg^{-1}). Other important traits of canopy like leaf angle distribution and leaf clumpiness determination can be done by on site observations only as they need to be carried out in undisturbed canopies. Direct methods are normally unmanageable, tedious and time demanding. Hence, the indirect methods have been largely used for estimating canopy structure. These methods use the relationship with other more easily measurable parameters, such as canopy transmittance, green cover estimate or correlation with dry or green biomass. In addition, gap-fraction inversion is adequately used to estimate LAI (Welles and Norman, 1991). Gaps are obtained from specially designed devices such as LAI-2000 or hemispherical photographs. The LAI-2000 finds the gap fraction at five angles, making it possible to estimate the average leaf angle also. Ceptometers are also used to estimate LAI through gap fraction (López-Lozano *et al.,* 2009). Hemispherical photographs are usually taken looking upward (Rich, 1990) and the digital processing allows the estimation not only the total gap fraction but also to partition these gaps in different angles. Hemispherical photographs are the most robust technique, from many standpoints of evaluation, the least sensible to illumination conditions and thus can be applied for a large range of canopy structures (Garrigues *et al.,* 2008). However, the indirect methods based on light extinction through the canopy can be affected by canopy structure assumptions, exceptionally regarding the clumping of leaves (Larsen & Kershaw, 1996).

Fig. 2. Canopy structure modification for enhancing light interception

Fig. 3. Hemispherecal photograph of mango tree.

LIGHT INTERCEPTION

Plant life depends upon the conversion of light energy into chemical energy via the process of photosynthesis. Farmers are dependent on this process to convert the light energy falling on their crops into economic yield. Over the last two deacdes, considerable attention has been focused on the relationship between growth rate and the amount of solar radiation intercepted by agricultural crops and tree stands (Dewar 1996). Although, solar radiation is basic need for plant growth but up to which extent unit solar radiation will support plant growth is a fundamental question which has been addressed by several researcher (Loomis and Amthor, 1999; Long *et al.,* 2006; Zhu *et al.,* 2008), still not completely resolved. The fraction of incident solar radiation intercepted by a plant community depends on the leaf area and orientation. Sparse canopies intercept little radiation; dense canopies may intercept it all. Intercepted radiation is absorbed, transmitted or reflected. Canopies fully intercepting incident solar radiation might absorb 90–95% of PAR in the solar spectrum (Hipps *et al.,* 1983). However, some of that PAR may be absorbed by pigments that do not contribute to photosynthesis. Approximate values for inactive absorption of intercepted PAR have been given to be 4–10% (Long *et al.,* 2006). The issue of inactive absorption, especially at the canopy scale, is unresolved and sometimes ignored. However, it is assumed that a full canopy can absorb 93% of incident PAR and that 92% of that absorption can be by photosynthetic pigments (Amthor, 2010).

Despite the atmospheric attenuation, the distance between the earth and the sun determines the intensity of solar radiation that is received by the earth. The elliptic motion of the earth round the sun causes the seasonal variation in the solar radiation as received by the earth. With increasing latitude, the annual mean value of the daily radiation decreases approximately linearly. However, relative amplitude of the seasonal variation in the daily radiation increases approximately linearly with increasing latitude. By utilizing this relationship the average daily radiation on any particular day at a site of known latitude can be predicted. The path length of the radiant energy through the earth's atmosphere changes with the time of the day, season and latitude. The proportion of the radiant energy that is absorbed or scattered during its passage through the atmosphere will depend upon the path length and the physical characteristics of the atmosphere such as the amount of dust in the air, the amount of water vapors etc. Variations in the density of dust particles and the density and extent of cloud cover contribute to the day to day noise observed in the measurements of the downward radiant energy flux density at ground level at any particular site.

Radiant energy in the 400 - 700 nm wave band only is used by plants for the photosynthetic conversion of gaseous carbon dioxide to the carbohydrate. The energy acquired by an atom or molecule in absorbing one quantum depends upon the frequency of the irradiating light. The energy associated with the 400 - 700 nm wave band depends upon the spectral composition of the irradiating light. Approximately 45% of the total radiation flux density incident upon an unobstructed surface at ground level is associated with light in the 400 – 700 nm wave band. Any molecule activated by light absorbs one quantum of the light that causes the activation. According to Planck's equation [$E = h\upsilon$, where h is Planck's constant and υ is the frequency of the irradiating light] the energy acquired by the molecule depends upon the frequency of the light. The energy absorbed per mole of substance will be Avogadro's number times $h\upsilon$, or $E = nh\upsilon$. This quantity of energy is an Einstein. Measurements of total radiation at the earth's surface include all wavelengths from the shorter, ultraviolet wavelengths to the longer, infrared wavelengths. There is negative correlation between the light energy (400 - 700 nm) to the total radiant energy upon a horizontal unobstructed surface. In the early morning and late evening, when the sun angle is low, more incident radiant energy flux density will be in the 400 - 700 nm wavebands. At around noon, when sun remains near its zenith, the ratio decreases, although the total incident flux density increases several fold. The changing spectral composition of the radiant energy flux with time of day will also reflect the changes in path length of the radiant energy through the earth's atmosphere with time of day. The Transmission and scattering of radiation through and in the atmosphere depends upon the wavelength of the radiation. The mean value of the average daily air temperature decreases more or less linearly with increase in latitudes. On the contrary relative amplitude of the seasonal variation in the average daily air temperature increases approximately linearly with latitude. There is a close relationship between intercepted light and dry matter production from a range of crops (Palmer, 1999). The light environment is dependent on both the latitude, as this determines the incoming solar radiation above the atmosphere and the cloudiness of the particular location (Palmer, 1999). Seed/ fruit growth is affected by temperature also which varies with locations,. hence while recording light interception data this should be taken into consideration (Palmer, 1999). Horticultural tree crop yields tend to be linearly correlated with the percentage of photosynthetically active radiation (PAR) intercepted by the canopies (Olsen *et al.*, 2007). Excessive shading causes reduced flowering (Jackson 1980), increased fruit abscission (Byers *et al.* 1991), reduction in fruit size (Warrington *et al.*1996), lower anthocyanin biosynthesis, poor colour development in the peel (Jackson 1970) and a reduction in internal fruit quality (Warrington *et al.* 1996).

Fig. 4. Light interception by plant (Source: http://www.sportsfieldmanagementmagazine.com)

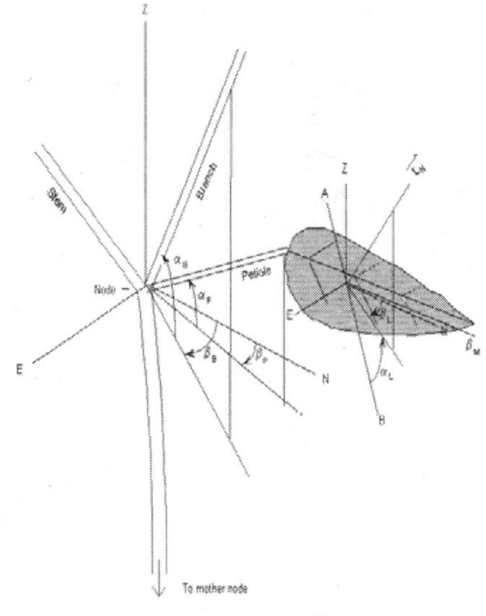

Fig. 5. Different leaf angles for enhancing light interception (Source: http://www.imagebank. osa.org).

DIRECT VS. DIFFUSED RADIATION

Solar radiation available at leaf surface may be direct or diffused; characteristics of both of these are different. Hence, their ratio at leaf surface affects plant productivity. Diffuse radiation enhances photosynthesis of terrestrial plants (Gu *et al.,* 1999, 2002; Urban *et al.,* 2007; Dengel and Grace, 2010) conversely; direct solar radiation may cause saturation of photosynthesis at the top of the canopy and possibly photo-inhibition (Powles, 1984; Krause, 1988; Long *et al.,* 1994). Furthermore, unsaturated photosynthesis during direct solar radiation is possibly occurring within the canopy and under storey region as a result of shading (Kanniah *et al.,* 2012).

The spectral distribution of radiation is important for plant growth and morphogenesis (Endler, 1993; Escobar-Gutuérrez *et al.,* 2009). However, substantially more energy in the photosynthetically active wave bands penetrates the canopy in case of diffuse radiation. Optimal photosynthetic activity of the canopy is achieved under diffuse radiation (cloudy and overcast) conditions, when scattered light penetrates throughout the canopy, illuminating all the leaves to some extent and providing a more uniform distribution of light between the leaves (Urban *et al.,* 2007, 2012; Dengel and Grace, 2010). Under cloudy skies more blue light reaches within the canopy, possibly a result of multiple reflections and scattering (Reicosky and Hanover, 1978; Cape and Percy, 1993). Blue enrichment may have important implications for stomatal control of photosynthesis and water use. Blue wavelengths are more effective in causing stomatal opening than red wavelengths (Morison and Jarvis, 1983). At low PPFD stomata open only in response to blue light, red light being ineffective (Smith, 1982). Conditions of diffuse radiation are especially conducive to stomatal opening in the lower regions of the canopy. Under clear skies the red: far-red ratio in the canopy remains much lower (Hertel *et al.,* 2011), indicating a lower photomorphogenical light quality (Smith, 1982) than under diffuse conditions. Realistic estimation of radiation profiles in plant canopies is important in order to correctly understand processes occurring at the leaf-level, such as photosynthesis and evaporation (Bodin and Franklin, 2012). It is important to study direct and diffuse constituents of solar radiation because a high direct: diffused solar radiation ratio has been shown to enhance gross primary production (Bodin and Franklin, 2012).

RADIATION UTILIZATION EFFICIENCY

Radiation use efficiency (RUE) is the ratio of dry matter produced per unit of radiant energy used in its production (Monteith 1977). Values of RUE for tree species are generally smaller than those for C_3 herbaceous species (Kiniry *et al.* 1989), because of the high energy cost of woody biomass and the respiration of supporting organs. Radiation-use efficiency is dependent on light (Connor and Sadras 1992, Villalobos *et al.* 1992), temperature (McMurtrie and Wang 1993), vapor pressure deficit (Stockle and Kiniry 1990, Landsberg and Hingston 1996) and factors inherent to plant species. For example, RUE increases with increasing rate of leaf photosynthesis (Sinclair and Horie 1989). Likewise, RUE decreases with increasing leaf age, respiration (Russell *et al.* 1989) and with the higher energy costs of some plant constituents (Trapani *et al.* 1992).

Rate of light saturated leaf photosynthesis, F_{max} is one of the most obvious physiological factors of the crop light utilization efficiency. The rate of light saturated photosynthesis per unit leaf area (F_{max}) usually increases as the mean daily light flux density experienced by the leaf during growth increase (Charles-Edwards, 1979). There is clear difference between solar energy flux (irradiance) and photon flux density. Energy of photon is inversely related to wavelength, so that the potential efficiency of energy use is greater with longer wavelength photons. The maximal spectral solar irradiance may occur at 450–500 nm, whereas the maximal spectral photon flux density occurs more broadly at longer wavelengths i.e., 550–700 nm (Amthor 2010). In the closed crop system, the instant rate of net photosynthesis per unit of ground area by a crop canopy is a function of leaf photosynthetic characters i.e., canopy architecture, light flux density incident upon the crop and the proportion of the incident light flux density intercepted by crop canopy.

There may be variation in F_{max} between cultivars which is related with the differences in their leaf thickness. As a general rule, thicker leaved cultivars have higher values for F_{max} than the thinner ones (Charles-Edward, 1979). Thicker leaves show lower rates of leaf area expansion. Thick leaved cultivars of a crop take longer time period to attain full ground cover. Genotypes with a high F_{max} may therefore be offset by a reduction in the amount of light energy intercepted by a crop of these genotypes during the early stages of growth and development. F_{max} may also depend upon the nitrogen content of the leaf (Charles-Edwards, 1981, Gulman and Chu, 1981). Leaf nitrogen content will depend upon the amount of nitrogen available to the plant. Hence, fertility may also have a direct effect on the light use efficiency of crops. Usually, species with lower leaf nitrogen contents have lower specific leaf area (thicker leaves). The plants with lower specific leaf photosynthetic activities have larger photosynthetic systems beneath unit leaf area. Measurement of the proportion of elemental nitrogen in the leaf and specific area of fully expanded leaves at the top of the canopy may provide useful information on factors affecting the rate of light saturated leaf photosynthesis and the efficiency with which new dry matter is produced. For crops consisting of prostrate plant types, the canopy light extinction will be high. In a canopy of erect leaves light is distributed more evenly throughout the canopy.

Smith (1982) indicated that the blue absorbing photoreceptor present in plants acts to measure light quantity and that the pigment phytochrome can act to detect the red:far-red ratio as an indicator of light quality. Blue light may have important implications for stomatal control, causing stomatal opening (Morison and Jarvis, 1983), while the red:far-red ratio is known to influence photomorphogenesis, heat regulation, as well as stem elongation and chlorophyll synthesis (Smith, 1982; Wherley *et al.,* 2005; Casal, 2013). In growing canopies, foliar traits (such as leaf area index and leaf mass per unit area) are the important factors in leaf light harvesting capacity and photosynthetic potentials (Niinemets and Sack, 2006). The radiation regime of plants depends on the incoming radiation. Variations in solar radiation inside the canopy can be reduced by examining the changes in the incident radiation above the canopy caused by the following three factors viz., meteorological changes (clouds, sun position etc.), variations in tree density (cutting and planting of trees) and size (growth of trees). Light utilization efficiency of the plant is proportional to the rate of light saturated photosynthesis. If increase in the light utilization efficiency is related with increase in specific leaf area or leaf thickness, plants with higher light use efficiencies may have lower rates of leaf area development. Simultaneously, at a particular leaf area plants with a higher light use efficiency may be more productive, this advantage may be counter balanced by the time it takes them to develop an effective light intercepting canopy. Usually, early crop production is of economic importance. This may be benefited from a rapid rate of leaf area expansion even if this rapid rate of leaf area expansion does not turn out to be associated with thin leaves and low light utilization efficiency. Comparison of two canopies absorbing the same amount of light energy shows that if the average light flux density on the illuminated leaves within one canopy is less than the average incident upon leaves in the second; first canopy will make more efficient use of the light incident upon them than the leaves in the second canopy. Hence, it may be anticipated that the first canopy have higher light utilization efficiency than the second. However, if surrounding plants (guard plants) from around a block of plants are removed and light is allowed to penetrate the crop canopy from the side; although the efficiency of the leaves in this block of plants may be reduced, because each leaf is receiving more light, the total amount of light energy absorbed by the block may have increased because the crop is now exposed to light previously absorbed by the guard plants. For comparisons of crop growth rates, crop area should be sufficiently large or sufficiently well guarded so that edge effects can be ignored. With the help of amount of light energy that is incident upon a particular crop and available to it for the production of new dry matter, the efficiency with which the crop uses absorbed light energy for the production of new dry matter can be calculated. The amount and quality (direct or diffused) of radiant energy at ground level is important for crop production. Although, data for seasonal and geographical variation in the radiant solar energy incident at the top of the atmosphere is available but the radiant energy incident at ground level is not so readily available (Charles-Edward, 1981).

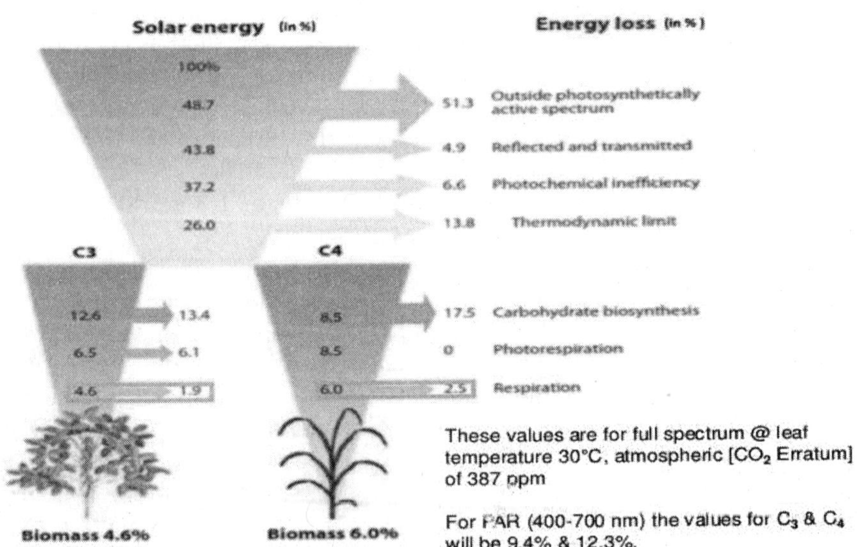

Source: Annual review of plant Biology. 2010.61, 235-261.

Fig. 6. Radiation use efficiency of C_3 and C_4 plants.

DRY MATTER PRODUCTION

The amount of energy accumulated in the plant depends on the amount of solar energy captured and the efficiency of its use (Dohleman and Long, 2009). Monteith (1972) showed that phytomass production under tropical climate conditions is correlated with the amount of photosynthetically active radiation (PAR) absorbed by plants. Although several internal and external factors play decisive role in the growth and development of a plant but physiological processes are central in determining the rate and extent of plant growth. In a crop growing in the field, each plant is subjected to complex competition from its neighbors. Conversely, a single well spaced plant, growing free from competition with its neighbors is a simplest crop system. Such plant may be presumed as free from any shading effect by its neighbors or self shading by its own leaves. In such condition the amount of light intercepted by the plant is directly proportional to its leaf area projected downward onto a horizontal surface beneath it. The amount of light energy intercepted by the plant each day equals the light energy incident per unit ground area.

The rate of leaf net photosynthesis does not increases linearly with increasing light flux densities incident upon the leaf surface. As the light flux density incident upon the leaf increases, the rate of leaf photosynthesis increases less and less quickly, until it reaches a maximum value i.e., the rate of light saturated leaf net photosynthesis. In other words, as the light flux density incident upon the leaf surface increases, the efficiency with which it uses that light energy in carbon assimilation (photosynthesis) decreases. Although the relationship between total dry matter production and light interception is often very tight, the relationship between yield of crop and light interception is variable. This is to be expected as the partitioning of carbohydrate into economic part depends upon crop load and the light distribution within the canopy (Palmer, 1999).

In a plant canopy, the above ground biomass determines the productivity. Solar radiant energy absorbed by plants is converted to biomass through photosynthesis. For a crop, supply of radiant energy sets a limit to potential production, however, if light is not a limiting factor then other variables may be determinant for the actual production (Oker-Blom, 1991). Hence, a plant stand may be appropriately considered as a solar energy trap i.e., higher the amount of solar energy it captures, the greater is its photosynthetic capacity. For perennial crops like fruit crop orchards, the capacity for harvest of solar radiation depends upon planting pattern of trees as well as tree density (Myneni, 1991). Reduced photosynthetic activity in old, large trees could be caused by low vascular system hydraulic conductance, which in turn causes reduced stomatal conductance and CO_2 uptake (Yoder *et al.,* 1994; Ryan *et al.,* 1997). Slow photosynthesis in old trees might also be caused, in part, by limited N availability to leaves or other factors (Landsberg and Gower, 1997).

CROP DURATION (VEGETATIVE VS REPRODUCTIVE PHASE)

Duration of the production of the harvestable component of the crops is one of the important physiological determinants of crop yields. In certain pulse crops like mungbean, soybean the seed yield of crops increases linearly with the duration of the period of productive growth. The grain yield of cereals like wheat is also a function of the duration of the period of grain filling. Proper understanding of the effects of physiological and environmental factors on the phenology of the crop is decisive for the prediction of the duration of the vegetative as well as reproductive growth phases of a crop. During the vegetative phase of crop growth there is an initial exponential phase of growth i.e., period when the leaf canopy of the plants is being established by the crop. Subsequently, the standing crop dry matter increases as a linear function with respect to time. In general, the longer the phase of vegetative crop growth the greater the yields of the vegetative parts of the crop will be. If the seasonal change in daily light and average air temperature is known, the potential length of the cropping season for a particular site and the potential total dry matter yield of a vegetative crop at the site can be predicted. For an indeterminate crop, the duration of fruit or seed production will most probably depend upon the partition of assimilate between the reproductive and vegetative plant parts. This partitioning may be determined by environmental factors, such as temperature, day length or water availability. The change from vegetative to reproductive growth will also be determined by environmental factors such as day length and temperature. There are multiple switches between vegetative and reproductive phase (King and Roughgarden, 1982).

Understanding and studies on the changes in physiological state of plants as they move from vegetative to reproductive growth is still inadequate and this is a fertile area of research. In fact, physiological studies of the transition from vegetative to reproductive growth have concentrated on those plants which respond rapidly to some external stimulus. Such experiments easily identify the proportion of plants in a particular population that flower in response to the stimulus. However, such experiments provide little information on the changing physiological state of plant as it approaches and then crosses, the boundary between vegetative and reproductive growth states. There is a clear boundary between vegetative and reproductive states by the ratio of the concentration of vegetative enzyme and flowering enzyme (Wang, *et. al.*, 2016). The boundary is defined by a particular numerical value for this ratio. Stimuli can upset the system and cause progressive changes in the ratio. At some point the critical ratio is attained, and the system changes from the vegetative to the reproductive state. For example wheat crop becomes reproductive when the flower is initiated on terminal apical meristem. However, newly formed leaves continue to expand for some time after flower initiation and the far greater proportion of dry matter is partitioned to the leaves than to the developing flower. Only about, anthesis does leaf expansion ceases. Hence, physiologically the time of anthesis will reflect more accurate boundary between the vegetative and reproductive growth phases of the crop.

RESPIRATION (GROWTH VS MAINTENANCE RESPIRATION)

Plant respiration is the metabolic link between gross primary productivity and net primary productivity. It is also a large component of a plant's carbon budget; perhaps typically 50-70% of carbon assimilated in gross primary productivity is released back to the atmosphere as CO_2 during subsequent plant respiration (Amthor and Baldocchi, 2001). Growth is the conversion of temporary pools of substrates such as carbohydrates and amides into new structures including enzymes and permanent storage such as starch in seeds. Growth includes monomer synthesis, polymerization of monomers into polymers such as cellulose and proteins, organization of polymers into organelles and cells and tool maintenance or the turnover of molecules catalyzing growth (Penning de Vries *et al.,* 1974). Substrate requirement for growth is the amount of sucrose needed to provide carbon skeletons, ATP and NADPH for synthesis of a compound, or whole plant, including polymerization reactions. For whole tissues/ plants, substrate requirements for individual classes of compounds are summed in proportion to phytomass composition (Thornley and France, 2007). To derive the approximate whole plant substrate requirement, therefore, representative values are needed for cellulose, hemicelluloses, lignins, proteins, lipids, organic acids and in some plants, pectins and/or storage carbohydrates (Penning de Vries *et al.,* 1974, 1983). In addition, the cost of mineral uptake is also needed.

Furthermore, life requires maintenance energy to acclimate to environmental fluctuations. Maintenance respiration is the CO_2 and energy release associated with maintenance processes (Penning de Vries, 1975). Maintenance respiration includes the energy and carbon used for turnover of labile cellular constituents, active intracellular transport to counteract membrane leaks and ion imbalances and repair as well as acclimation processes that result from a stressful or variable environment (Penning de Vries, 1975). This includes the regular replacement of enzyme and lipid population with different population better suited to new conditions and developmental states, repair of damage from, for example, endogenous and exogenous oxidants and repair/replacement of compounds subject to spontaneous breakdown. A large fraction of the carbon assimilated in higher plant photosynthesis is released back to the atmosphere as CO_2 during subsequent plant respiration. Thus, plant respiration is a large negative component of the carbon budget of plants. It contributes to the control of net primary production (NPP) because NPP is gross primary production (GPP) minus plant respiration. The relationship between NPP and GPP is controlled by respiration. Respiration is regulated in large part by needs for and uses as well as turnover of its products (Dry *et al.,* 1987). A range of carbon skeleton intermediates generated all along the respiratory pathways, ATP and the reductants (NADH and NADPH) are important products. Respiratory products accumulate when they are not used by other processes, thus they retard respiration through negative feedback mechanisms. Furthermore, without the use/ turnover of ATP, NADH and NADPH the required respiratory substrates ADP, NAD^+ and $NADP^+$ remain unavailable for further respiration. When respiratory products are used rapidly, negative feedbacks on respiration are released and associated respiratory substrates are regenerated with the result that respiration rate increases.

McCree (1970) has related the respiratory losses of carbon dioxide by plants to both the photosynthetic activity and to their net dry weight. McCree analysis is a robust relationship and useful for an analysis of crop dry matter production. McCree's analysis has led to the recognition of two components of plant /crop respiration activity. One of these components, called the maintenance respiration component is generally related to the net weight of the plant or crop. However, it does not necessarily follow that the size of the metabolically active component of a plant or crop is related to its total dry weight. Gradually, metabolically active component of the crop becoming an increasingly smaller proportion of the total crop biomass. There may be relation between the size of the metabolically active component of the crop to the amount of light energy currently being absorbed by the crop. In turn, the amount of light energy absorbed by a crop does not increase linearly with standing crop dry weight. It approaches a maximum value with increasing leaf area index. The maintenance respiration rate is very dependent upon the temperature. Usually, for every 10 °C increase in temperature, respiration rate doubles (Q10 value) (Ryan 1991). Nevertheless, Q10 values ranges from 1.4 to 4.0 (Azcón-Bieto 1992), this is highly variable. It depends on measurement temperature, species (Larigauderie and Körner 1995), season (Stockfors and Linder 1998, Atkin *et al.* 2000), growth temperature (Atkin *et al.* 2000) and leaf metabolic state (Berry and Raison 1981). In field condition, respiration rate is a function of a plant's physiological history and is subject to acclimation, adaptation or both (Amthor 1989, Atkin *et al.* 2000). Moreover, the change in protein content of the plant during growth will also depend upon the growth temperature of the plant.

Moreover, at ambient carbon dioxide concentrations there is a substantial rate of photorespiration by the leaves of C_3 plants. However, photorespiration is absent in the leaves of C_4 plants. Photorespiration takes place due to the photo-oxidation of ribulose-1,5-bisphosphate. If the ambient oxygen concentration in the air surrounding the leaf of a C_3 plant is reduced, photorespiration decreases with simultaneous increase in the net rate of leaf photosynthesis. In contrast, reduced ambient oxygen concentrations have meager or no effect on the net rate of photosynthesis by leaves of C_4 plants. Growth rates of plant depend equally on both the ability to intercept the incident light energy as well as plants capacity to use that light energy in the production of new dry matter. However, there is ample difference in the rates of light saturated leaf net photosynthesis within both C_3 and C_4 plant species. The major source of the genetic variation in the difference lies in the size of photosynthetic system beneath unit leaf area rather than in the specific activity of the system.

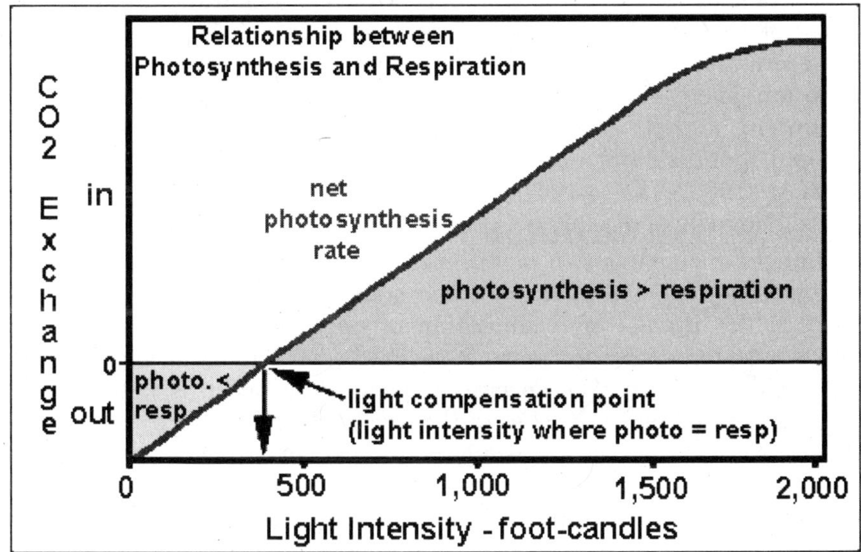

Fig. 7. Relationship between photosynthesis and respiration.

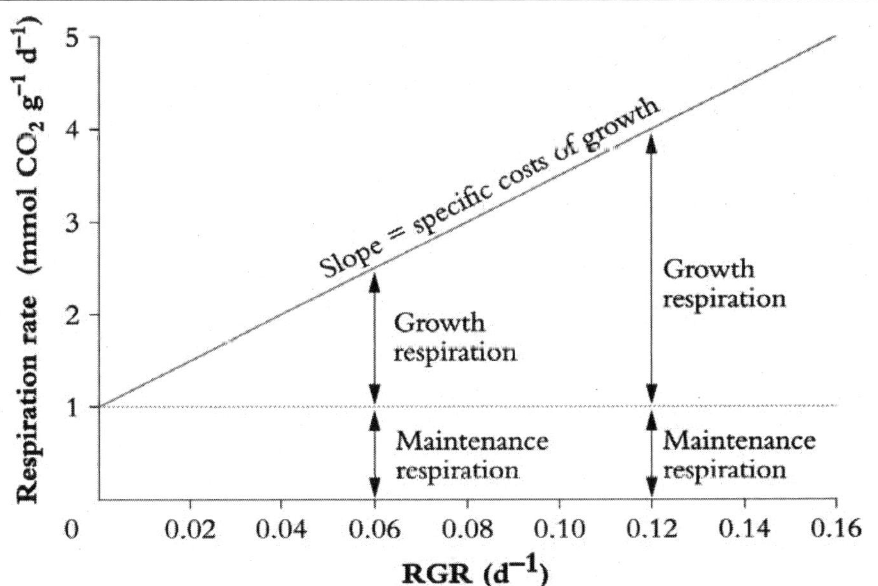

Fig. 8. Growth vs. maintenance respiration.

DRY MATTER PARTITIONING

For phloem translocation, chloroplastic starch is converted to sucrose via maltose (Chia *et al.,* 2004; Smith *et al.,* 2005). The mobilization of chloroplastic starch and the translocation of sucrose via phloem require ATP, which comes from respiration. The ATP required for phloem translocation depends on the number of active membrane crossings and the cost of each crossing (Amthor, 1994; Patrick, 1997). Minimal values may be one active crossing and one ATP per crossing. For plants that require temporary storage between sources and (future) sinks (e.g. starch and protein stored in stems and later mobilized for seed growth), additional mobilization and translocation costs are needed (Penning de Vries *et al.,* 1983). The cost of starch mobilization depends on the reactions involved (Noguchi *et al.,* 2001).

Ontogenetic changes in partitioning patterns may also be attributed to ontogenetic changes in the specific activities of root and shoots with respect to the uptake or assimilation of particular elements. Davidson (1969) suggested that root mass × specific root activity = shoot mass × specific shoot activity. The specific activity of an organ is defined as its activity expressed per unit mass of the organ. Specific activity of the shoot with respect to carbon uptake would be the rate of carbon assimilation per unit of shoot mass. Specific activity of the shoot with respect to carbon assimilation is directly proportional to the light flux density incident upon the plant. The specific growth rate of the plant is related hyperbolically to the available nutrient concentration and the light level incident upon the plant. Analysis of the effect of changes in partitioning attributed due to changes in the plant's aerial environment, particularly changes in the air temperature is complex and need to be integrated with other related factors also.

At reproductive phase of a crop, consideration of the number of reproductive organs (pods or fruits) set per unit of ground area by the crop and their average size is important. During the initiation of each reproductive unit some critical rate of supply of assimilates is required. Some plants behave in an indeterminate manner. After commencement of flowering and fruit set, they partition a constant proportion of their dry matter to the reproductive parts, the remainder being partitioned to new vegetative growth. Other plants behave in determinate manner. On flowering, they cease vegetative growth and photosynthate is diverted towards reproductive organs. However, if both the stages of plant growth (vegetative and reproductive) are under consideration then the boundary between vegetative and reproductive growth stage should be well demarcated i.e., at what point of time the plant shifts from vegetative to reproductive growth. If moment of shift is known then new pattern of dry matter partitioning during reproductive growth should be obvious. However, this depends upon the plant type whether it is determinate plant or an indeterminate one. During the early stages of growth both the dry weight and leaf area of the plant increases approximately exponentially. For an agriculturally determinate crop a proportion 'a' of the dry matter previously partitioned to the leaves, stems and roots is partitioned to the reproductive parts and remainder (1- a) is partitioned to the roots or to a storage organ. Different crops may have similar rates of above ground dry matter production. If they have closed green surface, i.e., they are intercepting the entire light energy incident upon them. Similarly, crop growth rates suggest a similarity in the light utilization efficiencies of the different crops. The light distribution within a canopy can affect the efficiency with which it is able to use that light energy to produce new dry matter (Medlyn, 1997).

HARVEST INDEX (HI)

Harvest index depends on phenological, physiological and environmental factors. Harvest index adds little to our understanding of the crop performance hence a better and more meaningful understanding can be obtained by the direct analysis of the main determinants of grain yield. The harvest index of a crop is the ratio of its economic yield (grain yield) to the total standing above ground dry matter of the crop at harvest. The total above ground dry matter of the crop is often called its biological yield. There is a strong association between harvest index and grain yield (Nass 1973; Riggs *et al.* 1981) and the use of measurement of harvest index and biological yield have been believed as an instrument that would make an important contribution to our understanding and advancement of crop performance (Donald and Hamblin, 1976).

The harvest index will be hyperbolically dependent on the proportion of the light energy intercepted during grain production, the duration of grain production and the proportion of the dry matter partitioned to the grain during grain production. In addition, it will be inversely related to the proportion of the incident light intercepted during vegetative growth, the duration of the vegetative growth period, the proportion of dry matter partitioned to stem during vegetative growth and the proportion of the dry matter partitioned to chaff during grain production. Harvest index declines in cultivars that reach anthesis late. Usually, high yielding varieties have high harvest index, these are shorter and reach anthesis earlier than the older varieties. The harvest index is a function of both phenological and physiological characters of the crop. Hence, for improving grain yield it seems to be more logical and practical to look directly at the phenological, physiological and environmental determinants of grain yield. However, for tuberous and bulbous crops in which underground biomass is of economic importance, criteria of harvest index is quite different as compared to cereals, pulses or oil seed crops (i.e., crops in which above ground plant parts are of economic importance).

INFERENCE AND APPLICATIONS

Above mentioned determinants have decisive role in crop yield however, dry matter production is affected by five main determinants viz., (i) the efficiency with which light energy is used in the production of new dry matter (ii) the amount of light energy intercepted by the crop/plant during each 24 h (day/night) period (iii) the proportion of one daily increment in new dry matter partitioned to the different crop/plant parts (iv) the daily loss of dry matter (v) the duration of production of that part of the crop for which crop is grown (economic part). Detailed information of crop production with respect to these determinants is pivotal for understanding of the genetic, environmental and managerial factors of growth through their effects on these determinants of growth. Availability of accurate information and reliable data affects management of cropping system.

For proper management of a crop, there are three distinct phases of crop growth viz., (i) the period between showing and establishment of the crop growth (ii) the period of vegetative crop growth (iii) the period of reproductive crop growth. These three separate growth phases define four prime times in the life of a crop viz., sowing, establishment, initiation of reproductive phase and maturity. Although there may be ontogenetic or environmentally induced differences in some or all of the determinants of growth during each of the three growth phases, however, there are discontinuous changes in one or more of them between phases. Analysis of the standing crop dry matter at these prime times helps to understand these determinants. The inference of gross crop growth rates during reproductive growth is more difficult with an indeterminate plant. For a determinate crop, leaf dry matter production during reproductive growth may be subtracted and gross increment in the above ground dry weight may be estimated directly as the increment in standing stem and reproductive parts dry weight.

REFERENCES

Acreche, M. M., Briceno-Felix, G., Sanchez, J. A. M. and Slafer, G. A. (2009). Radiation interception and use efficiency as affected by breeding in mediterranean wheat. *Field Crop Res.* **110:** 91 – 97.

Amthor, J. S. (2010). From sunlight to phytomass: on the potential efficiency of converting solar radiation to phyto-energy. *New Phytol.*, 188: 939 – 959.

Amthor, J. S. and Baldocchi, D. D. (2001). Terrestrial Higher Plant Respiration and Net Primary Production. In: Terrestrial Global Productivity. p. 33 – 59.

Amthor, J. S. (1989). Respiration and crop productivity. SpringerVerlag, New York, p. 215.

Angelopoulus, K., Dichio, B. and Xiloyannis, C. (1996). Inhibition of photosynthesis in olive trees (*Olea europaea* L.) during water stress and re-watering. *J. Exp. Bot.* **49:** 1093–1100.

Asner, G.P., Scurlock, J. M. O. and Hicke, J. A. (2003). Global synthesis of leaf area observations: implications for ecological and remote sensing studies. *Global Ecol. Biogeograph.*, **12 (3):** 191 – 205. OnlineISSN: 1466 – 8238.

Atkin, O.A., Holly, C. and Ball. M. C. (2000). Acclimation of snow gum (*Eucalyptus pauciflora*) leaf respiration to seasonal and diurnal variations in temperature: the importance of changes in the capacity and temperature sensitivity of respiration. *Plant Cell Environ.* **23:** 15 – 26.

Azcón-Bieto, J. (1992). Relationships between photosynthesis and respiration in the dark in plants. In Trends in Photosynthesis Research. Eds. J. Barber, M.G. Guerrero and H. Medrano. Intercept Ltd., Anover, Hampshire, U.K., p. 241 – 253.

Berry, J. A. and Raison, J. K. (1981). Responses of macrophytes to temperature. In Physiological Plant Ecology I. Responses to the Physical Environment. Eds. O.L. Lange, P.S. Nobel, C.B. Osmond and H. Zeigler. Springer-Verlag, Berlin, p. 277 – 238.

Bloom, A. J., Sukrapanna, S. S. and Warner, R. L. (1992). Root respiration associated with ammonium and nitrate absorption by barley. *Plant Physiol.* **99:** 1294 – 1301.

Bodin, P. and Franklin, O. (2012). Efficient modeling of sun/shade canopy radiation dynamics explicitly accounting for scattering. *Geosci. Model Dev.*, **5:** 535 – 541.

Bouma, T. J., de Visser, R., van Leeuwen, P. H., de Kock, M. J. And Lambers H. (1995). The respiratory energy requirements involved in nocturnal carbohydrate export from starch-storing mature source leaves and their contribution to leaf dark respiration. *J. Exp. Bot.* **46:** 1185 – 1194.

Breda, N. J. J. (2003). Ground-based measurements of leaf area index: a review of methods, instruments and current controversies. *J. Exp. Bot.* **54 (392):** 2403 - 2417, ISSN 0022-0957.

Byers, R. E., Carbaugh, D. H., Presley, C. N., Wolf, T. K. (1991). The influence of low light on apple fruit abscission. *J. Hort. Sci.*, **66:** 7 – 17.

Campbell, G. S. and Norman, J. M. (1989). The description and measurement of plant canopy structure In: Plant canopies: their growth, form and function (Eds. Russell, G., Marshall, B. and Jarvis, P.G.), p. 1 - 19, Cambridge University Press, ISBN 0 521 39563 1, Cambridge, UK.

Cape, J. N. and Percy, K. E. (1993). Environmental influences on the development of spruce needle cuticles. *New Phytol.*, **125:** 787 – 799.

Casal, J. J (2013). Photoreceptor signaling networks in plant responses to shade, *Ann. Rev. Plant Biol.*, **64:** 403 – 427.

Charles-Edward, D. A. (1979). Photosynthesis and crop growth. In: Photosynthesis and Plant Development (Ed. Marcelle, R., Clijsters, H. and van Pouke, M.). p.111 – 124. Junk. The Hague.

Charles-Edward, D. A. (1981). The Mathematics of Photosynthesis and Productivity. Academic Press, London.

Chia, T., Thorneycroft, D., Chapple, A., Messerli, G., Chen, J., Zeeman, S. C., Smith, S. M. and Smith, A. M. (2004). A cytosolic glucosyltransferase is required for conversion of starch to sucrose in Arabidopsis leaves at night. *Plant J.*, **37:** 853 – 863.

Connor, D. J. and V. O. Sadras (1992). Physiology of yield expression in sunflower. *Field Crops Res.* **30:** 333–389.

Davidson, R. L. (1969). Effect of root/ leaf temperature on root shoot ratios in some pasture grasses and clover. *Ann. Bot.*, **33:** 561 – 569.

Dengel, S. and Grace, J. (2010). Carbon dioxide exchange and canopy conductance of two coniferous forests under various sky conditions, *Oecologia* **164:** 797 – 808.

Dewar, R. C. (1996). The Correlation between Plant Growth and Intercepted Radiation: An Interpretation in Terms of Optimal Plant Nitrogen Content. *Ann. Bot.*, **78:** 125 – 136.

Dohleman, F. G., and Long, S. P. (2009). More productive than maize in the Midwest: how does Miscanthus do it? *Plant Physiol.* **150:** 2104 – 2115.

Donald, C. M. and Hamblin, J. (1976). Biological yield and harvest index of cereals as agronomic and plant breeding criteria. *Adv. Agron.*, **28:** 361 – 405.

Dry, I. B., Bryce, J. H., and Wiskich, J. T. (1987). Regulation of mitochondrial respiration. In: The Biochemistry of Plants, Vol. **11**, Biochemistry of Metabolism" (Ed. D. D. Davies), p. 213 - 252. Academic Press, San Diego.

Endler, J. A. (1993). The Color of Light in Forests and Its Implications, *Ecol. Monogr.*, 63 : 1 – 27.

Escobar-Gutiérrez, A. J., Combes, D., Rakocevic, M., de Berranger, C., Eprinchard-Ciesla, A., Sinoquet, H., and Varlet-Grancher, C. (2009). Functional relationships to estimate Morphogenetically Active Radiation (MAR) from PAR and solar broadband irradiance measurements: The case of a sorghum crop, *Agr. Forest Meteorol.*, **149:** 1244 – 1253.

Garrigues, S., Shabanov, N. V., Swanson, K., Morisette, J. T., Baret, F. and Myneni, R. B. (2008). Intercomparison and sensitivity analysis of Leaf Area Index retrievals from LAI2000, AccuPAR, and digital hemispherical photography over croplands. *Agric. Forest Meteor.*, **148 (8-9):** 1193 – 1209, ISSN 0168-1923.

Geiger, D. R. (1975). Phloem loading. (In: Transport in plants. I Phloem loading, Eds., Zimmerman, M. H., and Milburn J. A.) *Encyclo. Plant Physiol. (New Series),* **1:** 395–431, Berlin: Springer-Verlag.

Gu, L., Baldocchi, D., Verma, S. B., Black, T. A., Vesala, T., Falge, E. M., and Dowty, P. R. (2002). Advantages of diffuse radiation for terrestrial ecosystem productivity, *J. Geophys. Res.*, **107:** 2 – 23.

Gu, L., Fuentes, J. D., Shugart, H. H., Staebler, R. M., and Black, T. A. (1999). Responses of net ecosystem exchanges of carbon dioxide to changes in cloudiness: Results from two North American deciduous forests. *J. Geophys. Res.*, **104:** 31421 – 31434.

Gulman, S. L., and Chu, C. C. (1981). The effects of light and nitrogen on photosynthesis, leaf characteristics and dry matter accumulation in the chaparral shrub *Diplacus auranticas. Oecology (Berl.),* **49:** 207 – 212.

Hipps, L. E., Asrar, G., Kanemasu, E. T. (1983). Assessing the interception of photosynthetically active radiation in winter wheat. *Agric. Meteor.* **28:** 253 – 259.

Hue, L. (1982). Futile cycles and regulation of metabolism. In: Metabolic compartmentation (Ed., Sies, H.). New York: Academic Press, 71 – 97.

Jackson, J. E. (1970). Aspects of light climate within apple orchards. *J. Appl. Ecol.,* **7:** 207 – 216. doi: 10.2307/2401373.

Jackson, J. E. (1980). Light interception and utilization by orchard systems. *Hort. Rev.,* **2:** 208 – 267.

Kanniah, K. D., Beringer, J., North, P. and Hutley, L. (2012). Control of atmospheric particles on diffuse radiation and terrestrial plant productivity A review. *Prog. Phys. Geogr.*, **36:** 209 – 237.

King, D. and Roughgarden, J. (1982). Multiple switches between vegetative and reproductive growth in annual plants. *Theo. Popul. Biol.* **21 (2):** 194 – 204.

Kiniry, J.R., C.A. Jones, J.C. O'Toole, R. Blanchet, M. Cabelguenne and D.A. Spanel. 1989. Radiation-use efficiency in biomass accumulation prior to grain filling for five grain-crop species. *Field Crops Res.*, **20:** 51 – 64.

Krause, G. H. (1988). Photoinhibition of photosynthesis, An evaluation of damaging and protective mechanisms. *Physiol. Plant.*, **74:** 566 – 574.

Lambers, H. (1997). Respiration and the alternative pathway. In: A molecular approach to primary metabolism in plants (Eds., Foyer, C. H. and Quick, P.), Taylor and Francis, London. 295 – 309.

Landsberg, J. J., and Gower, S. T. (1997). Applications of Physiological Ecology to Forest Management". Academic Press, San Diego.

Landsberg, J. J. and F. J. Hingston. (1996). Evaluating a simple radiation/dry matter conversion model using data from Eucalyptus globulus plantations in Western Australia. *Tree Physiol.*, **16:** 801 – 808.

Larigauderie, A. and Körner, C. (1995). Acclimation of leaf dark respiration to temperature in alpine and lowland plant species. *Ann. Bot.*, **76:** 245 – 252.

Long, S. P., Zhu, X-G, Naidu, S. L. and Ort, D. R. (2006). Can improvement in photosynthesis increase crop yields? *Plant, Cell Environ.* **29:** 315 – 330.

Long, S. P., Humphries, S. and Falkowski, P .G. (1994). Photo-inhibition of Photosynthesis in Nature. *Ann. Rev. Plant Physiol. Plant Molecul. Biol.*, **45:** 633 – 662.

Loomis, R. S. and Amthor, J. S. (1999). Yield potential, plant assimilatory capacity, and metabolic efficiencies. *Crop Sci.*, **39:** 1584 – 1596.

López-Lozano, R., Baret, F., Cortázar-Atauri, I. G., Bertrand, N. and Casterad, M. A. (2009). Optimal geometric configuration and algorithms for LAI indirect estimates under row canopies: The case of vineyards. *Agric. Forest Meteor.*, **149 (8):** 1307 – 1316, ISSN 0168-1923.

Maddonni, G. A., Otegui, M. E. And Cirilo, A. G. (2001). Plant population density, row spacing and hybrid effects on maize canopy architecture and light attenuation. *Field Crop Res.,* **71:** 183 – 193.

McCree, K. J. (1970). An equation for the rate of respiration of white clover plants grown under controlled conditions. In: Prediction and Measurement of Photosynthetic Productivity." (Proc. IBP/PP Technical Meeting, Trebon), Centre for Agricultural Publishing, Wageninger.

McMurtrie, R. E. and Wang, Y. P. (1993). Mathematical models of the photosynthetic response of tree stands to rising CO2 concentrations and temperatures. *Plant Cell Environ.*, **16:** 1 – 13.

Medlyn, B. E. (1997). Physiological basis of the light use efficiency model. *Tree. Physiol.*, **18:** 167 – 176.

Monteith JL. 1972. Solar radiation and productivity in tropical ecosystems. *J. Appl. Ecol.*, **9:** 747 – 766.

Monteith, J. L. (1977). Climate and the efficiency of crop production in Britain. *Philosoph. Trans. Royal Soc. London B.* **281:** 277 – 294.

Morison, J. I. L. and Jarvis, P. G. (1983). Direct and Indirect Effects of Light on Stomata .1. in Scots Pine and Sitka Spruce. *Plant Cell Environ.,* **6:** 95 – 101.

Myneni, R. B. (1991). *Agric. Forest Meteor.,* **55:** 323.

Nass, H. G. (1973). Determination of characters for yield selection in spring wheat. *Can. J. Plant Sci.*, **53:** 755 – 762.

Niinemets, U. and Sack, L. (2006). Structural determinants of leaf light-harvesting capacity and photosynthetic potentials. *Prog in Bot.*, **67:** 385 – 419.

Nobel, P. S., Forseth, I. N. and Long, S. P. (1993). Canopy structure and light interception. In: Photosynthesis and production in a changing environment: a field and laboratory manual. (Eds., Hall, D. O., Scurlock, J. M. O., Bolhàr-Nordenkampf, H. R., Leegood, R. C. and Long, S.P.). p. 79 – 90. Chapman & Hall, ISBN 0 412 42900 4, London, UK.

Noguchi, K., Go C. S., Miyazawa, S. I., Terashima, I., Ueda, S., and Yoshinari, T. (2001). Costs of protein turnover and carbohydrate export in leaves of sun and shade species. *Aust. J. Plant Physiol.* **28:** 37 – 47.

Oker-Blom, P., Lappi, J. and Smolander, H. (1991). "Radiation regime and photosynthesis of coniferous stands." In: "Photon-Vegetation Interactions: Applications in Plant Physiology and Optical Remote Sensing" (Eds., Myneni, R. B. and Ross, J.). Springer-Verlag, 469.

Palmer, J. W. (1999). Light, Canopies, Fruit and Dollars. 42nd Annual IDFTA Conference, February 20 - 24, 1999, Hamilton, Ontario.

Penning de Vries, F. W. T., Brunsting, A. H. M. and van Laar, H. H. (1974). Products, requirements and efficiency of biosynthesis: a quantitative approach. *J. Theo. Biol.*, **45:** 339 – 377.

Penning de Vries, F. W. T., van Laar, H. H. and Chardon, M. C. M. (1983). Bioenergetics of growth of seeds, fruits and storage organs. In: Potential productivity of field crops under different environments (Eds., Smith, W. H. and Banta, S. J.). Los Banos, Philippines: International Rice Research Institute, 37–59.

Penning de Vries, F. W. T. (1975). The cost of maintenance processes in plant cells. *Ann. Bot.*, **39:** 77 - 92.

Powles, S. B. (1984). Photoinhibition of Photosynthesis Induced by Visible Light, *Ann. Rev. Plant Physiol.*, **35:** 15 – 44.

Reicosky, D. A. and Hanover, J. W. (1978). Physiological Effects of Surface Waxes: I. Light Reflectance for Glaucous and Nonglaucous Picea pungens, *Plant Physiol.*, **62:** 101 – 104.

Rich, P.M. (1990). Characterizing plant canopies with hemispherical photographs. *Remote Sens. Rev.,* **5 (1):** 13 - 29, ISSN 0275-7257.

Rigg, T. J., Hanson, P. R., Start, N. D., Miles, D. M., Morgan, C. L., and Ford, M. A. (1981). Comparisons of spring barley varieties grown in England and Wales between 1880 and 1890. *J. Agric Sci. Camb.*, **97:** 599 – 610.

Russell, G., Jarvis, J. P. and Monteith, J. L. (1989). Absorption of radiation by canopies and stand growth. In Plant Canopies: Their Growth, Form and Function. (Eds. Russel, G., Marshall, B. and Jarvis, P.G.). Cambridge University Press, London, UK. p 21 – 40.

Ryan, M. G., Binkley, D., and Fownes, J. H. (1997). Age-related decline in forest productivity: Pattern and process. *Adv. Ecol. Res.,* **27:** 213 - 262.

Ryan, M. G. (1991). Effects of climate change on plant respiration. *Ecol. Appl.*, **1:** 157 – 167.

Simpson FB. 1987. Hydrogen reactions of nitrogenase. *Physiol. Plant.*, **69:** 187 – 190.

Sinclair, T. R. and Horie, T. (1989). Leaf nitrogen, photosynthesis, and crop radiation-use efficiency: a review. *Crop Sci.*, **29:** 90 – 98.

Smith, A. M., Zeeman, S. C., Smith, S. M. (2005). Starch degradation. *Ann. Rev. Plant Biol.,* **56:** 73–98.

Smith, H. (1982). Light Quality, Photoperception, and Plant Strategy. *Ann. Rev. Plant Physiol.,* **33:** 481 – 518.

Stockfors, J. and Linder. S. (1998). The effect of nutrition on the seasonal course of needle respiration in Norway spruce stands. *Trees,* **12:** 130–138.

Stockle, C. O. and Kiniry, J. R. (1990). Variability in crop radiation-use efficiency associated with vapor-pressure deficit. *Field Crops Res.* **25:** 171 – 181.

Thornley, J. H. M. and France, J. (2007). Mathematical models in agriculture, 2nd edn. Wallingford, UK: CABI.

Thornley, J. H. M. and Johnson, R. L. (1990). Plant and crop modeling. A mathematical approach to plant and crop physiology. Oxford: Oxford Science Publications.

Trapani, N., Hall, J. A., Sadras, V.O. and Villalobos, F. J. (1992). Ontogenic changes in radiation-use efficiency of sunflower (*Helianthus annus* L.) crops. *Field Crops Res.*, **29:** 301 – 316.

Urban, O., Janous, D., Acosta, M., Czerny, R., Markova, I., Navrátil, M., Pavelka, M., Pokorný, R., Šprtová, M., Zhang, R., Špunda, V., Grace, J., and Marek, M. (2007). Ecophysiological controls over the net ecosystem exchange of mountain spruce stand, Comparison of the response in direct vs. diffuse solar radiation, *Glob. Change Biol.,* **13:** 157 – 168.

Urban, O., Klem, K., Aˇc, A., Havránková, K., Holišová, P., Navrátil, M., Zitová, M., Kozlová, K., Pokorný, R., Šprtová, M., Tomášková, I., Špunda, V., and Grace, J. (2012). Impact of clear and cloudy sky conditions on the vertical distribution of photosynthetic CO2 uptake within a spruce canopy. *Funct. Ecol.,* **26:** 46 – 55.

Vierstra, R. D. (1993). Protein degradation in plants. *Ann. Rev. Plant Physiol. Plant Mol. Biol.,* **44:** 358 – 410.

Villalobos, F. J., Soriano, A. and Fereres, E. (1992). Effects of shading on dry matter partitioning and yield of field-grown sunflower. *Eur. J. Agron.*, **1:** 109 – 115.

Wang, Q., Hasson, A., Rossmann, S. and Theres, K. (2016). Divide et impera: boundaries shape the plant body and initiate new meristems. *New Phytol.*, **209:** 485 – 498.

Warrington I.J., Stanley C.J., Tustin. D.S., Hirst, P. M. and Cashmore, W. M. (1996). Light transmission, yield distribution, and fruit quality in six tree canopy forms of 'Granny Smith' apple. *J. Tree Fruit Prod.*, **1:** 27 – 54. doi: 10.1300/J072v01n01 03.

Welles, J.M. and Norman, J. M. (1991). Instrument for indirect measurement of canopy architecture. *Agron. J.,* **83 (5):** 818 – 825, ISSN 0002-1962.

Wherley, B.G., Gardner, D.S. and Metzger, J.D. (2005). Tall Fescue Photomorphogenesis as Influenced by Changes in the Spectral Composition and Light Intensity. *Crop Sci.,* **45:** 562 – 568.

Woodward, F.I. (1983). Instruments for the Measurement of Photosynthetically Active Radiation and Red, Far-Red and Blue Light. *J. Appl. Ecol.,* **20:** 103 – 115.

Yoder, B.J., Ryan, M.G., Waring, R.H., Schoettle, A.W. and Kaufmann, M.R. (1994). Evidence of reduced photosynthetic rates in old trees. *For. Sci.,* **40:** 513 - 527.

Zhu, X.G., Long, S.P. and Ort, D.R. (2008). What is the maximum efficiency with which photosynthesis can convert solar energy into biomass? *Curr. Opin. Biotech.*, **19:** 153 – 159.

■■■■

Section V:

Nutraceuticals, Medicinal & Aromatic Plant Wealth

20

MICROALALGAE AS A POTENTIAL SOURCE OF NUTRACEUTICALS AND PHARMACEUTICALS

Surendra Singh*, Ankita Kachhwaha and Rishibha Dixit

Algal Biotechnology Laboratory,
Department of Post Graduate Studies and Research in Biological Science,
Rani Durgavati University, Jabalpur (M.P.), India
E-mail: singhbiosci@yahoo.co.in

INTRODUCTION

Microalgae are microscopic photosynthetic organisms that are found in both marine and freshwater environments. They occupy the bottom of the food chain in aquatic ecosystems and possess the intrinsic ability to take up H_2O and CO_2 in presence of sunlight to form complex organic compounds (Guedes *et al.*, 2011). Microalgae appeared then as a good source of protein and has continued as such, but with an increased interest due to the unique bioactive ingredients recently found in these small microorganisms, which gives them great potential as a food source and as a source of functional molecules (Chacon-Lee and Gonzalez-Marino, 2010). They are a rich and varied sustainable source of pharmacologically active natural products and nutraceuticals and over 15,000 novel compounds originated from algal biomass have been identified. While nutraceutical and pharmaceutical content in the microalgal strain is very small, current market values for these products are extremely high. The major products being commercialized or under consideration for commercial extraction include carotenoids, phycobilins, fatty acids, polysaccharides, vitamins, sterols and biologically active molecules for use in human and animal health (Talero *et al.*, 2015).

Some of the microalgae have been consumed as food or health food due to their high nutritional value. *Nostoc* has been used in Asia and *Spirulina* by certain tribes in Africa. Microalgae such as *Spirulina, Chlorella, Dunaliella, Haematococcus, Schizochytrium* etc. are classified as food sources falling into the GRAS (Generally Regarded as Safe) category by the U.S. Food and Drug Administration and are gaining acceptance in the food and health-food market (Chacon-Lee and Gonzalez-Mari, 2010). Many high-value compounds produced by microalgae can be administered as a powder of dried or freeze-dried biomass with no extraction undertaken (Walker *et al.,* 2005).

This review article focuses upon the general composition of value of identified and established microalgae which are the sources of nutraceuticals and pharmaceuticals based on pigments, protein and several bioactive compounds essential for human health and nutrition as well as elucidating the side effects, if any.

GENERAL COMPOSITION OF MICROALGAE

Microalgae have been harvested for food and are capable of producing necessary vitamins including: A (Retinol), B1 (Thiamine), B2 (Riboflavin), B3 (Niacin), B6 (Pyridoxine), B9 (Folic acid), B12 (Cobalamin), C (L-Ascorbic acid), D, E (Tocopherol), and H (Biotin). Also, microalgae concentrate essential elements including: Potassium, Zinc, Iodine, Selenium, Iron, Manganese, Copper, Phosphorus, Sodium, Nitrogen, Magnesium, Cobalt, Molybdenum, Sulfur and Calcium. Microalgae are also high producers of essential amino acids and Omega 6 (Arachidonic acid) and Omega 3 (Docosahexaenoic acid, Eicosapentaenoic acid) fatty acids. Due to their abundant production of beneficial compounds and nutritive contents, the market for increased microalgae production for nutraceuticals is profitable (Bishop and Zubeck, 2012; Abd El-Baky *et al.,* 2004).

Microalgae can be incorporated in food to gain the benefits such as improving cardiovascular systems (Khan *et al.,* 2005), their slimming properties (Lyons and O'Brien, 2002), energizing properties (Desai and Sivakami, 2004), antioxidant capabilities (Lyons and O'Brien, 2002), and cholesterol and triglyceride lowering effects (Geppert *et al.,* 2006), the most popular way to consume microalgae is as a dietary supplement in the form of tablet, capsule, powder (Molina *et al.,* 2003). Microalgae are also added to pasta, snack foods or drinks either as nutritional supplements or natural food colorants (Becker, 2004).

Table 1: General composition (N% of dry matter) of different microalgae and comparison values with traditional sources of food:

Species	Protein	Carbohydrates	Lipids	Reference
Spirulina platensis	63	15	11	Becker, 2007
	61.32 to 64.43	15.09 to 15.81	7.09 to 8.03	Tokusoglu and Unal, 2003
Arthrospira maxima	60 to 71	13 to 16	6 to 7	Becker, 2007
Chlorella vulgaris	51 to 58	12 to 17	14 to 22	Becker, 2007
	47.82	8.08	13.32	Tokusoglu and Unal, 2003
Chlorella pyreinodosa	57	26	2	Becker, 2007
Dunaliella salina	57	32	6	Becker, 2007
Porphyridium cruentum	28 to 39	40 to 57	9 to 14	Becker, 2007
	34.1	32.1	6.53	Rebolloso-Fuentes, 2000
Scenedesmus obliquus	50 to 56	10 to 17	12 to 14	Becker, 2007
Aphanizomenon flos-aquae	62	23	3	Becker, 2007
Chlamydomonas rheinhardtii	48	17	21	Becker, 2007
Anabaena cylindrical	43 to 56	25 to 30	4 to 7	Becker, 2007
Euglena gracilis	39 to 61	14 to 18	14 to 20	Becker, 2007
Spirogyra sp.	6 to 20	33 to 64	11 to 21	Becker, 2007
Synechococcus sp.	46 to 63	8 to 14	4 to 9	Becker, 2007
Nannocloropsis spp.	28.8	35.9	18.36	Rebolloso-Fuentes, 2001
Haematococcus pluvialis	48	27	15	Gouveia, 2008a
Isochrisis galbana	26.99	16.98	17.16	Tokusoglu and Unal, 2003
Bakers yeast	39	38	1	Gouveia *et al.*, 2008a
Meat	43	1	34	Gouveia *et al.*, 2008a
Egg	47	4	41	Gouveia *et al.*, 2008a
Milk	26	38	28	Gouveia *et al.*, 2008a
Rice	8	77	2	Gouveia *et al.*, 2008a
Soya	37	30	20	Gouveia *et al.*, 2008a

Values expressed as percent dry matter.

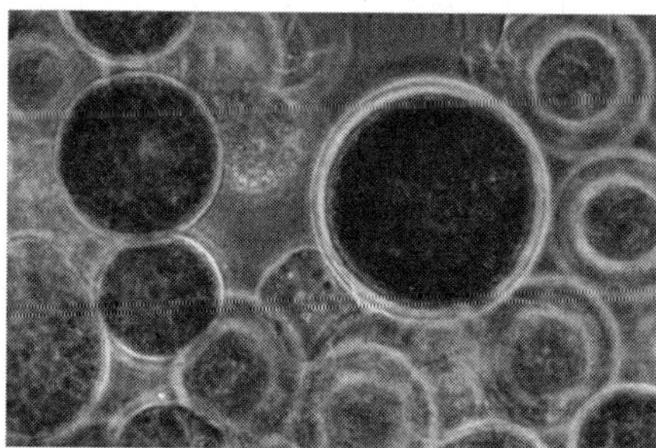

Fig. 1: *Chlorella* sp.

Fig. 2: *Chlorella* Powder

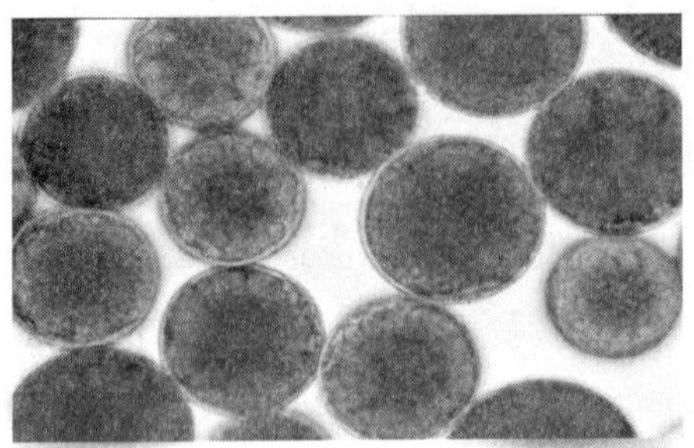

Fig. 3: *Haematococcus* sp.

Fig. 4: Astaxanthin Powder

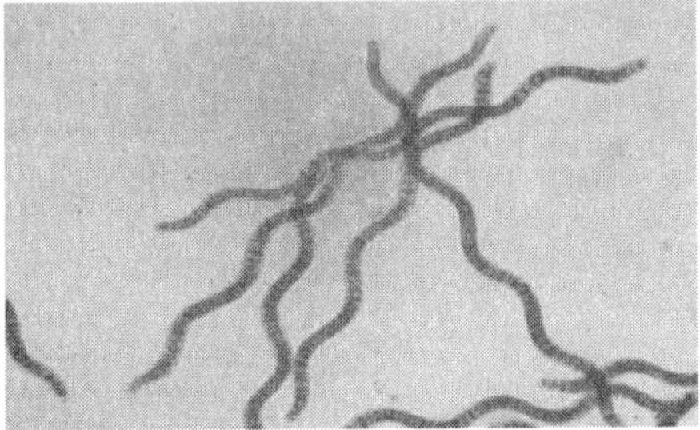

Fig. 5: *Spirulina* sp.

Fig. 6: *Spirulina* Powder

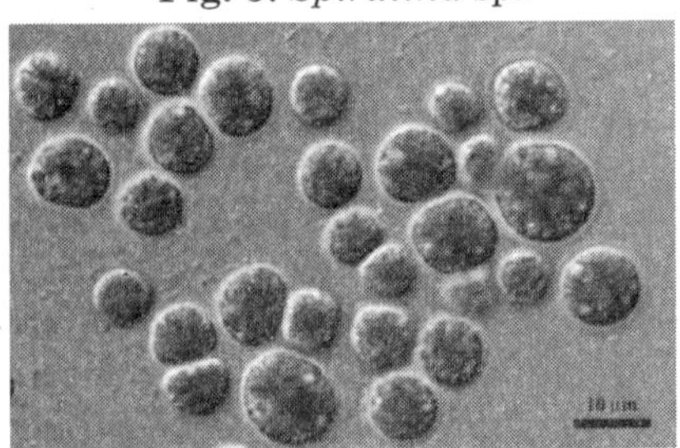

Fig. 7: *Porphyridium* sp.

Fig. 8: Phycoerythrin Powder

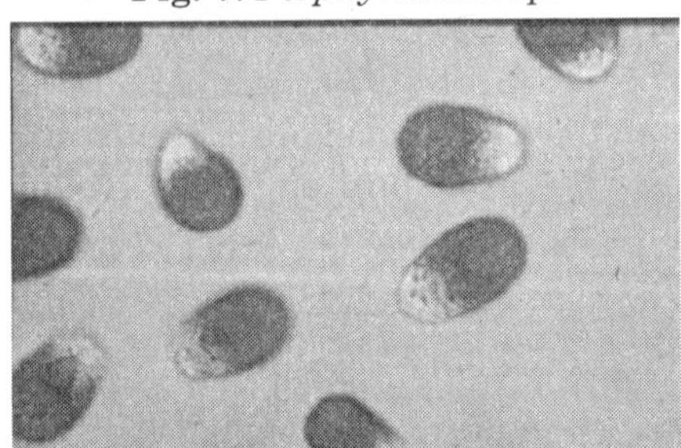

Fig. 9: *Dunaliella* sp.

Fig. 10: Beta Carotene Powder

MICROALGAE AS A SOURCE OF NUTRACEUTICALS

Nutraceuticals are nutrients from food or food products that not only supplement the diet but also facilitate the prevention or treatment of a disease (Kalra, 2003). An increased global interest has arisen in these substances due to their documented role in health enhancement. With the increase in demand for nutraceuticals and food supplements, organisms that can rapidly produce nutritional compounds are desired (Rajasekaran *et al.,* 2008). The growing use of microalgal biomass for nutraceutical purposes is expected to provide an attractive revenue stream for microalgae producers. While nutraceutical content in microalgae strain is very small, current market values for these products are extremely high. Physiologically-active nutraceuticals from microalgae include food supplements, dietary supplements in the form of tablets, soft gels, capsules etc (Ciferri and Tiboni, 1985; Abdulquader *et al.,* 2000).

MICROALGAE AS A SOURCE OF PHARMACEUTICALS

Microalgae possess physiological, biochemical and molecular strategies to cope with stress thus are capable of synthesising a variety of bioactive chemicals (Tandeau-de-Marsac, 1993). The bioactive compounds are secondary metabolites which include various types of substances ranging from organic acids, carbohydrates, amino acids and peptides, vitamins, growth substances, antibiotics, enzymes to toxic compounds (Guedes *et al.,* 2011). The need for development of alternative antibiotic agent was investigated since the emergence of antibiotic resistant microbes. Microalgae have potential pharmaceutical applications (Katircioglu *et al.,* 2006). There are a range of pharmaceutical products derived from microalgae like antimicrobials, antivirals, antifungals, neuroprotective products, therapeutic proteins and drugs. The biological activities of the microalgae may be attributed to the presence of volatile compounds, some phenols, free fatty acids and their oxidized derivatives (Amaro *et al.,* 2011).

PIGMENTS

Microalgal pigment has commercial uses as a natural food coloring and cosmetic ingredient Carotenoids are fat-soluble pigments which decrease the risk of cancer. Beta-carotene, lycopene, lutein and zeaxanthin act as antioxidants. Beta-Carotene may have added benefits due its ability to be converted to vitamin A. Carotenoids are found to be powerful anti-oxidants. β-Carotene and astaxanthin are attractive microalgal products as they command a high market price and are present in high concentrations in *Dunaliella* and *Haematococcus* respectively (Chen *et al.,* 1993). Beta carotene is used as a food coloring (with a major application in providing the yellow color to margarine), as a food additive to enhance the color of the flesh of fish and the yolk of eggs, and to improve the health and fertility of grain-fed cattle Carotene is effective in controlling cholesterol and in reducing risks of heart disease.

Microalgae are a polyphyletic and biochemically diverse assemblage of chlorophyll-a containing microorganisms capable of oxygenic photosynthesis that are predominantly found in aquatic environments with observed high levels of ultraviolet (UV) radiation. Microalgae such as *Haematococcus pluvialis, Chondrus crispus, Mastocarpus stellatus, Ascophyllum nodosum, Alaria esculenta, Spirulina platensis, Nannochloropsis oculata, Chlorella vulgaris* and *Dunaliella salina* produce organic metabolites, such as sporopollenin, scytonemin and mycosporine-like amino acids, to protect themselves from UV radiation and are used in cosmetics as thickening agents, water-binding agents and antioxidants (Priyadarshani and Rath, 2012).

β-Carotene

ß-Carotene is a natural pigment used as a yellow-orange food coloring and prevent certain types of cancers. *Dunaliella* have the potential to accumulate the pigment to up to 10% of the algal dry weight. ß-Carotene production is proportional to the light intensity with the effective range of light wavelengths overlapping the photosynthetically active radiation range (Ben-Amotz and Avron, 1989). ß-Carotene protect the skin from damage, boost immunity, promote healthy growth and development, eye and vision health, lower inflammation etc. (Cakmak *et al.,* 2014).

Astaxanthin

Astaxanthin protect against oxidation of essential polyunsaturated fatty acids, cancer, UV light effects and modify immune function (Rao *et al.,* 2013). Astaxanthin can protect skin from UV radiation effects, protect against chemically induced cancers, and enhance the immune system (Cysewski and Lorenz, 2004).

Lutein

Lutein is a yellow oxycarotenoid or xanthophyll containing two cyclic end groups (one beta and one epsilon-ionone ring). In green microalgae, lutein protects cells from ROS damage under stress conditions. Lutein has been widely used as a feed additive and a food coloration agent in industry (Fernandez-Sevilla *et al.,* 2010). *Chlorella sorokiniana* is also a good source for lutein, and the yield can be improved by managing several nutritional and environmental factors and by random mutagenesis (Pasquet *et al.,* 2011). *Chlorella prothecoides* has also been proposed as a potential source of lutein (Shi *et al.,* 2002).

Violaxanthin and Zeaxanthin

Violaxanthin is a natural xanthophyll pigment with an orange color found in a variety of plants and microalgae. This product has been isolated from several microalgae including *Dunaliella tertiolecta* (Fernandez-Sevilla *et al.,* 2010) and *Chlorella ellipsoidea* (Soontornchaiboon *et al.,* 2012). Zeaxanthin together with lutein accumulate in the central retina. Insufficient dietary lutein and zeaxanthin intake or lower serum zeaxanthin levels are associated with increased risk for age-related macular degeneration (Bian *et al.,* 2012). Zeaxanthin has been obtained from the *Nannochloropsis oculata* and *Chlorella saccharophila* (Singh *et al.,* 2013).

VITAMINS

Dunaliella is rich in β-carotene, tocopherol, thiamin and folic acid (Fabregas and Herrero, 1990). Biotin and α-tocopherol (Vitamin E) is found in *Euglena gracilis* (Li *et al.,* 2008). Microalgae have been shown to be an important source of tocopherol, a fat-soluble phenolic compound with varying degrees of vitamin E antioxidant activity showing anti-inflammatory and antitumoral effects (Goh *et al.,* 2009; Carballo-Cardenas *et al.,* 2003; Trevithick *et al.,* 1992). Microalgae could provide a rich source of ascorbic acid in our diet, despite the differences in the composition found in microalgae species (0.11%–1.62% ascorbic acid) (Brown and Miller, 1992). Ascorbic acid (Vitamin C) in *Chlorella* sp. Ascorbic acid and its derivatives have been shown to protect against sunburn in mice, delay the onset of skin tumors and reduce UV-induced skin wrinkling (Besaratinia *et al.,* 2007).

POLYUNSATURATED FATTY ACID

The average lipid content in microalgae varies between 1% and 40% and, according to growing conditions, can be as high as 85% of dry weight. Algal lipids are typically composed of glycerol, sugars, or bases esterified to fatty acids, with carbon numbers in the range of C12 to C22 (Chacon-Lee and Gonzalez-Marino, 2010). Long-chain polyunsaturated fatty acids of the omega 3 and omega 6 series are straight chain carboxylic acids of 20 or more carbon atoms that contain 3 or more double bonds. Microalgae can be easily cultivated at different stress conditions and are effective in the prevention and treatment of several diseases like atherosclerosis, thrombosis, arthritis, cancers, etc. (Benemann *et al.,* 1987).

Spirulina is rich in γ-linolenic acid (GLA) (Mahajan and Kamat, 1995). *Chlorella minutissima* is a eukaryotic species with a fast growth rate and high PUFA content (Seto *et al.,* 1984) and could be another important source of a PUFA-rich nutraceutical supplement. The red microalgae, *Porphyridium* contain more than 30% of total fatty acids as AA (Arachidonic Acid) and equally high concentrations of EPA (Arad and Richmond, 2004). *Dunaliella* can also be used as natural antimicrobials and antioxidants in the food/feed and pharmaceutical industry because of its high unsaturated fatty acid content (Cakmak *et al.,* 2014). Another of the main active compounds in marine microalgae includes the sterols, such as fucosterol which showed antioxidant activity and cytotoxicity against several cancer cell lines (Khanavi *et al.,* 2012).

Table 2: Commonly used microalgae: their potential benefits for human health.

Genus	Product/application area	Activity claimed	References
Spirulina	Biomass γ -linolenic acid Phycocianin Health foods Cosmetics Infant formulas Nutritional supplements Human & animal nutrition Vitamin B12	Immunoenhancing prevention and treatment of heart diseases, obesity, manic depression, antitumor effect, antioxidant activity, nutritional supplements to inhibit replication and infectivity of virus including HIV, CMV, HSV and influenza A	Pulz and Gross, 2004; Spolaore *et al.,* 2006
Chlorella	Biomass Carbohydrate extract Phycobiliproteins Health foods Food supplement Feed surrogates Cosmetics Human & animal nutrition Aquacultured	Improves the immune system Reduces blood lipid Antioxidant Health promoting effects like : Efficacy on gastric ulcers, wounds, and constipation, preventive action against atherosclerosis and hypercholesterolemia and antitumor action	Pulz and Gross, 2004; Spolaore *et al.,* 2006
Dunaliella	Carotenoids B-carotene Glycerol Health food Food supplement Feed Pigments DHA oil Human & animal nutrition Protein	Vitamin A precursor Anticancer, antioxidant, antiviral Positively influences the energy metabolism of the skin Antipertensive Bronchodilator Analgesic Muscle relaxant Antiedema activity	Pulz and Gross, 2004; Dufosse *et al.,* 2005
Haema-tococcus	Carotenoids Astaxanthin Pigment Health foods Pharmaceuticals Feed additives Nutraceuticals Human nutrition Aquaculture Cosmetics	Antiinflammatory Antioxidant Treatment of carpal syndrome Treats muscle soreness Coloring muscles in fish Used to prevent arteriosclerosis, coronary artery disease, and ischemic brain development	Pulz and Gross, 2004; Dufosse *et al.,* 2005; Spolaore *et al.,* 2006
Odontella	Fatty acids Pharmaceuticals Cosmetics Baby food EPA and DHA	Anti-inflammatory	Pulz and Gross, 2004

Genus	Product/application area	Activity claimed	References
Porphyridium	Polysaccharides Pharmaceuticals Cosmeticsa Infant formulas Nutrition Soil conditioning B-phycoerythin C-phycocyanin	Nutrition Antioxidant activity	Pulz and Gross, 2004; Dufosse *et al.*, 2005; Spolaore *et al.*, 2006
Shizochytrium	DHA oil, biomass, health foods, cosmetics, baby foods, and infant formulas Food additives Aquaculture	Treats brain and heart disorders Helps brain development	Pulz and Gross, 2004
Isochrysis	DHA oil Polyunsaturated fatty acids Health care Food supplement Animal nutrition	Treats brain and heart disorders Helps brain development, mass oil production, animal nutrition Aquaculture	Pulz and Gross, 2004; Spolaore *et al.*, 2006

PROTEIN

Microalgae are neuroprotective agents and promote nerve cell survival. *Spirulina* is found to partially prevent neurotoxicity and oxidative stress, suggesting it could be a possible alternative in experimental therapy (Hu, 2004). Expression of recombinant proteins in green algal chloroplast holds substantial promise as a platform for the production of human therapeutic proteins. *Chlamydomonas reinhardtii* is rich in therapeutic proteins. Several therapeutic bioactivities such as hepatoprotective, anti-inflammatory, immunomodulating, antioxidant and anticancer effects have been described for phycobiliproteins from red microalgae, such as *Porphyridium* sp. (McCarty *et al.*, 2015; Yang *et al.*, 2014).

POLYSACCHARIDE

A polysaccharide found in *Chlorella* has been linked to antitumor effects (Iwamoto, 2004). *Chlorella* cells contain ß-1,3-glucan, an active immune-stimulator, which acts as a free-radical scavenger and as a reducer of blood lipids (Becker, 2004).

GLYCOLIPID

Microalgae glycolipids are effective against skin inflammation and cancer (Colombo *et al.*, 2011). Anti-inflammatory properties of sulphated polysaccharides derived from red microalgae also have been reported effective in the treatment of skin inflammation (Matsui *et al.*, 2003).

Table 3: Antibacterial activity of selected compounds from microalgae:

Algal species	Extract source	Target bacteria	References
Pithophora oedogonium	Ethanol extract	*Salmonella, Staphylococcus* sp., 4978	Danyal *et al.,* 2013
Rivularia bullata, Nostoc spongiaeforme, Codium fragile, Colpomenia peregrina Sauvageau, *Cystoseira barbata, Zanardinia typus*	Methanol Chlorofom Diethyleter Dichloromethne Ethanol	Gram negative and Gram positive bacteria	Akgul *et al.,* 2013
Sargassum wightii, Chaetomorpha linum, Padina gymnospora.	Acetone, methanol	*P. aeruginosa* (ATCC27853), *S. typhi*-B, *Erwinia amy-lovora* (MTCC2760) (*E. amylovora*), Enterobacter aerogenes (MTCC111) (*E. aerogenes*), *Proteus vulgaris* (MTCC1771) (*P. vulgaris*), *Klebsiella pneumonia* (ATCC15380) (*K. pneumonia*) and *E. coli* (ATCC25922). gram-positive bacterial strains were Methicillin resistant *S. aureus,*	Rosaline *et al.,* 2012
Asparagopsis taxiformis	Ethanol extract	*Vibrio alginolyticus, Vibrio vulnificus* and *Aeromonas salmonicida* subsp. *salmonicida Photobacterium damselae* subsp. *Damselae* and *Photobacterium damselae* subsp. *piscicida, Salmonella* sp., *Vibrio cholerae, Vibrio harveyi* and *Vibrio parahaemolyticus*	Genovese *et al.,* 2012
Chlorococcum humicola	Bioactive compounds	*Escherichia coli, Pseudomonas aeruginosa, Salmonella typhi-murium, Klebsiella pneumoniae, Vibreo cholerae, Staphylococcus aureus, Bacillus subtilis, Candida albicans, Aspergillus niger* and *Aspergillus flavus.*	Bhagavathy *et al.,* 2011
Gloeocapsa sp. *Synechocystis* sp. *Anabaena* sp. *Aphanizomenon flos-aquae. Nostoc* sp. *Nostoc entophytum, Nostoc muscorum, Scytonema ocellatum, Arthrospira fusiformis* (Voronich), *Scenedesmus obliquus, Coelastrella* sp. *Chlorella* sp. *Rhodella violacea, Porphyridium cruentum* (AG.) NAG	Ethanol extract	*Staphylococcus aureus* 209, *Streptococcus pyogenes* 981, *Bacillus cereus* 2421 *Escherichia coli* 3702, *Pseudomonas aeruginosa* 1396, *Salmonella typhimurium* 123, and *Yersinia enterocolitica* 623	Najdenski *et al.,* 2013

Table 4: Antiviral activities of selected compounds from microalgae:

Algal species	Extract source	Target Virus	References
Haematococcus Pluvialis and *Dunaliella salina*	Pressurized liquid extraction	Herpes simplex virus type 1	Santoyo *et al.*, 2012
*Gyrodinium Impudium (*sulfated polysaccharide, p-KG03)	Sulfated polysaccharide	Influenza virus	Kim *et al.*, 2012
Navicula directa	Polysaccharide	HSV1 & 2, Influenza A virus	Lee *et al.*, 2006
Gyrodinium impudicum	p-KG03 exopolysaccharides	Encephalomyocarditis Virus	Yim *et al.*, 2004

Table 5: Antifungal activities of selected compounds from microalgae:

Algal species	Extract source	Target fungi	References
Pithophora oedogonium	Ethanol extract	*Penicillium viridicatum* 1101, *Fusarium solini* 1127	Danyal *et al.*, 2013
Gloeocapsa sp.	Exopolysaccharides	*Candida albicans*	Najdenski *et al.*, 2013
Haematococcus pluvialis	Butanoic acid and methyl lactate	*Candida albicans*	Santoyo *et al.*, 2009

SIDE EFFECTS

Chlorella consumption as a nutritional product may cause side effects such as nausea, diarrhoea, vomiting and gastrointestinal problems. *Chlorella* tablets have been reported in the causation of acute tubulointerstitial nephritis often resulting in renal failure (Yim *et al.*, 2007; Tiberg *et al.*, 1995). Some question has been raised on the digestion capacity and bioavailability of nutrients like vitamin B12 from *Spirulina* (Watanabe *et al.*, 2002). However, efficient processing claims to release nutrients into a digestible form for humans. Some *Spirulina* supplements have been shown to contain traces of the toxin microcystin due to invasion of other algae (like *Microcystis* spp.) in culturing areas, which acts primarily on decreasing liver function and could cause liver cancer and other diseases (Gilroy *et al.*, 2000). Also, since *Spirulina* can concentrate elemental constituents it may contain elevated amounts of unnecessary elements like mercury and create potential risks, depending on the growth environment. Also, allergic reactions, though rare, may result it insomnia and anxiety (Griffith, 1998; Bratman and Kroll, 1999).

Hence the relevance of microalgae as described in the preceding sections, the article needs a brief summarization before coming to a precise conclusion. Overall microalgae has been produced for commercial exploitation, with biotechnological applications. Many species of microalgae, *Chlorella, Dunaliella, Haematococcus, Porphyridium, Shizochytrium* etc., are able to produce valuable metabolites like antioxidants, carotenoids, polyunsaturated fatty acids, vitamins, anticancer and antimicrobial drugs. They are autotrophic organisms with simple and rapid growth characteristics capable of producing secondary metabolites when exposed to stress conditions like nutrient deprivation, variations in light intensity, temperature, salinity and pH. The major products currently being commercialized include cyanovirin, oleic acid, linolenic acid, palmitoleic acid, vitamin E, B12, β-carotene, astraxanthin, phycocyanin, phycoerythrin, lutein, zeaxanthin etc. Due to the emerging infectious diseases, viral infections and raise in antibiotic resistant bacteria, there is an urgent need for development of alternative treatment therapies against infectious diseases. With the increasing demand of natural products, biologically active molecules from microalgae are exceedingly desired for use in nutraceuticals and pharmaceuticals having antimicrobial, antioxidant and anti-inflammatory properties and have been found effective in the treatment of cardiovascular diseases, cancer and microbial infections. Some of the keywords for this article are: **Microalgae, Pharmaceuticals, Nutraceuticals, Health benefits.**

CONCLUSION

As the human population continues to increase, demand for nutritive food and health products increases. Microalgae have shown viability in meeting nutritive demands due to their rapid growth, health benefits and enriched compounds they produce. Microalgae such as *Spirulina, Chlorella, Dunaliella, Haematococcus, Schizochytrium, Isochyris* etc. are potential sources of high-value products, including nutraceuticals, and bioactive molecules that may lead to the discovery of new drugs. They have a wide range of physiological and biochemical characteristics and contain high content of protein, lipid, glycerol, vitamins, pigments, antioxidants etc. compared to other plants or animals. They prove to be cost-effective sources and are effective in the treatment of cardiovascular diseases, cancer etc. However, more research and development is required for greater production to meet the ever-growing nutritional demand. Selection of high-yield strains, optimization of cultivation conditions and use of genetic engineering to modify the strains to get high-value-added products is needed. There is great potential to further exploit microalgae various biotechnological applications.

REFERENCES

Abd-El-baky, H.A., Farouk, K., Baz-Gamal, E. and El-baroty, S. (2004). Production of antioxidant by the green alga D*unaliella salina*. International Journal of Agriculture and Biology. 6(1): 49-57.

Abdulquader, G., Barsanti, L. and Tredici, M. (2000). Harvest of Arthrospira platensis from lake Kossoram and its household usage among the Kanembu. *J. Appl. Phycol.* 12: 493-49.

Akgul, R., Suerdem, T.B. and Akgul, F. (2013). Antimicrobial Activities of Some Marine Algae and Some Cyanobacteria from Canakkale *J. Algal. Biomass. Utln.* 4: 35-40.

Amaro, H.M., Catarina-Guedes, A. and Xavier-Malcata, F. (2011). Antimicrobial activities of microalgae: an invited review. *Science against microbial pathogens: communicating current research and technological advances.* 1272-1280.

Arad, S. and Richmond, A. (2004). Industrial production of microalgal cell-mass and secondary products—mayor industrial species. *Porphyridium* spp. In: RichmondA, editor. Handbook of microalgal culture. *Biotechnology and applied phycology. Oxford , U.K. : Blackwell Science.*1-289.

Becker, W. (2004). Microalgae in human and animal nutrition. In: Richmond A, editor. Handbook of microalgal culture. Biotechnology and applied phycology. Oxford, U.K.: Blackwell Science. 312-51.

Ben-Amotz, A. and Avron, M. (1989). The biotechnology of mass culturing *Dunaliella* for products of commercial interest. 91–114. In: R. C. Cresswell, T. A. V. Rees, and N. Shah (eds.) *Algal and Cyanobacterial Biotechnology*, Longman Scientific & Technical, London, UK.

Benemann, J.R., Tillett, D.M. and Weissman, J.C. (1987). Microalgae biotechnology. *Trends in Biotechnol.* 5: 47-53.

Besaratinia, A., Kim, S.I., Bates, S.E. and Pfeifer, G.P. (2007). Riboflavin activated by ultraviolet A1 irradiation induces oxidative DNA damage-mediated mutations inhibited by vitamin C. Proc. Natl. Acad. Sci. USA. 104: 5953-5958.

Bhagavathy, S., Sumathi, P. and Bell, J.S. (2011). Green algae Chlorococcum humicola-a new source of bioactive compounds with antimicrobial activity. *Asian Pacific Journal of Tropical Biomedicine.* 1: 1-7.

Bian, Q., Gao, S., Zhou, J., Qin, J., Taylor, A., Johnson, E.J., Tang, G., Sparrow, J.R., Gierhart, D. and Shang, F. (2012). Lutein and zeaxanthin supplementation reduces photooxidative damage and modulates the expression of inflammation-related genes in retinal pigment epithelial cells. *Free Radic Biol Med.* 53(6): 1298-307.

Bishop, W.M. and Zubeck, H.M. (2012). Evaluation of Microalgae for use as Nutraceuticals and Nutritional Supplements. *J Nutr Food Sci.* 2(147): 1 6.

Bratman, S. and Kroll, D. (1999). Natural health Bible. Prima Publishing, Rocklin, CA, USA.

Brown, M.R. and Miller, K.A. (1992). The ascorbic acid content of eleven species of microalgae used in mariculture. *J. Appl. Phycol.* 4: 205-215.

Cakmak, Y.S., Kaya, M. and Asan-Ozusaglam, M. (2014). Biochemical composition and bioactivity screening of various extracts from *Dunaliella salina*, a green microalga. *Excli J.* 13: 679–690.

Carballo-Cardenas, E.C., Tuan, P.M., Janssen, M. and Wijffels, R.H. (2003). Vitamin E (alpha-tocopherol) production by the marine microalgae *Dunaliella tertiolecta* and *Tetraselmis suecica* in batch cultivation. *Biomol. Eng.* 20: 139-147.

Chacon-Lee, T.L. and Gonzalez-Mari, G.E. (2010). Microalgae for "Healthy" Foods-Possibilities and Challenges. *Comprehensive Reviews in Food Science and Food Safety.* 9: 655-675.

Chen, B.H., Chuang, J.R., Lin, J.H. and Chiu, C.P. (1993). Quantification of provitamin A compounds in Chinese vegetables by high-performance liquid chromatography. *J Food Prot.* 56(1): 51-54.

Ciferri, O. and Tiboni, O. (1985). The biochemistry and industrial potential of *Spirulina. Annu Rev Microbiol.* 39: 503-526.

Colombo, D., Tringali, C., Franchini, L., Cirillo, F. and Venerando, B. (2011). Glycoglycerolipid analogues inhibit PKC translocation to the plasma membrane and downstream signaling pathways in PMA-treated fibroblasts and human glioblastoma cells, U87MG. *Eur. J. Med. Chem.* 46: 1827-1834.

Cysewski, G.R. and Lorenz, R.T. (2004). Industrial production of microalgal cell-mass and secondary products-mayor industrial species. Haematococcus. In: Richmond, A., editor. Handbook of microalgal culture. Biotechnology and applied phycology. Oxford, U.K.: Blackwell Science. 281–8.

Danyal, A., Mubeen, U. and Malik, K.A. (2013). Investigating Two Native Algal Species to Determine Antibiotic Susceptibility Against some Pathogens. *Curr Res J Biol Sci.* 5: 70-74

Desai, K. and Sivakami, S. (2004). *Spirulina* the wonder food of the 21st century. *APBN.* 8: 1298–1302.

Dufosse, L., Galaup, P., Yaron, A., Arad, S.M., Chidambara, Murthy, K.N. and Ravishankar, G.A. (2005). Microorganisms and microalgae as sources of pigments for food use: a scientific odditiy or an industrial reality? *Trends Food Sci Technol.* 16: 289-406.

Fabregas, J. and Herrero, C. (1990). Vitamin content of four marine microalgae. Potential use as source of vitamins in nutrition. *Journal of Industrial Microbiology.* 5: 259-264.

Fernandez-Sevilla, J.M., Acien-Fernandez, F.G., Molina-Grima, E. (2010). Biotechnological production of lutein and its applications. *Appl Microbiol Biotechnol.* 86: 27-40.

Genovese, G., Faggio, C., Gugliandolo, C., Torre, A. and Spano, A. (2012). In vitro evaluation of antibacterial activity of Asparagopsis taxiformis from the Straits of Messina against pathogens relevant in aquaculture. *Mar Environ Res.* 73: 1-6.

Geppert, J., Kraft, V., Demmelmair, H. and Koletzko, B. (2006). Microalgal docosahexaenoic acid decreases plasma triacylglycerol in normolipidaemic vegetarians: a randomised trial. *Br J Nutr.* 95: 779–86.

Gilroy, D.J., Kauffman, K.W., Hall, R.A., Huang, X. and Chu, F.S. (2000). Assessing potential health risks from microcystin toxins in blue-green algae dietary supplements. *Environ Health Perspect.* 108: 435-439.

Goh, L.P. and Loh, S.P. (2009). Fatimah, M.Y.; Perumal, K. Bioaccessibility of Carotenoids and Tocopherols in Marine Microalgae, *Nannochloropsis* sp. and *Chaetoceros* sp. *Malays J. Nutr.* 15: 77–86.

Gouveia, L., Batista, A.P., Sousa, I., Raymundo, A. and Bandarra, N.M. (2008a). Microalgae in novel food products. In:PapadopoulosKN, editor. Food chemistry research developments. New York: Nova Science Publishers, Inc. 75–111.

Griffith, H.W. (1998). Vitamins, herbs, minerals & supplements: the complete guide. Fisher Books, Tucson, AZ, USA.

Guedes, A.C., Amaro, H.M. and Malcata, F.X. (2011). Microalgae as Sources of Carotenoids. *Mar. Drugs.* 9: 625-644.

Guedes, A.C., Barbosa, C.R., Amaro, H.M., Pereira, C.I. and Malcata, F.X. (2011). Microalgal and cyanobacterial cell extracts for use as natural antibacterial additives against food pathogens. *International Journal of Food Science and Technology*. 46: 862-870.

Hu, Q.H. (2004). Industrial production of microalgal cell-mass and secondary products-mayor industrial species. *Arthrospira (Spirulina) platensis* sp. In: Richmond A, editor. Handbook of microalgal culture. Biotechnology and applied phycology. Oxford, U.K.: Blackwell Science. 254-72.

Iwamoto, H. (2004). Industrial production of microalgal cell-mass and secondary products-mayor industrial species. Chlorella. In: Richmond, A, editor. Handbook of microalgal culture. Biotechnology and applied phycology. Oxford, U.K.: Blackwell Science. 255-63.

Kalra, E.K. (2003). Nutraceutical-definition and introduction. AAPS Pharm. Sci. 5: E25.

Katircioglu, H., Beyatli, Y., Aslim, B., Yuksekdag, Z. and Atici, T. (2006). Screening for antimicrobial agent production in fresh water. *Internet J. Microbiol.* 2(2). 1-9.

Khan, M., Varadharaj, S., Ganesan, L.P., Shobha, J.C. and Naidu, M.U. (2005). C-phycocyanin protects against ischemia-reperfusion injury of heart through involvement of p38 MAPK and ERK signalling. *Am J Physiol-Heart Circulatory Physiol.* (5): 2136-2145.

Khanavi, M., Gheidarloo, R., Sadati, N., Ardekani, M.R.S. and Nabavi, S.M.B. (2012). Cytotoxicity of fucosterol containing fraction of marine algae against breast and colon carcinoma cell line. *Pharmacogn. Mag.* 8: 60-64.

Kim, M., Yim, J.H., Kim, S.Y., Kim, H.S. and Lee, W.G. (2012). In vitro inhibition of influenza A virus infection by marine microalga-derived sulfated polysaccharide p-KG03. *Antiviral Res.* 93: 253-259.

Lee, J.B., Hayashi, K., Hirata, M., Kuroda, E. and Suzuki, E. (2006). Antiviral sulphated polysaccharide from *Navicula directa*, a diatom collected from deep-sea water in Toyama Bay. *Biol Pharma Bull.* 29: 2135-2139.

Li, Y., Horsman, M., Wu, N., Lan, C.Q. and Dubois-Calero, N. (2008). Biofuels from Microalgae. *Biotechnol. Prog.* 24: 815-820.

Lyons, N.M. and O'Brien, N.M. (2002). Modulatory effects of an algal extract containing astaxanthin on UVA-irradiated cells in culture. *J Dermatol Sci.* 30: 73–84.

Mahajan, G. and Kamat, M. (1995). γ-Linolenic production from *Spirulina platensis. Applied Microbiology and Biotechnology*. 43(3): 466-469.

Matsui, M.S., Muizzuddin, N., Arad, S. and Marenus, K. (2003). Sulfated polysaccharides from red microalgae have antiinflammatory properties in vitro and in vivo. *Arppl Biochem Biotechnol.* 104(1): 13-22.

McCarthy, M., Bane, V., García-Altares, M., Van-Pelt, F.N.A.M., Furey, A. and O'Halloran, J. (2015). Assessment of emerging biotoxins (pinnatoxin G and spirolides) at Europe's first marine reserve: Lough Hyne. *Toxicon.* 108: 202-209.

Molina, G.E., Belarbi, E.H., Acién-Fernández, F.G., Robles-Medina and A., Chisti, Y. (2003). Recovery of microalgal biomass and metabolites: process options and economics. *Biotechnol Adv.* 20: 491–515.

Najdenski, H.M., Gigova, L.G., Iliev, I.I., Pilarski, P.S. and Lukavsky, J. (2013) Antibacterial and antifungal activities of selected microalgae and cyanobacteria. *International Journal of Food Science and Technology.* 48: 1533-1540.

Pasquet, V., Morisset, P., Ihammouine, S., Chepied, A., Aumailley, L., Berard, J.B., Serive, B., Kaas, R., Lanneluc, I., Thiery, V., Lafferriere, M., Piot, J.-M., Patrice, T., Cadoret, J.P. and Picot, L. (2011). Antiproliferative Activity of Violaxanthin Isolated from Bioguided Fractionation of *Dunaliella tertiolecta* Extracts. *Mar. Drugs.* 9: 819-831.

Priyadarshani, I. and Rath, B. (2012). Commercial and industrial applications of micro algae - A review. *J. Algal Biomass Utln.* 3(4): 89-100.

Pulz, O. and Gross, W. (2004). Valuable products from biotechnology of microalgae. *Appl Microbiol Biotechnol.* 65: 635-48.

Rajasekaran, A., Sivagnanam, G. and Xavier, R. (2008). Nutraceuticals as therapeutic agents: A review. *Res J Pharm Tech.* 1: 328-340.

Rao, A.R.., Sindhuja, H.N., Dharmesh, S.M., Sankar, K.U., Sarada, R. and Ravishankar, G.A. (2013). Effective inhibition of skin cancer, tyrosinase, and antioxidative properties by astaxanthin and astaxanthin esters from the green alga *Haematococcus pluvialis. J. Agric. Food Chem.* 61: 3842–3851.

Rebolloso-Fuentes, M.M., Acién-Fernández, G.G., Sánchez-Pérez, J.A., Guil-Guerrero, J.L. (2000). Biomass nutrient profiles of the microalga *Porphyridium cruentum. Food Chem.* 70: 345–53.

Rosaline, X.D., Sakthivel, S.K., Rajendran, K. and Janarthanan, S. (2012). Screening of selected marine algae from the coastal Tamil Nadu, South India for antibacterial activity. *Asian Pacific Journal of Tropical Biomedicine.* 140-146.

Santoyo, S., Jaime, L., Plaza, M., Herrero, M. and Rodriguez-Meizoso, I. (2012). Antiviral compounds obtained from microalgae commonly used as carotenoid sources. *Journal of Applied Phycology.* 24: 731-741.

Seto, A., Wang, H.I. and Hesseltine, C.W. (1984). Culture conditions affect eicosapentaenoic acid content of *Chlorella minutissima. Journal of the American Oil Chemists Society.* 61(5): 892-894.

Shi, X.M., Jiang, Y. and Chen, F. (2002). High-yield production of lutein by the green microalga *Chlorella protothecoides* in heterotrophic fed-batch culture. *Biotechnol. Prog.* 18: 723-727.

Singh, D., Puri, M., Wilkens, S., Mathur A.S. and Tuli, D.K. (2013). Characterization of a new zeaxanthin producing strain of *Chlorella saccharophila* isolated from New Zealand marine waters. *Bioresource Technol.* 143: 308-314.

Soontornchaiboon, W., Joo, S.S., Kim, S.M. (2012). Anti-inflammatory effects of violaxanthin isolated from microalga Chlorella ellipsoidea in RAW 264.7 macrophages. *Biol Pharm Bull.* 35(7): 1137-44.

Spolaore, P., Joaniss-Cassan, C., Duran, E. and Isambert, A. (2006). Commercial applications of microalgae. *J Biosci Engr.* 101: 87–96.

Talero, E., Ávila-Román, S.J., Rodríguez-Luna, A., Alcaide, A. and Motilva, V. (2015). Bioactive Compounds Isolated from Microalgae in Chronic Inflammation and Cancer. *Mar Drugs.* 13(10): 6152-6209.

Tandeau-de-Marsac, H.J. (1993). Adaptation of cyanobacteria to environmental stimuli: new steps towards molecular mechanisms. *FEMS Microbiology Reviews.* 104: 119-190.

Tiberg, E., Dreborg, S. and Bjorksten, B. (1995). Allergy to green algae (*Chlorella*) among children. *J Allergy Clin Immunol.* 96: 257-259.

Tokusoglu, Ö. and Ünal, M.K. (2003). Biomass nutrient profiles of three microalgae: *Spirulina platensis*, *Chlorella vulgaris*, and *Isochrisis galbana*. *J Food Sci.* 68: 1144-48.

Trevithick, J.R., Xiong, H., Lee, S., Shum, D.T., Sanford, S.E., Karlik, S.J., Norley, C. and Dilworth, G.R. (1992). Topical tocopherol acetate reduces post-UVB, sunburn-associated erythema, edema, and skin sensitivity in hairless mice. *Arch. Biochem. Biophys.* 296: 575-582.

Walker, T.L., Purton, S., Becker, D.K. and Collet, C. (2005). Microalgae as bioreactors. *Plant Cell Rep.* 24: 629-641.

Watanabe, F., Takenaka, S., Kittaka-Katsura, H., Ebara, S. and Miyamoto, E. (2002). Characterization and bioavailability of vitamin B12-compounds from edible algae. *J Nutr Sci Vitaminol.* 48: 325-331.

Yang, F., Wong, K.H., Yang, Y., Li, X., Jiang, J., Zheng, W., Wu, H. and Chen, T. (2014). Purification and in vitro antioxidant activities of tellurium-containing phycobiliproteins from tellurium-enriched *Spirulina platensis*. *Drug. Des. Devel. Ther.* 8: 1789-1800.

Yim, H.E., Yoo, K.H., Seo, W.H., Won, N.H. and Hong, Y.S. (2007). Acute tubulointerstitial nephritis following ingestion of *Chlorella* tablets. *Pediatr Nephrol.* 22: 887-888.

Yim, J.H., Kim, S.J., Ahn, S.H., Lee, C.K. and Rhie, K.T. (2004). Antiviral effects of sulfated exo-polysacchride from the marine microalga *Gyrodinium impudicum* strain KG03. *Marine Biotechnol.* 6: 17-25.

■■■■

Comprehensive Review

21

MEDICINAL AND AROMATIC PLANTS: WEALTH OF INDIA AT HIGH AND LOW ALTITUDES

Dhiman Mukherjee

Department of Agronomy, Bidhan Chandra Krishi Viswavidyalaya, Kalayani, Nadia, W.B.-741235, India.

INTRODUCTION

India is home to a great variety of ethno-medicinally important plant species, and is ranked sixth among 12 mega-diversity countries of the world. Medicinal and Aromatic Plants (MAPs) as a group comprise approximately 8000 species and account for around 50% of all higher flowering plant species of India (Mukherjee, 2016). The Himalayas is recognised as one of the global biodiversity hotspots, where ecological, phyto-geographical and evolutionary factors favour high species diversity (Mukherjee, 2008). A bio-geographically unique region, it has the highest degree of species endemism in the Asian region - it supports about 18,440 species of plants, of which 25.3% are endemic to the region, a high proportion of which are of use for medicinal purposes. Because of the problem of **SAFETY** with **modern system** of medicine, there is increasing global interest in traditional and herbal medicines and essential oils. Medicinal and aromatic plants (MAPs) are traded both as raw materials and as processed final products. Demand for a wide variety of species is increasing as these markets expand and new end-uses are developed. The price trends of most of the Himalayan species of medicinal plants traded in market have been upwards in the last 3 years. However, about 90% of medicinal plants used by the industries are collected from the wild. While over 800 species are used in the industry, less than 20 species of plants are under commercial cultivation. The increase in demand increases the threat of depletion, and there are currently seventeen Himalayan medicinal plant species listed in the Indian government Red Data Book of Indian Plants. The collection and, more recently, marketing of these plants has provided an important source of income for communities living in mountain areas.

Glimpses of growing **MAPs in india** may be had from the following facts:

- Approx. 10 million floral spp.
- Approx. 2 million scientifically named
- 5-15% properly studied
- One among 12 mega- biodiverse countries
- 16 agro climatic zones
- 2 hotspots: Western- Ghats and North-East
- 2.5% of total world's area and 8% of recorded global floral spp.
- 45000 plant spp.-15000 spp. of medicinal plants (Charak *samhita and Susruta samhita)*
- 9000 spp. are used in traditional system
- Indian System of Medicine& Homeopathy (ISM&H), Min. of Health & Family Welfare, Govt of India. :
    - 1500 spp. of medicinal plants identified
    - 500 spp. Commonly used in the preparations of herbal drugs
    - 150 spp. Endangered
    - 90% drugs used in ISM&H is plant based
    - 60-70% population use medicinal plants for primary health care needs
    - 90% medicinal plants used come from the wild
    - India is one of leading exporter after China & USA
    - USA, Hong Kong & Germany are major markets

CLASSIFICATION OF MEDICINAL PLANTS

Of the 2,50,000 higher plant species on earth, more than 80,000 species are reported to have at least some medicinal value and around 5000 species have specific therapeutic value. They are classified according to the part used, habit, habitat, therapeutic value etc, besides the usual botanical classification.

1. Based on part used

i) Whole plant : *Boerhaavia diffusa, Phyllanthus neruri*

ii) Root : *Dasamula*

iii) Stem : *Tinospora cordifolia, Acorus calamus*
iv) Bark : *Saraca asoca*
v) Leaf : *Indigofera tinctoria, Lawsonia inermis, Aloe vera*
vi) Flower :*Biophytum sensityvum, Mimusops elenji*
vii) Fruit : *Solanum species*
viii) Seed : *Datura stramonium*

2. Based on habit

i) Grasses : *Cynodon dactylon*
ii) Sedges : *Cyperus rotundus*
iii) Herbs : *Vernonia cineria*
iv) Shrubs : *Solanum species*
v) Climbers : *Asparagus racemosus*
vi) Trees : *Azadirachta indica*

3. Based on habitat

i) Tropical : *Andrographis paniculata*
ii) Sub-tropical : *Mentha arvensis*
iii) Temperate : *Atropa belladona*

4. Based on therapeutic value

Antimalarial : *Cinchona officinalis, Artemisia annua*
Anticancer : *Catharanthus roseus, Taxus baccata*
Antiulcer : *Azadirachta indica, Glycyrrhiza glabra*
Antidiabetic : *Catharanthus roseus, Momordica charantia*
Anticholesterol : *Allium sativum*
Antiinflammatory : *Curcuma domestica, Desmodium gangeticum*
Antiviral : *Acacia catechu*
Antibacterial : *Plumbago indica*
Antifungal : *Allium sativum*
Antiprotozoal : *Ailanthus sp., Cephaelis ipecacuanha*
Antidiarrhoeal : *Psidium gujava, Curcuma domestica*
Hypotensive : *Coleus forskohlii, Alium sativum*
Tranquilizing : *Rauvolfia serpentina*
Anaesthetic : *Erythroxylum coca*
Spasmolytic : *Atropa belladona, Hyoscyamus niger*
Diuretic : *Phyllanthus niruri, Centella asiatica*
Astringent : *Piper betle, Abrus precatorius*
Anthelmentic : *Quisqualis indica, Punica granatum*
Cardiotonic : *Digitalis sp., Thevetia sp.*

Antiallergic : *Nandina domestica, Scutellaria baicalensis*
Hepatoprotective : *Silybum marianum, Andrographis paniculata*

5. Based on Ayurvedic formulations in which used

a) The ten roots of the Dasamoola (*Dasamoolam*)

i) Desmodium gangeticum (Orila)
ii) Uraria lagopoides (Cheria orila)
iii) Solanum jacquinii (Kantakari)
iv) Solanum indicum (Cheruchunda)
v) Tribulus terrestris (Njerinjil)
vi) Aegle marmelos (Koovalam)
vii) Oroxylum indicum (Palakapayyani)
viii) Gmelina arborea (Kumizhu)
ix) Steriospermum suaveolens (Pathiri)
x) Premna spinosus (Munja)

b) The ten flowers of the Dasapushpa (*Dasapushpam*)

i) Biophytum sensitivum (Mukkutti)
ii) Ipomea maxima (Thiruthali)
iii) Eclipta prostrata (Kayyuniam)
iv) Vernonia cineria (Poovamkurunnil)
v) Evolvulus alsinoides (Vishnukranthi)
vi) Cynodon dactylon (Karuka)
vii) Emelia sonchifolia (Muyalcheviyan)
viii) Curculigo orchioides (Nilappana)
ix) Cardiospermum halicacabum (Uzhinja)
x) Aerva lanata (Cherula)

c) The four trees of the Nalpamara (*Nalpamaram*)

i) Ficus racemosa (Athi)
ii) Ficus microcarpa (Ithi)
iii) Ficus relegiosa (Arayal)
iv) Ficus benghalensis (Peral)

d) The three fruits of the Triphala (*Thriphalam*)

i) Phyllanthus emblica (Nellikka)
ii) Terminalia bellerica (Thannikka)
iii) Terminalia chebula (Kadukka)

6. Botanical classification

This is the most comprehensive and scientific classification. The various medicinal plants are grouped according to their Class, Series, Order, Family, Genus and Species.

Most of the medicinal and aromatic plants belong to the following families:

Compositae, Labiatae, Umbelliferae, Leguminosae, Roseaceae, Rutaceae, Solanaceae, Cruciferae, Liliceae, Caryophyllaceae, Boraginaceae, Ranunculaceae, Papaveraceae, Malvaceae, Cucurbitaceae, Verbenaceae, Scrophulariceae and Phytolaccaceae

DEVELOPMENT OF MEDICINAL AND AROMATIC PLANTS (MAPS)

India with its varied agro-climatic conditions provides an ideal habitat for rich repository of number of medicinal and aromatic plants which provide raw material for Indian system of medicinal and cosmetic industry since ancient times. With greater realization of health hazard and side effect of synthetic drugs, the interest for herbal drug is increasing worldwide. An estimated amount of US $ 3000 Million per annum is being earned from medicinal and aromatic plants from India. Indiscriminate commercial exploitation of these medicinal plants, which are derived from the wild forests, is posing a great threat to bio-diversity and endangering a number of valuable species (Mukherjee and Roy, 2005). Despite the progress made in synthetic organic chemistry and in biotechnology, higher plants are still an indispensible source of well-known drugs and are used as raw material for many galenic preparations. The hope that cell culture and fermentation processes would replace the whole plant has not yet materialized.

Nowadays, one quarter of all prescribed drugs in the USA are of higher plants origin, exactly as it was 20 years ago. Many other species are used as galenic products in the health food markets, herbal teas, cosmetics and other similar products. It seems likely that higher plants will continue to play a considerable role in the pharmaceutical industry as well as in the health food business.

IMPORTANCE AND SCOPE

Most of the MAPs belong to important herb category. Herbs are staging a comeback and herbal 'renaissance' is happening all over the globe. The herbal products today symbolise safety in contrast to the synthetics that are regarded as unsafe to human and environment. Although herbs had been priced for their medicinal, flavouring and aromatic qualities for centuries, the synthetic products of the modern age surpassed their importance, for a while. However, the blind dependence on synthetics is over and people are returning to the naturals with hope of safety and security. Over three-quarters of the world population relies mainly on plants and plant extracts for health care. More than 30% of the entire plant species, at one time or other, were used for medicinal purposes. It is estimated that world market for plant derived drugs may account for about Rs.2,00,000 crores. Presently, Indian contribution is less than Rs.2000 crores. Indian export of raw drugs has steadily grown at 32 % to Rs.312 crores in 2014-15 from Rs.130 crores in 1991-'92. The annual production of medicinal and aromatic plant's raw material is worth about Rs.200 crores. This is likely to touch US $2535 by the year 2000 and US $5 trillion by 2050 (Mukherjee, 2009). It has been estimated that in developed countries such as United States, plant drugs constitute as much as 25% of the total drugs, while in fast developing countries such as China and India, the contribution is as much as 80%. Thus, the economic importance of medicinal plants is much more to countries such as India than to rest of the world (Mukherjee, 2006). These countries provide two third of the plants used in modern system of medicine and the health care system of rural population depend on indigenous systems of medicine.

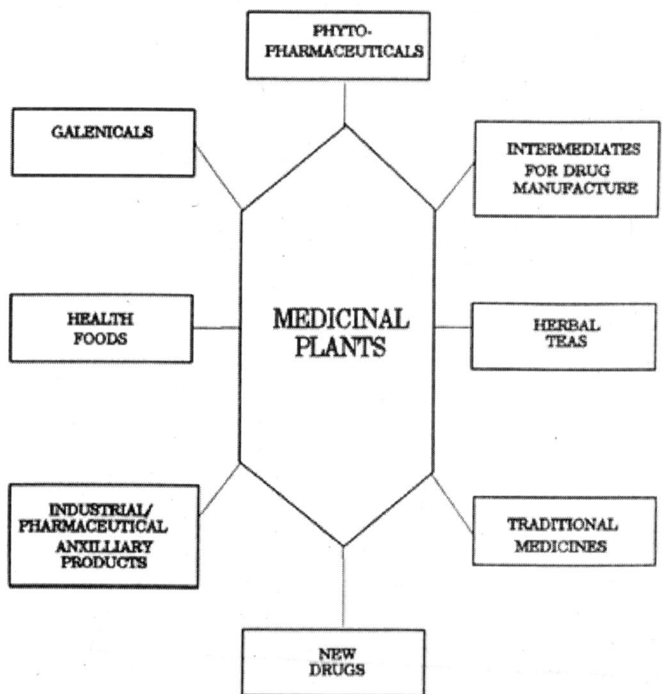

Fig. 1: Industrial uses of Medicinal Plants

In India, drugs of herbal origin have been used in traditional systems of medicines such as *Unani* and *Ayurveda* since ancient times. The *Ayurveda* system of medicine uses about 700 species, *Unani* 700, *Siddha* 600, Amchi 600 and modern medicine around 30 species (Chakraborty *et al.,* 2015). The drugs are derived either from the whole plant or from different organs, like leaves, stem, bark, root, flower, seed, etc. Some drugs are prepared from excretory plant product such as gum, resins and latex. Even the Allopathic system of medicine has adopted a number of plant-derived drugs which form an important segment of the modern pharmacopoeia. Some important chemical intermediates needed for manufacturing the modern drugs are also obtained from plants (Eg. diosgenin, solasodine). Not only, that plant-derived drug offers a stable market world wide, but also plants continue to be an important source for new drugs. Brazil, Indonesia, China and India are the leading world producers of essential oils. However, as the consumption of essential oils in India is very high, very little is made available for export. The important essential oils produced in India are from sandalwood, lemon grass, palmarosa, eucalyptus (*Eucalyptus citriodora* Hook. and *Eucalyptus globulus* Labill.), mentha, khus and linaloe (Mukherjee, 2009 a).

Following species of Medicinal Plants have commercial potentials at all levels :

Table 1: Highly demanded top 20 medicinal plants traded in India

S.N.	Trade Name	Botanical Name	Part used
1	ATIS	*Acanitum heteorphyllum,* Wall. ex Royle	Tuberous roots
2	MEETHA TELIA/ BACHNAG	*Aconitum violoceum* (Jacq.) Stapf	Tuberous roots
3	SAFED MUSALI	*Chlorophytum arundinaceum* Baker	Tubers
4	GUGGUL	*Commiphora wightii* (Arn.) Bhandari	Resin
5	NANURA/MISHMIBITTER	*Coptis teeta* Wall	Roots
6	SALAM PANJA/SALEP	*Dactylorrhiza hattagirea* (D.Don) Soo	Tubers
7	VIDANGA/BAIBARANGA	*Embelia ribes* Burm f.	Fruits
8	NAGKESAR	*Mesua nagasssrum* (Burm f.) Kostern	Dry stamens
9	RAMPATRI/BOMBAY MACE	*Myristica malabarica* Lam	Aril & Mace
10	JATAMANSI	*Nardostachys jatamansi* DC	Rhizomes & Roots
11	GAOZAGAN	*Onosma bractearum* Wall	Flowers
12	KUTKI	*Pycrorrhiza kurooa* Royle ex Benth	Roots & Rhizomes

S.N.	Trade Name	Botanical Name	Part used
13	KAKRA-SINGI	*Pistacia integerrina* Stewart ex Brandis	Galls
14	SARPAGANDHA	*Rauvolfia serpentina* Benth. ex Kurz	Roots
15	MANJISHTA	*Rubia cordifolia* Linn	Roots
16	CHANDANA/ SANDALWOOD	*Santalum album* Linn	Heartwood
17	CHOBCHINI GULABI	*Smilax glabra* Roxb	Roots
18	CHIRAIYITA	*Swertia chirayita* Karst.	Whole Plant
19	TAGGAR/ MUSHKBALA	*Valeriana Hardwickia* Wall	Rhizomes & Roots
20	BANAFSHA	*Viola pilosa* Blume	Flowers

About 121 (45 tropical and 76 subtropical) major plant drugs have been identified for which no synthetic one is currently available (Table 2). The scientific study of traditional medicines, derivation of drugs through bioprospecting and systematic conservation of the concerned medicinal plants are thus of great importance.

Table 2. Major plant drugs for which no synthetic one is currently available (Kumar *et al*, 1997).

Drug Plant Use

- Vinblastine *Catharanthus roseus* Anticancer
- Vinblastine *Catharanthus roseus* Anticancer
- Ajmalacine *Catharanthus roseus* Anticancer, hypotensive
- Rescinnamine *Rauvolfia serpentina* Tranquilizer
- Reserpine *Rauvolfia serpentina* Tranquilizer
- Quinine *Cinchona sp.* Antimalarial, amoebic dysentery
- Pilocarpine *Pilocarpus jaborandi* Antiglucoma
- Cocaine *Erythroxylum coca* Topical anaesthetic
- Morphine *Papaver somniferum* Painkiller
- Codeine *Papaver somniferum* Anticough
- Atropine *Atropa belladonna* Spasmolytic, cold
- Atropine *Hyoscyamus niger* Spasmolytic, cold
- Cardiac glycosides *Digitalis sp.* For congestive heart failure
- Artemisinin *Artemesia annua* Antimalarial,
- Taxol *Taxus baccata, T. Brevifolia* Breast and ovary cancer, antitumour
- Berberine *Berberis* For leishmaniasis
- Pristimerin *Celastrus paniculata* Antimalarial

- Quassinoids *Ailanthus* Antiprotozoal
- Plumbagin *Plumbago indica* Antibacterial, antifungal
- Diospyrin *Diospyros montana*
- Gossypol *Gossypium sp.* Antispermatogenic
- Allicin *Allium sativum* Antifungal, amoebiasis
- Ricin *Ricinus communis*
- Emetine *Cephaelis ipecacuanha* Amoebiasis
- Glycyrrhizin *Glycyrrhizia glabra* Antiulcer
- Nimbidin *Azadirachta indica* Antiulcer
- Catechin *Acacia catechu* Antiulcer
- Sophoradin *Sophora subprostrata* Antiulcer
- Magnolol *Magnolia bark* Peptic ulcer
- Forskolin *Coleus forskohlii* Hypotensive, cardiotonic
- Digitoxin, Digoxin *Digitalis, Thevetia* Cardio tonic
- Thevenerin, *Thevetia* Cardio tonic
- Nerrifolin *Thevetia* Cardio tonic
- Podophyllin *Podophyllum emodi* Anticancer
- Indicine N-oxide *Heliotropium indicum* Anticancer
- Elipticine *Ochrosia* Anticancer
- Homoharringtonine *Cephalotaxus* Anticancer
- Camptothecine *Camptotheca acuminata* Anticancer

UTILIZATION & MANAGEMENT - CURRENT STATUS

As mentioned above 85% to 90% of medicinal and aromatic plants come from the wild, that is, forests. The forests of the state are largely under the control of Autonomous District Councils (ADC). These forests are subjected to very little or no scientific management. In the government controlled reserved and protected forests, which constitutes about 3.9 % of the total geographical area. Medicinal and Aromatic Plants occur in these forests as herbs, shrubs and trees. These plants are generally collected in an unorganized and unscientific manner and used mostly in herbal formulations. In view of fast growing demand for the MAPs for the use in Pharmaceuticals, Neutraceuticals, Cosmatologicals, Dietary Supplements etc. both at National as well as International level, implementation of the projects and the schemes meant for cultivation, collection, R&D, awareness, commercial supply of the medicinal plants for generation of income and employment while ensuring conservation of endemic and threatened species through the sustainable use is the urgent need of the hour.

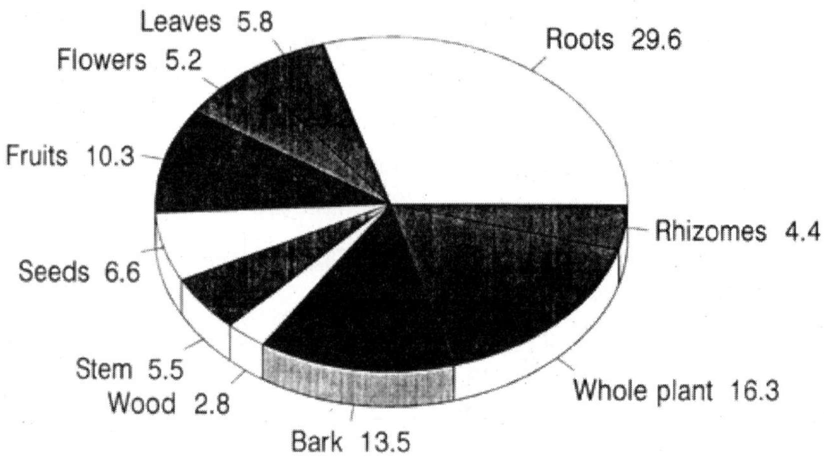

Figure 2: Destructive collections: distribution of medicinal plants by parts used (72% destructive and 28% non-destructive) (Source: Chakraborty *et al.*, 2016)

The following Medicinal and Aromatic plants are grown by the farmers, in their farmlands considering the area specific Agro-climatic Zones and Geo-morphological conditions in different parts of the State. The cultivation of the following species of MAPs have been taken up by the farmers in their farmland under Contractual Farming under State Medicinal Plants Board(SMPB), Sikkim funded by the National Medicinal Plants Board(NMPB), Department of Ayurveda, Yoga & Naturopathy, Unani, Siddha and Homoeopathy (AYUSH), Ministry of Health and Family Welfare Department, Government of India.

Table 3. Some important wild plant used in traditional system of medicine by local vaidays and procure in their farm land.

Botanical name	Vernacular name	Part used/ mode of application	Uses
Aconitum atrox	Mitha bish	Paste of rhizome fried in ghee (clarified butter) is applied externally	Rheumatism, neuralgia, paralysis, rheumatic and puerperal fever.
A. heterophyllum	Atibish	Root powder mixed with honey and galls of *Pistacia ichinjuk* (kakarsingi) is applied externally.	Bronchial inflammation
Actaea acuminata	Mamira	Decoction of roots	Flatulence, colic

Botanical name	Vernacular name	Part used/ mode of application	Uses
Asparagus filichinus	Jhirni	Powder of dried tuberous roots	Sexual debility and urinogenital disorder
Bergenia straqcheyi	Shilpari	Root decoction and juice of leaves	Kidney stone, sores, swellings and jaundice
Dactylorrhiza hatagirea	Salampanja	Powder of roots	Cuts and wounds
Dioscorea bulbifera	Genthi	Tuber	Bronchial cough
Diocorea deltoids	Tairu	Rhizomes	Spermetonorrhoea
Euphorbia hirta	Dudhibari	Entire plant with curd	Pile
Hedychium accuminatum	Kapoorkachri	Purified root powder	Dyspepsia and piles
Megacarpea polyandra	Barmoola	Roots	Fever, stomach disorder
Picrorrhiza kurroo	Kutki	Root powder	Severe coughing, fever and stomach disorder.
Swertia chirayita	Chirota	Whole plant	Fever, malaria remedies
Valerianal jatamansi	Sameva	Roots, rhizome	Epilepsy, hysteria
Zanthozylum acanthopodium	Timru	Seed powder and stem bark	Toothache, tooth decay.

SYSTEM OF MEDICINE PRACTICES IN INDIA

India is a unique country and proud to possess a well documented and well established system of medicine which has been practised for centuries, flourishing from 2500 to 500 B.C. There are two levels of medicine system in India, the classical well documented systems of Ayurveda, Siddha and Unani, which have text books, are taught in colleges, grant degrees to practitioners and are incorporated into the official health care system. The other system of folk medicine is informal and exists in communities. It is passed orally from generation to generation. These systems have survived for thousands of years because of their common use and associated cultural beliefs of the populations. The Ayurveda, Siddha and Unani systems of medicine have coexisted with homeopathy for centuries in India (WHO, 2001). The Ayurveda system of medicine was taught in ancient universities in India, established as early as 700 B.C. (Takshila) and 500 B.C. (Nalanda). Ayurveda is believed to have matured between 2,500 and 500 B.C. and is deeply rooted in the Indian culture. During the medieval period, it suffered first because of the Mughals and then the British who patronized their own systems of medicine. However, it survived because of its inherent strength, cultural support and use by common people. There are over 2,000 medicinal plants used in the Ayurvedic medicine system.

The Siddha system of medicine was developed in the South of India simultaneously with the Ayurveda. Siddha is one of the oldest systems of medicine practised in India. The Ayurveda and Siddha systems are being integrated with the modern system in the national health care programmes and patient care. The Unani medicine system introduced by Mughal, is based on Greek philosophy and Hippocrates (460- 377 B.C.), the father of modern medicine, was its founder. It was then adopted by the Islamic rulers of the Arabian countries and flourished and developed during the medieval period. It was introduced to India by the Arabs in 1351 A.D. Homeopathy, founded by Dr Samuel Hahnemann (1755-1843), originated in Germany and entered India in 1839 when Dr John Honigberger successfully treated Maharaja Ranjit Singh, the ruler of the Punjab for paralysis of the vocal cords. It is now one of the popular medicine systems practised in India.

Some of the Indian population on the Indo-China border area practises the Tibetan system of medicine, which is based on herbs. The Tibetan Medicine and Astrological Institute at Dharamsala in the state of Himachal Pradesh offers a six-year degree course in Tibetan medicine and has 30 branches throughout India and Nepal.

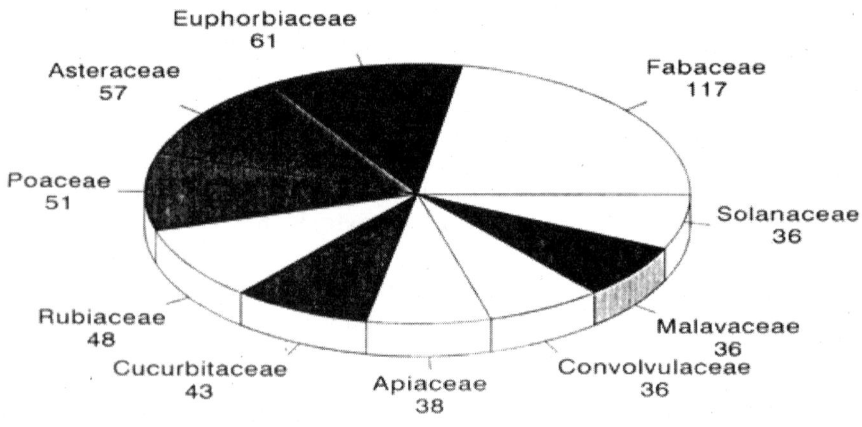

Figure 3: Distribution of Ayurvedic medicinal plants among the ten most represented families (Source : WHO 201).

The Indian system of medicine and homeopathy including Ayurveda, Siddha, Unani and drugless therapies such as yoga and naturopathy are widely practised in India. These systems are well integrated into the national health care system and are officially recognized by the central and state governments. These are safe, easy to use, economically viable and are widely accepted by the masses (Mukherjee, 2008 a). The Vaidyas and Hakims had been practising these systems on the strength of their own merit. Each of these systems of medicine is based on its own individual philosophy and principles and has its own core areas of strengths and weaknesses. In post independence India, the government recognized the merit of each of these medicine systems and made attempts to develop them as viable systems of medicine for the health care needs of its people.

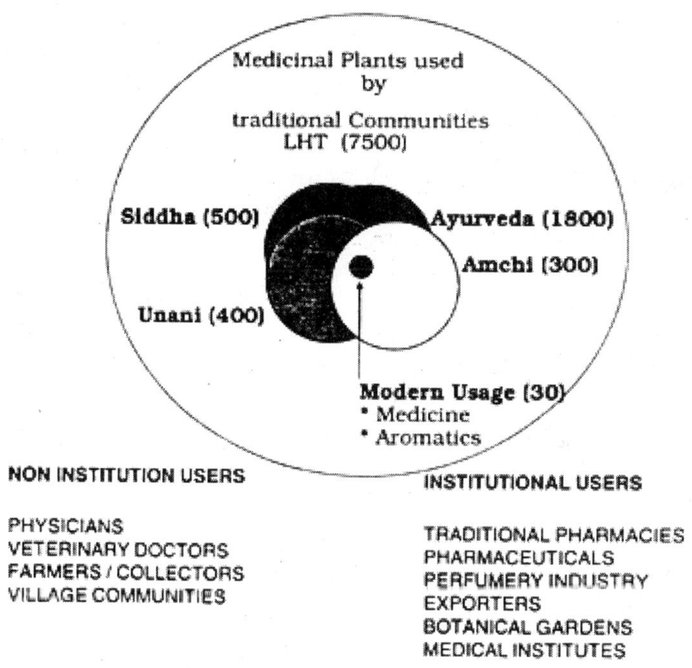

Figure 4: Resource base of traditional medicine (Source : WHO 2001)

TRADE AND MARKETING

Most of the medicinal raw materials produced in the country are used locally. However, with the introduction of cultivation of high demand plants, export is expected to rise in the coming years. The present export volume of crude drugs from India 140 stands at 36,200 tonnes worth about US$24 million. India, apart from the request for medicinal plants for internal consumption, is one of the major exporters of crude drugs, mainly to six developed countries: France, Germany, Japan, Switzerland, the UK and the USA, who share 75 to 80% of the total export of crude drugs from India. Despite the potential for producing raw material within the country, some crude drugs are imported from countries such as Afghanistan, Bangladesh, Bhutan, China, Nepal, Pakistan and Singapore. The most prominent are *Glycyrrhiza glabra* L. from Afghanistan, Iran and Pakistan; *Atropa belladonna* L. from Germany; *Hedychium spicatum* from China; *Commiphora wightii* (Arn.) Bhandari Engl. from Pakistan and *Swertia chirayita* (Roxb. ex Fleming) H. Karst. from Nepal (Moktan and Mukherjee, 2008). India exported finished Ayurvedic and Unani medicines worth US $512 million to countries including Germany, Hong Kong, Malaysia, Russia, the UK and the USA in the year 2009-2010. There are 7,843 licensed pharmacies of the Indian System of Medicine in addition to 857 of homeopathy and a number of unlicensed small scale processing units engaged in the manufacture of medicines to meet the requirement of 0.46 million registered practitioners of AYUSH and other users in the country. These pharmacies range from large Indian drug houses such as Baidyanath, Dabur, Himalaya Drugs, Zandu, Charak, etc. that employ modern/sophisticated equipment and methods for the production of drugs on a mass scale to small companies which manufacture drugs generally following the traditional prescriptions of ancient medical texts in the form of Nighantus (Lexicons) and texts on Bhaisaj Kalpana (Pharmacy) that specifically deal with plants and plant products.

Table 4: List of Medicinal and Aromatic plants having high export market demand :

Sl.No.	Medicinal Plants
1.	*Picrorhiza kurroa*
2.	*Nardostachys jatamansi*
3.	*Swertia chirata*
4.	*Valeriana jatamansi*
5.	*Aconitum heterophyllum*
6.	*Sassurea lappa*
7.	*Asparagus racemosus*
8.	*Rauvolfia serpentina*
9.	*Dactylorhiza hatagarea*
10.	*Withania somnifers*
11.	*Costus specious*
13.	*Bacopa monnieri*
14.	*Taxus wallichiana*
15.	*Digitallis purpurea*
Aromatic Plants	
1.	*Rosa amascene*
2.	*Lavandula intermedia*
3.	*Pelargonium graveolen*
4.	*Citronella*

However if we go zone-wise of their demand, we saw that most of the plant herb confined to high to low altitude range and these have high market demand (Mukherjee, 2009).

High Altitude/ Sub-Alpine areas above, 2500 Meter.

- *Aconitum heterophyllum* (Atees),
- *Aconitum ferox , palmatum* (Bikh),
- *Picrorhiza kurooa* (Kutki),
- *Nardostachys jatamansi* (Jatamansi),
- *Digitalis purpurea* (Fox glove),
- *Podophyllum hexandrum* (Patch patey),
- *Panax pseudoginseng* (Ginseng),
- *Rheum emodi* (Padamchal),
- *Selinum tenuifolium* (Bhut kesh),
- *Rhododendron anthopogon* (Sunpati),
- *Juniperus macropoda* (Shukpa),

Middle and Lower Hills/Temperate and Sub-Tropical areas:

- *Swertia chiraita* (Chiraita),
- *Acorus calamus* (Bojho),
- *Aloe barbadensis* (Ghiu Kumari),
- Asparagus racemosus (Kurilo),
- *Glorisa superba* (langarey tarul),
- *Piper longum* (Pipla),
- *Valeriana hard wickii* (Nakali jatamansi),
- *Solanum nigrum* (Bokshikara),
- *Ocimum sanctum* (Tulsi)
- *Terminalia belerica,T. chebula, Emblica officinalis* (Triphala)
- *Azadiracta indica* (Neem)

All the Medicinal and Aromatic plants (Maps) listed above are having high efficacy and have medicinal values which have been widely used by the different Indian System of Medicines like Ayurveda, Homeopathy, Unani, Siddha, Amchi, and Allopathy. Some of the species are also used for Cosmetic, Neutraciticals, Food products, Beverages etc. and have tremendous demands from pharmaceuticals and many other Herbal Based Industries (Mukherjee, 2008 b).

Table 5: List of plants traded to National and International markets.

S. No	Botanical Name	Local/Common Name	Exported To
1	*Acorus calamus*	Aorch	National Use
2	*Adiantum incisum*	Parshoshan	Germany, Scotland, Iran
3	*Adiantum capillus-veneris*	Parshoshan	Germany, Scotland and Iran
4	*Bergenia ciliata*	Qamar Panra	China
5	*Colchicum luteum*	Soranjan talakh	Germany, Scotland, Iran, and Afghanistan.
6	*Dioscoria deltoids*	Kaneez	Japan
7	*Diospyrus lotus*	Tor Amlok	National use
8	*Morchalla esculenta*	Guchi	Switzerland, France, Germany, Scotland, Austria, Belgium.
9	*Myrtus communis*	Maroon	National use
10	*Paeonia emodi*	Mamekh	Europe and India.
11	*Pistacia integrimma*	Shanai	China
12	*Podophyllun hexandrum*	Kakora	Europe and East Asia.
13	*Bistorta amplexicaulis*	Anjabar	Europe and South America.
14	*Valeriana wallachii*	Mushk Bala	Europe and Iran.
15	*Viola betonicifolia*	Banafsha	Germany and Scotland.
16	*Viola serpens*	Banafsha	Germany and Scotland.

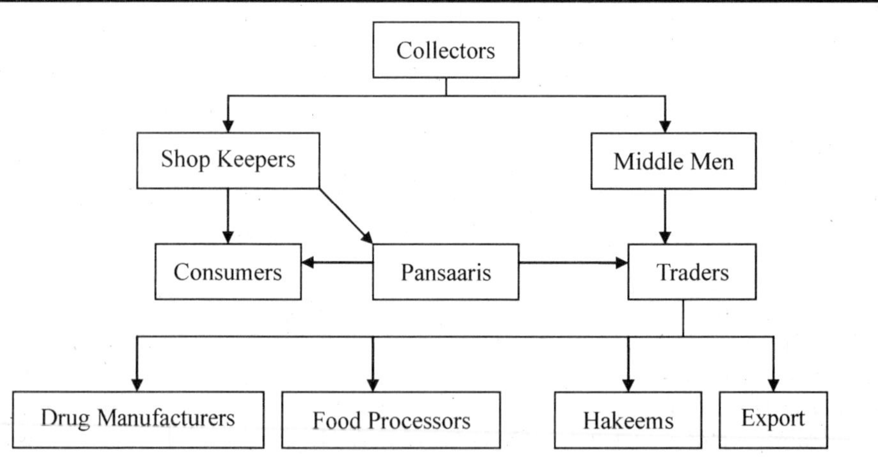

Figure 5: Chain of people involved in the medicinal plant trade in India.

Table 6: Marketing of available medicinal plant in Himalaya range.

No	Botanical Name	Part Used	Availa-bility	Collection Method	Plant Status	L.Pr. (Rs./Kg)
1	*Aconitum hetrophyllum*	Roots	Dec.	Digging	V	830
2	*Aconitum violaceum*	Roots	Dec.	Digging	V	342
3	*Acorus calamus*	Rhizome	Dec.	Digging	E	265
4	*Artemissia vulgaris*	Rizome, Leaves	Inc.	Digging	-	254
5	*Berberis lycium*	Roots	Dec.	Digging	V	214
6	*Berberis vulgaris*	Roots	Dec.	Digging	E	98
7	*Bergenia ciliata*	Leaves	Pers.	Plucking	R	116
8	*Bistorta amplexicaulis*	Rhizome	Dec.	Digging	E	185
9	*Cuminum cyminum*	Seeds	Pers.	Picking	R	198
10	*Calendula arvensis*	Roots	Dec.	Digging	-	245
11	*Caltha alba*	Plant	Pers.	Plucking	-	89
12	*Cichorium intybus*	Leaves	Inc.	Plucking	-	187
13	*Colchicum luteum*	Rhizome	Dec.	Digging	V	69
14	*Coriandrum sativum*	Fruits	Inc.	Picking	-	145
15	*Bunium persicum*	Seeds	Pers.	Picking	V	210
16	*Dioscoria deltoides*	Root	Dec.	Digging	E	285
17	*Diospyrus lotus*	Fruit	Pers.	Picking	-	198
18	*Ephedra intermedia*	Bark, Leaves	Dec.	Cutting	V	412
19	*Foeniculum vulgare*	Fruit	Pers.	Picking	-	312
20	*Geranium wallichianum*	Roots	Pers.	Digging	-	145
21	*Hyoscyamus niger*	Seeds	Dec.	Picking	-	245
22	*Hypericum perforatum*	Shoot	Pers.	Plucking	-	316
23	*Morchella esculenta*	Plant	Dec.	Picking	R	925
24	*Peganum harmala*	Seeds	Inc.	Picking	-	96
25	*Pistacia integrimma*	Pod	Dec.	Picking	V	298
26	*Plantago lanceolatum*	Seeds	Pers.	Picking	R	358
27	*Podophyllum hexandrum*	Rhizome	Dec.	Digging	E	875
28	*Skimmia laureola*	Leaves	Pers.	Plucking	R	121
29	*Swertia chirayita*	Whole plant	Nov-Dec	Plucking	E	460
30	*Valeriana jatamansi*	Rhizome	Dec.	Digging	E	109

Abbreviations
Availability: Dec. stands for decreased, Pers. for persistent, Inc. for increased.
Plant Status: E stands for endangered, V for vulnerable, R for rare.

PROBLEMS AND CONSTRAINTS

Some of the major constraints in commercial exploitation of the medicinal and aromatic plant sector in India include: major dependence on collection from natural sources; poor harvesting (indiscriminate) and post harvest treatment practices; domestication; improper storage; unorganized trade practices; and lack of coordination of the plant based drug industry. There is insufficient data on the demand and supply situation of medicinal and aromatic palnts. Farmers have been taking initiatives to cultivate medicinal and aromatic plants but price instability affects the level of confidence of farmers necessary to take up large scale cultivation. However, for some crops such as senna, plantago, withania, asparagus, chlorophytum, mentha and lemongrass, farmers have developed confidence.

Constraints Associated with Processing of MAPs

Some of the constraints associated with the processing of medicinal plants which may result in reducing their competitiveness in global markets and which have to be remedied are:

- Poor agricultural practices
- Poor harvesting (indiscriminate) and post-harvest treatment practices
- Lack of research on development of high-yielding varieties, domestication etc.
- Poor propagation methods
- Inefficient processing techniques leading to low yields and poor quality products
- Poor quality control procedures
- High energy losses during processing
- Lack of current good manufacturing practices
- Lack of R & D on product and process development
- Difficulties in marketing
- Lack of local market for primary processed products
- Lack of trained personnel and equipment
- Lack of facilities to fabricate equipment locally
- Lack of access to latest technological and market information

Systematic cultivation of many medicinal plants needs specific cultural practices and agronomical requirements. These are species-specific and are not only dependent on soil, water and climatic conditions. Hence research and development work has to be done to formulate Good Agricultural Practices which will include proper selection and identification, propagation methods, cultivation techniques, harvesting, step-wise quality control of raw material up to processing stage, post-harvest treatment, storage and safety. These aspects have to be incorporated into protocols for the cultivation of medicinal plants.

Medicinal Plant confined to High Altitude (> 1500 ft a.s.l)

Podophyllum hexandrum (Plate 1)

Picrorrhiza kurroo (Plate 2)

Coptis teeta (Plate 3)

Coptis teeta flowring stage (Plate 4)

Barbaris aristata (Plate 5)

Valeriana jatamasi (Plate 6)

Bergenia ciliata (Plate 7)

Aleocarpus granites (Rudraksha) (Plate 8)

Gentiana kurro (Plate 9)

Swertia chirayita (Plate 10)

Aloe barbadensis (Plate 11)

Artemesia annua (Plate 12)

Cinchona (Plate 13)

Citronella (Plate 14)

Costus specious (Plate 15)

Mentha citrata(Plate 16)

Berginia rigulata (Plate 17)

Phaichinan ciliata(Plate 18)

Vetiver spp. (Plate 19)

Hysamum niger (Plate 20)

Organic farming is another practice that is gaining wide acceptance as world demand particularly in developed countries for organically grown crops is rapidly on the increase. Farmers have to be trained in all aspects of organic farming of medicinal plants and herbs including obtaining certification from associations that do the monitoring starting from cultivation to final harvesting. Organic farming which is labour-intensive gives the developing countries the comparative advantage to be competitive.

PROCESSING AND UTILIZATION

Medicinal principles are present in different parts of the plant like root, stem, bark, heartwood, leaf, flower, fruit or plant exudates. These medicinal principles are separated by different processes; the most common being extraction. Extraction is the separation of the required constituents from plant materials using a solvent. In the case of medicinal plants, the extraction procedure falls into two categories (Paroda, 1993).

a) Where it is sufficient to achieve within set limits equilibrium of concentration between drug components and the solution, e.g., Tinctures, decoction, teas, etc.

b) Where it is necessary to extract the drug to exhaustion, i.e., until all solvent extractable are removed by the solvent.

Both the methods are employed depending on the requirement although in industry the latter method is mostly used. In all industrial procedures, the raw material is pre-treated with solvent outside the extractor before changing the latter. This prevents sudden bulk volume changes (which are the main cause of channelling during extraction) and facilitates the breaking up of the cell walls to release the extractable. To facilitate the extraction, the solvent should diffuse inside the cell and the substance must be sufficiently soluble in the solvent. The ideal solvent for complete extraction is one that is most selective, has the best capacity for extraction and is compatible with the properties of the material to be extracted. These parameters are predetermined experimentally. The cost and availability of the solvent are also taken into account. Alcohol, though widely used, because of its great extractive power it is often the least selective, in that it extracts all soluble constituents. Alcohol in various ratios is used to minimise selectivity. The ideal alcohol ratio for woody or bark material is 75%. For leafy material, it is often less than 50% thus avoiding extraction of the chlorophyll which makes purification difficult. Some materials such as alkaloids being soluble in acids, their extraction is facilitated by adjusting the pH in the acidic range. A number of alkaloids can be extracted easily with hydrocarbons after they have been released from combination with organic acids by grinding with alkali. It is first ground with moist calcium oxide and extracted with chloroform. A large number of alkaloids can be extracted directly with aqueous acids, organic or inorganic acids, and the alkalised extracts counter extracted with hydrocarbons or other apolar solvents. Experiment used for extraction with solvents usually comprise an extraction vessel with a heating jacket for steam heating or fitted with electrical devices, a condenser in reflux position, a solvent reservoir, a facility to convert to reboiler position or a separate reboiler and a short column for solvent recovery. Sometimes, sophisticated and costly equipment like the Carousel or the Inoxa extractor is employed. Technology for the manufacture of standardised extracts and phytochemicals is available and there are many extracts already in the international market as drugs. A drug such as an extract of *Centella asiatica* can be manufactured as an extract containing a standard quantity of asiaticoside. Similarly for senna a standardised extract of which, containing a standard quantity of sennosides a and b could easily be produced with equipment that can be designed and constructed in most developing countries (Wijesekera, 1991).

The promotion and development of processing of medicinal and aromatic plants have gained momentum recently in many developing countries. Green consumerism and resurgence of interest for plant based products, liberalised and free market economy, increasing awareness about biodiversity conservation and sustainable use of natural resources coupled with poor socio-economic conditions of native populations are ground realities for planning and harnessing the low-cost and purpose oriented process technologies. UNIDO has developed a Polyvalent Pilot Plant with a view to enabling developing countries to upgrade their technology for the processing of medicinal and aromatic plants. This plant incorporates all salient features of a low cost, efficient, small capacity factory which can carry out solvent extraction, solvent percolation, concentration of miscella, solvent recovery, steam distillation and oil separation.

FORMULATION AND INDUSTRIAL UTILIZATION

Medicinal plants are used as raw materials for extraction of active constituents in pure form (eg. alkaloids like quinine and quinidine from cinchona bark, emetine from ipecacuanha root, glycosides from digitalis leaves, sennosides from senna leaves), as precursors for synthetic vitamins or steroids, and as preparations for herbal and indigenous medicines. Products such as ginseng, valerian and liquorice roots are part of the herbal and health food market, as well as the food flavours, fragrance and cosmetic industries. Certain plant products are industrially exploited like liquorice in confectionery and tobacco, papaine as meat tenderiser, quinine as soft drink tonic and cinchona as wine flavour. A large quantity of medicinal plant material is used in the preparation of herbal and medicinal teas, eg. chamomile. These herbal and food uses are of great importance, also to the exporters from developing countries. Hundreds of medicinal plants are items of commerce, however relatively small countries are used in formulated herbal remedies. Several formulations like herbal teas, extracts, decoctions, infusions, tinctures, etc are prepared from medicinal plants (Kraisintu, 1997).

1. **Herbal teas, Herbal remedies**: herbal tea or infusion mixtures are mixture of unground or suitably ground medicinal plants to which drug plant extracts, ethereal oils or medicinal substances can be added. Infusion mixtures should be as homogenous as possible.
2. **Drug extracts**: They are preparations obtained by extracting drugs of a certain particle size with suitable extraction agents (menstrua). The extract obtained after separation of the liquid from the drug residue is called miscella. It may already represent the final liquid dose form eg. as a so called fluid extract, or be used as an intermediary product which is to be further processed as quickly as possible.
3. **Aqueous drug extracts**: The following degrees of comminution are used for the extract depending on the type of plant parts. Leaves, flowers and herbs shredded (4000 mm); woods, barks and roots shredded (2800 mm); fruits and seeds (2000 mm). Alkaloid containing drugs powdered (700 mm).
   a. **Decoctions**: The drug in the prescribed combination is put in to water at a temperature above 90^{0}C. The container is suspended in a water bath and maintained at this temperature for 30 minutes, with repeated stirring. The mixture is then strained while still hot.

b. **Infusions**: One part of the commented drug is kneaded several times in a mortar with 3-5 parts of water and left to stand for 15 minutes. The rest of the boiling water is then poured on to the mixture, which is suspended in a container in a water bath and kept for 5 minutes, with repeated stirring at a temperature above 90^{0}C. The mixture is covered and left to stand until cool.

c. **Macerates**: The comminute drug is left to stand, with occasional stirring, for 30 minutes after the required quantity of water has been poured on to it at room temperature. The extract is then strained and made up to the prescribed weight with rinsing.

d. **Tinctures**: Tinctures are extracts from drug plants prepared with ethanol of varying concentration, ether or mixtures of these, perhaps with certain additives, in such a way that one part of drug is extracted with more than two parts, but at most ten parts, of extraction liquid.

e. **Fluid extracts:** Like tinctures, they are liquid preparations, the difference being that they are more concentrated.

f. **Dry extracts:** They are usually very hygroscopic and should therefore be ground and mixed under conditions which exclude moisture as much as possible. Intermediate and end product must also be stored under dry conditions. There are also liquid, semisolid, solid and controlled release formulations or preparations. The other dose forms are injections, implants, ocular preparations, inhalations and transdermal systems. Liquid formulations may be solutions, emulsions, colloids or suspensions in the increasing order of particle size. They may be intended for administration parenterally, orally or topically including administration into body cavities. Homogeneity for the formulations is very important, particularly where the active ingredient is present in lower concentration.

GENERAL SCHEME FOR QUALITY ASSURANCE OF CRUDE DRUGS AND RAW MATERIALS

In general, during the drug production process, the raw materials are subjected to macroscopic and microscopic examination and physicochemical parameters such as ash values, analysis of ash for major elements such as Sodium, Potassium and Calcium, alcohol soluble and water soluble extractive values and fluorescence analysis, quantitative estimation of phytoconstituents such as total tannins, total glycosides, total alkaloids, total resins and total sugars of the raw materials as well as the formulations was carried out. The formulations were also evaluated for the general parameters such as organoleptic properties, pH, viscosity, specific gravity, optical rotation and refractive index. High Performance Thin Layer Chromatography (HPTLC) technique was employed to obtain characteristic HPTLC *fingerprints* of the individual raw materials and formulations. Using the spectral patterns of the separated components, the presence of certain raw materials in the formulations could also be established. Batch to batch variation was also studied using HPTLC *fingerprinting* technique.

The general scheme for quality assurance of crude drugs and raw materials as suggested by Pei-Gen and Hui-zhen (1997) is given below.

1. Importance of quality assurance of crude drugs and raw material:
   - Guarantee the best final pharmaceutical products.
   - Environmental protection
   - Sustainable utilisation and development of natural resource
2. Criteria of good quality:
   - Good efficacy- high active ingredient, high yield
   - Good safety-less toxicity and side effects, minimum pesticide residues, minimum heavy metals
   - Purity
   - Stability
3. Gene bank conservation
   - Biodiversity conservation
   - To store plant germplasm for future uses
   - To make germplasm available to create new cultivars
   - *Insitu* conservation- genebanks of medicinal and aromatic plants in Asia
   - *Invitro* conservation
   - Breeding
4. Biotechnology
   - Plant cell culture eg. *Digitalis, Catharanthus*
   - Hairy root culture eg. *Salvia, Glycyrrhiza uralensis, Datura stramoniun, Artemesia annua*
   - Tissue culture eg. *Aloe, Crocus sativa, Mentha*
   - Genetic engineering: Isolation and purification of an antifungal protein from *Phytolacca americana* against American ginseng pathogens and synthesis of its gene and expression in *E. coli.*
5. Suitable growth region: In order to get higher quality of crude drugs and raw materials selection of the most suitable growth region for relevant medicinal plant is quite important.

According to the ecological conditions, flora and other criteria, several regions of crude drug development have been identified.

6. GAP: Good Agro technological Practices. Large cultivation of medicinal plants relies upon strong and continuing research (Mukherjee *et al.,* 2009). Plant varieties with an abundance of desired constituents can be reproduced and improved upon under cultivation even in an entirely different area, e.g., cultivation of American ginseng (*Panax quinquefolia*) in China. Attempt should be made to select appropriate region based on similar ecological conditions to introduce good cultivated variety, improve yield of the desired secondary metabolite and reduce the undesirable constituents (Mukherjee and Chakraborty, 2010).
7. Non polluted cultivation: In order to protect the environment, to sustainably utilise the resources and to get a good quality of crude drug, non-polluted agrotechnology is rapidly developed in recent years. These products are commonly called as "Green crude drugs" This involve biological control of insects and pathogens and use of botanical pesticides for the control of pest and diseases.
8. Post harvest technology: Right time harvesting, good processing, good storage, extraction or distillation, quality control.

ROLE OF BIOTECHNOLOGY IN MAPs

Biotechnological tools are important for multiplication and genetic enhancement of the medicinal plants by adopting techniques such as *in-vitro* regeneration and genetic transformations. It can also be harnessed for production of secondary metabolites using plants as bioreactors. Tissue culture is useful for multiplying and conserving the species, which are difficult to regenerate by conventional methods and save them from extinction. The production of secondary metabolites can be enhanced using bioreactors. Bioreactors offer a great hope for the large-scale synthesis of therapeutically active compounds in medicinal plants. Since the biosynthetic efficiency of populations varies, a high yielding variety is recommended as a starting material. *In-vitro* propagation of plants holds tremendous potential for the production of high-quality plant-based medicines. This can be achieved through different methods including micropropagation. Micropropagation has many advantages over conventional methods of vegetative propagation, which suffer from several limitations. With micropropagation, the multiplication rate is greatly increased. It also permits the production of pathogen-free material. Numerous factors are reported to influence the success of *in-vitro* propagation of different medicinal plants. The effects of auxins and cytokinins on shoot multiplication of various medicinal plants have been reported.

Callus-Mediated Organogenesis

The induction of callus growth and subsequent differentiation and organogenesis is accomplished by the differential application of growth regulators and the control of conditions in the culture medium. With the stimulus of endogenous growth substances or by addition of exogenous growth regulators to the nutrient medium, cell division, cell growth and tissue differentiation are induced. There are many reports on the regeneration of various medicinal plants via callus culture. Satheesh and Bhavanandan (1988) have reported the regeneration of shoots from callus of *Plumbago rosea* using appropriate concentrations of auxins and cytokinins. Successful plant regeneration was reported from stem and leaf-derived callus of *Centella asiatica* on MS medium supplemented with 4.0 mg/L BA, 2.0 mg/L kinetin, 0.25 mg/L NAA (Rout, 1997), and 20 mg/L adenine sulphate (Patra and Rai, 1998). Rapid regeneration of *Plumbago zeylanica* in callus culture was achieved on MS medium supplemented with 2.0 mg/L BA and 0.5 mg/L IAA21. The efficient systems for *in-vitro* regeneration of *Solanum laciniatum* and *Echinacea pallida* have been established from leaf explants on medium supplemented with BA and NAA35, 36. Pande *et al.* (2002) have also reported the *in-vitro* regeneration of *Lepidium sativum* from various explants on medium supplemented with BA and NAA.

Regeneration Through Somatic Embryogenesis

Somatic embryogenesis is a process where groups of somatic cells/tissues lead to the formation of somatic embryos which resemble the zygotic embryos of intact seeds and can grow into seedlings on suitable medium. Plant regeneration via somatic embryogenesis from single cells, that can be induced to produce an embryo and then a complete plant, has been demonstrated in many medicinal plant species.

Conservation through Cryopreservation

The cryopreservation of *in-vitro* cultures of medicinal plants is a useful technique. Cryopreservation is long-term conservation method in liquid nitrogen (–196°C) in which cell division and metabolic and biochemical processes are arrested. A large number of cultured materials can be stored in liquid nitrogen (Bajaj, 1991). Since whole plants can regenerate from frozen culture, cryopreservation provides an opportunity for conservation of endangered medicinal plants. For example, low temperature storage has been reported to be effective for cell cultures of medicinal and alkaloidproducing plants such as *Rauvollfia serpentine, D. lanalta, A. belladonna, Hyoscyamus* spp (Pezzato, 1996). When plants are regenerated and no abnormality is seen either in fertility or in alkaloid content, the materials can be stored using cryopreservation methods. Cryopreservation has been used successfully to store a range of tissue types, including meristems, anthers/pollens, embryos, calli and even protoplasts. However, the system will depend on the availability of liquid nitrogen methods.

Genetic Transformation

The recent advances and developments in plant genetics and recombinant DNA technology have helped to improve and boost research into secondary metabolite biosynthesis. A major line of research has been to identify enzymes of a metabolic pathway and then manipulate these enzymes to provide better control of that pathway. Transformation is currently used for genetic manipulation of more than 120 species of at least 35 families, including the major economic crops, vegetables, ornamental, medicinal, fruit, tree and pasture plants, using *Agrobacterium*mediated or direct transformation methods (Birch, 1997). However, *Agrobacterium*-mediated transformation offers several advantages over direct gene transfer methodologies (particle bombardment, electroporation, etc), such as the possibility to transfer only one or few copies of DNA fragments carrying the genes of interest at higher efficiencies with lower cost and the transfer of very large DNA fragments with minimal rearrangement (Shibata and Liu, 2000). The gram-negative soil bacteria, *Agrobacterium* Genetic transformation would be a powerful tool for enhancing the productivity of novel secondary metabolites of limited yield. Hairy roots, transformed with *Agrobacterium rhizogenes*, have been found to be suitable for the production of secondary metabolites because of their stable and high productivity in hormone-free culture conditions. A number of plant species including many medicinal plants have been successfully transformed with *Agrobacterium rhizogenes*. The hairy root culture system of the medical plant *Artemisia annua* L. was established by infection with *Agrobacterium rhizogenes* and the optimum concentration of artimisin was 4.8 mg/L88. Giri *et al.* (1997) induced the development of hairy roots in *Aconitum heterophyllum* using *Agrobacterium rhizogenes*. Pradel *et al.* (1997) developed a system for producing transformed plants from root explants of *Digitalis lanata*64. They evaluated different wild strains of *Agrobacterium rhizogenes* for the productions of secondary products obtained from hairy roots and transgenic plants. They reported higher amounts of anthraquinones and flavonoids in the transformed hairy roots than in untransformed roots. An efficient protocol for the development of transgenic opium poppy (*Papaver somniferum* L.) and California poppy (*Eschscholzia californica* Cham.) root cultures using *Agrobacterium rhizogenes* is reported (Park and Facchini, 2000).

NEED FOR GOVERNMENT INTERVENTION FOR MAPs DEVELOPMENT

The Department of Ayurveda, Yoga & Naturopathy, Unani, Sidha and Homoeopathy (AYUSH), formerly the Department of Indian Systems of Medicine & Homeopathy (ISM&H) was created in March 1995 with the primary work areas of education, standardization of medicines, enhancement of availability of raw materials, research and development, information dissemination, communication and the involvement of traditional medicine and homeopathy in national health care. In 2000, a separate National Medicinal Plant Board under the control of the Department of AYUSH was established to coordinate all aspects of medicinal and aromatic plants.

MAPs' Sector offers immense potential for economic activity by providing income-generating opportunities to a very large section of the rural population through the conservation & sustainable use of the natural resources. Success stories of Medicinal Plants based economic enterprise of China and other South-East Asian countries are well known. Northeast India, including all seven state has better growing conditions harbors rich natural plant species. There are other factor endowments like the traditional skill of people in use of Medicinal and aromatic plants, available labour, huge internal and export market and a good transport network.

ACTIVITIES AND PROGRAMMES TO BE IMPLEMENTED

(1) Assessment and Periodic Monitoring of the Resource:
Assessment of growing stock
Mapping of Medicinal Plants forests including Sacred Groves
Monitoring changes in the status of the resource base
Create and maintain a database on the resource
Establishment of Herbal Gardens

(2) Development of Resource:
Sustainable Management of Medicinal Plants forests
Silvicultural practice based on working scheme
Enhancement of productivity
Adoption of modern nursery practices
Introduction of suitable exotic species
Clonal multiplication
Creation of Medicinal Plants Plantations
In-situ and Ex- situ conservation of endemic and threatened species
The Forest Department will raise plantations

(3) Utilization & Economic Enterprise
Development of cottage and small scale industries
Promotion of medium to large scale industries
Market Information Support
Developing marketing infrastructure
Export promotion

(4) Extension & Training

Organize training for processing of raw material and use of machines

Training for plantation and its management

Setting up of demonstration centers

Awareness programmes

(5) Miscellaneous

At a later stage, provision for levying a cess on the MAPs based trade may be made for generating revenue for the Forest Department

REFERENCES

Bajaj YPS. 1991. Biotechnology in Agriculture and Forestry 15. *Medicinal and aromatic PlantsIII*. Berlin: Springer-Verlag.

Birch RG. 1997. Plant transformation: Problems and strategies for practical application. In Annual Rivew meeting of plant genetic resources. *Plant Molecular Biology*, . 48: 297- 326.

Chakraborty, S., Mukherjee, Dhiman and Baskey, S. 2016. Morphological diversity and nomenclature of *Swertia chirayita* (Gentianaceae)- Recovery of endangered medicinal plant population in north eastern Himalaya. *American Journal of Plant Sciences*, 7 : 741-755.

Chakraborty, S., Mukherjee, Dhiman and Baskey, S. 2015. Paradigm of demographic stochasticity way to extinction of *Valeriana jatamansi* Jones, a valuable medicinal plant in North Eastern Himalaya region. *Ecology, Environment and Conservation*, 21 (1) : 521-528.

Giri A, Banerjee S, Ahuja PS, Giri CC. 1997. Production of hairy roots in *Aconitum heterophyllum* Wall. using Agrobacterium rhizogenes. Invitro Cell Dev Biol-Plant. 33: 293-306.

Kraisintu, K. 1997. Industrial exploitation of indigenous medicinal and aromatic plants: Formulation and industrial utilisation. UNDP .pp 56-78.

Kumar, N., M. Abdul Khader, J. B. M., Rangaswami, P. and Irulappan, I. 1997. *Introduction to Spices,Plantation Crops, Medicinal and Aromatic Plants*. Oxford and IBH Publishing Co. Pvt. Ltd., New Delhi.

Moktan, M.W. and Mukherjee, Dhiman. 2008. Trade and marketing strategy for spice crops in Darjeeling District. *Environment and Ecology* 26 (3A) : 1302-1305.

Mukherje, Dhiman and Chakrabarty, Soumendra. 2010. Effect of hormonal treatment on seed germination of *Valeriana jatamansi*. *Journal of Medicinal and Aromatic Plant Sciences,* 32 (2) : 145-147.

Mukherjee, Dhiman and Roy, Amar. 2005. Ethics and values of medicinal plant in the context of hilly tracts of Darjeeling district and Sikkim Region for entrepreneurship development. *(In) Abstract Seminar on value addition to agro-horti-medicinal produce and its marketing*. October 17-19, College of Agriculture, Nagpur. p. 42.

Mukherjee, Dhiman. 2006. Values of Medicinal plant in the context of hilly tract of Darjeeling District. *Proceedings of 5th Refresher Course in Agricultural*

Sciences conducted by UGC Academic Staff College, July 29- August 18, 2006. Banaras Hindu University, Varanasi pp 5-7.

Mukherjee, Dhiman 2008. Potential of herbals (medicinal plant) in context of India. *MAPs Dew,* 4 (3) : 1-5.

Mukherjee, Dhiman. 2008 a. Association of Medicinal Plants with important Tree species in hills of Darjeeling. *Environment and Ecology,* 26 (4A) : 1697-99.

Mukherjee, Dhiman. 2008 b. Cultivation of *Swertia chirayita* : a high value medicinal herb in high altitude. *Journal of Medicinal & Aromatic Plant Sciences,* 30 (4) : 350-355.

Mukherjee, Dhiman. 2009. Medicinal plant in Darjeeling hills. In : KRISHI SANDESH (eds. Rai, S., Mokthan, M.W. and Ali, S). Miazik International Volunteer Center, Japan. pp 118-121.

Mukherjee, D. 2009 a. Study on distribution pattern of medicinal plant in Darjeeling hills. 4th World Congress on Conservation Agriculture. February, 4-7, 2009, Indian Agricultural Research Institute, New Delhi. p 357.

Mukherjee, Dhiman., Chakraborty, Soumendra., Roy, Amar and Mokthan, M.W. 2009. Differential approach of germplasm conservation of high value of medicinal plants in north eastern himalayan region. *International Journal of Agriculture,. Environment.and Biotechnology.* 2 (4) : 332-340.

Mukherjee, Dhiman. 2016. Studies on ecology and outcome of various seed and organic input treatments on yield attribute and economic yield of *Valeriana jatamansi. The Himalayan Review,* 1 (2) : 193-200.

Pande D, Malik S, Bora M, Srivastava PS. 2002. Rapid protocol for in-vitro-micropropagation of *Lepidium sativum* Linn and enhancement in the yield of lepidine. In-vitro Cell Dev *Biology of Plant,* 38: 451-5.

Park SU, Facchini PJ. 2000. Agrobacterium rhizogenesmediated transformation of opium poppy, *Papaver somniferum* L. and California poppy, *Eschscholzia californica* cham, root cultures. *Journal of Experimental Botany,* 51: 1005-16.

Paroda, R. S. 1993. Medicinal and aromatic plants based cropping systems in South Asia. In FAO.

Patra A, Rai B. 1998. Successful plant regeneration from callus cultures of *Centella asiatica* (Linn.) urban. Plant Growth Reg. 24: 13-6

Pei-Gen, X. and Hui-shen, C. 1997. Quality Assurance of Crude Drugs and Raw Materials. In UNDP, 1997.

Pezzuto J. 1996. Taxol® production in plant cell culture comes of age. *Nature Biotechnol,* 14: 1083.

Pradel H, Dumkelehmann U, Diettrich B, Luckner M. 1997. Hairy root cultures of Digitalis lanata. Secondary metabolism and plant regeneration. *Journal of Plant Physiology,* . 151: 209 15.

Rout GR, Das P. 1997. In-vitroorganogenesis in ginger (*Zingiber officinale* Rosc.). *Journal of Herbs, Spices and Medicinal Plants,* 4: 41-51.

Satheesh Kumar K, Bhavanandan KV. 1988. Micropropagation of *Plumbago rosea* Linn. *Plant Cell Tissue Organ Culture,* . 15: 275- 8.

Shibata D, Liu YG. 2000. Agrobacterium-mediated plant transformation with large DNA fragments. *Trends Plant Sciences,* 5: 354-7.

WHO, 2001. Legal Status of Traditional Medicine and Complementary/Alternative Medicine: A Worldwide Review, WHO/EDM/TRM/2001.2, WHO, Geneva, p. 188

Wijesekera, R.O.B. (Ed.) 1991. *The medicinal plant industry*. CRC Press, Boca Raton. 1-17.

■■■■

SUBJECT INDEX

1-aminocyclopropane-1-carboxylic acid synthase genes 110
2-acetyl-1-proline (2-AP) 108
3D imaging 230
4-hydroxy-2- nonenal (HNE) 261
AaSMO1 13
ABA synthesis 136
ABA transport system 28
Abelmoschus esculentus 3
Abiotic determinants 119
Abiotic stress 37
Abscisic-acid signaling pathways 247
Absicic acid (ABA) 27
Adenosine triphosphatase (ATPase) 362
ADP-glucose pyrophosphorylase (AGPase) 106
Agatea 305
AHK1 9
AHK2 242, 246
AHK3 246
Alcoholic fermentation pathway 135
Alfalfa 156
Alkaline phosphatases 332, 333, 362
Allium sativum lectin (ASA lectin) 362
Altered gene expression 13
Aluminium 275
Amaranthus retroflexus 279
Amino adipic semialdehyde 261
Aminoketobutyrate 261
Anabaena cylindrical 407
Anaerobic stress 135, 136
Anastatica hierochuntica 7
Anion exchange capacity (AEC) 317
Antioxidant defense 88
Antioxidant defense system 2
Antioxidant system 147
Antioxidants mechanism 5
Antioxidative defense system 108, 256, 264
Antioxidative enzymes 147, 269
Antioxidative system 262
AP2/ERF factors 35
Aphanizomenon flosaquae 407
Aphanothece halophytica 13
Aphidius ervi 365
Aphis gossypii 363
Apoplastic compartment 263
Apoplastic pathway 268
APPF 218
Aquaporins 275
Aqueous drug extracts 447
Arabidopsis 365
Arabidopsis halleri 150
Arabidopsis histidine kinase (HK) 9
Arabidopsis thaliana 2, 162, 278
Areal leaf mass (ALM) 102
AREB 10
Arginine 261
Argophyllum 305
Arsenate 274
Arsenic 274
Arsenite uptake 275
Artemisia annua 13
Arthrospira maxima 407
Ascorbate 265
Ascorbate peroxidase (APX) 265
Ascorbate-glutathione cycle 265
Asparagopsis taxiformis 414
Assimilate translocation (partitioning) 89

Astaxanthin 410
Asteracea 278
Aulacorthum solani 365
Automated phenotyping platforms 234
Aux/IAA genes 7
Aux/IAA proteins 244
Aux/IAA reporter 244
AUX1 243
Auxin 7
Auxin signaling 243, 244
Bacterial symbionts 188
BADH 13
b-carotene 410
BcZAT12 gene 53
Behaviour 210
Berkheya coddi 279
betA 12
Biological nitrogen fixation 187
Biomass 88
Biotechnology 450
Boron 276
Brassica juncea 278, 279
Brassicaceae 278
Brassinosteroids 249
Brown Plant Hopper (BPH) 363
Ca^{2+}permeable channels 268
Cabbage 150
Cadmium 267
Cadmium enrichment 141
Cadmium enrichment in soil 143
Cadmium toxicity 268
Cadmium uptake 145
Calcium 242
Callus-mediated organogenesis 451
Camellia sinensis 3
Canopy structure 381
Canopy temperature 6, 88, 232
Car NAC5 13
Carbon exchange rate (CER) 98
Carbon gain hypothesis 98
CarNAC5 12
Carotenoids 263
Casearia 305
Casperian strips 268
CAT 109
Catalase (CAT) 265, 271
Catharanthus roseus 10
CAT-peroxidase 263
Cd absorption 149
Cd accumulation ecotype 151
Cd transport 152
Cd-carbonic anhydrase 150
Cell wall 151
Cellular transduction 31
Chenopodium album 279
Chlamydomonas rheinhardtii 407
Chlorella 406
Chlorella minutissima 411
Chlorella powder 407
Chlorella pyreinodosa 407
Chlorella vulgaris 407
Chlorococcum humicola 414
Chlorophyll 159
Chlorophyll content 85, 88, 232
Chlorophyll flourescence 231
Chloroplasts 262
Chlorosis 272
Chromatin 209
Chromium 269
Cicer arietinum 12
CIMMYT 51
CIPK25 8
Citric acid 306
CKI1 246
Clays 317
Cleidion 305
Cnaphalocrocis medinalis 365
CO_2 assimilation 274
Coleoptile elongation 134
Computed tomography 230
Copper 270
Craterostigma pumilum 2
CRE1/AHK4 246
Crop duration 390
Crop improvement programmes 136
Crop yield 380

CrRLKs 10
Cryopreservation 451
CT 215
Cunoniaceae 305
Cyclic adenosine monophosphate (cAMP) 242
Cynodon dactylon 279
Cyperus rotundus 279
Cysteine 152, 261
Cytochromes 271
Cytokinin (CK) 6
Cytokinin mediated signaling 246
Data analysis 218
Defense mechanisms 164
Dehydration stress 1, 5
Dehydration stress tolerance 11
Dehydrin 32
Dehydroascorbate reductase (DHAR) 265
DELLA proteins 245, 251
DELLAs 245
Deoxymugineic acid 152
Desiccation tolerance 1, 32
Developmental stages 82
DHA oil 413
Diabrotica undecimpunctata 365
Diacyl glycerol (DAG) 242
Diatraea saccharalis 365
Diffused radiation 385
Dimensionality 206
Di-metallic active site 298
Dolichos biflorus 367
DPPN 218
DREB 10
DREB A-5 13
DREB1 genes 31
DREB1A 33
DREB1B 34
DREB1C 34
Drought 46, 191
Drought stress 50, 52
Drought tolerance 52
Drug extracts 447
Dry matter accumulation 105
Dry matter partitioning 394
Dry matter production 389
Dunaliella 406, 409
Dunaliella salina 407
Early ground cover 88
Early light inducible proteins (ELIPs) 111
Early maturation 88
Echinochloa crussgali 279
Economic yield 395
Effective tiller per plant 88
Eichornia crassipes 279
EIN4 248
Eipgenomics 204
ELISA 213
Endoplasmic reticulum 263
Enzymatic antioxidants 86
Enzymic antioxidants 265
Eoreuma loftini 365
Epicuticular wax 88
Epigenomics 208
EPPN 218
ERS1 248
ERS2 248
Escalloniaceae 305
Escherichia coli 12
Ethylene induced signaling 248
Ethylene responding transcription factors 35
ETR1 248
ETR2 248
Euglena gracilis 100, 407
Euphorbiaceae 278, 305
Fabaceae 278
Fenugreek 279
Ferredoxin, aconitase 271
Fe-SOD 154
Festuca arundinacea 105
Flacourtiaceae 305
Fluorescence imaging 215, 230
Flux measurements 150
Forward phenomics 207
Free radicals 86

Fructose-1, 6-bisphosphatase (FBPase) 51
Functional traits 122
Fv/Fm 85
Fv/Fm chlorophyll fluorescence 88
Galactosyldiacylglyceral 86
Galanthus nivalis agglutinin (GNA) 363
Galanthus nivalis lectin (GNA) 362
GCR2 247
GDH 13
Geissois 305
Gene expression 244
Genetic transformation 452
Genetic variation 137
Genomics 204
Genotype loop 219
Genotype-phenotype map 203
Genotyping 202
Germination percentage 133
Gibberellin signaling 245
Glucose 6 phosphate dehydrogenase 268
Glutamate 87
Glutathione 265
Glutathione reductase (GR) 265
Glycine max 162
Glycine soja 367
Glycine-betaine 87
Glycolate-oxidase 263
Glycolipid 413
Glyoxalase I 302
Glyoxalase I in plants 304
Glyoxysomes 263
Gossypium 278
Grain filling duration (GFD) 88
Grain weight 88
Growscreen 217
Growscreen fluoro 211
Growth 15, 391
Growth analysis 98
Growth pattern 233
GSMT 13
H^+-ATPase 151
Haematococcus 406, 409
Haematococcus pluvialis 407
Halliwell-Asada pathway 265, 266
Harvest index (HI) 395
Harvesting complex II 86
Heat tolerance 81
Heat tolerance breeding 89
Heavy metals 192
Helianthus annuus 7, 162, 278
Helicoverpa zea 366
Heliothis virescens 365
Herbal remedies 447
Herbal teas 447
High throughput phenotyping 210, 228
High-throughput phenotyping techniques 207
Hill reaction 274
Histidine 306
Histidine kinase-dependent phosphotransfer 248
Homalium 305
Hordeum vulgare 279
Hormone signaling 250
HSP70 90
HSP90 90
HTPheno 217
HVCB4 34
Hybanthus 305
Hydrilla verticillata 271
Hydrogen peroxide (H_2O_2) 160
Hydroxyl radical (·OH) 261
Hydroxyl radical (HO^-) 160
Hyper-spectral imaging 215
Hyperspectral imaging 230
Image-based Plant Phenotyping 227
Imaging 210
Imaging techniques 229
Inositol triphosphate (IP_3) 242
Iodine 314
Iodine concentration 318
Iodine content 317
Iodine transfer factor in soil 319
Iodine volatilization 324
IPPN 217
Iron 271

Iron sulphur proteins 271
Isochrisis galbana 407
Isocitrate dehydrogenase 268
Jatropha curcus 279
Juncus maritimus 279
Keto fatty acids 261
Lacanobia oleracea 365
Lactate dehydrogenase (LDH) 362
Lamiaceae 278
Laodelphax striatellus 365
Laser imaging 230
Laser sterior imaging 214
Lasiochlamys 305
Late senescence 83
LEA 32
Lea promoter 37
Lead 268
Leaf anatomy 99
Leaf area index (LAI) 98
Leaf area ratio (LAR) 98, 125
Leaf glaucousness 88
Leaf hooper 364
Leaf rolling 88
Leaf senescence 3
Leaf sucrose-phosphate synthase (SPS) 106
Lectins 360
Leghemoglobin 271
Legume symbiosis 186
Lemna minor 278
LemnaTec 217
Light harvesting antenna complex of PSII (LHCII) 101
Light interception 383
Lipaphis erysimi 365
Lipid peroxidation 147
Lipoprotein complexes 159
Lotus corniculatus 367
Low light intensity 96
Low light tolerance 97
Lutein 410
Lycopersicon esculentum 304
Magnetic resonance images 230
Magnetic resonance imaging (MRI) 211
Maintenance respiration 391
Maize 365
Malic enzyme 268
Malondialdehyde (MDA) 261
Malondialdehyde (MDA) content 109
Management strategies 91
Manganese 271
Mannitol 87
MAPK 248
Medicago sativa 278
Medicinal and Aromatic Plants (MAPs) 425
Membrane thermostability 86, 88
Metabolic Alterations 256
Metabolomics 90, 205, 208
Metal Stressed Plants 264
Metal tolerance 151
Metal toxicity 256
Metalloenzyme active sites 293
Methionine 261
Microalalgae 405
Microalgae 406
miR1026 14
miR1030 14
miR1035 14
miR1050 14
miR1088 14
miR1125 14
miR1126 14
miR1432 14
miR1450 14
miR156 13, 14
miR158 13
miR159 13, 14, 15
miR165 13
miR166 14
miR167 13
miR168 13, 14
miR169 13, 14
miR170 14
miR171 13, 14
miR172 14
*MiR*172c 15
miR1867 14

miR1881 14
miR319 13, 14
miR393 13, 14
miR394 13
miR395 14
miR396 14, 15
miR397 14
miR398 14
miR408 14
miR474 14
miR528 14
miR529 14
miR845 14
miR851 14
miR854 14
miR894 14
miR901 14
miR903 14
miRNAs 13
Mn transport 147
Mn-SOD 154
Model of Heat-adaptive traits 84
Molecular oxygen 262
Molecular physiology 1, 15
Monodehydroascorbate reductase (MDHAR) 265
Monogalactosyldiacylglyceral 86
Monomer synthesis 391
Morphological traits 88
MRI 215
Multi spectral imaging 214
Mustard 365
Mustard aphids 364
MYB factors (OsMYB2) 36
MYB15 13
MYB15 gene 12
Mycorrhizal transpiration 268
Myzus persicae 363, 365
NADH dehydrogenase 53
NADH oxidase 263
NADP-malic enzyme (NADP-ME) 51
Nannocloropsis 407
NCED gene 28
Near infra red imaging 230
Nephotettix virescens 363, 364, 365
Net assimilation rate (NAR) 98, 125
Net photosynthetic rate 102
Net primary production (NPP) 391
N-fixing legumes 188
Ni Metalloenzymes 294
Nickel 273, 291
Nickel abundance 291
Nickel hyperaccumulating plants 304
Nickel toxicity 308
Nicotiana tabacum 13, 278
Nicotianamine 307
Ni-enzyme 294
Nilaparvata lugens 362, 365
Nitrate assimilation 190
Nitrogen fixation 110
Nodule development 189
Nodule system 156
Non-enzyme proteins 136
Nonenzymic antioxidants 265
Nongenomic 248
NRAMP protein 272
Number of grains 88
Nutraceuticals 405, 408
Nutrient availability 124
OAT 12
Oconthecaceae 305
Oncontheca 305
Ontogenetic changes 394
Organic matter 317
Oryza sativa 2, 12, 13, 49, 162, 275
OsAKT1 8
OsDREB1A 31
OsDREB1F 31
OsDREB1G 31
OsDREB2A 31
OsDREB2A1 31
OsDREB2A2 31
OsDREB2B 31
Osmolyte accumulation 87
Ostrinia nubilaris 365
Oxidative damage 52
Oxidative stress 160, 256

Oxidized glutathione (GSSG) 266
Oxygenase 262
P deficiency 147
P5CR 12
P5CS 12
Palisade cells 102
PAM fluorescence 215
Pancheria 305
Parallorhogas pyralophagus 365
Peroxidases (POX) 265, 271
Peroxidation 261
Peroxisome 263
pH 317
Pharmaceuticals 405, 413
Phaseolus vulgaris 147
Phenodyn 217
Phenolics 265
Phenome 218
Phenomic projects 216
Phenomic scale measurements 209
Phenomic tools 208
Phenomics 200, 205, 206
Phenomics facilities 235
Phenopsis 217
Phenotyping 202, 229
Phenotyping system 205
Pheophytin 263
Phloem transport 152
Phosphatase 268
Phosphoenolpyruvate carboxylase (PEPCase) 51
Phospholipid 270
Photooxidative damage 86
Photoprotection 263
Photoprotective metabolites 86
Photosynthesis 51, 85, 158
Photosynthetic nitrogen-use efficiency (PNUE) 125
Photosynthetic pigments 100
Photosynthetic rate 88
Photosynthetically active radiation (PAR) 97
Photosystems I and II 4
Phototropism 241
Phragmites australis 162
Phyllantus 305
Physiological breeding 200
Physiological traits 88
Phytohaemagglutinin 369
Phytohormones 241
Phytolacca acinosa 279
Phytometallophores 270
Phytoremediation 256, 278
Phyto-therapy 325
Phytotoxic Metals 267
PIN1 243
Pisum sativum 153, 156, 162
Pityrogamma calomelanos 279
Planck's equation 384
Plant disease symptoms 213
Plant growth 153
Plant hormones 136
Plant nutrition 52, 318
Plant phenomics 213, 218
Plant phenotyping 206
Plant phenotyping platforms 216
Plant productivity 96
Plant Purple Acid Phosphatases 331, 336
Plant system 319
Plant-Insect interaction 212
PlantScan 3D meshes 217
Plasma membrane 151, 263
POD 109
Polyols 87
Polysaccharides 413
Polyunsaturated fatty acid 411, 413
Populus canescens 162
Populus tremula 279
Populus trichocarpa 13
Porphyridium cruentum 407
Portulaca oleracea 7
Potato 365
Proline 87, 261
Proline-rich extensin-like receptor kinase (PERKs) 10
Protein 413
Protein content 159

Protein phosphatase-2C (PP2C) 29
Protein sulphydryl groups 275
Proteomics 89, 204, 208
PSI electron transport 262
PSII/PSI ratio 102
Psychotria 305
Pteris vittata 279
Purple acid phosphatases 334, 339, 342
Pyrabactin resistance (PYR) 29
Pyrabactin resistance like (PYR1-L) 29
Pyridoxine 109
Pyruvate orthophosphate dikinase (PPDK) 51
Quantitative traits loci (Qtl) analysis 87
Radiation use efficiency (RUE) 84, 388
Radiation utilization efficiency 386
Radical variation 379
RCB imaging 214
Reactive oxygen species (ROS) 53, 261, 262
Reactive oxygen species 53
Red kidney bean 336
Redox signaling 86
Regulatory component of ABA response (RCAR) 29
Regulatory elements 36
Relative growth rate (RGR) 125
Relative water content (RWC) 6
Reproductive phase 390
Resin 431
Resolution 206
Respiration 391
Respiratory electron transport chain 270
Resurrection plants 2
Reverse phenomics 207
RFLP (Restriction Fragment Length Polymorphism) 202
Ribonuclease 268
Ribulose-1,5 bisphosphate carboxylase-oxygenase (Rubisco) 51, 262
Rice 365
RLK1-like family 10
RLK-mediated signaling pathways 10
Root phenotyping 233
Root plasmolysis 275
RootReader3D 217
ROS 15
ROS generation 160
ROS metabolism 4
ROS production 2
ROS–antioxidant interactions 4
ROS-responsive regulatory genes 4
Rubiaceae 305
Rubisco activity 101
Saccharomyces cerevisiae 151
Saccharum officinarum 162
Salicornia brachiata 13
Salinity 191
Salinity stress 30
Salinity tolerance 36
Salix nigra 278
Sapotaceae 305
Scendesmus 104
Scenedesmus obliquus 407
Schizochytrium 406
Scirpophaga incertulas 365
Scrophulariaceae 278
Sebertia 305
Secale cereale 273, 278
Seedling survival 133
Senescence 3
Sesquioxides 317
Shade tolerance 97, 103
Signal 31
Signal transduction 241
Single perception 242
Singlet oxygen 86
Sinorhizobium meliloti 156
Sitobion avenae 366
Skeletonoma 104
Small heat shock proteins 71
SNAC1 11
SOD 109
Soil acidity 193
Soil factors 144
Soil metal toxicity 260

Soil-plant-air-system 314
Solanum tuberosum 13
Somatic embryogenesis 451
Sorbitol 87
Sorghum halepense 279
Soybean 342
SPAD chlorophyll content 88
SPAD measurements 85
Specific leaf weight (SLW) 98
Spectroscopic 297
Spinach 150
Spinacia oleracea 278
Spirogyra 407
Spirulina 406
Spirulina platensis 407
Starch synthase activity 106
Stay-green 83, 88
Stem reserve 88
Stomatal conductance 88
Stress-responsive NAC1 11
Sucrose non fermenting protein kinase 2(SnRK2) 29
Sucrose synthase (SS) 106
Sucrose-starch metabolic pathway 106
Sugarcane 365
Superoxide dismutase (SOD) 265, 271
Superoxide radical ($O_2^{\cdot-}$) 160, 261
Sustainability 46
Sweet potato 339
Symbiosis 187
Symbiotic nitrogen fixation 189
Synechococcus 407
Syntrichia caninervis 13
Systems biology 205
Tabebuia chrysotricha 104
TaMYB33 6
TaNAC69 6
Tartrate-resistant acid 335
TaWLIP19 6
TaWRKY10 6
TaWRKY16 6
TaWRKY17 6
TaWRKY19-C 6
TaWRKY24 6
TaWRKY59 6
TaWRKY61 6
TaWRKY82 6
Temperature 192
Tetraethylenepentamine 270
Thalspi caerulescens 278
Thermal imaging 214, 230
Thermal stress managers 71
Thousand grains weight (TGW) 88
Threonine 261
Throughput 206
Thylakoid membranes 86
Time of flight imaging 214
Tithonia diversifolia 278
Tobacco 365
Tolerant varieties 137
Toxicity 146
Traditional phenotyping 228
Trait-based physiological breeding 213
TraitMill 217
Transcription 6
Transcription factors (TFs) 10, 32
Transcriptomics 5, 204, 208
Transgenic 37
Transpirational cooling 88
Tree seedling 119
Trigonella foenumgraecum 279
Triticum aestivum 3, 13, 151
Triticum duram 162
Triticum turgidum 151
Triticum vulgare 273
Tropical dry forests (TDFs) 119
Tuberous roots 431
UDP-glucose pyrophosphorylase (UGPase) 106
Urease 295, 298
Urease inhibitors 301
V. aconitifolia 12
Violaceae 305
Violaxanthin 410
Visible light imaging 230
Vitamins 411
Volatilization 323
Wall-associated kinases (WAKs) 10

Water use efficiency (WUE) 26
Wet chemistry destructive sampling 215
Wheat 81
WinCELL 217
WinFOLIA 217
WinRHIZO 217
WinSEEDLE 217
WRKY mediated stress tolerance 37
Xerocomus chrysenteron 366
X-ray 215
Xylem sap 152
Xylem transport 152
Xylosma 305
Yellow lupin 341
Yield 88
Yield variation 379
Zea mays 12, 49, 273, 279
Zeaxanthin 410
Zinc (Zn) hyperaccumulator 151
Zinc 274
Zinc-binding sustances 153
ZIP family 151
Zn/Cd hyperaccumulator 150
ZNT1 cDNA 151
α-tocopherol 265
β-carotene 265
γ-glutamyl semialdehyde 261